Signal
Processing *for*
Intelligent
Sensor Systems

David C. Swanson

**The Pennsylvania State University
University Park, Pennsylvania**

MARCEL DEKKER, INC.   NEW YORK · BASEL

Signal Processing *for* Intelligent Sensor Systems

ISBN: 0-8247-9942-9

This book is printed on acid-free paper.

Headquarters
Marcel Dekker, Inc.
270 Madison Avenue, New York, NY 10016
tel: 212-696-9000; fax: 212-685-4540

Eastern Hemisphere Distribution
Marcel Dekker AG
Hutgasse 4, Postfach 812, CH-4001 Basel, Switzerland
tel: 41-61-261-8482; fax: 41-61-261-8896

World Wide Web
http://www.dekker.com

The publisher offers discounts on this book when ordered in bulk quantities. For more information, write to Special Sales/Professional Marketing at the headquarters address above.

Series Introduction

Over the past 50 years, digital signal processing has evolved as a major engineering discipline. The fields of signal processing have grown from the origin of fast Fourier transform and digital filter design to statistical spectral analysis and array processing, and image, audio, and multimedia processing, and shaped developments in high-performance VLSI signal processor design. Indeed, there are few fields that enjoy so many applications—signal processing is everywhere in our lives.

When one uses a cellular phone, the voice is compressed, coded, and modulated using signal processing techniques. As a cruise missile winds along hillsides searching for the target, the signal processor is busy processing the images taken along the way. When we are watching a movie in HDTV, millions of audio and video data are being sent to our homes and received with unbelievable fidelity. When scientists compare DNA samples, fast pattern recognition techniques are being used. On and on, one can see the impact of signal processing in almost every engineering and scientific discipline.

Because of the immense importance of signal processing and the fast-growing demands of business and industry, this series on signal processing serves to report up-to-date developments and advances in the field. The topics of interest include but are not limited to the following:

- Signal theory and analysis
- Statistical signal processing
- Speech and audio processing
- Image and video processing
- Multimedia signal processing and technology
- Signal processing for communications
- Signal processing architectures and VLSI design

I hope this series will provide the interested audience with high-quality, state-of-the-art signal processing literature through research monographs, edited books, and rigorously written textbooks by experts in their fields.

K. J. Ray Liu

iii

Preface

Signal Processing for Intelligent Sensor Systems covers a broad range of topics that are essential to the design of intelligent autonomous computing systems. A unified approach is presented linking data acquisition, system modeling, signal filtering in one and two dimensions, adaptive filtering, Kalman filtering, system identification, wavenumber processing, pattern recognition, sensor systems, and noise cancellation techniques. Together these topics form the technical basis for the state of the art in radar, sonar, medical and machinery health diagnosis and prognosis, and "smart" sensor systems in general. Applications are given throughout the book in the areas of passive remote sensing, active radar and sonar, digital image processing, tracking filters, acoustic imaging and diagnostics, and wavenumber filters. Additional references and example problems are provided in each topic area for further research by the reader. This book presents adaptive signal processing from a physical, rather than mathematical, point of view, with emphasis on application to intelligent sensor systems. Engineers and scientists working on the development of advanced sensor and control systems should find this text useful in bringing together the required topics. Unifying these topics in a single book allows the uncertainties from the basic sensor detection elements to be propagated through each adaptive signal processing topic to produce meaningful metrics for overall system performance. Many signal processing applications require knowledge of a wide range of adaptive signal processing topics for developing such systems. This text pays specific attention to the underlying physics behind the signal processing application, and, where appropriate, examines the signal processing system as a physical device with physical laws to be examined and exploited for science.

The text is well suited for senior undergraduate and graduate students in science and engineering as well as professionals with a similar background. Some prior knowledge of digital signal processing, statistics, and acoustics/field theory would be helpful to readers, but an undergraduate level of understanding in complex matrix algebra, field theory, and Fourier-LaPlace transforms should be sufficient background. From this starting point, the book develops basic adaptive signal processing principles and applies them to the problems such as adaptive beamforming, system identification, and data tracking. The benefit of using this book is that one comes away with a more global view of adaptive signal processing in the context of a "smart" sensor and/or control system as well as a detailed understanding of many state-of-the-art techniques. While adaptive algorithms extract information from input data according to optimization schemes such as least-squared error, they can also be used to "intelligently" adapt the sensor system to the environment. Adaptive systems that optimize themselves based on command inputs and the sensor and/or actuator environment represent the most advanced "sentient" systems under development today. The term "sentient" means "having the five senses" and the associated awareness of the environment. Sensor technology in the 21st century will no doubt achieve sentient processing and this text is aimed at providing an interdisciplinary groundwork toward this end.

A simple example of an environment-sensitive adaptive algorithm is automatic exposure, focus, and image stabilization on many commercially available video cameras. When coupled to a digital frame-grabber and computer vision system, adaptive image processing algorithms can be implemented for detecting, say, product defects in an assembly line. Adaptive systems designed to detect specific "information patterns" in a wide range of environments are often referred to as automatic target recognition (ATR) systems, especially in defense applications. The ATR problem has proved to be one of the most difficult and intriguing problems in adaptive signal processing, especially in the computer vision area. Acoustic ATRs have enjoyed some degree of success in the areas of machine condition vibration monitoring, sound navigation and ranging (SONAR), and speech recognition. In medical imaging, adaptive image processing systems can produce remarkable measurements of bone density, geometry, or tissue properties, but the actual end recognition is currently done by people, not by machines. It is curious to note that while a toddler can easily find a partially obscured toy in a full toybox, it's not an easy task for even the most sophisticated computer vision system, due to the complexity of the signal separation problem. It is likely that people will always be "in the loop" in making critical decisions based on information that intelligent sensor systems provide because of the inherent intelligence and personal responsibility human beings can display. But as technology progresses, we should certainly expect many noncritical sensor-controller applications to be fully automated at significant levels.

What we can do today is build adaptive systems in which the accepted laws of physics and mathematics are exploited in computer algorithms that extract information from sensor data. The information is then used to do useful things such as: optimize the sensor configuration for maximum resolution; recognize primitive patterns in the input data and track the patterns and information to determine statistical trends; and logically assemble the information to test and score hypotheses. The results of the adaptive processing can in some instances lead to a completely automatic recognition system. However, since many of the current ATR applications are life-critical (such as military targeting, medicine, and machine or structure prognosis), the idea of eliminating the "man-in-the-loop" is being replaced by the idea of making the man-in-the-loop smarter and more efficient. This text is concerned with the mathematics of the adaptive algorithms and their relationship to the underlying physics of the detection problems at hand. The mechanical control systems on automobile engines (choke, spark-plug timing, etc.), mechanical audio recordings, and many household appliances changed to digital systems only in the last decade or so. Most industrial process controls and practically all military control and communication systems are digital, at least in part. The reasons for this unprecedented proliferation of basic digital system technology are not just the increased precision and sophistication of digital systems. Digital control systems are now far cheaper to manufacture and have better repeatability and reliability than their analog computers. One could argue that a new industrial revolution is already underway, in which the machines and practices from the previous revolution are being computerized, optimized, and reinvented with "machine intelligence."

The text is organized into five main parts. Fundamentals, frequency domain processing, adaptive filtering, wavenumber systems, and signal processing applications. Together, these five parts cover the major system operations of an intelligent sensor system. However, the emphasis is on applied use, physics, and data confidence rather than just applied mathematical theory. There are many available texts listed in each chapter's bibliography that can supply the reader with sufficient detail beyond that offered here. By focusing on the essential elements of measurement, filtering, detection, and information confidence metrics, this book provides a good foundation for an intelligent sensor system. In the future, sensor system engineering will need to be more accessible to the nonelectrical engineering discipline, while electrical engineers will need a stronger applied background in physics and information processing. Bridging this gap in an interdisciplinary approach has been the main challenge of preparing this text. The student with a strong signal processing background should be advised that even in the fundamentals chapter there are subtle physical application points to be learned. Furthermore, students relatively new to signal processing should not be intimidated by the advanced signal processing topics such as lattice filters and adaptive beamforming. These advanced topics are explained in a very straightforward and practical manner for the essential techniques, leaving many of the more narrow techniques to presentations given in other texts.

The prime objective of this book is to organize the broad scope of adaptive signal processing into a practical theory for the technical components of smart machines. The longer a normal human being works in the area of developing a computing system's eyes, ears, motor control, and brains, the more incredible biological life appears to be. It's almost funny how our most powerful supercomputers, with a throughput of over billions of operations per second, have the real-time neural network capacity of a slug (okay, a smart slug). Computer programs and algorithms teach us a great deal about our own thought processes, as well as how our imaginations lead us to both inventiveness and human error. Perhaps the most important thing building machine intelligence can teach us is what an amazing gift we all have: to be human with the ability to learn, create, and communicate.

David C. Swanson

Acknowledgments

Preparation of this text began in earnest in the fall of 1993 and took more than five years to complete. There are many reasons that projects such as creating a textbook take longer than originally envisioned, not the least of which is the fact that it is far more interesting to create a textbook than to read and use one. I can acknowledge three main factors that contributed enormously to the content and the required time spent preparing this text. In order of importance, they are my family, my students and colleagues, and my research sponsors. At Penn State's Applied Research Laboratory I have been extremely fortunate to have very challenging applied research projects funded by sponsors willing to pay me to learn. Teaching graduate students with backgrounds in electrical and mechanical engineering, physics, and mathematics in Penn State's Graduate Program in Acoustics not only has been a privilege but has shaped the interdisciplinary approach of this text. As a result of the time demands of applied research and teaching, this text was written in its entirety on a home PC during evenings and weekends. That time reasonably belongs to a very patient wife, Nadine, and three children, Drew, age 6, Anya, age 3, and Erik, age 3 months at the time of this writing. They made things far easier on me than my absences were on them. In particular, I owe Nadine enormous love, gratitude, and respect for tolerating the encumbrance the creation of this book placed on our home life. If you find this text interesting and useful, don't just consider the author's efforts—consider the role of a strong and loving family, outstanding students and colleagues, and challenging applied research projects. Without any one of these three important factors, this text would not have been possible. Finally, the creators of Matlab, WordPerfect, and Corel Draw also deserve some acknowledgment for producing some really outstanding software that I used exclusively to produce this book.

Contents

Part I

Fundamentals of Digital Signal Processing

1

Sampled Data Systems

Figure 1 shows a basic general architecture which can be seen to depict most adaptive signal processing systems. The number of inputs to the system can be very large, especially for image processing sensor systems. Since adaptive signal processing system is constructed using a computer, the inputs generally fall into the categories of analog "sensor" inputs from the physical world and digital inputs from other computers or human communication. The outputs also can be categorized into digital information such as identified patterns, and analog outputs which may drive actuators (active electrical, mechanical, and/or acoustical sources) to instigate physical *control* over some part of the outside world. In this chapter we examine the basic constructs of signal input, processing using digital filters, and output. While these very basic operations may seem rather simple compared to the algorithms presented later in the text, careful consideration is needed to insure a high fidelity adaptive processing system. For example, Figure 1 shows the adaptive process controlling the analog input and output gains. This technique is relatively straightforward to implement and allows high fidelity signal acquisition and output over a wide dynamic range. With a programmed knowledge-base of rules for acceptable input and output gains, the adaptive system can also decide if a transducer channel is broken, distorted, or operating normally. Therefore, we will need to pay close attention to the fundamentals of sampling analog signals and digital filtering. The next chapter will focus on fundamental techniques for extracting information from the signals.

Consider a transducer system which produces a voltage in response to some electromagnetic or mechanical wave. In the case of a microphone, the transducer sensitivity would have units of volts/Pascal. For the case of a video camera pixel sensor, it would be volts per lumen/m^2, while for an infrared imaging system the sensitivity might be given as volts per °K. In any case, the transducer voltage is conditioned by filtering and amplification in order to make the best use of the A/D convertor system. While most adaptive signal processing systems use floating-point numbers for computation, the A/D convertors generally produce fixed-point (integer) digital samples. The integer samples from the A/D are con-

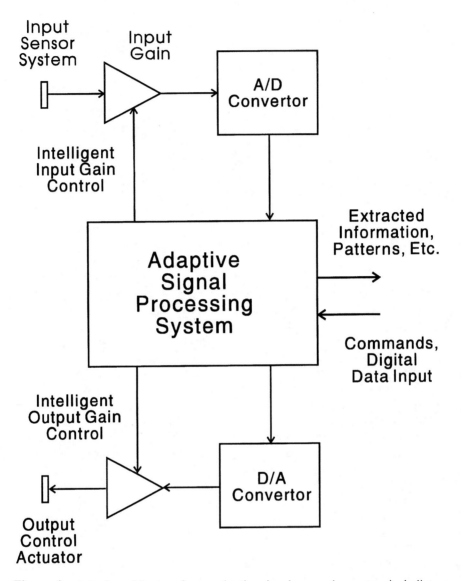

Figure 1 A basic architecture for an adaptive signal processing system including sensory inputs, control outputs, and information inputs and outputs for human interaction.

verted to floating-point by the signal processor chip before subsequent processing. This relieves the algorithm developer from the problem of controlling numerical dynamic range to avoid underflow or overflow errors in fixed-point processing unless less or more expensive floating-point processors are used. If the processed signals are to be output, then floating-point samples are simply re-converted to integer and an analog voltage is produced using a digital-to-analog (D/A) convertor system and filtered and amplified.

1.1 A/D CONVERSION

Quite often, adaptive signal processing systems are used to dynamically calibrate and adjust input and output gains of their respective A/D and D/A convertors. This extremely useful technique requires a clear understanding of how most data acquisition systems really work. Consider a generic successive approximation 8-bit A/D convertor as seen in Figure 2 below. The operation of the A/D actually involves an internal D/A convertor which produces an analog voltage for the "current" decoded digital output. A D/A convertor simply sums the appropriate voltages corresponding to the bits set to 1. If the analog input to the A/D does not match the internal D/A converted output, the binary counter counts up or down to compensate. The actual voltage from the transducer must be sampled and held constant (on a capacitor) while the successive approximation completes. Upon completion, the least significant bit (LSB) of the digital output number will randomly toggle between 0 and 1 as the internal D/A analog output voltage converges about the analog input voltage. The "settling time" for this process increases with the number of bits quantized in the digital output. The shorter the settling time, the faster the digital output sample rate may be. The toggling of the LSB as it approximates the analog input signal leads to a low level of uniformly-distributed (between 0 and 1) random noise in the digitized signal. This is normal, expected, and not a problem so long as the sensor signal strengths are sufficient enough that the quantization noise is small compared to signal levels. It is important to under-

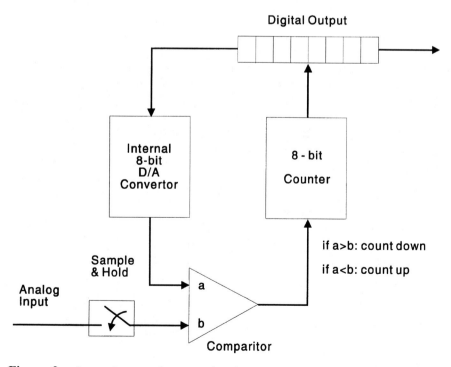

Figure 2　A generic successive approximation 8-bit A/D convertor actually has an internal D/A sub-system.

stand how transducer and data acquisition systems work so that the adaptive signal processing algorithms can exploit and control their operation.

While there are many digital coding schemes, the binary number produced by the A/D are usually *coded in either offset binary or in two's complement formats.* Offset binary is used for either all-positive or all-negative data such as absolute temperature. The internal D/A convertor in Figure 2 is set to produce a voltage V_{min} which corresponds to the number 0, and V_{max} for the biggest number or 255 (11111111), for the 8-bit A/D. The largest number produced by an M-bit A/D is therefore 2^M-1. The smallest number, or (LSB) will actually be wrong about 50% of the time due to the approximation process. Most data acquisition systems are built around either 8, 12, or 16-bit A/D convertors giving maximum offset binary numbers of 255, 4095, and 65535, respectively. If a "noise-less" signal corresponds to a number of, say 1000, on a 12-bit A/D, the signal-to-noise ratio (SNR) of the quantization is 1000:1, or approximately 60 dB.

Signed numbers are generally encoded in two's complement format where the most significant bit (MSB) is 1 for negative numbers and 0 for positive numbers. This is the normal "signed integer" format in programming languages such as "C". If the MSB is 1 indicating a negative number, the magnitude of the binary number is found by complementing (changing 0 to 1 or 1 to 0) all of the bits and adding 1. The reason for this apparently confusing coding scheme has to do with the binary requirements of logic-based addition and subtraction circuitry in all of today's computers. The logic simplicity of two's complement arithmetic can be seen when considering that the sum of two two's complement numbers N_1 and N_2 is done exactly the same as for offset binary numbers, except any carry from the MSB is simply ignored. Subtraction of N_1 from N_2 is done simply by forming the two's complement of N_1 (complementing the bits and adding 1), and then adding the two numbers together ignoring any MSB carry. Table 1 below shows two's complement binary for a 3-bit ± 3.5 V A/D and shows the effect of subtracting the number $+2$ (010 or $+2.5$ V) from each of the possible 3-bit numbers. Note that the complement of $+2$ is (101) and adding 1 gives the "two's complement" of (110), which is equal to numerical -2, or -1.5 V in the table.

As can be easily seen in Table 1, the numbers and voltages with an asterisk are rather grossly in error. This type of numerical error is the single most reason to use floating-point rather than fixed-point signal processors. It is true that fixed-point

Table 1 The Effect of Subtracting 2 from the Range of Numbers From a 3-bit Two's Complement A/D

Voltage N	Binary N	Binary N-2	Voltage N-2
+3.5	011	001	+1.5
+2.5	010	000	+0.5
+1.5	001	111	−0.5
+0.5	000	110	−1.5
−0.5	111	101	−2.5
−1.5	110	100	−3.5
−2.5	101	011*	+1.5*
−3.5	100	010*	+0.5*

signal processor chips are very inexpensive and sometimes faster at fixed-point arithmetic. However, a great deal of attention must be paid to insuring that no numerical errors of the type in Table 1 occur in a fixed-point processor. Fixed-point processing severely limits the numerical dynamic range of the adaptive algorithms used. In particular, algorithms involving many divisions, matrix operations, or transcendental functions such as logarithms or trigonometric functions are generally not good candidates for fixed-point processing. All of the subtractions are off by at least 0.5 V, or half the LSB. A final point worth noting from Table 1 is that while the analog voltages of the A/D are symmetric about 0 V, the coded binary numbers are not, giving a small numerical offset from the two's complement coding. In general, the design of analog circuits with nearly zero offset voltage is a difficult enough task that one should always assume some non-zero offset in all digitized sensor data. The maximum M-bit two's complement positive number is $2^{M-1}-1$ and the minimum negative number is -2^{M-1}. Even though the A/D and analog circuitry offset is small, it is good practice in any signal processing system to numerically remove it. This is simply done by recursively computing the mean of the A/D samples and subtracting this time-averaged mean from each A/D sample.

1.2 SAMPLING THEORY

We now consider the effect of the periodic rate of A/D conversion relative to the frequency of the waveform of interest. There appear to be certain advantages to randomly spaced A/D conversions or "dithering" (1), but this separate issue will not be addressed here. According to Fourier's theorem, any waveform can be represented as a weighted sum of complex exponentials of the form $A_m e^{j\omega m t}$; $-\infty < m < +\infty$. A low frequency waveform will have plenty of samples per wavelength and will be well represented in the digital domain. But, as one considers higher frequency components of the waveform relative to the sampling rate, the number of samples per wavelength declines. As will be seen below for a real sinusoid, at least two equally-spaced samples per wavelength are needed to adequately represent the waveform in the digital domain. Consider the arbitrary waveform in Eq. (1.2.1).

$$x(t) = A\cos(\omega t) = \frac{A}{2}e^{j\omega t} + \frac{A}{2}e^{-j\omega t} \qquad (1.2.1)$$

We now sample $x(t)$ every T sec giving a sampling *frequency* of f_s Hz (samples per sec). The digital waveform is denoted as $x[n]$ where n refers to the nth sample in the digitized sequence.

$$x[n] = x(nT) = A\cos(\omega nT)$$
$$= A\cos\left(\frac{2\pi f}{f_s}n\right) \qquad (1.2.2)$$

Eq. (1.2.2) shows a "digital frequency" of $\Omega = 2\pi f/f_s$, which has the same period as an analog waveform of frequency f so long as f is less than $f_s/2$. Clearly, for the real sampled cosine waveform, a digital frequency of 1.1π is basically indistinguishable from 0.9π except that the period of the 1.1π waveform will actually be longer than the analog frequency f! Figures 3 and 4 graphically illustrate this phenomenon well-known as aliasing. Figure 3 shows a 100 Hz analog waveform

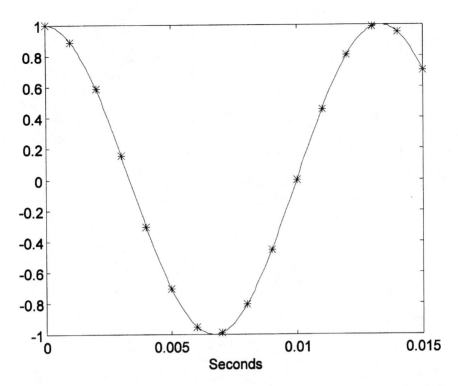

Figure 3   A 75 Hz sinusoid (−) is sampled at 1000 Hz giving a digital signal (*) which clearly represents the analog signal.

sampled 1000 times per sec. Figure 4 shows a 950 Hz analog signal with the same 1000 Hz sample rate. Since the periods of the sampled and analog signals match only when $f \leq f_s/2$, these frequency components of the analog waveform are said to be *unaliased*, and therefore adequately represented in the digital domain.

Restricting real analog frequencies to be less than $f_s/2$ has become widely known as the Nyquist sampling criterion. This restriction is generally implemented by a low-pass filter with −3 dB cutoff frequency in the range of $0.4\,f_s$ to insure a wide margin of attenuation for frequencies above $f_s/2$. However, as will be discussed in the rest of this chapter, the "anti-aliasing" filters can have environment-dependent frequency responses which adaptive signal processing systems can intelligently compensate.

It will be very useful for us to explore the mathematics of aliasing to fully understand the phenomenon, and to take advantage of its properties in high frequency bandlimited A/D systems. Consider a complex exponential representation of the digital waveform in Eq. (1.2.3) showing both positive and negative frequencies.

$$
\begin{aligned}
x[n] &= A\cos(\Omega n) \\
&= \frac{A}{2}e^{+j\Omega n} + \frac{A}{2}e^{-j\Omega n}
\end{aligned}
\tag{1.2.3}
$$

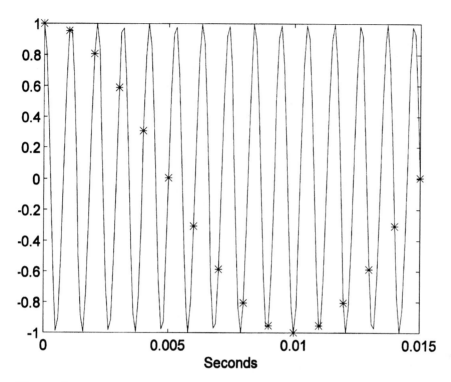

Figure 4 A 950 Hz analog signal sampled 1000 times per sec actually appears "aliased" as a 50 Hz signal.

While Eq. (1.2.3) compares well with (1.2.1) there is a big difference due to the digital sampling. Assuming no anti-aliasing filters are used, the digital frequency of $\Omega = 2\pi f/f_s$ (from the analog waveform sampled every T sec), could represent a multiplicity of analog frequencies.

$$A\cos(\Omega n) = A\cos(\Omega n \pm 2\pi m); \quad m = 0, 1, 2\ldots \tag{1.2.4}$$

For the real signal in Eq. (1.2.3), both the positive and negative frequencies have images at $\pm 2\pi m$; $m = 0, 1, 2, \ldots$ Therefore, if the analog frequency f is outside the Nyquist bandwidth of $0 - f_s/2$ Hz, one of the images of $\pm f$ will appear within the Nyquist bandwidth, but at the wrong (aliased) frequency. Since we want the digital waveform to a linear approximation to the original analog waveform, the frequencies of the two must be equal. One must always suppress frequencies outside the Nyquist bandwidth to be sure that no aliasing occurs. In practice, it is not possible to make an analog signal filter which perfectly passes signals in the Nyquist band while completely suppressing all frequencies outside this range. One should expect a transition zone near the Nyquist band upper frequency where unaliased frequencies are attenuated and some aliased frequency "images" are detectable. Most spectral analysis equipment will implement an anti-alias filter with a -3 dB cutoff frequency about $1/3$ the sampling frequency. The frequency range from $1/3 f_s$ to $1/2 f_s$ is usually not displayed as part of the observed spectrum so the user does not notice the the anti-alias filter's transition region and the filter very effectively suppresses frequencies above $f_s/2$.

Figure 5 shows a graphical representation of the digital frequencies and images for a sample rate of 1000 Hz and a range of analog frequencies including those of 100 Hz and 950 Hz in Figures 3 and 4, respectively. When the analog frequency exceeds the Nyquist rate of $f_s/2$ (π on the Ω axis), one of the negative frequency images (dotted lines) appears in the Nyquist band with the wrong (aliased) frequency, violating assumptions of system linearity.

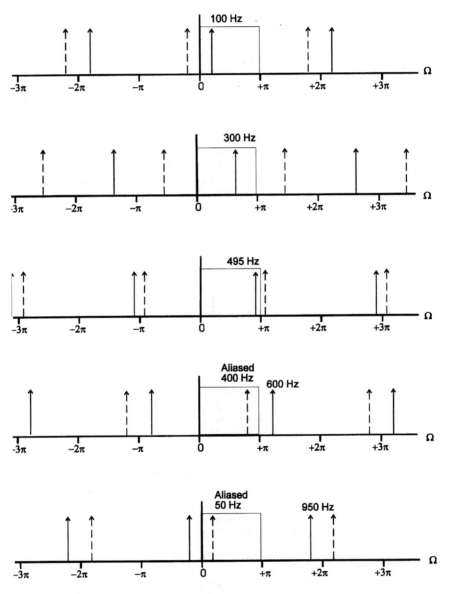

Figure 5 A graphical view of 100, 300, 495, 600 and 950 Hz sampled at 1000 Hz showing aliasing for 600 and 950 Hz.

1.3 COMPLEX BANDPASS SAMPLING

Bandpass sampling systems are extremely useful to adaptive signal processing systems which use high frequency sensor data but with a very narrow bandwidth of interest. Some excellent examples of these systems are active sonar, radar, and ultrasonic systems for medical imaging or non-destructive testing and evaluation of materials. These systems in general require highly directional transmit and receive transducers which physically means that the wavelengths used must be much smaller than the size of the transducers. The transmit and receive "beams" (comparable to a flashlight beam) can then be used to scan a volume for echoes from relatively big objects (relative to wavelength) with different impedances than the medium. The travel time from transmission to the received echo is related to the object's range by the wave speed.

Wave propagation speeds for active radar and sonar vary from a speedy 300 m/μsec for electromagnetic waves, to 1500 m/sec for sound waves in water, to a relatively slow 345 m/sec for sound waves in air at room temperature. Also of interest is the relative motion of the object along the beam. If the object is approaching, the received echo will be shifted higher in frequency due to Doppler, and lower in frequency if the object is moving away. Use of Doppler, time of arrival, and bearing of arrival, provide the basic target tracking inputs to active radar and sonar systems. Doppler radar has also become a standard meteorological tool for observing wind patterns. Doppler ultrasound has found important uses in monitoring fluid flow both in industrial processes as well as the human cardio-vascular system.

Given the sensor system's need for high frequency operation and relatively narrow signal bandwidth, a digital data acquisition system can exploit the phenomenon of aliasing to drastically reduce the Nyquist rate from twice the highest frequency of interest down to the bandwidth of interest. For example, suppose a Doppler ultrasound system operates at 1 MHz to measure fluid flow of approximately ± 0.15 m/sec. If the speed of sound is approximately 1500 m/sec, one might expect a Doppler shift of only ± 100 Hz. Therefore, if the received ultrasound is bandpass filtered from 999.9 kHz to 1.0001 MHz, it should be possible to extract the information using a sample rate on the order of 1 kHz, rather than the over 2 MHz required to sample the full frequency range. From an information point of view, bandpass sampling makes a lot of sense because only 0.01% of the over 1 MHz frequency range is actually required.

We can show a straightforward example using *real* aliased samples for the above case of a 1 MHz frequency with Doppler bandwidth of ± 100 Hz. First the analog signal is bandpass filtered attenuating all frequencies outside the 999.9 kHz to 1.0001 MHz frequency range of interest. By sampling at a rate commensurate with the signal bandwidth rather than absolute frequency, one of the aliased images will appear in the baseband between 0 Hz and the Nyquist rate. As seen in Figure 5, as the analog frequency increases to the right, the negative images all move to the left. Therefore, one of the positive images of the analog frequency is sought in the baseband. Figure 6 depicts the aliased bands in terms on the sample rate f_s.

So if the 1 MHz, ± 100 Hz signal is bandpass filtered from 999.9 kHz to 1.0001 MHz, we can sample at a rate of 1000.75 Hz putting the analog signal in the middle

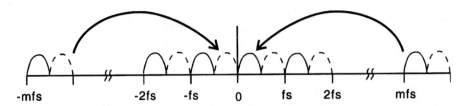

Figure 6 Analog frequencies bandpass filtered for the mth band will appear from 0 to $f_s/2$ Hz after sampling.

of the 999th positive image band. Therefore, one would expect to find a 1.0000 MHz signal aliased at 250.1875 Hz, 1.0001 MHz aliased at 350.1875 Hz, and 999.9 kHz at 150.1875 Hz in the digital domain. The extra 150 Hz at the top and bottom of the digital baseband allow for a transition zone of the anti-aliasing filters. Practical use of this technique requires precise bandpass filtering and selection of the sample rate. However, Figure 6 should also raise concerns about the effects of high-frequency analog noise "leaking" into digital signal processing systems at the point of A/D conversion. The problem of aliased electronic noise is particularly acute is systems where many high-speed digital signal processors operate in close proximity to high impedance analog circuits and the A/D subsystem has a large number of resolution bits.

For the case of a very narrow bandwidth at a high frequency it is obvious to see the numerical savings and it is relatively easy to pick a sample rate where only a little bandwidth is left unused. However, for wider analog signal bandwidths a more general approach is needed where the bandwidth of interest is not required to lie within a multiple of the digital baseband. To accomplish this we must insure that the negative images of the sampled data do not mix with the positive images for some arbituary bandwidth of interest. The best way to do this is to simply get rid of the negative frequency and its images entirely by using complex (real plus imaginary) samples.

How can one obtain complex samples from the real output of the A/D convertor? Mathematically one can describe a "cosine" waveform as the real part of a complex exponential. But, in the real world where we live (at least most of us some of the time), the sinusoidal waveform is generally observed and measured as a real quantity. Some exceptions to this are simultaneous measurement of spatially orthogonal (e.g. horizontal and vertical polarized) wave components such as polarization of electromagnetic waves, surface Rayleigh waves, or orbital vibrations of rotating equipment, all of which can directly generate complex digital samples. To generate a complex sample from a single real A/D convertor, we must tolerate a signal phase delay which varies with frequency. However, since this phase response of the complex sampling process is known, one can easily remove the phase effect in the frequency domain.

The usual approach is to gather the real part as before and to subtract in the imaginary part using a $T/4$ delayed sample.

$$x^R[n] = A\cos(2\pi fnT + \phi)$$
$$jx^I[n] = -A\cos\left(2\pi f\left[nT + \frac{T}{4}\right] + \phi\right)$$

(1.3.1)

The parameter ϕ in Eq. (1.3.1) is just an arbitrary phase angle for generality. For a frequency of $f = f_s$, Eq. (1.3.1) reduces to

$$x^R[n] = A\cos(2\pi n + \phi)$$
$$jx^I[n] = -\cos\left(2\pi n + \phi + \frac{\pi}{2}\right) \qquad (1.3.2)$$
$$= A\sin(2\pi n + \phi)$$

so that for this particular frequency the phase of the imaginary part is actually correct. We now have a usable bandwidth of f_s, rather than $f_s/2$ as with real samples. But, each complex sample is actually two real samples, keeping the total information rate (number of samples per sec) constant! As the frequency decreases towards 0, a phase error bias will increase towards a phase lag of $\pi/2$. However, since we wish to apply complex sampling to high frequency bandpass systems, the phase bias can be changing very rapidly with frequency, but it will be fixed for the given sample rate. The complex samples in terms of the digital frequency Ω and analog frequency f are

$$x^R[n] = A\cos(\Omega n + \phi)$$
$$jx^I[n] = -A\cos\left(\Omega n + \phi + \frac{\pi f}{2f_s}\right) \qquad (1.3.3)$$

giving a sampling phase bias (in the imaginary part only) of

$$\Delta\theta = -\frac{\pi}{2}\left(1 - \frac{f}{f_s}\right) \qquad (1.3.4)$$

For adaptive signal processing systems which require phase information, usually two or more channels have their relative phases measured. Since the phase bias caused by the complex sampling is identical for all channels, the phase bias can usually be ignored if relative channel phase is needed. The scheme for complex sampling presented here is sometimes referred to as "quadrature sampling" or even "Hilbert transform sampling" due to the mathematical relationship between the real and imaginary parts of the sampled signal in the frequency domain.

Figure 7 shows how any arbitrary bandwidth can be complex sampled at a rate equal to the bandwidth in Hz, and then digitally "demodulated" into the Nyquist baseband. If the signal bandwidth of interest extends from f_1 to f_2 Hz, an analog bandpass filter is used to band-limit the signal and complex samples are formed as seen in Figure 7 at a sample rate of $f_s = f_2 - f_1$ samples per sec. To move the complex data with frequency f_1 down to 0 Hz and the data at f_2 down to f_s Hz, all one needs to do is multiply the complex samples by $e^{-j\Omega_1 n}$, where Ω_1 is simply $2\pi f_1/f_s$. Therefore, the complex samples in Eq. (1.3.1) are demodulated as seen in Eq. (1.3.5).

$$x^R[n] = A\cos(\Omega n + \phi)e^{-j\Omega_1 n}$$
$$jx^I[n] = -A\cos\left(\Omega\left[n + \frac{1}{4}\right] + \phi\right)e^{-j\Omega_1 n} \qquad (1.3.5)$$

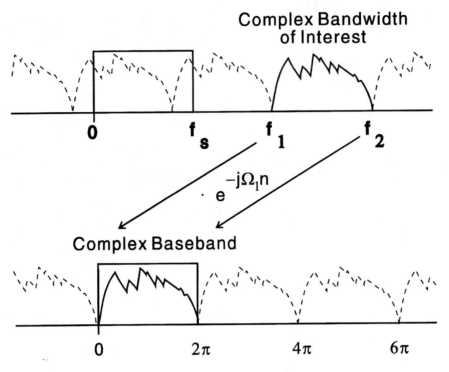

Figure 7 An arbitrary signal bandwith may be complex sampled and demodulated to baseband.

Analog signal reconstruction can be done by re-modulating the real and imaginary samples by f_1 in the analog domain. Two oscillators are needed, one for the $\cos(2\pi f_1 t)$ and the other for the $\sin(2\pi f_1 t)$. A real analog waveform can be reconstructed from the analog multiplication of the D/A converted real sample times the cosine minus the D/A converted imaginary sample times the sinusoid. As with the complex sample construction, some phase bias will occur. However, the technique of modulation and demodulation is well established in amplitude modulated (AM) radio. In fact, one could have just as easily demodulated (i.e. via an analog heterodyne circuit) a high frequency signal, band-limited it to a low-pass frequency range of half the sample rate, and A/D converted it as real samples. Reconstruction would simply involve D/A conversion, low-pass filtering, and re-modulation by a cosine waveform. In either case, the net signal information rate (number of total samples per sec) is constant for the same signal bandwidth. It is merely a matter of algorithm convenience and desired analog circuitry complexity from which the system developer must decide how to handle high frequency band-limited signals.

1.4 SUMMARY, PROBLEMS, AND BIBLIOGRAPHY

This section has reviewed the basic process of analog waveform digitization and sampling. The binary numbers from an A/D convertor can be coded into offset-binary or into two's complement formats for use with unsigned or signed integer arithmetic, respectively. Floating-point digital signal processors subsequently convert the integers from the A/D to their internal floating-point format for processing, and then back to the appropriate integer format for D/A conversion. Even though floating-point arithmetic has a huge numerical dynamic range, the limited dynamic range of the A/D and D/A convertors must always be considered. Adaptive signal processing systems can, and should, adaptively adjust input and output gains while maintaining floating-point data calibration. Adaptive signal calibration is straightforwardly-based on known transducer sensitivities, signal conditioning gains, and the voltage sensitivity and number of bits in the A/D and D/A convertors. The least-significant bit (LSB) is considered to be a random noise source, both numerically for the A/D convertor, and electronically for the D/A convertor. Given a periodic rate for sampling analog data and reconstruction of analog data from digital samples, analog filters must be applied before A/D and after D/A conversion to avoid unwanted signal aliasing. For real digital data, the sample rate must be at least twice the highest frequency which passes through the analog "anti-aliasing" filters. For complex samples, the complex-pair sample rate equals the bandwidth of interest, which may be demodulated to baseband if the bandwidth of interest was in a high frequency range. The frequency response of D/A conversion as well as sophisticated techniques for analog signal reconstruction will be discussed in the subsection on reconstruction filters later in the text.

PROBLEMS

1. An accelerometer with sensitivity 10 mV/G (1.0 G is 9.801 m/sec^2) is subjected to a ± 25 G acceleration. The electrical output of the accelerometer is amplified by 11.5 dB before A/D conversion with a 14-bit two's complement encoder with an input sensitivity of 0.305 mV/bit.

   (a) What is the numerical range of the digitized data?
   (b) If the amplifier can be programmed in 1.5 dB steps, what would be the amplification for maximum SNR? What is the SNR?

2. An 8-bit two's complement A/D system is to have no detectable signal aliasing at a sample rate of 100,000 samples per sec. An 8th-order (-48 dB/octave) programmable cutoff frequency low-pass filter is available.

   (a) What is a possible cutoff frequency fc?
   (b) For a 16-bit signed A/D what would the cutoff frequency be?
   (c) If you could tolerate some aliasing between fc and the Nyquist rate, what is the highest fc possible for the 16-bit system in part b?

3. An acceptable resolution for a medical ultrasonic image is declared to be 1 mm. Assume sound travels at 1500 m/sec in the human body.

   (a) What is the absolute minimum A/D sample rate for a receiver if it is to detect echoes from scatterers as close as 1 mm apart?
   (b) If the velocity of blood flow is to be measured in the range of ± 1m/sec (we don't need resolution here) using a 5 Mhz ultrasonic sinusoidal

burst, what is the minimum required bandwidth and sample rate for an A/D convertor? (hint: a Doppler-shifted frequency fd can be determined by $fd = f(1 + v/c)$, $c < v < + c$; where f is the transmitted frequency, c is the wave speed, and v is the velocity of the scatterer towards the receiver).

4. A microphone has a voltage sensitivity of $12 \, mV/Pa$ (1 Pascal = 1 Nt/m^2). If a sinusoidal sound of about 94 dB (approximately 1 Pa rms in the atmosphere) is to be digitally recorded, how much gain would be needed to insure a "clean" recording for a 10 V 16-bit signed A/D system?

5. A standard television in the United States has 525 vertical lines scanned in even and odd frames 30 times per sec. If the vertical field of view covers a distance of 1.0 m, what is the size of the smallest horizontal line thickness which would appear unaliased?

6. A new car television commercial is being produced where the wheels of the car have 12 stylish holes spaced every 30° around the rim. If the wheels are 0.7 m in diameter, how fast can the car move before the wheels start appearing to be rotating backwards?

7. Suppose a very low frequency high signal-to-noise ratio signal is being sampled at a high rate by a limited dynamic range 8-bit signed A/D convertor. If one simply adds consecutive pairs of samples together one has 9-bit data at half the sample rate. Adding consecutive pairs of the 9-bit samples together gives 10-bit data at 1/4 the 8-bit sample rate, and so on.

   (a) If one continued on to get 16-bit data from the original 8-data sampled at 10,000 Hz, what would the data rate be for the 16-bit data?

   (b) Suppose we had a very fast device which samples data using only 1-bit, 0 for negative and 1 for positive. How fast would the 1-bit A/D have to sample to produce 16-bit data at the standard digital audio rate of 44,100 samples per sec?

8. An "offset binary" encoded signal assumes single-signed input voltage where one adds a dc offset (if necessary) which forces the A/D input to be between 0 and the maximum positive voltage. The offset binary range for an M-bit A/D is 0 to 2^M-1, or 0 to 65535 for a 16-bit A/D.

   (a) For a sine wave with its zero crossing at 32767 and a 16-bit offset binary A/D, plot the wave response for a few cycles if the A/D data is misinterpreted as two's complement.

   (b) Suppose we have a zero mean two's complement A/D sampled sine wave. Plot the response of this digital signal if it is misinterpreted as offset binary encoded.

BIBLIOGRAPHY

P. M. Aziz, et al. An Overview of Sigma-Delta Converters, IEEE Sig Proc Mag, Jan 1996, pp. 6183.

A. Gill, Machine and Assembly Language Programming of the PDP-11, Englewood Cliffs: Prentice-Hall, 1978.

K. Hwang, Computer Arithmetic, Wiley: New York, 1979, p. 71.

A. V. Oppenheim, R. W. Schafer, Discrete-Time Signal Processing, Englewood Cliffs: Prentice-Hall, 1973.

REFERENCE

1. N. S. Jayant, P. Noll, Digital Coding of Waveforms, Englewood Cliffs: Prentice-Hall, 1984.

2

The Z-Transform

Given a complete mathematical expression for a discrete time-domain signal, why transform it to another domain? The main reason for time-frequency transforms is that many mathematical reductions are much simpler in one domain than the other. The z-transform in the digital domain is the counterpart to the Laplace transform in the analog domain. The z-transform is an extremely useful tool for analyzing the stability of digital sequences, designing stable digital filters, and relating digital signal processing operations to the equivalent mathematics in the analog domain. The Laplace transform provides a systematic method for solving analog systems described by differential equations. Both the z-transform and Laplace transform map their respective finite difference or differential systems of equations in the time or spatial domain to much simpler algebraic systems in the frequency or wavenumber domains, respectively. We begin by assuming time t increases as life progresses into the future, and a general signal of the form e^{st}, $s = \sigma + j\omega$, is *stable* for $\sigma \le 0$. A plot of our general signal is seen in Figure 1.

The quantity $s = \sigma + j\omega$, is a complex frequency where the real part σ represents the damping of the signal ($\sigma = -10.0$ Nepers/sec and $\omega = 50\pi$ rad/sec, or 25 Hz, in Figure 1). All signals, both digital and analog, can be described in terms of sums of the general waveform seen in Figure 1. This includes transient characteristics governed by σ. For $\sigma = 0$, one has a steady-state sinusoid. For $\sigma < 0$ as seen in Figure 1, one has an exponentially decaying sinusoid. If $\sigma > 0$, the exponentially increasing sinusoid is seen as unstable, since eventually it will become infinite in magnitude. Signals which change levels over time can be mathematically described using piecewise sums of stable and unstable complex exponentials for various periods of time as needed.

The same process of generalized signal modeling is applied to the *signal responses* of systems such as mechanical or electrical filters, wave propagation "systems," and digital signal processing algorithms. We define a "linear system" as an operator which changes the amplitude and/or phase (time delay) of an input signal to give an output signal with the same frequencies as the input, independent of the input signal's amplitude, phase, or frequency content. Linear systems can

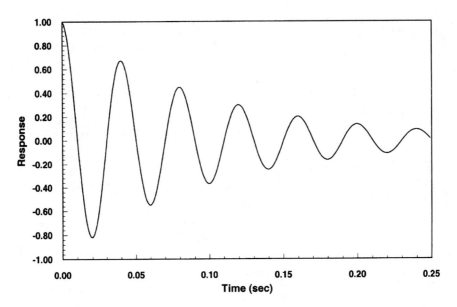

Figure 1 A "general" signal form of $e^{(\sigma+k\omega)t}$ where $\sigma \leq 0$ indicates a stable waveform for positive time.

be dispersive, where some frequencies travel through them faster than others, so long as the same system input–output response occurs independent of the input signal. Since there are an infinite number of input signal types, we focus on one very special input signal type called an impulse. An impulse waveform contains the same energy level at all frequencies including 0 Hz (direct current), and is exactly reproducible. For a digital waveform, a digital impulse simply has only one sample non-zero. The response of linear systems to the standard impulse input is called the *system impulse response*. The impulse response is simply the system's response to a Dirac delta function (or the unity amplitude digital-domain equivalent), when the system has zero initial conditions. The impulse response for a linear system is unique and a great deal of useful information about the system can be extracted from its analog or digital domain transform.

2.1 COMPARISON OF LAPLACE AND Z-TRANSFORMS

Equation (2.1.1) describes a general integral transform where $y(t)$ is transformed to $Y(s)$ using the kernal $K(s, t)$.

$$Y(s) = \int_{-\infty}^{+\infty} K(s, t)u(t)dt \tag{2.1.1}$$

The LaPlace transform makes use of the kernal $K(s, t) = e^{st}$, which is also in the form of our "general" signal as seen in Figure 1 above. We present the LaPlace transform as a pair of integral transforms relating the time and "s" domains.

$$Y(s) = \mathcal{L}\{y(t)\} = \int_0^{+\infty} y(t)e^{-st}\,dt$$

$$y(t) = \mathcal{L}^{-1}\{Y(s)\} = \frac{1}{2\pi j} \int_{\sigma-j\infty}^{\sigma+j\infty} Y(s)e^{+st}\,ds$$

(2.1.2)

The corresponding z-transform pair for discrete signals is seen in Eq. (2.1.3) where t is replaced with nT and denoted as $[n]$, and $z = e^{st}$.

$$Y[z] = Z\{y[n]\} = \sum_{n=0}^{+\infty} y[n]z^{-n}$$

$$y[n] = Z^{-1}\{Y[z]\} = \frac{1}{2\pi j} \oint_\Gamma Y[z]z^{n-1}\,dz$$

(2.1.3)

The closed contour Γ in Eq. (2.1.3) must enclose all the poles of the function $Y[z]\,z^{n-1}$. Both $Y(s)$ and $Y[z]$ are, in the most general terms, ratios of polynomials where the zeros of the numerator are also zeros of the system. Since the system response tends to diverge if excited near a zero of the denominator polynomial, the zeros of the denominator are called the system poles. The transforms in Eq. (2.1.2) and (2.1.3) are applied to signals, but if these "signals" represent system impulse or frequency responses, our subsequent analysis will refer to them as "systems," or "system responses."

There are two key points which must be discussed regarding the LaPlace and z-transforms. We are presenting what is called a "one-sided" or "causal" transform. This is seen in the time integral of Eq. (2.1.2) starting at $t = 0$, and the sum in Eq. (2.1.3) starting at $n = 0$. Physically, this means that the current system response is a result of the current and past inputs, *and specifically not future inputs*. Conversely, a current system input can have no effect on previous system outputs. Time only moves forward in the real physical world (at least as we know it in the 20th Century), so a distinction must be made in our mathematical models to represent this fact. Our positive time movement mathematical convention has a critical role to play in designating stable and unstable signals and systems mathematically. In the LaPlace transform's s-plane ($s = \sigma + j\omega$), only signals and system responses with a $\sigma \leq 0$ are mathematically stable in their causal response (time moving forward). Therefore, system responses represented by values of s in the left-hand plane ($j\omega$ is the vertical Cartesian axis) are stable causal response systems. As will be seen below, the nonlinear mapping from s-plane (analog signals and systems) to z-plane (digital signals and systems) maps the stable causal left-half s-plane to the region inside a unity radius circle on the z-plane, called the *unit circle*.

The comparison of the LaPlace and z-transforms is most useful when considering the mapping between the complex s-plane and the complex z-plane, where $z = e^{sT}$, T being the time interval in seconds between digital samples of the analog signal. The structure of this mapping depends on the digital sample rate and whether real or complex samples are used. An understanding of this mapping will allow one to easily design digital systems which model (or control) real physical systems in the analog domain. Also, adaptive system modeling in the digital domain of real physical systems can be quantitatively interpreted and related to other information processing in the adaptive system. However, if we have an analytical expression for a signal or system in the frequency domain, it may or may not be realizable as a stable causal signal or system response in the time domain (digital or analog). Again, this is due to the obliviousness of time to positive or negative direction. If we are mostly concerned with the magnitude response, we can generally adjust the phase (by adding time delay) to realize any desired response as a stable causal system. Table 1 below gives a partial listing of some useful LaPlace transforms and the corresponding z-transforms assuming regularly sampled data every T seconds ($f_s = 1/T$ samples per sec).

One of the more subtle distinctions between the LaPlace transforms and the corresponding z-transforms in Table 1 are how some of the z-transform magnitudes scale with the sample interval T. It can be seen that the result of the scaling is that the sampled impulse responses may not match the inverse z-transform if a simple

Table 1 Some Useful Signal Transforms

Time Domain	s Domain	z Domain
$1 \; t \geq 0$ $0 \; t < 0$	$\dfrac{1}{s}$	$\dfrac{z}{z - 1}$
$e^{s_0 t}$	$\dfrac{1}{s - s_0}$	$\dfrac{z}{z - e^{s_0 T}}$
$t \, e^{s_0 t}$	$\dfrac{1}{(s - s_0)^2}$	$\dfrac{T z e^{s_0 T}}{(z - e^{s_0 T})^2}$
$e^{-at} \sin \omega_0 t$	$\dfrac{\omega_0}{s^2 + 2as + a^2 + \omega_0^2}$	$\dfrac{z e^{-aT} \sin \omega_0 T}{z^2 - 2z e^{-aT} \cos \omega_0 T + e^{-2aT}}$
$e^{-at} \cos(\omega_0 t - \theta)$	$\dfrac{\cos \theta (s + a) + \omega_0 \sin \theta}{(s + a)^2 + \omega_0^2}$	$\dfrac{z \cos \theta (z - \alpha) - z \beta \sin \theta}{(z - \alpha)^2 + \beta^2}$ $\alpha = e^{-aT} \cos \omega_0 T$ $\beta = e^{-aT} \sin \omega_0 T$
$\dfrac{1}{ab} + \dfrac{e^{-at}}{a(a - b)} + \dfrac{e^{-bt}}{b(b - a)}$	$\dfrac{1}{s(s + s)(s + b)}$	$\dfrac{(Az + B)z}{(z - e^{-aT})(z - e^{-bT})(z - 1)}$ $A = \dfrac{b(1 - e^{-aT}) - a(1 - e^{-bT})}{ab(b - a)}$ $B = \dfrac{ae^{-aT}(1 - e^{-bT}) - be^{-bT}(1 - e^{-aT})}{ab(b - 1)}$

direct s-to-z mapping is used. Since adaptive digital signal processing can be used to measure and model physical system responses, we must be diligent to eliminate digital system responses where the amplitude depends on the sample rate. However, in the next section, it will be shown that careful consideration of the scaling for each system resonance or pole will yield a very close match between the digital system and its analog counterpart. At this point in our presentation of the z-transform, we compare the critical mathematical properties for linear time-invariant systems in both the analog LaPlace transform and the digital z-transform.

The LaPlace transform and the z-transform have many mathematical similarities, the most important of which are the properties of linearity and shift invariance. Linear shift-invariant system modeling is essential to adaptive signal processing since most optimization are based on a quadratic squared output error minimization. But even more significantly, linear time-invariant physical systems allow a wide range linear algebra to apply for the straightforward analysis of such systems. Most of the world around us is linear and time-invariant provided the responses we are modeling are relatively small in amplitude and quick in time. For example, the vibration response of a beam slowly corroding due to weather and rust is linear and time-invariant for small vibration amplitudes over a period of, say, days or weeks. But, over periods of years the beam's corrosion is changing the vibration response, therefore making it time varying in the frequency domain. If the forces on the beam approach its yield strength, the stress-strain relationship is no longer linear and single frequency vibration inputs into the beam will yield nonlinear multiple frequency outputs. Nonlinear signals are rich in physical information but require very complicated models. From a signal processing point of view, it is extremely valuable to respect the physics of the world around us, which is only linear and time-invariant within specific physical constraints, and exploit linearity and time-invariance wherever possible. Nonlinear signal processing is still something much to be developed in the future. Below is a summary comparison of LaPlace and z-transforms.

Linearity

Both the LaPlace and z-transforms are linear operators.

$$\mathcal{L}\{af(t) + bg(t)\} = aF(s) + bG(s)$$
$$Z\{af[k] + bg[k]\} = aF[z] + bG[z]$$

(2.1.4.)

The inverse LaPlace and z-transforms are also linear.

Delay Shift Invariance

Assuming one-sided signals $f(t) = f[k] = 0$ for $t, k < 0$, (no initial conditions)

$$\mathcal{L}\{f(t - \tau)\} = e^{-s\tau}F(s)$$
$$Z\{f(k - N)\} = z^{-N}F[z]$$

(2.1.5)

Convolution

Linear shift-invariant systems have the property a multiplication of two signals in one domain is equivalent to a convolution in the other domain.

$$\mathcal{L}\{f(t) * g(t)\} \doteq \mathcal{L}\left\{ \int_0^t f(\tau)g(t - \tau)d\tau \right\} = F(s)G(s) \tag{2.1.6}$$

A more detailed derivation of Eq. (2.1.6) will be presented in the next section. In the digital domain, the convolution integral becomes a simple summation.

$$Z\{f[k] * g[k]\} \doteq Z\left\{ \sum_{k=0}^m f[k]g[m - k] \right\} = F[z]G[z] \tag{2.1.7}$$

If $f[k]$ is the impulse response of a system and $g[k]$ is an input signal to the system, the system output response to the input excitation $g[k]$ is found in the time domain by the convolution of $g[k]$ and $f[k]$. However, the system must be both linear and shift invariant (a shift of k samples in the input gives a shift of k samples in the output), for the convolution property to apply. Eq. (2.1.7) is fundamental to digital systems theory and will be discussed in great detail later.

Initial Value

The initial value of a one-sided (causal) impulse response is found by taking the limit as s or z approaches infinity.

$$\lim_{t \to 0} f(t) = \lim_{s \to \infty} sF(s) \tag{2.1.8}$$

The initial value of digital impulse response can be found in an analogous manner.

$$f[0] = \lim_{z \to \infty} F[z] \tag{2.1.9}$$

Final Value

The final value of a causal impulse response can be used as an indication of the stability of a system as well as to determine any static offsets.

$$\lim_{t \to \infty} f(t) = \lim_{s \to 0} sF(s) \tag{2.2.10}$$

Equation (2.1.10) holds so long as $sF(s)$ is analytic in the right-half of the s-plane (no poles on the $j\omega$ axis and for $\sigma \geq 0$). $F(s)$ is allowed to have one pole at the origin and still be stable at $t = \infty$. The final value in the digital domain is seen in Eq. (2.1.11).

$$\lim_{k \to \infty} f[k] = \lim_{z \to 1} (1 - z^{-1})F[z] \tag{2.1.11}$$

$(1 - z^{-1})F[z]$ must also be analytic in the region on and outside the unit circle on the z-plane. The region $|z| \geq 1$, on and outside the unit circle on the z-plane, corresponds to the region $\sigma \geq 0$, on the $j\omega$ axis and in the right-hand s-plane. The s-plane pole $F(s)$

is allowed to have at $s = 0$ in equation maps to a z-plane pole for $F[z]$ at $z = 1$ since $z = e^{sT}$. The allowance of these poles is related to the restriction of causality for one-sided transforms. The mapping between the s and z planes will be discussed in some more detail on pages 26–28.

Frequency Translation/Scaling

Multiplication of the analog time-domain signal by an exponential leads directly to a frequency shift.

$$\mathcal{L}\{e^{-at}f(t)\} = F(s + a) \tag{2.1.12}$$

In the digital domain, multiplication of the sequence $f[k]$ by a geometric sequence α^k results in scaling the frequency range.

$$Z\{\alpha^k f[k]\} = \sum_{k=0}^{\infty} f[k]\left(\frac{z}{\alpha}\right)^{-k} = F[z/\alpha] \tag{2.1.13}$$

Differentiation

The LaPlace transform of the derivative of the function $f(t)$ is found using integration by parts.

$$\mathcal{L}\left\{\frac{\partial f}{\partial t}\right\} = sF(s) - f(0) \tag{2.1.14}$$

Carrying out integration by parts as in Eq. (2.1.14) for higher-order derivatives yields the general formula

$$\mathcal{L}\left\{\frac{\partial f^N}{\partial t^N}\right\} = s^N F(s) - \sum_{k=0}^{N-1} s^{N-1-k} f^{(k)}(0) \tag{2.1.15}$$

where $f^{(k)}(0)$ is the kth derivative of $f(t)$ at $t = 0$. The initial conditions for $f(t)$ are necessary to its LaPlace transform just as they are necessary for the complete solution of an ordinary differential equation. For the digital case, we must first employ a formula for carrying forward initial conditions in the z-transform of a time-advanced signal.

$$Z\{x[n + N]\} = x^N X[z] - \sum_{k=0}^{N-1} z^{N-k} x[k] \tag{2.1.16}$$

For a causal sequence, Eq. (2.1.16) can be easily proved from the definition of the z-transform. Using an approximation based on the definition of the derivatives, the first derivatives of a digital sequence is seen as

$$\dot{x}[n + 1] = \frac{1}{T}(x[n + 1] - x[n]) \tag{2.1.16}$$

where T is the sample increment. Applying the time-advance formula in Eq. (2.1.16) gives the z-transform of the first derivative.

$$Z\{\dot{x}[n+1]\} = \frac{1}{T}\{(z-1)X[z] - zx[0]\} \tag{2.1.18}$$

Delaying the sequence by 1 sample shows the z-transform of the first derivative of $x[n]$ at sample n.

$$Z\{\dot{x}[n]\} = \frac{1}{T}\{(1-z^{-1})X[z] - x[0]\} \tag{2.1.19}$$

The second derivative can be seen to be

$$Z\{\ddot{x}[n]\} = \frac{1}{T^2}\{(1-z^{-1})^2 X[z] \\ - [(1-2z^{-1})x[0] + z^{-1}x[1]]\} \tag{2.1.20}$$

The pattern of how the initial samples enter into the derivatives can be more easily seen in the third derivative of $x[n]$, where the polynomial coefficients weighting the initial samples can be seen as fragments of the binomial polynomial created by the triple zero at $z = 1$.

$$Z\{\dddot{x}[n]\} = \frac{1}{T^3}\{(1-z^{-1})^3 X[z] \\ - (1-3z^{-1}+3z^{-2})x[0] \\ - (z^{-1}-3z^{-2})x[1] - z^{-2}x[2]\} \tag{2.1.21}$$

Putting aside the initial conditions on the digital-domain definitive, it is straightforward to show that the z-transform of the Nth definitive of $x[n]$ simply has N zeros at $z = 1$ corresponding to the analogous N zeros at $s = 0$ in the analog domain.

$$Z\{x^{(N)}[n]\} = \frac{1}{T^N}\{(1-z^{-1})^N X[z] - initial\ conditions\} \tag{2.1.22}$$

Mapping Between the s and z Planes

As with the aliased data in section 1.1 above, the effect of sampling can be seen as a mapping between the series of analog frequency bands and the digital baseband defined by the sample rate and type (real or complex). To make sampling useful, one must band-limit the analog frequency response to a bandwidth equal to the sample rate for complex samples, or low-pass filter to half the sample rate for real samples. Consider the effect of replacing the analog t in $z^n = e^{st}$ with nT, where n is the sample number and $T = 1/f_s$ is the sampling interval in seconds.

$$z^n = e^{(\sigma+j\omega)nT} = e^{\left(\frac{\sigma}{f_s}+j\frac{2\pi f}{f_s}\right)n} \tag{2.1.23}$$

As can be seen in Eq. (2.1.23), the analog frequency repeats every multiple of f_s (a full f_s Hz bandwidth is available for complex samples). For real samples (represented by a phase shifted sine or cosine rather than a complex exponential), a frequency band f_s Hz wide will be centered about 0 Hz giving an effective signal bandwidth of only $f_s/2$ Hz for positive frequency. The real part of the complex spectrum is symmetric for positive and negative frequency while the imaginary part is skew-symmetric (negative frequency amplitude is opposite in sign from positive frequency amplitude. This follows directly from the imaginary part of $e^{j\theta}$ being j $\sin\theta$. The amplitude of the real and imaginary parts of the signal spectrum are determined by the phase shift of the sine or cosine. For real time-domain signals sampled at f_s samples per sec, the effective bandwidth of the digital signal is from 0 to $f_s/2$ Hz. For $\sigma \pm 0$, a strip within $\leq \omega_s/2$ for the left-half of the complex s-plane maps into a region inside a unit radius circle on the complex z-plane. For complex sampled systems, each multiple of f_s Hz on the s-plane corresponds to a complete trip around the unit circle on the z-plane. In other words, the left-half of the s-plane is subdivided into an infinite number of parallel strips, each ω_s radians wide which all map into the unit circle of the z-plane. As described in Chapter 1, accurate digital representation requires that one band-limit the analog signal to one of the s-plane strips before sampling.

For real sampled systems, the upper half of the unit circle from the angles from 0 to π is a "mirror image" of the lower half circle from the angles of 0 to $-\pi$. Figure 2 shows a series of s-values (depicted as the complex values "A" through "I") and their corresponding positions on the complex z-plane for a real sampled system [2.2]. So long as the analog signals a filtered appropriately before sampling (and after D/A conversion for output analog signal reconstruction), the digital representation on the complex z-plane will be accurate. The letters "A" through "I" depict mappings between the analog s-plane and the digital z-plane in Figure 2.

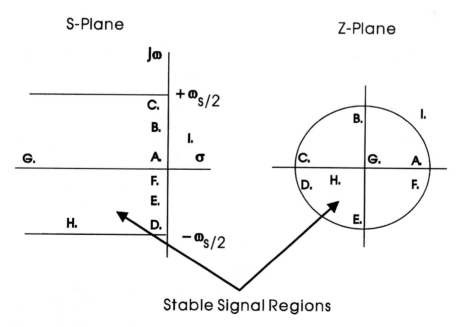

Figure 2 The region between $\pm\omega_s/2$ (and its images) on the left-hand s-plane maps to a region inside the unit circle on the z-plane.

The mapping between the s and z planes can be seen as a straightforward implementation of periodic sampling. If one had an mathematical expression for the LaPlace transform $H(s)$ of some system impulse response $h(t)$, the poles and zeros of $H(s)$ can be directly mapped to the z-plane as seen in Figure 2, giving a digital domain system response $H[z]$ with inverse z-transform $h[n]$. The sampled system impulse response $h[n]$ should simply be the analog domain impulse response $h(t)$ sampled every nT sec, provided of course that the analog signal is first appropriately band-limited for the sample rate.

While the mapping given here is similar to bi-linear transformation, the mapping from the s-plane $j\omega$ axis onto the z-plane unit circle in Figure 2 is actually linear in frequency. Bilinear transformations include a nonlinear mapping to the "w-plane" where the frequency part of the mapping follows a tangent, rather than linear, relationship. Only at very low frequencies on the w-plane do the two frequency scales approximate one another. The w-plane is typically used for stability analysis using techniques like the Routh–Hurwitz criterion or Jury's test. In our work here, we simply map the poles and zeros back and forth as needed between the z-plane and a band-limited s-plane primary strip bounded by $\pm \omega_s/2$. The interior of the z-plane unit circle corresponds to the left-half ($\sigma < 0$) of the s-plane primary strip.

The approach of mapping between the s and z planes does have an area where some attention to physics is needed. The mapping of the s-plane and z-plane poles and zeros insures that the frequency response of the two systems is relatively accurate at low frequencies. But, an accurate digital system representation of a physical analog-domain system must have an impulse response that is identical in both the digital and analog domains. As will be seen below, the digital-domain impulse response will differ from the analog impulse response by a scale factor for any system with more than a single pole mapped directly between the s and z-planes. The scale factor is found in a straightforward manner by comparing the analog and digital impulse responses from the inverse LaPlace and inverse z-transforms, respectively. Each system "mode" can be isolated using a partial fraction expansion and scaled independently to make the analog and digital domain systems very nearly identical. Because $z = e^{sT}$ is not a linear function, the transition from analog to digital domain is not a perfect match. One can match the system modes which will give a nearly identical impulse response but will not match the spectral zeros (or nulls in the frequency response), or one can match the frequencies of the poles (spectral peaks or resonances) and zeros, but the modal amplitudes and impulse response will differ between the digital and analog system responses. The next section illustrates the process for a simple mass-spring-damper mechanical oscillator and its corresponding digital system model.

2.2 SYSTEM THEORY

Common systems such as the wave equation, electrical circuits made up of resistors, capacitors, and inductors, or mechanical systems made up of dampers, springs, and masses are described in terms of 2nd-order linear differential equations of the form

$$M \frac{\partial^2 y(t)}{\partial t^2} + R \frac{\partial y(t)}{\partial t} + Ky(t) = Ax(t) \qquad (2.2.1)$$

where $Ax(t)$ is the applied input to the system for $t > 0$ and $y(t)$ is the resulting output assuming all system initial conditions are zero. If Eq. (2.2.1) describes a mechanical oscillator, $Ax(t)$ has units of force, $y(t)$ is displacement in meters m, M is mass in Kg, K is stiffness in Kg/s^2, and R is damping in Kg/s. Note that each of the three terms on the left side of Eq. (2.2.1) has units of force in nT, where $1\ nT = 1\ Kg\ m/s^2$. Assuming the input force waveform $Ax(t)$ is a force impulse $f_0\delta(t)$, the displacement output $y(t)$ of Eq. (2.2.1) can be seen as a system impulse response only if there are zero initial conditions on $y(t)$ and all its time derivatives. Taking LaPlace transforms of both sides of Eq. (2.2.1) reveals the general system response.

$$M\{s^2 Y(s) - sy(0) - \dot{y}(0)\} + R\{s Y(s) - y(0)\} + KY(s) = f_0 \tag{2.2.2}$$

Since the LaPlace transform of $f(t) = f_0\delta(t)$ is $F(s) = f_0$,

$$\begin{aligned} Y(s) &= \frac{F(s) + (Ms + R)y(0) + M\dot{y}(0)}{\{Ms^2 + Rs + K\}} \\ &= H(s)G(s) \end{aligned} \tag{2.2.3}$$

where $H(s)$ is called the system function

$$H(s) = \frac{1}{\{Ms^2 + Rs + K\}} \tag{2.2.4}$$

and $G(s)$ is called the system excitation function.

$$G(S) = F(s) + (Ms + R)y(0) + M\dot{y}(0) \tag{2.2.5}$$

Usually, one separates the initial conditions from the system response

$$Y(s) = F(s)H(s) + \frac{(Ms + R)y(0) + M\dot{y}(0)}{\{Ms^2 + Rs + K\}} \tag{2.2.6}$$

since the effect of the initial conditions on the system eventually die out relative to a steady state excitation $F(s)$. Figure 3 depicts a mass-spring-damper mechanical oscillator and its LaPlace transform system model equivalent.

Taking inverse LaPlace transforms of Eq. (2.2.6) gives the system displacement response $y(t)$ to the force input $f(t)$ and the initial displacement $y(0)$ and velocity $\dot{y}(0)$. The inverse LaPlace transform of the s-domain product $F(s)H(s)$ brings us to a very important relationship called the *convolution integral*. We now examine this important relationship between the time and frequency domains. Applying the definition of the LaPlace transform in Eq. (2.1.2), to $F(s)H(s)$ one obtains

$$F(s)H(s) = \int_0^\infty f(\tau)e^{-s\tau}d\tau \int_0^\infty h(\beta)e^{-s\beta}d\beta \tag{2.2.7}$$

We can re-arrange Eq. (2.2.7) to give

$$F(s)H(s) = \int_0^\infty f(\tau)d\tau \int_0^\infty h(\beta)e^{-s(\beta+\tau)}d\beta \tag{2.2.8}$$

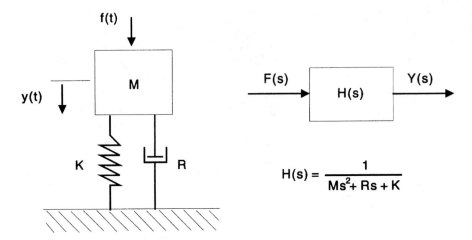

Figure 3 A mechanical oscillator and its equivalent LaPlace transform system model with force input and displacement output.

Equation (2.2.8) is now changed so $\beta = t - \tau$

$$F(s)H(s) = \int_0^\infty f(\tau)\,d\tau \int_\tau^\infty h(t - \tau)e^{-st}\,dt \tag{2.2.9}$$

Finally, interchanging the order of integration we obtain

$$F(s)H(s) = \int_0^\infty \left(\int_0^t f(\tau)h(t - \tau)\,d\tau \right)e^{-st}\,dt \tag{2.2.10}$$

where the middle integral in parenthesis in Eq. (2.2.10) is the convolution of $f(t)$ and $h(t)$. It can be seen that the product of two system functions in the frequency domain (s-plane or z-plane) results in the convolution of the corresponding time domain functions. Conversely, the product of two time domain functions is equivalent to the convolution of their corresponding frequency domain functions. Also note that the conjugate product in the frequency domain is equivalent to a cross correlation of the two time domain signals.

The inverse LaPlace transform of our mechanical oscillator system in Eq. (2.2.6) is therefore

$$y(t) = \int_0^t f(\tau)h(t - \tau)d\tau + y(0)\mathscr{L}^{-1}\left\{ \frac{Ms + R}{Ms^2 + Rs + K} \right\}$$
$$+ \dot{y}(0)\mathscr{L}^{-1}\left\{ \frac{M}{Ms^2 + Rs + K} \right\} \tag{2.2.11}$$

where $h(t)$ is the inverse LaPlace transform of the system function $H(s)$.

If the initial conditions are both zero the second two terms in Eq. (2.2.11) vanish, and if the force input to the system $f(t)$ is a unity amplitude Dirac delta function $f(t) = f_0 \delta(t); f_0 = 1$, the displacement output response $y(t)$ is exactly the same as the system impulse response $h(t)$. The system impulse response $h(t)$ can be seen as the solution to a homogeneous differential Eq. (2.2.1) while the first term of the right-hand side of Eq. (2.2.11) is seen as the forced response to a nonhomogeneous equation. The later two right-hand terms form the complete general solution to Eq. (2.2.1). In our chosen case of a unity impulse input force with zero initial conditions we can write the system response $H(s)$ as

$$H(s) = \frac{1}{Ms^2 + Rs + K} \qquad (2.2.12)$$

which can be written as a partial fraction expansion as

$$H(s) = \frac{1}{s_1 - s_2} \left\{ \frac{1}{s - s_1} - \frac{1}{s - s_2} \right\} \qquad (2.2.13)$$

where the two system poles s_1 and s_2 are simply

$$s_{1,2} = \frac{-R}{2M} \pm j \sqrt{\frac{K}{M} - \left(\frac{-R}{2M}\right)^2}$$

$$= -\zeta \pm j\omega_d \qquad (2.2.14)$$

The inverse LaPlace transform of Eq. (2.2.13) is therefore

$$h(t) = \frac{e^{-\zeta t}}{j 2\omega_d} \left[e^{+j\omega_d t} - e^{-j\omega_d t} \right]$$

$$= \frac{e^{-\zeta t}}{\omega_d} \sin \omega_d t \qquad (2.2.15)$$

Equation (2.2.15) is the analog domain system impulse response $h(t)$. This simple physical example of a system and its mathematical model serves us to show the importance of linearity as well as initial conditions. Given this well-founded mathematical model, we can transfer, or map, the model to the digital domain using z-transforms in place of the LaPlace transform. The physical relationship between physical system models in the digital and analog domains is extremely important to understand since we can use adaptive digital signal processing to identify physical models in the digital domain. Generating *information* about a physical system from the digital models requires a seamless transition from the z-plane back to the s-plane independent of digital system sample rates, number of A/D bits, and transducer sensitivities. In other words, if we build a processing algorithm which adapts to the input and output signals of an unknown system, and we wish to represent physical quantities (such as mass, stiffness, damping, etc) from their digital counterparts, we must master the transition between the digital and analog domains.

2.3 MAPPING OF *S*-PLANE SYSTEMS TO THE DIGITAL DOMAIN

We now examine the implications of mapping the *s*-plane poles in Eq. (2.2.14) to the *z*-plane using the bi-linear transformation $z = e^{sT}$. This transformation alone places the poles at the correct frequencies, but not necessarily the same amplitudes as the analog domain counterpart.

$$z_1 = e^{-\zeta T + j\omega_d T}$$
$$z_2 = e^{-\zeta T - j\omega_d T} \tag{2.3.1}$$

The mapped z-transform for the system is depicted as $H^m[z]$ and written as

$$H^m[z] = \frac{1}{(z - z_1)(z - z_2)} = z^{-2}\frac{1}{(1 - z_1 z^{-1})(1 - z_2 z^{-1})} \tag{2.3.2}$$

where the negative powers of *z* are preferred to give a causal z-transform using some additional time delay. Expanding Eq. (2.3.2) using partial fractions gives a sum of two poles which will simplify the inverse z-transform.

$$H^m[z] = \frac{z^{-1}}{(z_1 - z_2)}\left\{\frac{1}{(1 - z_1 z^{-1})} - \frac{1}{(1 - z_2 z^{-1})}\right\} \tag{2.3.3}$$

The two terms in the braces in Eq. (2.3.3) can be re-written as an infinite geometric series.

$$H^m[z] = \frac{z^{-1}}{(z_1 - z_2)}\sum_{k=0}^{\infty}(z_1^k - z_2^k)z^{-k} \tag{2.3.4}$$

By inserting the *z*-plane poles into Eq. (2.3.4) and substituting $n = k + 1$ we have

$$H^m[z] = \frac{1}{e^{-\zeta T}\sin(\omega_d T)}\sum_{n=1}^{\infty}e^{-\zeta T(n-1)}\sin(\omega_d T(n - 1))z^{-k} \tag{2.3.5}$$

From examination of Eqs (2.1.3) and (2.3.5) the inverse z-transform of the mapped poles is

$$h^m[n] = \frac{e^{-\zeta T(n-1)}\sin(\omega_d T(n - 1))}{e^{-\zeta T}\sin(\omega_d T)}; n > 0 \tag{2.3.6}$$

Clearly, the mapped system impulse response has a scale factor (in the denominator of Eq. (2.3.6)) which is dependent on the sample interval T. As ω_d approaches $\omega_s/2$ or 0, $h^m[n]$ will have a large amplitude compared to the sampled analog domain impulse response $h(nT)$. The time delay of 1 sample is a direct result of causality constraints on digital systems. Figure 4 below shows the impulse response of our system with a poles near ± 25 Hz and a damping factor of $\zeta = 10$ sampled at rates of 57 Hz, 125 Hz, and 300 Hz. At 57 samples per sec, $\omega_d T$ is nearly π and the mapped impulse response of Eq. (2.3.6) is larger in amplitude than the physical system. At 125 samples per sec, the pole at $\omega_d T$ is nearly $\pi/2$, making the sin function nearly unity and the two impulse responses to be a pretty good match. However, as we sample at faster and faster

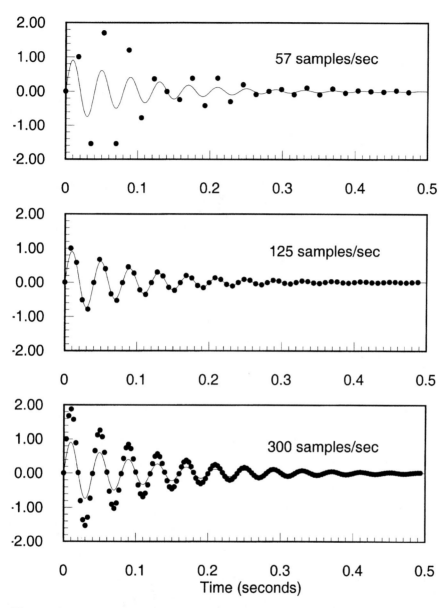

Figure 4   An unscaled digital impulse response may be accurate depending on sample rate.

rates, $\omega_d T$ approaches 0 and the unscaled digital impulse response becomes artificially huge. Conversely, given a digital measurement of a real physical system impulse response, extracting physical information from the digital domain model mapped to the analog domain requires the proper scaling which depends on the sample rate as seen in Eq. (2.3.6).

   Since we want the digital impulse response to match the true physical system's impulse response as perfectly as possible, we must scale the digital system to remove the sample rate dependence.

$$h([n-1]T) = \left\{ \frac{e^{-\zeta T} \sin(\omega_d T)}{\omega_d} \right\} h^m[n]; \; n > 0 \tag{2.3.7}$$

Figure 5 below shows that this scaling allows any sample rate to be used for our mechanical oscillator system's digital model. The scaling required is actually a result of the partial fraction expansions in the two domains. The time delay in the scaled mapped digital impulse response is unavoidable due to causality constraints on real

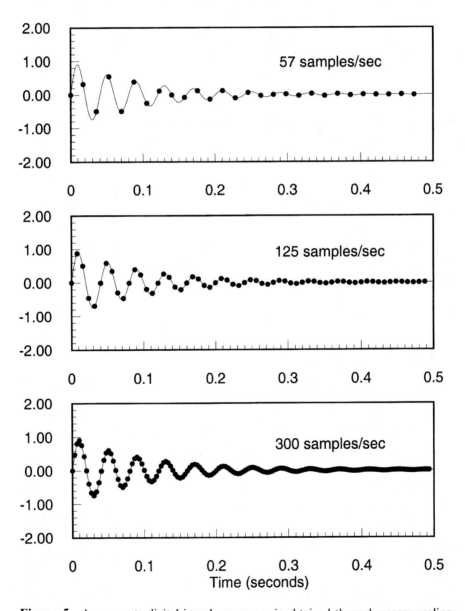

Figure 5   An accurate digital impulse response is obtained through proper scaling.

digital systems. Since the sample interval is usually quite small, the delay is generally of little consequence.

The scale factor in Eq. (2.3.6) allows the digital impulse response to very nearly match the analog system. For very

$$\frac{e^{-\zeta T}\sin(\omega_d T)}{\omega_d} = \frac{z_1 - z_2}{s_1 - s_2} \tag{2.3.8}$$

fast sample rates where ω_d is small compared to $2\pi f_s$, Eq. (2.3.8) can be approximated by $T = 1/f_s$ as presented in Ref. 3. The good news about the scaling between the z and s-planes is that it is always a constant factor (linear) regardless of the number of poles and zeros being mapped.

Comparing the frequency responses of the system LaPlace transform $H(s)$ evaluated for $j0 \le s \le j\omega_s/2$, and the scaled and mapped digital system $H[z]$ evaluated on the unit circle provides another measure of the importance of scaling. Evaluating the frequency response of a digital system $H[z]$ requires an additional scale factor of T, the digital sampling time interval which can be seen to be the counterpart of the differential "dt" in the LaPlace transform. The properly scaled frequency response of the digital domain $H[z]$ is seen to closely approximate that for the analog domain $H(s)$ for any non-aliased sample rate as shown in Eq. (2.3.9).

$$H(s)|_{s=j\omega} \approx H[z]|_{z=e^{j\omega T}} = T\left\{\frac{z_1 - z_2}{s_1 - s_2}\right\}\frac{1}{(z - z_1)(z - z_2)}|_{z=e^{j\omega T}} \tag{2.3.9}$$

Figure 6 below compares the frequency responses of $H(s)$ and $H[z]$ in terms of dB-magnitude and phase in degrees. The case shown in Figure 6 is for two 25 Hz conjugate poles with a damping factor of $\zeta = 10$ Nepers and a sampling frequency of 300 samples per sec. Except for the linear phase response due to the delay of one sample in $H[z]$, the magnitude and phase compare quite well, especially when considering the sensitivity of the log scale. For a high sampling frequency, or relatively low resonance, the net scale factor can be approximated by T^2. But, if the resonance is near the Nyquist frequency, the scaling in Eq. (2.3.9) should be employed.

As can be seen in Figure 6, the match is very close at low frequencies. If we remove the sample delay from the digital system phase frequency response, the two phase curves almost exactly match. It can be seen that even with careful scaling of digital systems, the transference to the digital domain is not perfect. At the upper end of the spectrum where system features are observed through signals which are sparsely sampled, the system response errors are larger in general. From a system fidelity point of view, oversampling is the most common way to drive the dominant system features to the low digital frequency range where very precise system modeling can occur. Figure 7 shows the same $H(s)$ system model but for a digital system $H[z]$ sampled at only 57 samples per sec. Clearly, the scaling shown in Eq. (2.3.9) succeeds in providing a close match near the system resonance, but high-frequency problems exist nonetheless. In the high frequency range, scaling by T^2, rather than that shown in Eq. (2.3.9), actually produces a very significant gain error.

Finding the precise scale constants becomes much more difficult for complicated systems with many system resonances, or modes. However, each mode

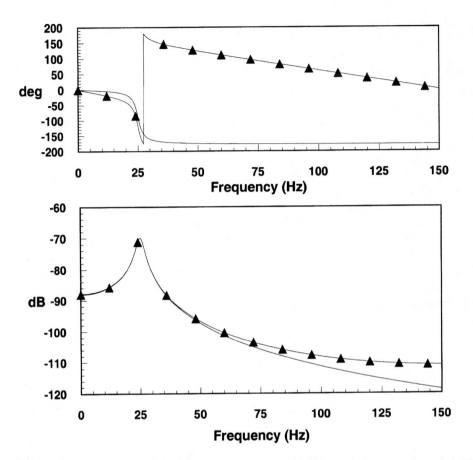

Figure 6   Comparison of the frequency responses of $H(s)(-)$ and the properly scaled $H[z]$ (▲) for a sample rate of 300 Hz.

(or system resonance) can be scaled in the same manner as the simple oscillator above to give a reasonable match for both the time-domain impulse responses, and the frequency-domain system responses. But for systems with both high frequency poles and zeros (resonances and anti-resonances, respectively), a design choice must be made. The choice is between either matching the frequencies of the poles and zeros with an error in system response magnitude, or matching the impulse response and system resonances with an error in the zero (system anti-resonance) frequencies. As will be seen below, the only recourse for precise analog and digital system and impulse response matching is to substantially "over-sample" the digital system, where all the poles and zeros are at very low frequencies.

Consider the following two-zero, four-pole system with a real impulse response requiring that the poles and zeros be conjugate pairs. Let the two zeros z_1^s be at $\sigma = +5$, $j\omega/2\pi = \pm 130$ Hz, the first pair of poles p_1^s at $\sigma = -20$, $j\omega/2\pi = \pm 160$ Hz, and the second pair of poles p_2^s at $\sigma = -10$, $j\omega/2\pi = \pm 240$ Hz on the s-plane. Note that the positive damping for the zeros does not cause an unstable system, but rather a *non-minimum phase* system, which will be discussed in more detail

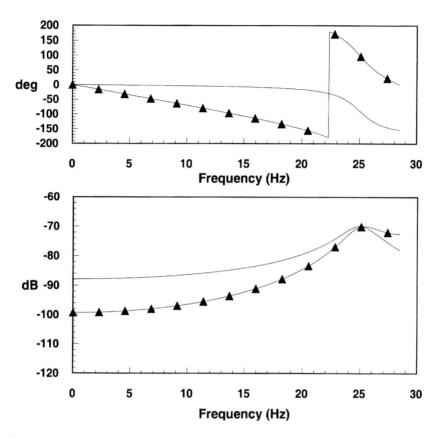

Figure 7 The same analog system $H(s)$ as in Figure 6, but for a digital system sampled at only 57 Hz showing the difficulties of spectral matching at high frequencies.

in the next section. Our s-plane system response function is

$$H(s) = \frac{(s - z_1^s)(s - z_1^{s*})}{(s - p_1^s)(s - p_1^{s*})(s - p_2^s)(s - p_2^{s*})} \tag{2.3.10}$$

where the impulse response is found using partial fraction expansions as seen in Eq. (2.3.11).

$$h(t) = A_s e^{p_1^s t} + B_s e^{p_1^{s*} t} + C_s e^{p_2^s t} + D_s e^{p_2^{s*} t} \tag{2.3.11}$$

where

$$A_s = \frac{(p_1^s - z_1^s)(p_1^s - z_1^{s*})}{(p_1^s - p_1^{s*})(p_1^s - p_2^s)(p_1^s - p_2^{s*})}$$

$$B_s = \frac{(p_1^{s*} - z_1^s)(p_1^{s*} - z_1^{s*})}{(p_1^{s*} - p_1^{s*})(p_1^{s*} - p_2^s)(p_1^{s*} - p_2^{s*})}$$

$$C_s = \frac{(p_2^s - z_1^s)(p_2^s - z_1^{s*})}{(p_2^s - p_1^{s*})(p_2^s - p_1^s)(p_2^s - p_2^{s*})} \tag{2.2.12}$$

$$D_s = \frac{(p_2^{s*} - z_1^s)(p_2^{s*} - z_1^{s*})}{(p_2^{s*} - p_1^{s*})(p_2^{s*} - p_1^s)(p_2^{s*} - p_2^s)}$$

Applying the mapping and modal scaling technique described previously in this section, the discrete impulse response which is seen to closely approximate Eq. (2.3.11) is

$$h[n-1] = A_s(p_1^z)^n + B_s(p_1^{z*})^n + C_s(p_2^z)^n + D_s(p_2^{z*})^n \tag{2.3.13}$$

where p_1^z is the z-plane mapped pole corresponding to p_1^s, etc. The discrete system frequency response is found following Eq. (2.3.9).

$$H^2[z] = T\left[\frac{A_s}{z-p_1^z} + \frac{B_s}{z-p_1^{z*}} + \frac{C_s}{z-p_2^z} + \frac{D_s}{z-p_2^{z*}}\right] \tag{2.3.14}$$

The spectral peaks in Eq. (2.3.14) will be well-matched to the analog system in Eq. (2.3.10) due to the modal scaling. However, the change in relative modal amplitudes causes the zero locations to also change as will be seen graphically below.

There appears to be little one can do to match both the poles and zeros in general at high frequencies with an algorithm other than empirical means. However, a fairly close match (much closer than using the mapped poles and zeros alone) can be achieved by writing $H[z]$ as a product of an all-zero system and an all-pole system. The two systems are then scaled separately where the zeros are each divided by T and the poles are scaled according to the modal scaling described previously in this section. A compensated system response with separate "linear" scaling is depicted as $H^c[z]$ in Eq. (2.3.15) below.

$$H^c[z] = T\frac{(z-z_1^z)(z-z_1^{z*})}{T}\left[\frac{A_c}{z-p_1^z} + \frac{B_c}{z-p_1^{z*}} + \frac{C_c}{z-p_2^z} + \frac{D_c}{z-p_2^{z*}}\right] \tag{2.3.15}$$

The partial fraction expansion coefficients A_c, B_c, C_c, and D_c are seen in Eq. (2.3.16) for the all-pole part of the system response.

$$\begin{aligned}
A_c &= \frac{1}{(p_1^s - p_1^{s*})(p_1^s - p_2^s)(p_1^s - p_2^{s*})} \\
B_c &= \frac{1}{(p_1^{s*} - p_1^s)(p_1^{s*} - p_2^s)(p_1^{s*} - p_2^{s*})} \\
C_c &= \frac{1}{(p_2^s - p_1^s)(p_2^s - p_1^{s*})(p_2^s - p_2^{s*})} \\
D_c &= \frac{1}{(p_2^{s*} - p_1^s)(p_2^{s*} - p_1^{s*})(p_2^{s*} - p_2^s)}
\end{aligned} \tag{2.3.16}$$

The compensated "linear" scaling shown in Eqs. (2.3.15) and (2.3.16) are seen as a compromise between matching the peak levels and maintaining consistent pole-zero frequencies. Another linear scaling technique seen in the literature applies to the z-plane mapped impulse response

$$h^c[n-1] = A_z(p_1^z)^n + B_z(p_1^{z*}) + C_z(p_2^z)^n + D_z(p_2^{z*})^n \tag{2.3.17}$$

where A_z, B_z, C_z, and D_z are the unscaled mapped z-plane counterparts to the s-plane coefficients given in Eq. (2.3.12) above. Linear scaling gives an approximate impulse response match which degrades as the system resonances increase in frequency. Consider the impulse responses shown in Figure 8 for the system sampled at 600 Hz. As can be seen in Figure 8, the effect of modal scaling is to give a much closer match to the real system than simple linear scaling. Figure 9 compares the frequency responses for the same system and sampling rate.

As clearly seen in Figures 8 and 9, modal scaling gives an accurate impulse response and spectral peaks but a significant error in the shifted location of the zero. Clearly, it is up to the designer to decide whether an accurate match of the frequencies of the poles and zeros is needed more than an accurate impulse response. The important lesson here is that mapping from the s-plane to the z-plane (and z-plane back to the s-plane), is not without some trade-offs. For systems where accurate transient response is needed, one would prefer modal mapping. If accurate steady-state response is needed in terms of resonance and anti-resonance frequencies, linear scaled mapping is preferred. However, if high precision matching between the z-plane and s-plane systems is required, it can best be achieved by "over-sampling" the system. Over-sampling can be seen as using a digital system bandwidth much greater than that required for the dominant system features (poles and zeros). As can be seen in Figures 10 and 11, where the previous Figure's system is over-sampled 5:1 at 3 kHz, an excellent spectral and impulse response match occurs for both digital systems using either type of scaling.

2.4 SUMMARY, PROBLEMS, AND BIBLIOGRAPHY

The z-transform is a useful tool for describing sampled-data systems mathematically. Its counterpart in the continuous signal analog domain is the

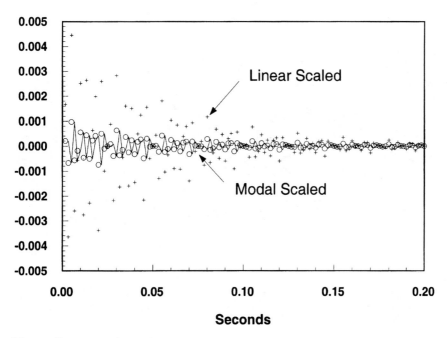

Figure 8 Comparison of the analog (–), modal scaled (○), and linear scaled (+) impulse responses sampled at 600 Hz.

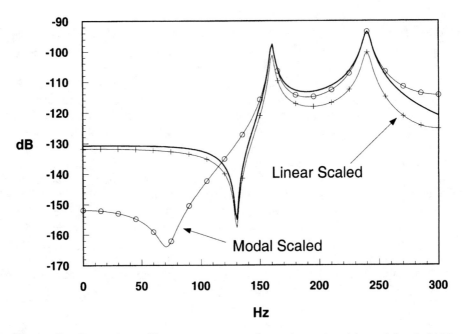

Figure 9 Comparison of frequency responses for analog system (–), modal scaled (○), and linear scaled (+) using a sample rate of 600 Hz.

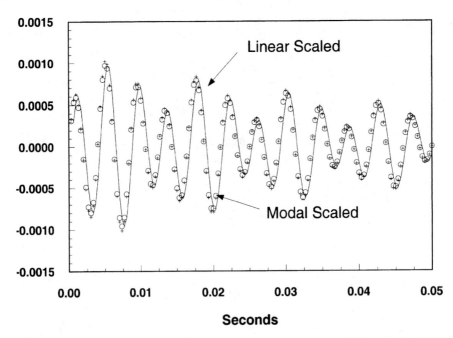

Figure 10 Comparison of the analog (–), modal scaled (○), and linear scaled (+) impulse responses sampled at 3000 Hz.

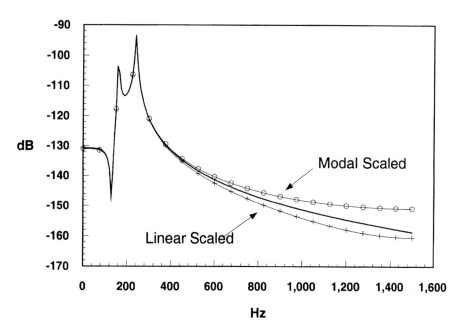

Figure 11   Comparison of frequency responses for analog system (–), modal scaled (○), and linear scaled (+) using a sample rate of 3000 Hz.

LaPlace transform. By using a linear frequency mapping between the analog s-plane of the LaPlace transform and the digital z-plane of the z-transform, one can model any physical system bandlimited to a frequency strip on the s-plane using a digital system. In general, the guarantee of a band-limited signal in the analog domain requires electronic filtering before A/D conversion at a suitable rate high enough for the bandwidth to be represented without frequency aliasing. The parameters of the digital system model can be determined by mapping the s-plane analog system poles and zeros directly to the z-plane for the digital system. However, this mapping is not without accuracy limitations. It can be seen in this section that the mapping is only precise at relatively low frequencies compared to the sampling rate fs. In general, an accurate match is attainable in the $0 < f < fs/4$ range, and only an approximate match is practical in the frequency range between $fs/4$ and the Nyquist rate $fs/2$ for real signals. For the frequency range approaching the Nyquist rate (fs for complex signals), one must decide whether the impulse response match is more important than the match of the zeros in the steady-state frequency response. If so, each "mode" or system resonance for the digital system can be scaled independently to insure a match to their analog system counterparts, giving a very accurate impulse response in the digital domain.

However, the modal scaling will affect the zero frequencies (or anti-resonances) in the digital domain since one can not control the phase of a digital pole-zero system at all frequencies. One could match both the magnitudes of the pole and zero frequencies in the digital system, but the phase at the system pole frequencies would have to be altered in order to have the zeros in the right place. Therefore, with the phases of the resonances adjusted, the digital impulse response would not match the true physical system. It should be clear that one can not simultaneously match the impulse and frequency responses of a digital and corresponding analog pole-zero system unless one "oversamples" at a very high rate driving the spectral features of interest to low frequencies.

PROBLEMS

1. Prove the properties of linearity and shift invariance for both the LaPlace transform and z-transform.

2. Show that for positive-moving time, the stable region for the z-transform is the region inside the unit circle and the stable region for the LaPlace transform is the left-half s-plane.

3. Show that for the frequency scaling in Eq. (2.1.13) where $|\alpha| = 1$, the digital domain scaling is actually the same as the analog domain frequency shifting in Eq. (2.1.12) where a is imaginary.

4. Assuming zero initial conditions, show the equivalence of differentiation in the analog and digital domains by mapping the LaPlace transform in Eq. (2.1.15) to the z-transform in Eq. (2.1.22).

5. Build a digital system model for a mechanical oscillator as seen in Figure 3 where the system resonance is 175 Hz, with a damping of 40 Nepers/sec, using a digital sample rate of 1000 real samples/sec. Find the z-plane poles and scaled impulse response as described in Eqs (2.3.3)–(2.3.7).

BIBLIOGRAPHY

W. E. Boyce, R. C. DiPrima Elementary Differential Equations and Boundary Value Problems, 3rd ed., New York: Wiley, 1977.

C. L. Phillips, H. T. Nagle Jr. Digital Control System Analysis and Design, Englewood Cliffs: Prentice-Hall, 1984.

H. F. VanLandingham Introduction to Digital Control systems, New York: MacMillan, 1985.

3

Digital Filtering

Digital filtering is a fundamental technique which allows digital computers to process sensor signals from the environment and generate output signals of almost any desired waveform. The digital computer in the 1990s has either replaced, or begun to replace, the analog electronics commonly found throughout the earlier part of the 21st Century. For electronic control systems, the advent of the digital computer has meant the replacement of analog circuitry with digital filters. Microprocessor-based digital filters have the advantages of incredible versatility, repeatability, reliability, small size, and low power requirements. During the last quarter of the 20th Century, digital signal processors have replaced even low-technology electronics and mechanical systems such as thermostats, clocks/timers, scales, and even the mechanical control systems on the internal combustion engine. Digital filters are generally used to detect important information from electronic sensors, provide precision output signals to actuators in the environment, or to form a control loop between sensors and actuators. The most sophisticated kind of signal processing system makes use of additional information, human knowledge, sensor data, and algorithms to adaptively optimize the digital filter parameters. Before we engage the full potential of adaptive signal processing, we will first examine the fundamentals of digital filtering and its application real physical systems.

In this section we first describe the two most fundamental types of digital filter: the finite impulse response (FIR); and infinite impulse response (IIR) filters. Using results from system theory, FIR and IIR filter design techniques are presented with emphasis on physical applications to real-world filtering problems. Of particular importance are the techniques for insuring stable IIR filters and the effects of delay on the digital system parameters. Digital systems can also be designed based on state variables and this relationship to IIR filtering is presented. Generally, state variables are used when the information of interest is directly seen from the system state equations (position, velocity, acceleration, etc.). For processing 2-dimensional (2D) data such as images or graphs, 2D FIR filters are presented. Image processing using 2D FIR filters (often referred to as convolution kernels) allows blurred or out-of-focus images to be sharpened. Finally, a set of popular applications of digital filters are given. Throughout the rest of this book, we will relate the more advanced adaptive processing techniques to some of these basic families of applications.

3.1   FIR DIGITAL FILTER DESIGN

There are many techniques for digital filter design which go far beyond the scope of this book. In this section we will concentrate on the fundamental relationship between a digital filter and its frequency response. In subsequent chapters, we will also examine several more sophisticated design procedures. Consider the z-transform of Eq. (2.1.3), but with z restricted to the unit circle on the z-plane ($z = e^{j2\pi f/f_s}$) as a means to obtain the frequency response of some digital signal $y[n]$.

$$Y[\Omega] = \frac{1}{N} \sum_{n=0}^{N-1} y[n]z^{-n}; \, z = e^{j\Omega}; \, \Omega = 2\pi\frac{f}{f_s} \qquad (3.1.1)$$

The infinite sum in Eq. (2.1.3) is made finite for our practical use where N is large. Eq. (3.1.1) is a discrete time Fourier transform (DTFT), the details of which will be explained in much greater detail in Chapter 5. If $y[n]$ is the output of a digital filter driven by a broadband input signal $x[n]$ containing every frequency of interest, the frequency response of the filter is found from system theory.

$$H[\Omega] = \frac{Y[\Omega]}{X[\Omega]} \qquad (3.1.2)$$

The impulse response of our digital filter is found most conveniently by computing the inverse discrete time Fourier transform (IDTFT) using the sum of the responses at K frequencies of interest.

$$h_n = \frac{1}{2\pi} \sum_{k=0}^{K-1} H[\Omega_k]e^{+j\Omega_k n} \qquad (3.1.3)$$

If $y[n]$ and $x[n]$ are real (rather than complex-sampled data), $h[n]$ must also be real. The DTFT of a real signal gives both real and imaginary frequency response components in what are known as Hilbert transform pairs. The real part of the frequency response is symmetric for positive and negative frequency; $\text{Re}\{H[\Omega]\} = \text{Re}\{H[-\Omega]\}$. The imaginary part of the frequency response is skew-symmetric; where $\text{Im}\{H[\Omega]\} = -\text{Im}\{H[-\Omega]\}$. It follows from the fact that $e^{j\Omega}$ has the same symmetry of its real and imaginary components. In order to have $h[n]$ real, both positive, and their corresponding negative, frequencies must be included in the IDTFT in Eq. (3.1.3). For complex time domain sampled data, the frequency range need only be from the lowest to highest positive frequency of interest. Any number of samples for $h[n]$ can be generated using the IDTFT in Eq. (3.1.3). The most important aspect of using the IDTFT to design a digital filter is that *any* frequency response may be specified and inverted to give the filter impulse response.

Finite Impulse Response (FIR) digital filters are usually designed by specifying a desired frequency response and computing an inverse Fourier transform to get the impulse response. This highly convenient approach to filter design is limited only by the number of samples desired in the impulse response and the corresponding spectral resolution desired in the frequency domain. The finer the desired spectral resolution is, the greater the number of samples in the impulse response will need to be. For example, if the desired resolution in the frequency domain is 8 Hz, the impulse response will have to be at least 1/8th sec long. At a sample rate of 1024 Hz, the impulse response would have at least 128 samples.

The method for implementing a digital filter in the time domain is to use a difference equation based on the inverse z-transform of Eq. (3.1.2).

$$Y[z] = X[z]\{h_0 + h_1 z^{-1} + \cdots + h_{N-1} z^{-N+1}\} \tag{3.1.4}$$

Noting the delay property in Eq. (2.1.5) where $z^{-1}X[z]$ transforms to $x[n-1]$, the inverse z-transform is seen to be

$$y[n] = h_0 x[n] + h_1 x[n-1] + \cdots + h_{N-1} x[n-N+1]$$
$$= \sum_{k=0}^{N-1} h_k x[n-k] \tag{3.1.5}$$

where the output $y[n]$ is the discrete convolution of the input $x[n]$ with the filter impulse response $h[n]$ as also seen in Eq. (2.1.7). The impulse response is finite in Eq. (3.1.5) due to the truncation at N samples. This type of filter is called a finite impulse response, or FIR, digital filter and has a convenient design implementation using the DTFT.

The question arises as to what penalty one pays for shortening the impulse response and simplifying Eq. (3.1.5). In Figure 1 the response of a mass-spring-damper system, with a damped resonance at 375 Hz and damping factor 10 Nepers, is modeled using an FIR filter and a sampling rate of 1024 samples per sec. This system is similar to the one in Figure 3 of the previous chapter. The frequency responses for 1024, 128, 16 sample impulse responses are given in Figure 1 to graphically show the effect of truncation. The 8 sample FIR filter has only 128 Hz resolution, while the 32 sample FIR has 32 Hz, and the 1024 sample filter has 1 Hz resolution. Clearly, the resolution of the FIR filter in the frequency domain should be finer than the underlying structure of the system being modeled, as is the case for the 1024 sample FIR filter. All one needs to do to design a FIR filter is to specify the magnitude and phase and the desired frequency resolution and sample rate will determine the length of the FIR impulse response. Given a long enough FIR filter, *any frequency response can be realized.*

However, if the desired frequency response contains only anti-resonances, or "dips", an FIR filter can model the response with absolute precision. Consider the case of multipath cancellation as shown in Figure 2. A distant transmitter radiates a plane wave to a receiver a distance "d" away from a reflecting boundary. The two different path lengths from source to receiver cause cancellation at frequencies where they differ by an odd multiple of half-wavelengths. From the geometry seen in Figure 2, it can be seen that the path difference for some angle θ measured from the normal to the reflecting boundary is

$$\Delta\ell = 2d\cos\theta \tag{3.1.6}$$

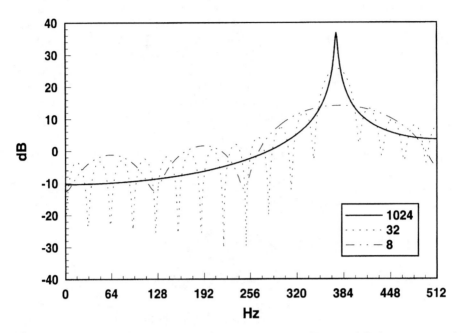

Figure 1 Comparison of 1024, 32, and 8 sample FIR filter models for a system with a damped resonance at 374 Hz.

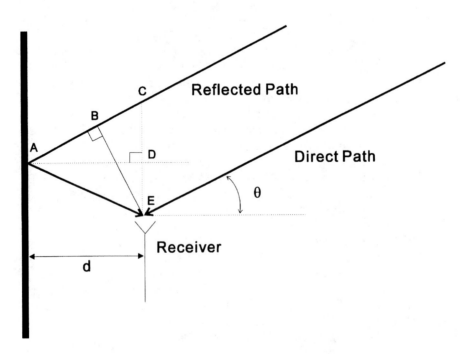

Figure 2 Multipath from a distant source gives a path difference seen along BAE which causes cancellation of some frequencies and enhancement of others.

The geometry in Figure 2 is straightforward to solve. The segment AD is "d" meters long. Therefore, segments AC and AE are each $d/\cos\theta$ long. The angles in BAD, DAE, and BED are all θ, making the segment CE be $2d\tan\theta$ long. Therefore, BC is $2d\tan\theta\sin\theta$ long and the path difference along BAE is $(2d/\cos\theta)(1-\sin^2\theta)$, or simply $2d\cos\theta$. If we consider an acoustic wave where d is 0.3 m, θ is 30°, and the reflecting boundary has a reflection coefficient of $+1$, the direct plus reflected waves combine to give the following analog system frequency response

$$H(\omega) = 1 + e^{jk\Delta\ell}; \quad k = \frac{2\pi f}{c} \tag{3.1.7}$$

where c is the speed of sound in air (about 343 m/sec at 20C). The path difference computes to be 0.5196 m giving a time delay between the direct and reflected paths of approximately 1.5 msec. If we are interested in frequencies up to 5 kHz, the digital sample rate is set to 10 kHz making the time delay approximately equal to 15 sample periods. The z-transform for a digital filter with a direct output plus a 15-sampled delayed output is simply

$$H[z] = 1 + z^{-15} \tag{3.1.8}$$

The digital filter model output $y[n]$ is found for the input $x[n]$ to be

$$y[n] = x[n] + x[n-15] \tag{3.1.9}$$

As can be seen in Eq. (3.1.9), there's no need to include more samples in the FIR filter! Eq. (3.1.8) is nearly an exact representation of the analog system its modeling in Eq. (3.1.7) and Figure 2. Figure 3 gives the frequency response of the system. Note that the acoustic pressure-squared doubles for all multiples of 666 Hz and cancels for odd multiples of 333 Hz. This frequency response structure is generally referred to as a comb filter, (affectionately named after the basic personal hygiene tool many signal processors used to be acquainted with). Comb filter effects are common ailments for sensors placed in multipath environments or near reflectors.

There is little one can do to correct comb filtering problems that is more effective than simply moving the sensor to a better location. However, the comb filter is the natural match for the FIR filter model since both systems contain only "dips" or "anti-resonances" in their frequency responses. The FIR polynomial in the z-domain can be factored into complex zeros which are located near the unit circle at the appropriate frequency angle. FIR filters are often referred to as "all-zero" filters, or even Moving Average (MA) filters since their outputs appear as a weighted average of the inputs.

3.2 IIR FILTER DESIGN AND STABILITY

Infinite Impulse Response (IIR) filters are essentially the spectral inverse of FIR filters. IIR filters are described by a ratio of polynomials in z and can have poles and zeros or only poles. The all-pole form of an IIR filter is often referred to as an Auto-Regressive (AR) filter. The pole-zero form of an IIR is called Auto-Regressive Moving Average (ARMA). We'll first examine the structure of

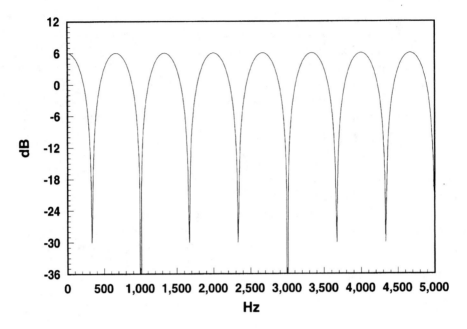

Figure 3 Frequency response of the multipath system in Figure 2 for $\theta = 30°$, $d = 0.3\,\text{m}$, and $c = 343\,\text{m/sec}$.

an AR filter and its difference equation to establish a straightforward rule for IIR stability. Consider the following Mth-order AR filter.

$$H[z] = \frac{1}{1 + a_1 z^{-1} + a_2 z^{-2} + \cdots + a_M z^{-M}} \tag{3.2.1}$$

For an IIR filter input $x[n]$ and output $y[n]$, taking z-transforms gives the relation

$$Y[z] = X[z]H[z] = X[z]\frac{1}{1 + a_1 z^{-1} + a_2 z^{-2} + \cdots + a_M z^{-M}} \tag{3.2.2}$$

Rearranging Eq. (3.2.2) one has

$$Y[z]\{1 + a_1 z^{-1} + a_2 z^{-2} + \cdots + a_M z^{-M}\} = X[z] \tag{3.2.3}$$

where taking inverse z-transforms gives the AR difference equation

$$y[n] = x[n] - \sum_{i=1}^{M} a_i y[n - i] \tag{3.2.4}$$

Equation (3.2.4) clearly shows that an infinite impulse response is generated for an arbitrary input $x[n]$. However, because of the feedback of past outputs back into the current output, we must be very careful not to create an unstable system. In addition, for the IIR filter to be realizable in the time-domain, the current output $y[n]$ must not contain any future inputs or outputs. A straightforward way to examine stability is write $H[z]$ as a partial fraction expansion.

$$H[z] = \frac{A_1}{1 - z_1 z^{-1}} + \frac{A_2}{1 - z_2 z^{-1}} + \cdots + \frac{A_M}{1 - z_M z^{-1}} \tag{3.2.5}$$

Applying the inverse transform techniques described in Eqs (2.3.3–2.3.4) it can be seen that the AR impulse response is a sum of M geometric series.

$$h[n] = A_1 \sum_{k=0}^{\infty} z_1^k z^{-k} + A_2 \sum_{k=0}^{\infty} z_2^k z^{-k} + \cdots + A_M \sum_{k=0}^{\infty} z_M^k z^{-k} \tag{3.2.6}$$

Clearly, for $h[n]$ to produce a stable output $y[n]$ from a stable input $x[n]$, the magnitude of the quotient z_i/z; $i = 0,1,2,...,M$ must be less than unity (i.e. $|z_i/z| < 1$). Therefore, it can be seen that the zeros of the denominator polynomial of $H[z]$, or the poles of $H[z]$, must all have magnitude less than unity requiring them to all lie within the unit circle on the complex z-plane to guarantee stability.

Consider the mass-spring-damper system modeled with FIR filters in Figure 1. We can derive an IIR filter with a very precise match to the actual system in both frequency response and impulse response using the s-plane to z-plane mapping techniques of the previous section. From examination of Eqs (2.3.1–2.3.7), the mapped and scaled IIR filter is

$$H[z] = \left\{ \frac{e^{-\zeta T} \sin(\omega_d T)}{\omega_d} \right\} \frac{z^{-2}}{1 + a_1 z^{-1} + a_2 z^{-2}} \tag{3.2.7}$$

where $\omega_d = 750\pi$, $T = 1/1024$ sec, and $\zeta = 10$ Nepers. The IIR coefficient $a_1 = -(z_1 + z_2) = 2e^{-\zeta T}\cos(\omega_d T)$ and $a_2 = e^{-2\zeta T}$. Numerical calculations reveal that the pole magnitudes are approximately 0.99 giving a stable IIR filter. The scale factor, defined as b_0, works out to be 3.19×10^{-4} while $a_1 = -1.32$ and $a_2 = 0.98$.

$$H[z] = \frac{b_0 z^{-2}}{1 + a_1 z^{-1} + a_2 z^{-2}} \tag{3.2.8}$$

The IIR finite difference equation is simply

$$y[n] = b_0 x[n - 2] - a_1 y[n - 1] - a_2 y[n - 2] \tag{3.2.9}$$

The three term sum in Eq. (3.2.9) is actually more accurate than the 1024 term sum shown in Figure 1. IIR filtering is very precise when the actual physical system being modeled is composed of spectral peaks which are easily modeled using poles in a digital filter. The same is true for pole-zero systems modeled by ARMA digital IIR filters. However, a serious question to be answered is how one can determine the ARMA numerator and denominator polynomial orders in the model. Most techniques for ARMA system identification provide a "best fit" for the model, given the number of poles and zeros to be modeled. Guessing the model orders wrong may still lead to a reasonable match between the desired and actual filter responses, but it may not be the best match possible. In general, a pole-zero difference equation for an ARMA model with Q zeros and P poles is seen in Eq. (3.2.10). ARMA models derived from mapped s-plane poles and zeros are also subject to the limitations of matching both the impulse and frequency responses simultaneously at high frequencies as described in detail in Chapter 2. Figure 4 gives a block diagram for an ARMA IIR filter.

$$y[n] = \sum_{i=0}^{Q} b_i x[n-i] - \sum_{j=1}^{P} a_j y[n-j] \tag{3.2.10}$$

3.3   WHITENING FILTERS, INVERTIBILITY, AND MINIMUM PHASE

The whitening filter for the ARMA IIR system $H[z] = B[z]/A[z]$ is its spectral inverse, $H^w[z] = A[z]/B[z]$. The spectral product of $H[z]$ and $H^w[z]$ is unity, or a spectrally "white" flat frequency response with unity gain. If both $H[z]$ and its inverse $H^w[z]$ exist as stable filters, both filters must have a property known as *minimum phase*. While IIR filters are far more flexible in modeling transfer functions of physical systems with spectral peaks than FIR filters, the major design concern is stability. This concern is due to the feedback of past IIR filter outputs into the current output as seen in the last term in Eq. (3.2.10). This feedback must be *causal*, meaning that the current output $y[n]$ must be a function of only current and past inputs and past outputs. From Eq. (3.2.6), it can be seen that the denominator polynomial $A[z]$ of an IIR system must have all of its zeros inside the unit circle on the z-plane for a stable impulse response. A polynomial with all its zeros inside the unit circle is said to be a *minimum phase* polynomial. For $H[z]$ to be stable, $A[z]$ must be minimum phase and for $H^w[z]$ to be stable $B[z]$ must also be minimum phase. For both $H[z]$ and $H^w[z]$ to be stable, both systems must be minimum phase in order to be invertible into a stable causal system.

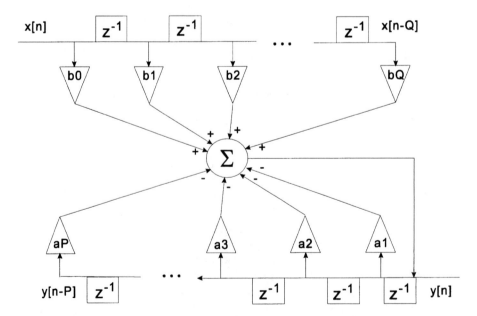

Figure 4   A block diagram of an ARMA IIR digital filter showing tapped delay lines for the input $x[n]$ and output $y[n]$.

Consider the effect of a delay in the form of $H[z] = z^{-d} B[z]/A[z]$ where both $B[z]$ and $A[z]$ are minimum phase. $H[z]$ is stable and causal, but $H^{w}[z] = z^{+d} A[z]/B[z]$ is not causal. Applying the ARMA difference Eq. (3.2.10) it can be seen that for output $x[n]$ and input $y[n]$, the whitening filter $H^{w}[z]$ is unrealizable because future inputs $y[n + d - j]$ are needed to compute the current output $x[n]$.

$$x[n] = \frac{1}{b_0} \left\{ \sum_{j=0}^{P} a_j y[n + d - j] - \sum_{i=1}^{Q} n_j x[n - j] \right\} \qquad (3.3.1)$$

Non-minimum phase numerator polynomials can be seen to contribute to the system delay. To examine this further, we conside a simple FIR system for real signals with a pair of conjugate-symetric zeros corresponding to 60 Nepers at 300 Hz. With a digital sample rate of 1000 Hz, the two zeros map to a magnitude of $e^{+0.06}$, or 1.062, and angles $\pm 0.6\pi$, or $\pm 108°$, on the z-plane. Since the zeros are slightly outside the unit circle, the inverse filter (which is an AR IIR filter) is unstable. However, we can minimize the FIR filter's phase without affecting its magnitude response by adding an *all-pass* filter $H^{ap}[z]$ in series. As seen in Figure 5, the all-pass filter has a pole to cancel each zero outside the unit circle, and a zero in the inverse conjugate position of the pole to maintain a constant frequency response.

The all-pass filter is by itself unstable due to the poles outside the unit circle. However, its purpose is to replace the non-minimum phase zeros with their minimum phase counterparts. The unstable poles are canceled by the non-minimum phase zeros, and therefore are of no consequence. The resulting minimum phase system $H^{min}[z]$, will have the same magnitude frequency response, but a very different phase response. We should also expect the impulse response to be significantly affected by the imposition of a minimum phase condition. Figure 6 compares the magnitude and phase of the original 2-zero FIR system, the all-pass filter, and the resulting 2-zero FIR minimum phase filter. As the non-minimum phase zeros move towards infinity, the amount of phase lag in the FIR filter increases. *In the extreme case with a pair of conjugate zeros at infinity, the all-pass filter will leave a pair of zeros at the origin which correspond to a delay of 2 samples in the minimum phase system.* Minimum phase filters are also referred to as minimum delay filters since phase

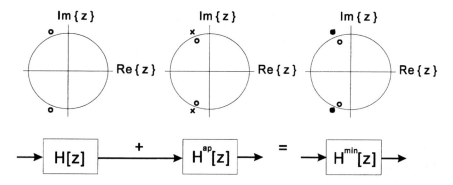

Figure 5   The all-pass filter $H^{ap}[z]$ replaces non-minimum phase zeros with their inverse conjugates maintaining the same magnitude response and minimizing the phase response.

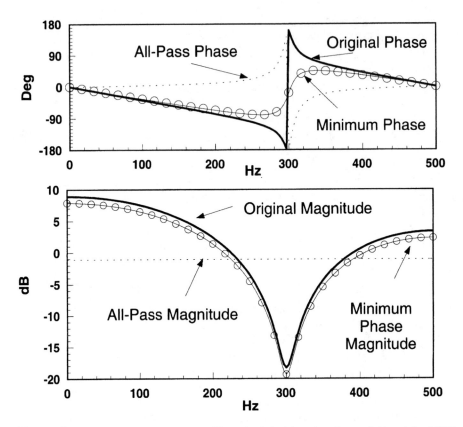

Figure 6 The all-pass filter has the effect of minimizing the phase of the original FIR system without causing a significant change to the magnitude response.

is also defined as radian frequency times delay. It follows then that constant delay filters have a linear phase response with negative slope.

As can be seen in Figures 5 and 6, the minimum phase system is invertible and differs only in phase from the non-minimum phase system (the less than 1 dB amplitude difference is of little consequence). If we wish to design an all-pole IIR filter to model an arbitrary frequency response where the spectral peaks are of primary interest, we could estimate an FIR model from the inverse Fourier transform of the spectral inverse of the IIR system. The FIR model is seen as a whitening filter because the product of its frequency response times the original frequency response of interest is a constant (spectrally flat or "white"). One would invert the spectrum of interest and compute the FIR whitening filter model from the inverse Fourier transform of the inverted spectrum. Because the phase of the original spectrum is arbitrary, the phase of the FIR whitening filter may or may not be minimum, giving one or more zeros outside the unit circle. This is typically the case when large numbers of zeros are modeled with long FIR filters. To invert the FIR whitening filter to get the desired stable IIR filter model, a minimum phase condition is imposed to insure stability of the IIR filter. One estimates the zeros of the FIR polynomial and "reflects" any non-minimum phase zeros into their respective inverse conjugate position to obtain an invertible minimum phase FIR filter. The phase of an FIR polynomial used as an FIR whitening filter can be arbitrarily defined in its Fourier transform, but the phase of an IIR filter is significantly restricted by the fact that its denominator polynomial must be minimum phase.

3.4 SUMMARY, PROBLEMS, AND BIBLIOGRAPHY

In this chapter, we presented two main linear digital filter structures in the form of FIR and IIR filter systems. There is far more to digital filter design than is presented here and the interested reader should refer to some of the excellent texts available on the subject. In order to achieve the best possible results in designing linear digital filters which model real physical systems, one must be driven by the underlying physical structure of the real physical system of interest. For example, if a relatively short impulse response is required where the system frequency response is composed mainly of spectral dips (antiresonances), an FIR filter design is generally the best choice. If the system of interest has a long impulse response, or has a frequency response characterized by dominant spectral peaks (resonances), a stabilized IIR filter should give the best results. However, the phase of the IIR filter will be very strictly constrained by the causality and minimum phase requirements for stability. If both the magnitude and phase of the physical system and its corresponding digital filter must be precisely matched with resolution Δf Hz, then one should specify an FIR filter with impulse response duration $1/\Delta f$ seconds. The main advantage for IIR digital filters is that they can represent spectral peaks accurately using a low number of coefficients. Furthermore, if one can specify the physical parameters and map them to the z-plane (as done in the previous section), one can obtain accurate impulse responses through modal scaling or oversampling techniques. Digital filters are the building blocks of real-time control systems and many signal processing operations.

PROBLEMS

1. For a sampling frequency of 44,100 Hz and real digital signals, design a FIR "notch" filter to reject 8000 Hz. Write the difference equation for computing the filtered output sequence.

2. How many coefficients are needed for a FIR filter to precisely model a system with sharp resonances approximately 10 Hz wide around 10 kHz where the sample rate is 50 kHz?

3. For modeling real physical systems with both poles and zeros, what is the best strategy for matching the frequency response as closely as possible.

    (a) scaling each mode appropriately
    (b) scaling the digital system by the sample period T
    (c) significantly oversampling the digital system and scaling appropriately

4. To match the impulse response of a physical system using a minimum bandwidth digital filter, which of the listed strategies in problem 3 makes the most sense?

5. Show that as a conjugate pair of zeros of an ARMA filter moves further outside the unit circle, the net system delay increases.

BIBLIOGRAPHY

T. Kailath, Linear Systems, Englewood Cliffs: Prentice-Hall, 1980.

K. Steiglitz, An Introduction to Discrete Systems, Wiley: New York, 1974.

F. J. Taylor, Digital Filter Design Handbook, Marcel-Dekker: New York, 1983, 4th printing.

J. Vlach, K. Singhal, Computer Methods for Circuit Analysis and Design, Van Nostrand Reinhold: New York, 1983.

4

Linear Filter Applications

Linear digital filters have as a typical application, temporal frequency filtering for attenuation of certain frequency ranges such as high-pass, low-pass, band-pass, and band-stop filters. In this section, we explore some important applications of digital filtering which will be used later in this book and are also widely used in the signal processing community. State variable theory is presented for applications where the system state (position, velocity, acceleration, etc.) is of interest. Any system (modeled as a mathematical function) can be completely described provided sufficient samples of the function are known, or if the function output and all its derivatives are known for a particular input. The mass-spring-damper oscillator is formulated in a state variable model for comparison to the IIR models developed in the previous sections. It will be shown that for discrete state variable filter, oversampling must be employed for an accurate system impulse response. Tracking filters are introduced for smoothing observed system output data and for examining unobservable system states using the non-adaptive α-β tracker.

Next we present 2-dimensional (2D) FIR filters which are widely used in image and video processing. Almost all image filtering involves FIR rather than IIR filters because the spatial causality constraints must be relaxed in order for a small 2D FIR convolution filter to process a much larger image by scanning and filtering. Perhaps the most common 2D filter in most households is an auto-focus system in a video camera. A high-pass 2D filter produces a maximum output signal when the image is in focus and giving many sharp edges. By constantly adjusting the lens system using a servo-motor to maximize the 2D high-pass filter output, the image stays in focus even as the camera moves about. In poor light or for very low contrast images, the high-pass filter approach to auto-focus has difficulty finding the maximum output point for the lens. Many other applications of 2D filters to image processing can be done including nonlinear operations such as brightness normalization, edge detection, and texture filtering.

One of the most important and often overlooked applications for digital filters is in the area of high fidelity analog signal reconstruction. Most systems where an analog signal is to be created simply use a straightforward digital-to-analog (D/A) conversion device to produce the corresponding analog voltage for each sample in the digital signal. The voltage is simply held constant between samples in what is known as a zero-order hold. The "staircase-like" shape of the analog signal is then smoothed using standard low-pass filters constructed from analog electronics. In the high frequency region near the Nyquist rate, the response of the zero-order hold and analog low-pass filters can be an issue if one desires low distortion. By creating new D/A samples (between the original samples), one can push the distortion and fidelity problems to higher frequencies which the analog filters suppress with greater efficiency. For example, one could simply compute the slope between samples and compute 3 additional samples to give a 4-times oversampled output with a first-order hold. Higher-order holds such as parabolic (2nd order hold), or even sinusoidal, are possible with very high oversampling rates to produce D/A signals which actually need no additional analog filtering. Oversampled D/A systems are quite common on modern compact disc audio recording playback systems and will likely be common for future digital television systems.

4.1 STATE VARIABLE THEORY

A popular formulation for digital systems is state variable formulation where the state variables completely describe the dynamics of the system. State variables can represent almost any physical quantity, so long as the set of state variables represent the minimum amount of information which is necessary to determine both future states and system outputs for the given inputs. State variable formulations in digital systems can be seen as a carryover from analog electronic control systems. Linear time-invariant systems can be described as parametric models such as ARMA systems, or as linear differential equations such as a state variable formulation. A function polynomial of order N can be completely described by $N+1$ samples of its input and output, or by the value at one sample point plus the values of N derivatives at that point. Early analog control systems used state variables in the form of derivatives because the individual state elements could be linked together using integrators (single pole low-pass filter).

The design of analog control systems is still a precise art requiring a great deal of skill an inventiveness from the designer. However, the digital age has almost completely eclipsed analog controllers due to the high reliability and consistent producability of digital controllers. The state variable in a digital control system can be some intermediate signal to the system, rather than specifically a representation of the system derivatives. The exceptions to this are the so-called alpha-beta non-adaptive position tracking filter and the adaptive Kalman tracking filter whose states are the derivatives of the underlying system model. Tracking filters will be covered in some detail later in the book with particular attention given to position tracking.

We now consider a general form of a state variable digital filter at iteration k with "r" inputs $\bar{u}(k)$, "m" outputs $\bar{y}(k)$ and "n" system states $\bar{x}(k)$. The system state at time $k+1$ has the functional relationship seen in Eq. (4.1.1).

$$\begin{aligned}\bar{x}(k+1) &= f[\bar{x}(k), \bar{u}(k)] \\ &= A\bar{x}(k) + B\bar{u}(k)\end{aligned} \qquad (4.1.1)$$

The system output also has a functional relationship with the state and input signals.

$$\bar{y}(k) = g[\bar{x}(k), \bar{u}(k)]$$
$$= C\bar{x}(k) + D\bar{u}(k) \qquad (4.1.2)$$

The dimensions of A is $n \times n$, B is $n \times r$, C is $m \times n$, and D is $m \times r$. For single-input single-output systems, B is a $n \times 1$ vector and C is a $1 \times n$ vector, while D becomes a scalar.

Consider the following ARMA filter system to be represented in a state variable model.

$$G[z] = \frac{b_0 + b_1 z^{-1} + \ldots + b_Q z^{-Q}}{1 + a_1 z^{-1} + \ldots + a_P z^{-P}} = \frac{Y[z]}{U[z]} \qquad (4.1.3)$$

The ARMA system in Eq. (4.1.3) will need P states making A $P \times P$, B $P \times 1$, C $1 \times P$, and D 1×1 in size. An expanded Eq. (4.1.1) is

$$\begin{bmatrix} x_1[k+1] \\ x_2[k+1] \\ \vdots \\ x_P[k+1] \end{bmatrix} = \begin{bmatrix} 0 & 1 & 0 & 0 & \ldots & 0 \\ 0 & 0 & 1 & 0 & \ldots & 0 \\ & & \vdots & & & \\ -a_P & -a_{P-1} & & \ldots & & -a_1 \end{bmatrix} \begin{bmatrix} x_1[[k] \\ x_2[k] \\ \vdots \\ x_P[k] \end{bmatrix} + \begin{bmatrix} 0 \\ 0 \\ \vdots \\ 1 \end{bmatrix} u[k]$$

$$(4.1.4)$$

and the system output is

$$y[k] = [b_Q b_{Q-1} \ldots b_0] \begin{bmatrix} x_1[k] \\ x_2[k] \\ \vdots \\ x_P[k] \end{bmatrix} \qquad (4.1.5)$$

Equation (4.1.5) is valid for $P = Q$ as written. If $P > Q$, $P - Q$ zeros follow to the right of b_0 in Eq. (4.1.5). If $Q > P$, $Q - P$ columns of zeros are added to the left of a_P in Eq. (4.1.4). State variable formulations are usually depicted in a signal flow diagram as seen in Figure 1 below for the ARMA system described in Eqs (4.1.4) and (4.1.5). The state variable implementation of the ARMA filter in Figure 1 is numerically identical to the IIR digital filter presented in the previous section. However, when the state vector elements represent derivatives rather than delay samples, the state variable formulation takes on an entirely different meaning.

Continuous State Variable Formulation

It is straightforward and useful for us to derive a continuous state variable system for the mass-spring-damper oscillator of Figure 3 in Chapter 2. Letting the force input $f(t)$ be denoted as $u(t)$ (so our notation is consistent with most publications on state variable filters), we have the 2nd-order differential equation

$$u(t) = M\ddot{y}(t) + R\dot{y}(t) + Ky(t) \qquad (4.1.6)$$

where $\ddot{y} = \partial^2 y / \partial t^2$ and $\dot{y} = \partial y / t$ write the 2nd-order system in Eq. (4.1.6), as well as

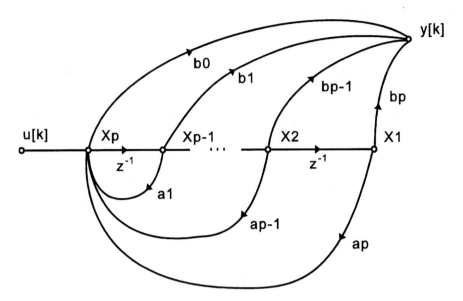

Figure 1 A flow diagram for the state variable formulation of an ARMA system shows a single tapped delay line propagating the state variables $x_i[k]$.

much higher order systems, as a first-order system by defining the necessary states in a vector which is linearly proportional to the observed output $y(t)$. The updates for the state vector are defined by the differential equation in $y(t)$, but the state vector itself is only required to be linearly proportional to $y(t)$.

$$\bar{x}(t) = \begin{bmatrix} x_1(t) \\ x_2(t) \end{bmatrix} = \begin{bmatrix} y(t) \\ \dot{y}(t) \end{bmatrix} \tag{4.1.7}$$

The 1st-order differential equation for the state vector is simply

$$\dot{\bar{x}}(t) = \begin{bmatrix} 0 & 1 \\ -\dfrac{K}{M} & -\dfrac{R}{M} \end{bmatrix} \begin{bmatrix} x_1(t) \\ x_2(t) \end{bmatrix} + \begin{bmatrix} 0 \\ \frac{1}{M} \end{bmatrix} u(t)$$
$$= A^c \bar{x}(t) + B^c u(t) \tag{4.1.8}$$

The state variable system's frequency and impulse responses are derived using LaPlace transforms

$$s\bar{X}(s) - \bar{x}(0^+) = A^c \bar{X}(s) + B^c U(s) \tag{4.1.9}$$

giving the state vector s-plane response as

$$\bar{X}(s) = [sI - A^c]^{-1} \bar{x}(0^+) + [sI - A^c]^{-1} B^c U(s) \tag{4.1.10}$$

The time-domain system response for arbitrary initial conditions $\bar{x}(0^+)$, and input forcing function $u(t)$ is

$$\bar{x}(t) = \bar{\phi}^{A^c}(t)\bar{x}(0^+) + \int_0^t \bar{\phi}^{A^c}(t-\tau)B^c u(\tau)d\tau \tag{4.1.11}$$

where $\bar{\phi}^{A^c}(t) = \mathcal{L}^{-1}\{[sI\ A^c]^{-1}\}$. If the initial position and velocity are both zero and $u(t)$ is a Dirac delta function, the system impulse response is therefore

$$\bar{h}(t) = B^c \int_0^t \bar{\phi}^{A^c}(t-\tau)\delta(\tau)d\tau = \bar{\phi}^{A^c}(t)B^c \tag{4.1.12}$$

Note that the impulse response signal is a 2×1 vector with the first element linearly proportional to the output $y(t)$ and second element linearly proportional to the output velocity. The proportionality constants are elements of the C matrix in Eq. (4.1.2) and are derived by equating the state vector impulse response to the actual response for the physical system.

$$\bar{h}(t) = \mathcal{L}^{-1}\left\{ \frac{\left[\begin{matrix} \left(s+\dfrac{R}{M}\right) & -\dfrac{K}{M} \\ 1 & s \end{matrix} \right] \left[\begin{matrix} 0 \\ \dfrac{1}{M} \end{matrix} \right]}{\left(s^2 + \dfrac{R}{M} + \dfrac{K}{M}\right)} \right\} = \mathcal{L}^{-1}\left\{ \frac{\left[\begin{matrix} -\dfrac{K}{M^2} \\ \dfrac{s}{M} \end{matrix} \right]}{\left(s^2 + \dfrac{R}{M} + \dfrac{K}{M}\right)} \right\} \tag{4.1.13}$$

The mass position given in Eq. (2.2.15) turns out to be $-M^2/K$ times $x_1(t)$ and the velocity of the mass is simply M times $x_2(t)$ where ω_d and ζ are defined in Eq. (2.2.14).

$$\bar{h}(t) = \left[\begin{matrix} x_1(t) \\ x_2(t) \end{matrix} \right] = \left[\begin{matrix} -\dfrac{K}{M^2\omega_d}e^{-\zeta t}\sin\omega_d t \\ \dfrac{1}{M}e^{-\zeta t}\cos\omega_d t - \dfrac{\zeta}{M\omega_d}e^{-\zeta t}\sin\omega_d t \end{matrix} \right] \tag{4.1.14}$$

Both an output position and velocity are available from the state variables using $\bar{y}(t) = C^c\bar{x}(t)$.

Discrete State Variable Formulation

To implement a discrete state variable digital filter and show the effects of sampling we start with the continuous system sampled every T seconds. The state vector elements are defined the same as for the continuous case.

$$\dot{\bar{x}}(t) = \left[\begin{matrix} 0 & 1 \\ -\dfrac{K}{M} & -\dfrac{R}{M} \end{matrix} \right] \left[\begin{matrix} x_1(kT) \\ x_2(kT) \end{matrix} \right] + \left[\begin{matrix} 0 \\ \dfrac{1}{M} \end{matrix} \right] u(kT)$$

$$= A\bar{x}_k + Bu_k \tag{4.1.16}$$

The discrete estimate for the derivative given in Eq. (2.1.17) is used to obtain

$$
\begin{aligned}
\bar{x}_{k+1} &= [TA + I]\bar{x}_k + TBu_k \\
&= \tilde{A}\bar{x}_k + \tilde{B}u_k
\end{aligned}
\tag{4.1.17}
$$

where $\tilde{A}$ and $\tilde{B}$ are the state transition and control input matrices for the digital system. To examine the digital frequency response, the z-transform is used to yield

$$
\bar{X}[z] = \frac{z}{[zI - \tilde{A}]}\bar{x}(0^+) + \frac{1}{[zI - \tilde{A}]}\tilde{B}U[z]
\tag{4.1.18}
$$

The digital time-domain state response to the input u_k is therefore

$$
\bar{x}_k = \tilde{A}^k\bar{x}(0^+) + \sum_{i=0}^{k-1}\tilde{A}^{k-i-1}\tilde{B}u_{k-i}
\tag{4.1.19}
$$

and the system impulse response is

$$
\bar{h}_k = \tilde{A}^{k-1}\tilde{B}
\tag{4.1.20}
$$

A plot of the true position impulse response for the case of $M = 1$ Kg, $R = 4$ Kg/s, and $K = 314$ Kg/s^2 where the sample rate f_s is 500 Hz is seen in Figure 2. For

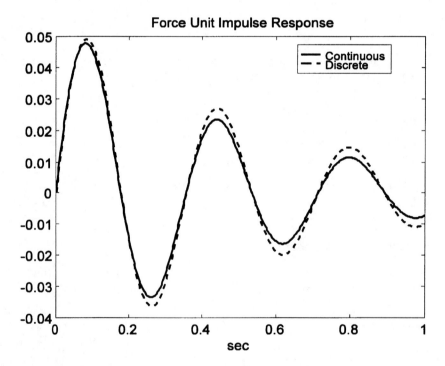

Figure 2 Comparison of state variable impulse responses in the analog and digital domain sampled at 500 Hz.

500 samples per sec, or $T = 2$ msec, there is clearly a reasonable, but not perfect, agreement between the continuous and digital state variable systems.

The reason for the difference between the two systems is that $\tilde{A}$ scales with the sample rate. As the sample rate decreases (T increases), the error between the two systems becomes much larger as seen in Figure 3 for a sample rate of only 75 Hz.

The instability in the example shown in Figure 3 illustrates an important design criteria for discrete state variable filters: *significant oversampling is required for an accurate and stable impulse response.* It can be shown that for the impulse response to be stable, the determinant of the discrete state transition matrix must be less than unity.

$$|\tilde{A}| = \begin{vmatrix} 1 & T \\ -\dfrac{K}{M}T & -\dfrac{TR}{M}+1 \end{vmatrix} = -T\frac{R}{M}+1+T^2\frac{K}{M} < 1 \tag{4.1.21}$$

After some algebra, $0 < T < R/K$, or since $T > 0$ always, $T < 2\zeta/\omega_0^2$, where $f_0 = \omega_0/2\pi$ is the undamped frequency of resonance in Hz. Clearly, the sampling frequency $f_s > \omega_0^2/2\zeta$ for stability. For the example in Figure 3, $f_0 = 27.65$ Hz and the minimum stable f_s is 78.75 Hz. It can be seen that while $f_s = 75$ Hz is easily high enough to represent the signal frequency properly, it is not high enough for

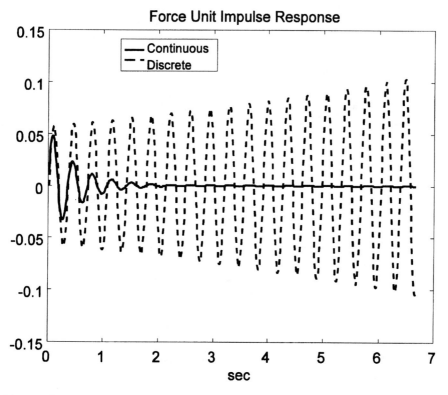

Figure 3   At a sample rate of 75 Hz the digital system is actually unstable.

a stable state variable system. Note that the sample rate must be even higher as the real physical system's damping decreases, and/or undamped frequency of resonance increases. An approximate guideline for *accurate* impulse responses can be seen to be $f_s > \omega_0^2/20\zeta$ or higher. The cause of the error is seen to be the finite difference approximation for the derivative of the state vector in Eq. (4.1.17). Contrast the state variable filter oversampling requirement to a properly-scaled IIR digital filter (which matches the impulse response for any unaliased sample rate) and one might wonder why one would use a digital state variable system. However, the reason state variable systems are important can be seen in the simple fact that in some adaptive signal processing applications, the state variables themselves are the main point of interest.

4.2   FIXED-GAIN TRACKING FILTERS

Perhaps the most important and popular use of digital state variable systems is in tracking systems. Common examples of tracking filter use include air-traffic control systems, stock/futures market programmed trading, autopilot controllers, and even some sophisticated heating, ventilation, and air conditioning (HVAC) temperature and humidity controllers. What makes a tracking filter unique from other digital filters is the underlying kinematic model. Rather than a state transition matrix based on the specific parameters of the physical system, such as the mass, stiffness, and damping in Eq. (4.1.16), a tracking filter's state transition matrix is based on the Newtonian relationship between position, velocity, acceleration, jerk, etc. The "jerk" is the term used for the rate of change of acceleration with time. However, our discussion here will be mainly limited to "position-velocity" states with no input signal u_k to simplify presentation of the tracking filter concept.

A tracking filter can have input signals u_k, but the more usual case involves just passive observations to be smoothed and predicted in the future. The position state is predicted based on last time's position and velocity and the Newtonian kinematic model. The prediction is then compared to an actual measurement of the position, and the resulting prediction error is weighted and used to correct, or "update" the state variables. The weights for the state updates are α for the position, β for the velocity, and γ for acceleration. The effect of the tracking filter is to "follow" the measurements and to maintain a kinematic model (position, velocity, acceleration, etc.) for predicting future positions. If the true kinematic system for the target has position, velocity, and acceleration components, and the tracking filter only has position and velocity, the α-β tracker will eventually "lose" a target under constant acceleration. However, if the target stops accelerating an α-β tracker will fairly quickly converge to the true target track. Clearly, the usefulness of tracking filters can be seen in the air traffic control system where collision avoidance requires estimates of target velocities and future positions.

An additional benefit of tracking filters is to reduce the noise in the measurements and estimated velocities, accelerations, etc. All measurement systems have inherent errors due to signal-to-noise ratio and unmodeled physics in the measurement environment. The underlying assumption for tracking filters of all types is that the measurement errors are zero-mean Gaussian (ZMG) with a known standard deviation of σ_w. This is a fairly broad and occasionally problematic assumption particularly when unmodeled environment dynamics produce occasional biases or periodic (chaotic), rather then ZMG random measurement errors. However, if the measurement is affected by a large number of random processes (such as from signal propagation through atmospheric turbulence), the measurement noise statistics will tend to be Gaussian, following the central limit theorem.

If one has an unbiased measurement system, the ZMG assumption is very practical for most applications. The amount of position noise reduction is equal to α, the weighting factor for the position state updates. Therefore, if $\alpha = 0.10$, one would expect a 90% reduction in noise for the predicted position state as compared to the raw measurements. For $\alpha > 1$ one would expect noise amplification. Let the raw measurements be depicted as z_k

$$z_k = Hx_k + w_k \tag{4.2.1}$$

where H is a matrix (analogous to C above) which relates the components of the state vector x_k to the measurements, while w_k is the ZMG measurement noise. Using the Newtonian kinematic model one can predict the state vector one time-step in advance

$$x_{k+1|k} = Fx_{k|k}; \qquad F = \begin{bmatrix} 1 & T & \dfrac{T^2}{2} \\ 0 & 1 & T \\ 0 & 0 & 1 \end{bmatrix} \tag{4.2.2}$$

where $x_{k|k}$ is a position-velocity-acceleration state vector updated at time step k. Examining the top row of the F matrix it can be clearly seen how the new position state is predicted for step $k + 1$ given updated data at step k from Newton's laws of motion.

$$x^p_{k+1|k} = x^p_{k|k} + x^v_{k|k} T + \frac{1}{2} x^a_{k|k} T^2 \tag{4.2.3}$$

The updated state vector elements from time step k are $x^p_{k|k}$ for position, $x^v_{k|k}$ for velocity, and $x^a_{k|k}$ for acceleration, and the time interval between steps is T seconds. The error between the predicted measurement from the predicted state vector and the actual measurement is

$$\epsilon_{k+1} = z_{k+1} - Hx_{k+1|k} \tag{4.2.4}$$

One then produces an "updated" state vector, separate from the "predicted" state vector, using the α-β-γ weights on the error.

$$x_{k+1|k+1} = x_{k+1|k} + W \epsilon_{k+1}; \qquad W = \left[\alpha, \frac{\beta}{T}, \frac{\gamma}{2T^2} \right]^T \tag{4.2.5}$$

The updated state vector will tend to follow the measurements slightly more closely than the predicted state vector. Hence, the predicted state $x_{k+1|k}$, is often referred to as the "smoothed" state estimate, although both state estimates will be less reactive to measurement noise (as well as target maneuvers) as α decreases.

The task of optimally setting α, β, and γ is yet to be presented. Choosing the tracking filter gains requires additional information in the form of the tolerable state vector *process noise*. Unlike the measurement noise, the process noise has nothing to do with the environment or measurement system. It is determined by some under-lying assumptions in the kinematic model. For example, for an α-β tracker there is an implicit assumption of a ZMG acceleration process with standard deviation σ_v. The underlying assumption of an unpredictable acceleration from step to step allows the tracking filter to follow changes in velocity from target maneuvers. Therefore, the state prediction equation is actually

$$x_{k+1|k} = Fx_{k|k} + v_k \tag{4.2.6}$$

where v_k is a ZMG random process with standard deviation σ_v substituting for the unmodeled dynamics.

For the α-β tracker, the process noise would nominally be set to be on the order of the maximum expected acceleration of the target. For the α-β-γ tracker, one assumes an unpredictable white jerk (not to be confused with an obnoxious Caucasian), to allow for ZMG changes in the rate of change in acceleration. Similar to the α-β tracker, the process noise for the α-β-γ tracker would be set on the order of the biggest jerk (no comment this time) expected from the target track. The process noise assumptions lead us to a very important parameter in fixed-gain tracking filters called the target maneuvering index λ_M for an α-β-γ tracker:

$$\lambda_M = \frac{\sigma_v}{\sigma_w} T^2 \tag{4.2.7}$$

The RMS position noise due to the assumed process noise σ_v is seen to be $\sigma_v T^2 / 2$. Choosing a smaller target maneuvering index (reducing the tolerated process noise for the available measurement noise) for the tracker will give a "sluggish" target track which greatly reduces track noise but is slow to respond to target maneuvers. Low target maneuvering indices are appropriate for targets such as large aircraft, ships, or very stable, slow moving, systems. Larger maneuverability indices may be appropriate for cases where one is not as interested in smoothing the states, but rather having low bias for real time dynamic state tracking.

In Chapter 10 we will present a derivation of the optimum least-squared error tracking gains based on the underlying statistics and kinematic model. Below we simply provide the solution for the tracking filter weights based on either of two simple design criteria: constant maneuvering index; and non-constant measurement noise. For both cases the target process noise is assumed to be constant and set based on the target kinematics. The constant maneuvering index case is therefore most appropriate when the measurement noise variance is assumed constant. One might then simply choose an α based on the amount of noise reduction desired. The smoothed position state will have a variance of $\alpha\sigma_w^2$. The β gain can be derived directly from α.

$$\beta = 2(2 - \alpha) - 4\sqrt{1 - \alpha} \tag{4.2.8}$$

The gain for γ is then set as β^2/α if an acceleration state exists. The target maneuvering index in terms of α and β is therefore

$$\lambda_M = \frac{\beta}{\sqrt{1 - \alpha}} = \frac{\sigma_v}{\sigma_w} T^2 \tag{4.2.9}$$

One can evaluate the resulting process noise σ_v from the choice of α, but the more typical way to design a fixed gain tracking filter is to choose σ_v based on the known target kinematics, determine σ_w objectively for the particular measurement sensor system in use, and compute the optimal α, β, and γ gains from the resulting target maneuvering index.

$$\beta = \frac{1}{4}(\lambda_M^2 + 4\lambda_M - \lambda_M\sqrt{\lambda_M^2 + 8\lambda_M}) \tag{4.2.10}$$

The α gain can then be determined directly from λ_M, or more conveniently from β.

$$\alpha = -\frac{1}{8}(\lambda_M^2 + 8\lambda_M - (\lambda_M + 4)\sqrt{\lambda_M^2 + 8\lambda_M}) = \sqrt{2\beta} - \frac{\beta}{2} \tag{4.2.11}$$

If an acceleration state is used, $\gamma = \beta^2/\alpha$ as previously noted. The optimal tracking filter fixed gains are determined from the algebraic solution of several nonlinear equations derived from the least-squared error solution. The solutions for α, β, and γ presented here assume a piecewise constant ZMG acceleration for the α-β tracker and piecewise constant ZMG acceleration increment (the jerk) for the α-β-γ tracker.

Consider the following example of an elevation tracking system for a small rocket which launches vertically from the ground, burns for 5 sec, and then falls back to earth. Neglecting changes in the mass of the rocket, the thrust would produce an acceleration of 15 m/sec^2 in a zero gravity vacuum. The rocket is also assumed to be subject to a drag deceleration of 0.25 times the velocity. During liftoff, the rocket's acceleration slowly decreases as the drag forces build up with velocity. At burnout, the maximum deceleration is imposed on the rocket from both gravity and drag. But as the rocket falls back to earth the drag forces again build up in the other direction, slowing the rocket's acceleration back towards the ground. Our measurement system provides altitude data 10 times per sec with a standard deviation of 3 m. It is estimated that the maximum deceleration is around 13 m/sec^2 which occurs at burnout. Therefore, we'll assume $\sigma_v = 13$, $\sigma_w = 3$, and $T = 0.1$ giving a target maneuvering index of $\lambda_M = 0.0433$, and tracking filter gains of $\alpha = 0.2548$, $\beta = 0.0374$, and $\gamma = 0.0055$. Figure 4 shows the results of the α-β-γ tracking filter.

Figure 4 clearly shows the benefits of tracking filters when one needs a good estimate of the target velocity. The velocity measurements shown in Figure 4 are computed by a simple finite difference and show significant error as compared to the tracking filter's velocity estimate. The measurement errors for acceleration based on finite difference are so great that they are omitted from the acceleration graph. Once the acceleration begins to settle into a nearly constant range, the acceleration states start to converge. The underlying assumption for the α-β-γ tracker is that the acceleration increment, or jerk, is ZMG. For our real physical problem this is indeed not the case, so it is not surprising that the acceleration losses track during changes in the target's acceleration (maneuvers). If one were to choose a process noise too small for the expected target maneuvers, such as $\sigma_v = 3$ rather than 13, the sluggish track ($\lambda_M = 0.01$) contains significant errors in position velocity and acceleration as seen in Figure 5. Choosing too large a process noise produces a very reactive track to target maneuvers, but the value of the estimated velocity and acceleration is greatly diminished by the huge increase in unnecessary tracking

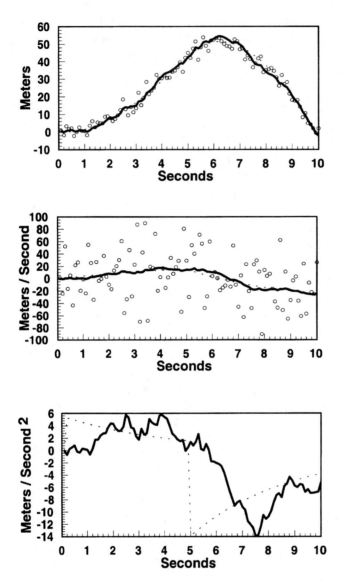

Figure 4 Typical α–β–γ results for a constant measurement error standard deviation of 3 m and a process noise set to be on the order of the maximum acceleration of 13 m. Time steps are 100 msec, true states are (- - -), measurements ($\bigcirc$), and tracking filter states are the solid line.

noise as seen in Figure 6 ($\sigma_v = 100$, $\lambda_M = 0.333$). The acceleration and velocity estimates from the tracking filter are almost useless with high track maneuverability but, they're still much better than the finite difference estimates.

The technique of determining the maneuvering index based on the target kinematics and measurement system noise can be very effective if the measurement error variance is not constant. Computing λ_M, α, β, and γ for each new estimated measurement noise σ_w allows the tracking filter to "ignore" noisy measurements and pay close attention to accurate measurements. In Figure 7, a burst of extra

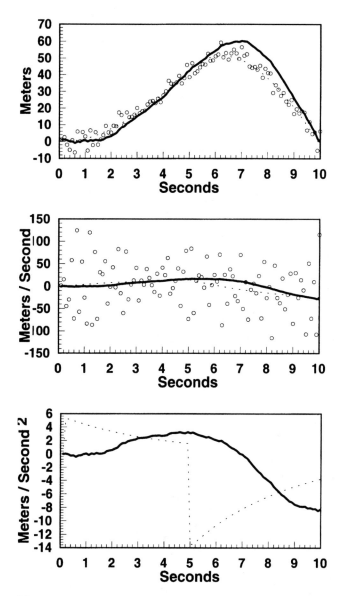

Figure 5 Tracking filter response using a sluggish $\sigma_v = 3$ rather than 13 for the same measurement data as seen in Figure 4.

measurement noise occurs between 7 and 8 sec during the simulation where σ_w goes from 3 m to 40 m and then back to 3 m. With the tracking filter assuming a constant measurement noise standard deviation of 3 m, a significant "glitch" appears in the state estimates. Figure 8 shows the performance possible when the target maneuvering index and filter gains are updated at every step to maintain optimality during changes in measurement noise.

It can be seen that the varying maneuvering index which follows the varying measurement noise is similar to the fully adaptive gain Kalman tracking

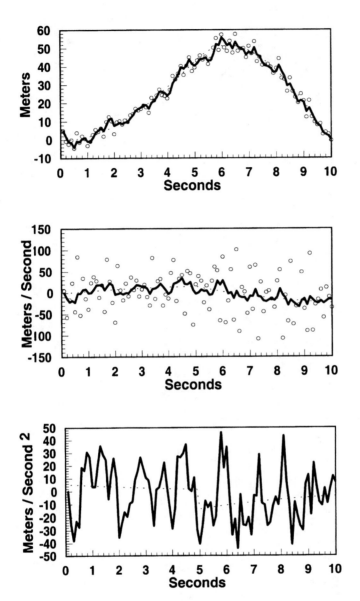

Figure 6 Tracking filter response using a hyperactive $\sigma_v = 100$ rather than 13 for the same measurement data as seen in Figure 4.

filter. However, the big difference in the Kalman filter (as seen in Chapter 10) is that the adaptive gains are determined from the state vector error variances as well as the measurement and process noises. In other words, the sophisticated Kalman filter considers its own estimates of the state error uncertainty when adapting to a new measurement. If the new measurement noise error variance is bigger than the estimated state error variance, the Kalman filter will place less emphasis on the new measurement than if the state error were larger then the measurement error. The much simpler fixed-gain tracker presented above

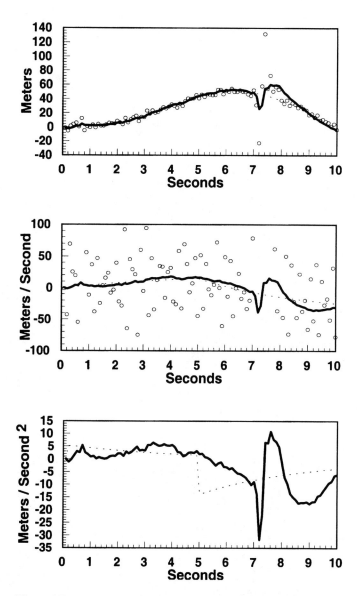

Figure 7 Tracking filter response for the same measurement data as seen in Figure 4 except a measurement noise burst of $\sigma_w = 40$ m occurs between 7 and 8 sec where the tracking filter assumes a constant $\sigma_w = 3$ m.

simply determines the tracking filter gains based on the assigned target maneuvering index.

4.3 2D FIR FILTERS

Two-dimensional (2D) FIR filters are most often used to process image data from camera systems. However, the techniques of sharpening, smoothing, edge detection,

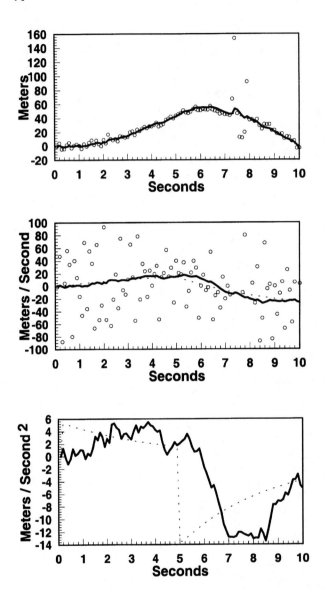

Figure 8 Tracking filter response for the same measurement data as seen in Figure 4 except a measurement noise burst of $\sigma_w = 40\,\text{m}$ occurs between 7 and 8 sec where the tracking filter follows the changes in σ_w and α, β, γ and λ_M.

and contrast enhancement can be applied to any 2D data such as level vs. frequency vs. time for example. Image processing techniques applied to non-image data such as acoustics, radar, or even business spreadsheets can be very useful in assisting the human eye in extracting important information from large complicated and/or noisy data sets. Data visualization is probably the most important benefit of modern computing in science. The amount of information passed through the human optic nerve is incredible but, the human brain's ability to rapidly process visual information completely eclipses all man-made computing machines. A robot can select, assemble, and manipulate nuts and bolts with efficiency well beyond that of a human. But if the robot drops a bolt into a pile of screws and has to recover it, the required computing complexity increases by an almost immeasurable amount. Even a child only a few years old could find the bolt with ease. One should never underestimate the value of even unskilled human labor in automated industries, particularly in the area of visual inspection.

Ultimately, one would like to develop algorithms constructed of relatively simple 2D processes which can enhance, detect, and hopefully recognize patterns in the data indicative of useful information. This is an extremely daunting task for an automatic computing system even for the simplest tasks. However, we can define a few simple operations using 2D FIR filters and show their usefulness by examining the effects on a digital 256-level grey-scale image of a house (the author's), as seen in Figure 9.

Each picture element (pixel) in Figure 9 is an 8-bit number representing the brightness of the image at the corresponding location on the camera focal plane. It is straightforward to see that if one were to combine adjacent pixels together, say in an weighted average, the sharpness, resolution, contrast, even texture of the image can be altered. Even more interesting is the idea that the signal processing system can apply a filter to the image in the form of a template detector to search for features such as windows, roof lines, walls, trees, etc. The template detector filter will have high output for the regions where there is a good match between the template and the local pixel distribution. These "vision" features can be used by a computer to recognize one house from another (depending on the diversity of house designs

Figure 9 Test image for 2D filtering tests consisting of a 256-level grey scale 640×480 pixels.

in the neighborhood). The process of developing computer algorithms which enable a signal processing system to have visual recognition capability is awesome in scope. The fundamental processing element for computer vision is the 2D digital filter, or convolver. We denote the grey-scale image brightness as $B(x,y)$, where x represents the row and y the column of the pixel. A 2D filter which combines regional pixels using weights $w_{i,j}$ to produce the filtered output image brightness $B'(x,y)$ can be seen in Eq. (4.3.1).

$$B'(x, y) = \sum_{i=-N}^{N} \sum_{j=-M}^{M} w_{i,j} B(x + i, y + j) \tag{4.3.1}$$

The process of computing a brightness output based on a weighted average of the adjacent pixels is likely where the term "moving average" for FIR digital filters originates. A wide range of visual effects can be implemented using simple 2D filters to alter the image focus, contrast, sharpness, and even color. 2D FIR filters can be designed to enhance or reduce various types of texture in an image. However, rarely would one try to implement an IIR filter since causality constraints would limit severely the direction of movement for some types of image patterns. For example, if the 2D filter moves in the $+x$ direction, "future-scanned" inputs $B(x+i,y)$ are used to compute the current output $B'(x,y)$. This is acceptable for an FIR filter. However, an IIR filter must only use past outputs to avoid a noncausal instability. In general, one only uses FIR filters for image processing since the direction of scan for the filter should not be a factor in the processed output image. To simply illustrate the 2D FIR filtering process, Figure 10 shows a process often referred to as "pixelation" where all the pixels in a block are replaced with the average of the higher-resolution original image pixels. Essentially the image sampling rate is reduced or *decimated*. This process can often be seen in broadcast television to obscure from view offensive or libelous material in a section of an image while leaving the rest of the image unchanged. Figure 10 shows an 8×8 pixelation of the original image in Figure 9.

Another interesting application of 2D filters to image processing is edge enhancement and detection. Edges in images can provide useful features for determining object size and dimension, and therefore are very useful elements in computer recognition algorithms. The most basic edge detector filter is a simple gradient which is shown in the positive x-direction (to the right) below. Since it is usually undesirable for a filtering operation to give an output image offset by ½ a pixel in the direction of the gradient, one typically estimates the gradient using a symmetric finite difference approximation.

$$\begin{aligned} B'(x, y) &= -\frac{\partial}{\partial x} B(x, y) \\ &= -[B(x + 1, y) - B(x - 1, y)] \end{aligned} \tag{4.3.2}$$

The negative gradient in the x-direction in Eq. (4.3.2) can be seen in Figure 11 below giving the digital image an "embossed" look. Examining the area between the tree trunk and house, one can clearly see that the transition from the black shadow to the white wall in the original image produces a dark vertical line along the edge of the house wall. The image is normalized to a mid-level grey which allows the

Figure 10 Test image after replacing each pixel in a 8 × 8 pixel group with the average brightness of the pixels in that group.

transition from bright to dark (seen in the left edges of the windows) to be represented as a bright vertical line. Note how the horizontal features of the image are almost completely suppressed. Figure 12 shows the test image with a negative gradient in the positive y-direction (upwards).

Derivatives based on a finite-difference operation are inherently noisy. The process of computing the difference between two pixels tends to amplify any random noise in the image while averaging pixels tends to "smooth" or suppress image noise. Generally, one can suppress the noise by including more pixels in the filter. A easy way to accomplish this is to simply average the derivatives in adjacent rows when the derivative is in the x-direction, and adjacent columns when the derivative is in the y-direction. A 2D FIR filter results with $N = M = 1$ for Eq. (4.3.1) where the filter weights for the negative gradient in the x-direction are given in Eq. (4.3.3).

$$-\nabla_x = \begin{bmatrix} w_{-1,+1} & w_{0,+1} & w_{+1,+1} \\ w_{-1,0} & w_{0,0} & w_{+1,0} \\ w_{-1,-1} & w_{0,-1} & w_{+1,-1} \end{bmatrix} = \begin{bmatrix} 1 & 0 & -1 \\ 2 & 0 & -2 \\ 1 & 0 & -1 \end{bmatrix} \qquad (4.3.3)$$

The 2D filter weights in Eq. (4.3.3) are known as a kernel because the weights can be rotated along with the direction of the desired gradient. For example, a negative gradient in the positive y-direction (upwards) can be realized by rotating the

Figure 11 Test image with a negative gradient applied in the x-direction to enhance vertical lines.

Figure 12 Test image with a vertical negative gradient filter applied to enhance horizontal image features.

weights in Eq. (4.3.3) counter-clockwise 90°.

$$-\nabla_y = \begin{bmatrix} -1 & -2 & -1 \\ 0 & 0 & 0 \\ 1 & 2 & 1 \end{bmatrix} \tag{4.3.4}$$

The kernel as written above can be conveniently rotated in 45° increments since the pixel in the center has eight neighbors. There are many possible kernels for a 2D gradient including larger, more complex, filters based on higher-order finite difference approximations. However, it is also useful to take the eight neighboring pixels and *estimate* the vector gradient for the local area. Gradient information from the 2D image data is very useful for simplifying the task of automated detection of geometrical features. These features of the image can subsequently be compared to a computer database of features as a means of automated detection of patterns leading to computer image recognition algorithms.

While directional derivatives can be very useful in detecting the orientation of image features such as edges, sometimes it is desirable to to detect all edges simultaneously. The geometry of the detected edges can then be used to identify important information in the image such as shape, relative size, and orientation. A straightforward edge detection method, known as Sobel edge detection, computes the spatial derivatives in the x and y directions, sum their squares, and compute the square-root of the sum as the output of the filter. A less complex operator, known as the Kirsh operator, accomplishes a more economical result without the need for squares and square-roots by estimating all eight gradients and taking the maximum absolute value as the edge detection output. The application of Sobel edge detection to our test image in Figure 9 can be seen in Figure 13 below.

While edge detection is useful for extracting various feature from the image for use in pattern recognition algorithms, it can also be used to enhance the visual quality of the image. The edge detector operator can easily be seen as a type of high-pass filter allowing only abrupt changes in spatial brightness to pass through to the output. If one could amplify the high frequencies in an image, (or attenuate the low frequencies), one could increase the sharpness and apparent visual acuity. Typically, sharpness control filtering is done using a rotationally-invariant LaPlacian operator as seen in Eq. (4.3.5).

$$\nabla^2 B = \frac{\partial^2 B}{\partial x^2} + \frac{\partial^2 B}{\partial y^2} \tag{4.3.5}$$

The LaPlacian, like the gradient, is approximated using finite differences. The sum of the x and y-direction second (negative) derivatives are seen in Eq. (4.3.6).

$$-\nabla^2 B = \begin{bmatrix} 0 & -1 & 0 \\ -1 & +4 & -1 \\ 0 & -1 & 0 \end{bmatrix} \tag{4.3.6}$$

Since we prefer to suppress noise by including all eight neighboring pixels in the edge detection operator, we simply add in the diagonal components to the LaPlacian as

Figure 13   Edge detection using a Sobel operator on the Test image from Figure 1.

seen in Eq. (4.3.7).

$$-\nabla^2 B = \begin{bmatrix} -1 & -1 & -1 \\ -1 & +8 & -1 \\ -1 & -1 & -1 \end{bmatrix} \tag{4.3.7}$$

Note that both Eqs (4.3.6) and (4.3.7) are normalized operators, that is, they do not cause a shift in the average brightness of the image. However, the operator will cause a slowly varying brightness to be nearly canceled since the sum of the eight neighbors will be nearly the same value as 8 times the central pixel. We can define a "sharpness operator" by simply subtracting the LaPlacian estimated at the central pixel from the central pixel value as seen with the 2D FIR filter operator in Eq. (4.3.8).

$$W^s = \begin{bmatrix} +1 & +1 & +1 \\ +1 & -7 & +1 \\ +1 & +1 & +1 \end{bmatrix} \tag{4.3.8}$$

Figure 14 has a sharpness operator applied to enhance the clarity of the test image.

The image can also be softened by a simple low pass moving average filter like the one used in Figure 10, but with the average value applied to each pixel rather than the 8 × 8 block, as seen in Figure 15.

As noted earlier in this section, 2D FIR filtering can be applied to non-image digital data such as the common spectrogram seen in Figure 16 below. A

Figure 14 A sharpness operator W^s applied to the test image.

Figure 15 Test image with low pass filtering to soften edges an enhance smoothness.

spectrogram is very common to speech signal analysis and sonar processing. Essentially, it is a plot of signal intensity displayed as a color vs. frequency vs. time. High signal intensities of a given frequency are seen in Figure 16 as bright vertical lines. Any "wiggles" in the bright lines indicate shifts in frequency over time. The data seen in Figure 16 is from the acoustic noise emitted by a large diesel engine on a slowly moving vehicle. The horizontal axis is frequency and the vertical axis is time. The spectrogram in Figure 16 is made by "stacking" Fourier transforms of the raw acoustic signal on successive time sections of the data. The many parallel lines are harmonics of the engine cylinder firing frequency and the variability of the frequencies gives some indication of the vehicle's movements.

However, the raw spectrogram is somewhat difficult to view and is also difficult for a computer to detect some of the more interesting patterns. Figure 17 shows the same spectrogram with a sharpening operator and re-scaling of the gray-scale data. The information content is basically the same in the two spectrograms except some of the noise randomness has been suppressed relative to the harmonic signals by the sharpening operator. The scaling operation simply adds contrast between the signal levels we want to see and the background noise (mainly from wind) which we don't care about.

Clearly, one can see how low pass and high pass 2D filters can be used for controlling image focus, signal-to-noise enhancement, and detecting orientation features and edges. The rather simple 2D filters presented here are the fundamental building blocks of much larger and more complex systems which comprise man's more recent attempts at computer vision. The 2D filter is presented here because

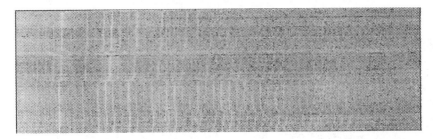

Figure 16   A spectrogram of acoustic noise from a large diesel engine of a construction vehicle where the horizontal axis is frequency and the vertical axis is time.

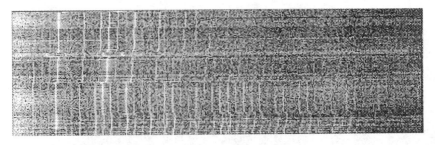

Figure 17   The spectrogram in Figure 4 with a sharpening operator and re-scaling of the grey-scale data to enhance the finer structures of the acoustic harmonics.

it is a fundamental tool in modern digital signal processing which extends well beyond image processing into the frontiers of computer visualization of complex data sets. The 2D filter can be made adaptive, it can be applied in the frequency domain as well as time or spatial domain, and in concert with a thorough understanding of the application's physics and geometry, can be made to extract the essential features necessary for computer vision.

Finally, a medical example of image enhancement is seen in Figure 18 for a magnetic resonance image (MRI) scan showing a medial slice along the spine. The spinal cord is easily seen as a tube structure running vertically alongside the vertebrae. In the original MRI image on the left the damage is barely visible. By applying some adjustments in brightness, contrast, and gray-scale adjustments, the obvious damage to the disc is seen on the right. The telltale feature of the ruptured disc is the break seen in the disc wall near the spinal canal. The damaged disc is the author's, giving the phrase "pain in the butt" a new and deeper meaning. Microsurgery corrected the problem brilliantly. A few "extra" images were obtainable from the MRI technician after explaining CAT scans to him as described in Section 7.3. An MRI device is quite clever and complicated. A high frequency electromagnetic pulse in a very intense static magnetic field energizes the tissue by inducing spin on the electrons of the tissue atoms. The various tissues re-radiate the electromagnetic energy at a frequency and damping factor unique to the individual molecules. With a high-frequency plane wave pulse incident on the body at some angle θ, the body re-radiates the energy differently in all directions depending on the distribution of the tissue molecules. By scanning many angles and inverse Fourier processing the re-radiation patterns, the spatial distribution

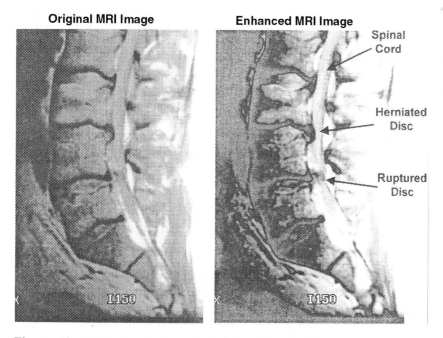

Figure 18 Original and enhanced detection MRI showing herniated and ruptured disc.

of tissue can be then reconstructed using a technique called back propagation as described in Section 7.3. Since the MRI transmitter and receivers can be "tuned" to enhance detection of a particular tissue type, body structures such as cartilage and nerves can be imaged, which is not easily done with low levels of X-rays.

4.4 D/A RECONSTRUCTION FILTERS

For signal processing and control systems where digital outputs are to be converted back to analog signals, the process of signal reconstruction is important to the fidelity of the total system. When an analog signal is sampled, its frequency range is band-limited according to the well-known rules for avoiding aliased frequencies. The analog-to-digital (A/D) process simply misses the signal information between samples (this is actually insignificant due to the anti-aliasing filter's attenuation of high frequencies). However, upon reconstruction of the output signal using digital-to-analog (D/A) conversion, no information is available to reconstruct a continuous signal in-between digital samples. Therefore, one must interpolate the continuous signal between samples. The simplest approach electronically is to simply hold the output voltage constant until the next digital sample is converted to analog. This holding of the voltage between samples is known as a *zero-order hold* D/A conversion and is the most straightforward way to produce analog signals from a sequence of digital numbers. In fact, if the frequency of the signal is very low compared to the sample rate, the reproduction is quite accurate.

One can visualize a D/A convertor as a simple bank of analog current sources, each allocated to a bit in the binary number to be converted and with corresponding current strength, where the total output current across a resistor produces the desired D/A voltage. A simple digital latch register keeps the D/A output voltage constant until the next digital sample is loaded into the register. This practical electronic circuitry for D/A conversion results in a "staircase" reproduction of the digital waveform as seen in Figure 19 below where the frequency is approximately $fs/5$.

The zero-order hold circuitry has the effect of low-pass filtering the analog output. If one were to reproduce each sample of the digital sequence as a short pulse of appropriate amplitude (analogous to a weighted Dirac delta function), both the positive and negative frequency components of a real sampled signal will have aliased components a multiples of the sampling frequency. For example, if the sample rate is 1000 Hz and the signal of interest is 400 Hz, D/A reproduction with impulses would produce the additional frequencies of ± 600 Hz (from ± 400 ± 1000), ± 1400 Hz, etc. With the zero-order hold in place, the higher aliased frequencies will be attenuated somewhat, but generally speaking, additional analog low pass filtering is required to faithfully reproduce the desired analog signal. In the early 1980s when digital music recording and playback equipment were in their commercial infancy, the quality of digital music reproduction came into controversy from an initial negative reaction by many hifi enthusiasts (many of whom still today swear by their vacuum tubes). There was little controversy about the elimination of tape hiss and phonograph rumble and the wear improvement for digital recordings. But many music enthusiasts could hear subtle effects of the aliased digital frequencies and low-pass filters in the D/A systems. By the mid 1980s most digital audio players employ a technique known as *oversampling* on the output which is

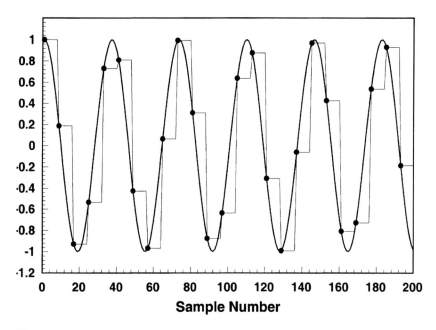

Figure 19 Unfiltered digital signal to analog signal conversion results in a "staircase" effect which contains many undesirable high frequency components.

an extremely useful method for achieving high fidelity without increasing the system sample rate.

Perhaps the most common example of sophisticated oversampling for analog signal reconstruction is in the modern audio compact disc player. Often these devices are marketed with features noted as "4 × oversampling" or even "8 × oversampling" Some ultra high fidelity systems take a direct digital output from a compact disc (CD) or digital audio tape (DAT) and using about 100 million floating-point operations per sec (100 MFLOPS), produce a 32 × oversampled audio signal. For a typical CD with 2 channels sampled at 44,100 Hz a 32 × oversampled system produces over 2,822,400 samples per sec! Where do the other 2,778,300 samples come from? To clearly illustrate what D/A oversampling is and how it works we will present below cases for 2 ×, 4 ×, and 8 × oversampling below.

Figure 20 shows the process of adding zeros to the digital sequence between the samples. One zero is added if 2 × oversampling is desired, 3 zeros if 4 × is desired, 7 zeros if 8 × oversampling is desired. The zero-filled digital sequence is then digitally processed by a *FIR interpolation filter*. The output sequence of the interpolation filter has the added zeros replaced by synthesized samples which "smooth out" the staircase effect seen with a zero-order hold. A 1st-order hold (connect the two actual samples with a straight line) is achieved by an FIR interpolation filter which moves the added zero sample to the average of the two adjacent actual samples for 2 × oversampling. A 3rd-order hold is achieved with an FIR filter which approximates a cubic function for 4 × oversampling, and so on. Figure 21 shows the FIR impulse responses for the zero, 1st, 3rd, and 7th order holds corresponding to 1 ×, 2 ×, 4 ×, and 8 × oversampling.

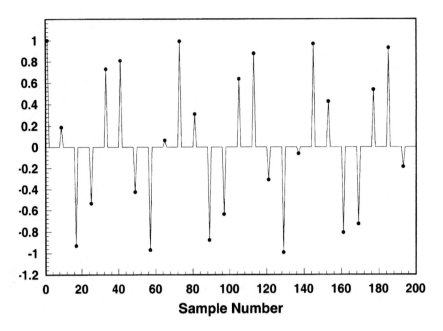

Figure 20   Adding zeros between actual output samples is the first step towards producing an oversampled D/A signal (the original samples are seen as black dots).

The symbols in Figure 21 depict the samples of the FIR impulse when the corresponding oversampled rate is chosen. Doing 8 × oversampling using a 1st-order filter can be done, but one would be better advised to use the 7th order FIR filter instead to get much better oversampling results. Note that once the over-sampled sequence is produced, the analog output is still produced by a analog electronic zero-order hold. However, since the D/A sample rate is increased to a very high rate, the aliased components are well out of the bandwidth of interest and almost completely eliminated by the low-pass filter properties of the interpolation filter and electronic zero-order hold. Figure 22 shows a comparison between the original waveform, a zero-order hold, and 1st-order hold with 2 × oversampling. Figure 23 compares the 3rd-order with 4 ×, and 7th-order holds with 8 × oversampling to the analog waveform.

Figures 22 and 23 show a significant improvement in the 4 × oversampled sequence smoothing when comparing to 1 × and 2 ×, but little improvement when going to 8 × oversampling with the 7th-order FIR interpolation filter. However, if one were to go to a higher frequency, the 8 × oversampled sequence would look much better than the 4 × sequence. The improvement in fidelity can also be seen in the higher-order FIR interpolating filters by examining their frequency responses as seen in Figure 24.

At first glance Figure 24 might indicate that the oversampling FIR filters leak more undesirable high frequencies in the range above the Nyquist rate (depicted with a dotted line). However, since these filters are processing oversampled sequences, the aliased components appear at much higher frequencies and are actually more attenuated than the zero-order hold case. Also of importance to high

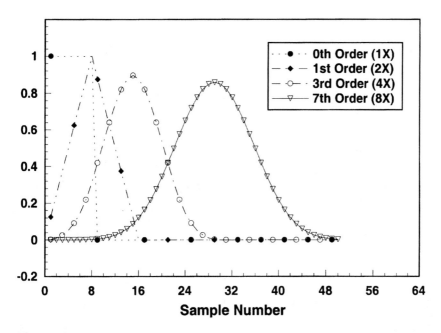

Figure 21 FIR interpolation filters for producing oversampled D/A output sequences from digital signals with added zeros.

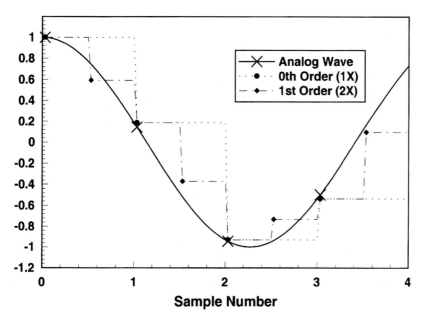

Figure 22 Comparison of the zeroth-order hold and the 2× oversampled 1st-order hold to the analog waveform.

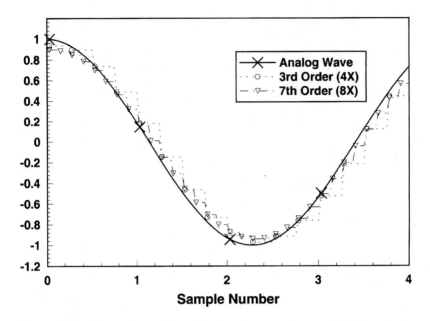

Figure 23   Comparison of the original analog waveform to the 3rd-order 4× oversampled and 7th-order 8× oversampled output.

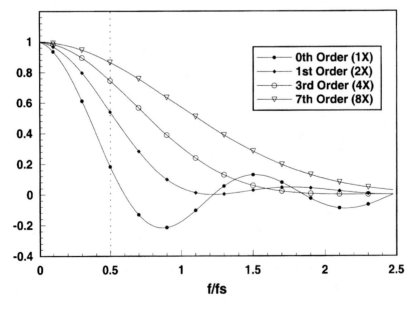

Figure 24   While the frequency responses of the 3rd and 7th-order filters are flatter in the pass-band, their aliased components are much higher in frequency and more attenuated.

fidelity D/A conversion is the observation that the higher oversampled filters are actually much flatter in frequency response within the pass band below the Nyquist rate. Clearly the zero-order hold is nearly 14 dB down at the Nyquist rate but has leakage above the Nyquist rate of about the same amount. The 8 × oversampled system is barely 1 dB down at the Nyquist rate and the nearest aliased frequency components will appear 8 times the sample rate away where they are well attenuated. The addition of a very simple low pass filter such as a capacitor and resistor completely suppresses the ultra-high frequency aliased components of the 8 × oversampled outputs. Clearly, oversampling for D/A conversion is very important to fidelity and may also have a significant impact on digital broadcast television as well as other systems.

4.5   SUMMARY, PROBLEMS, AND BIBLIOGRAPHY

Linear digital filters have many applications beyond simple frequency suppression of the time-domain signals. We have shown formulations of the digital state vector and the difference between a tapped delay line state vector and a differential state vector which can be used to model many physical systems. Conspicuously, one must have either a firm grasp of the underlying physics behind the modeling problem, or a firm grasp of a physicist to make sure the model inputs and outputs are reasonable. However, what is often overlooked in applications of digital filtering to real physical problems is the physics of the digital system's operation. Converting analog signals to digital data for computer processing, while extremely powerful and versatile, does carry with it some undesirable artifacts. For example, sampling without bandlimiting the analog signal leads to aliased frequencies. Mapping analog domain system poles and zeros to the digital domain requires modal scaling to obtain a reasonable, but not perfect, match between analog and digital impulse and frequency responses. Unless one significantly oversamples the signals, one cannot simultaneously match both poles and zeros for the same frequency response in the analog and digital domains. A similar problem arises in the digital state variable problem where the finite difference errors become significant as the frequencies of interest approach the Nyquist sampling rate. For all digital filters, we must maintain a stable causal response giving rise to the strict requirement for all system poles to be interior to the unit circle on the digital z-plane. In 2D filters we generally limit the structure of operators to be of the FIR type so that the stability of the filtering operation is not dependent on which direction the 2D filter is moved over the 2D data. These physical attributes of digital signal processing are perhaps the most important fundamental concepts to understand before moving on to frequency transforms and the adaptive processes presented in the next few chapters.

PROBLEMS

1. What is the discrete state variable flow diagram for the system $(b_0 + b_1 z^1)/(1 + a_1 z^1 + a_2 z^2)$?

2. For a series resonant tank circuit with $R = 10\Omega$, $C = 1\ \mu F$, and $L = 2\ mH$, what is the minimum sample period in seconds for a stable state variable model? What sample period is recommended?

3. A volatile stock is being smoothed by an α-β-γ type fixed-gain tracking filter to help remove random hourly fluctuations and improve the observability of market trends. If we wish to reduce the observed fluctuations to 10% of the original data, what are the values of α-β-γ?

4. We are interested in detecting lines in an image that runs along a $-45°$ line (from upper left to lower right). Define a 2D FIR filter to enhance detection of these lines.

5. Show that the 1st order ($2\times$) interpolation filter is the convolution of the 0th order filter with itself, the 3rd order is the convolution of the 2nd order with itself, and so on. That being the case, prove that the frequency response of the 1st order interpolation filter is the square of the 0th order (taking into account the doubled sample rate), 3rd order square of the 2nd order, and so on.

BIBLIOGRAPHY

Y. Bar-Shalom, X. Li, Estimation and Tracking: Principles, Techniques, and Software, Boston: Artech House, 1993.

A. V. Oppenheim, R. W. Schafer, Discrete-Time Signal Processing, Englewood Cliffs: Prentice-Hall, 1989.

J. C. Russ, The Image Processing Handbook, Boca Raton: CRC Press, 1992.

Part II

Frequency Domain Processing

Frequency domain processing of signals is an essential technique for extracting signal information with physical meaning as well as a filtering technique for enhancing detection of periodic signal components. The genesis of frequency domain dates back to the later half of the 19th century when Fourier (pronounced "4-E-A") published a theory that suggested any waveform could be represented by an infinite series of sinusoids of appropriate amplitudes and phases. This revolutionary thought led to the mathematical basis for many fundamental areas in physics such as diffraction theory and optics, field theory, structural vibrations and acoustics, just to name a few. However, it was the development of the digital computer in the 1950s and 1960s which allowed the widespread use of digital Fourier transformations to be applied to recorded signals. Now, in the last decade of the 20th century, real-time digital frequency transformations are commonplace using ever more astonishing rates of numerical computation for applications in almost every area of modern technology. The topic of frequency domain processing is of such extreme importance to modern adaptive signal processing that several chapters are dedicated to it here. One can do Fourier processing in the time, space, frequency and wavenumber domains for steady-state signals. However, one must adhere to the underlying physics and mathematical assumptions behind the particular frequency transformation of interest to be sure that the correct signal information is being extracted.

The Fourier transform mathematically is an integral of the product of the waveform of interest and a complex sinusoid with the frequency for which one would like to know the amplitude and phase of the waveform's sinusoidal component at that frequency. If one has an analytic function of the waveform in the time domain, $x(t)$, then one could analytically integrate $x(t)e^{j\omega t}dt$ over infinite time to obtain an equation for the frequency domain representation of the waveform $X(\omega)$. From Adam and Eve to Armageddon is too long a time integral. By truncating the time integral to t_1 to t_2, periodicity of the waveform in the signal buffer is implied over all time. This leads to what we call Fourier series. This has obvious mathematical utility because many differential equations are more easily solved algebraically in the frequency domain. However, in the straightforward applied mathematics of the signal processing world, one has a finite-length digital recording, $x[n]$, representing the waveform of interest. The analytic indefinite integral becomes a finite (truncated from t_1 to t_2) discrete sum which takes on the characteristics of spectral leakage, finite resolution, and the possibility of frequency aliasing. The limitations on the Fourier transform imposed by real-world characteristics of digital systems are manageable by controlling the size, or number of samples, of the discrete-time Fourier transform, the use of data envelope windows to control resolution and leakage, and by controlling and optimizing frequency resolution as needed for the application of interest. Our goal in this presentation of Fourier transforms is to demonstrate the effect of finite time (space) integration on frequency (wavenumber) resolution and spectral leakage.

Given the underlying physics of the waves of interest, the frequency domain data can be used to observe many important aspects of the field such as potential and kinetic energy densities, power flow, directivity, and wave coupling effects between media of differing wave impedance. When the sampled waveform represents spatial data, rather than time-sampled data, the frequency response represents the wavenumber spectrum which is a very important field parameter describing the wavelengths in the data. One can describe the physical man-made device of a pin-hole or lens-system camera as an inverse wavenumber Fourier transform system in 2 dimensions where the optical wavelengths are physically filtered in direction to reconstruct the spatial image data on a screen. The importance of wavenumber spectra will become evident when we use them to control focus in images as well as show radiation directivities (beamforming) by various transmitting arrays of sources. Perhaps what is most useful to keep in mind about Fourier transforms is the orthogonality of $e^{j\omega t}$ for time (t) and frequency (ω) transforms and e^{jkx} for space (x) and wavenumber ($k = 2\pi/\lambda$). Note that if t has units of seconds, ω has units of radians/second and if x has units of meters, k has units of radians/meter. Since the kernel sinusoids are orthogonal to each other, one obtains a discrete number of frequencies with amplitudes and phases independent of one another as the output of the "forward" (i.e. time to frequency or space to wavenumber) Fourier transform. The independence of the frequencies (or wavenumbers) is what allows differential equations in the time domain to be solved algebraically in the frequency domain.

It is interesting to note the many natural frequency transforms occurring all around us such as rainbows and cameras, musical harmony, speech and hearing, and the manner in which materials and structures radiate, transmit, and reflect mechanical, acoustic, and electromagnetic waves. In the most simple terms, a frequency transform can be seen as a wave filter, not unlike an audio graphic equalizer in many high fidelity audio systems. The frequency band volume slide controls on an audio equalizer can approximately represent the various frequency bands of the input signal for which the output volume is to be controlled by the slider position. Imagine for a moment thousands of audio equalizer slide controls and signal level meters for filtering and monitoring a given sound with very high frequency precision. While the analog electronic circuitry to build such a device would be extremely complicated, it can be achieved easily in a digital signal processing system in real-time (instant response with no missed or skipped data) using current technology.

Rainbows are formed from the white light of the sun passing through water droplets which naturally have a chromatic aberration (slightly different wave speed for each color in white light). The angle of a ray of white light entering a droplet changes inside the droplet differently for each color wavelength. Upon reaching the other side of the droplet, the curvature of the surface causes the rays to transmitted back into the air at slightly different angles, allowing one to see a rainbow after a rain storm. Chromatic and geometric distortions can be minimized in lens systems by elaborate design and corrective optics, but it is also possible to do so using signal processing on the digital image. In modern astronomy, imperfections in mirrors and lens systems are usually corrected in the digital wavenumber domain by characterization of the 2-dimensional (2-D) wavenumber response of the telescope (usually on a distant star near the object of interest) and "normalization" of the received image by the inverse of the wavenumber response of the telescope. The process is often referred to as "de-speckling" of the image because before the process a distant star appears as a group of dots due to the telescope distortions and atmospheric multipath due to turbulence. The "system" 2-D wavenumber transfer function which restores the distant star clarifies the entire image.

For musical harmony, the modes of vibration in a string, or acoustic resonances in a horn or woodwind instrument form a Fourier series of overtone frequencies, each nearly an exact integer multiple of a fundamental frequency. Western musical scales are based on the first 12 natural harmonics, where the frequency difference between the 11th and 12th harmonic is the smallest musical interval, or semitone, at the frequency of the 12th harmonic. In other words, there are 12 semitones in a musical octave, but the frequency difference between a note and the next semitone up on the scale is $^{12}\sqrt{2}$ (or approximately 1.059 times the lower note), higher in frequency. This limits the frequency complexity of music by insuring a large number of shared overtones in musical chords. Indeed, an octave chord sounds "rock solid" because all overtones are shared, while a minor 4th (5 semitones) and minor 5th (7 semitones) intervals form the basis for blues, and most rock and roll music. In many ancient eastern cultures, musical scales are based on 15 semitones per octave, and some are even based on 17-note octaves giving very interesting musical patterns and harmony. Chords using adjacent semitones have a very complex overtone structure as do many percussive instruments such as cymbals, snare drums, etc. In practical musical instruments, only the first few overtones are accurate harmonic multiples while the upper harmonics have slight "mis-tunings" because of the acoustic properties of the instrument, such as structural modes, excitation non-linearities, and even control by the musician. The same is certainly true for the human voice due to nasal cavities and non-linearities in the vocal chords.

In speech and hearing, vocal chords in most animal life vibrate harmonically to increase sound power output, but in human speech, it is largely the time rate of change of the frequency response of the vocal tract which determines the information content. For example, one can easily understand a (clearly heard) spoken whisper as well as much louder voiced speech. However, the frequency content of voiced speech also provides informational clues about the speaker's ages, health, gender, emotional state, and so on. Speech recognition by computers is a classic example of adaptive pattern recognition addressed in later chapters of this book. We are already talking to electronic telephone operators, office computers. and even remote controls for home videotape players! It is likely in the near future that acoustic recognition technology will be applied to monitoring insect populations, fish and wildlife, and perhaps even stress in endangered species and animals in captivity or under study.

The operation of the human ear (as well as many animals) can be seen as a frequency transformer/detector. As sound enters the inner ear it excites the Basilar membrane which has tiny hair-like cells attached to it, which when stretched, emit electrical signals to the brain. It is believed that some of the hair cells actually operate as little loudspeakers, receiving electrical signals from the brain and responding mechanically like high frequency muscles. The structure of the membrane wrapped up in the snail-shell shaped cochlea (about the size of a wooden pencil eraser), along with the active feedback from the brain, cause certain areas of the membrane to resonate with a very high Q, brain-controllable sensitivity, and a more distinct and adaptive frequency selectivity. The sensor hair cell outputs represent the various frequency bands (called critical bands) of hearing acuity, and are nearly analogous to 1/3 octave filters at most frequencies. At higher frequencies in the speech range, things become much more complex and the hair cells fire in response vibrations in very complex spatial patterns. Evolution has made our speech understanding abilities absolutely remarkable where the majority of the neural signal processing is done within the brain, not the ear. It is fascinating to point out that the "background ringing", or mild tinnitus, everyone notices in their hearing is actually a real sound measurable using laser vibrometry on the ear drum, and can even be canceled using a carefully phased sound source in the ear canal. Tinnitus is thought to be from the "active" hair cells, where the corresponding sensor hair cells have been damaged by disease or excessive sound levels, driving the acoustic feedback loop through the brain to instability. It is most often in the 4–16 kHz frequency region because this area of the Basilar membrane is closest to where the sound enters the cochlea. Severe tinnitus often accompanies deafness, and in extreme cases can lead to a loss of sanity for the sufferer. Future intelligent hearing aid adaptive signal processors will likely address issues of tinnitus and outside noise cancellation while also providing hearing condition information to the doctor. Based on current trends in hearing loss and increases in life expectancy, intelligent hearing aids will likely be an application of adaptive processing with enormous benefits to society.

Essential to the development of intelligent adaptive signal processing and control systems is a thorough understanding of the underlying physics of the system being monitored and/or controlled. For example, the mass loading of fluids or gasses in a reactor vessel will change the structural vibration resonances providing the opportunity to monitor the chemical reaction and control product quality using inexpensive vibration sensors, provided one has identified detectable vibration features which are causal to the reaction of interest. Given a good physical model for sound propagation in the sea or in the human body, one can adaptively optimize the pulse transmission to maximize the detection of the target backscatter of interest. Analogous optimizations can be done for radar, lidar (light detection and ranging), and optical corrective processing as mentioned earlier. The characterization of wave propagation media and structures is most often done in the frequency-wavenumber domain. An intelligent adaptive processor will encounter many waves and signals where useful information on situation awareness will be detected using practical frequency-wavenumber transforms. On a final note, it is interesting to point out that Fourier's general idea that any waveform, including sound waveforms, could be represented by a weighted sum of sinusoids was a very radical idea at the time. His application of the Fourier transform to heat transfer is considered one of the great scientific contributions of all time.

5

The Fourier Transform

The French scientist Joseph B. Fourier (1768–1830) developed the first important development in the theory of heat conduction presented to The Academy of Sciences in Paris in 1807. To induce Fourier to extend and improve his theory the Academy of Sciences in Paris assigned the problem of heat propagation as its prize competition in 1812. The judges were LaPlace, Lagrange, and Legendre. Fourier became a member of the Academy of Sciences in 1817. Fourier continued to develop his ideas and eventually authored the applied mathematics classic *Théorie Analytique de la Chaleur* (or Analytical Theory of Heat), in 1822. Fourier gets much due credit, for his techniques revolutionized methods for the solution of partial differential equations and has led us to perhaps the most prominent signal processing operation in use today.

Consider the Fourier transform pair

$$Y(\omega) = \int_{-\infty}^{+\infty} y(t)e^{-j\omega t}dt$$

$$y(t) = \frac{1}{2\pi} \int_{-\infty}^{+\infty} Y(\omega)e^{+j\omega t}d\omega$$

(5.0.1)

where $y(t)$ is a time domain waveform and $Y(\omega)$ is the frequency domain Fourier transform. Note the similarity with the LaPlace transform pair in Eq. (2.1.2) where the factor of "j" in the LaPlace transform is simply from the change of variable from "$j\omega$" to "s" We can eliminate the factor of "$1/2\pi$" by switching from "ω" to "$2\pi f$" as seen in Eq. (5.0.2), but this is not the historically preferred notation.

$$Y(f) = \int_{-\infty}^{+\infty} y(t)e^{-j2\pi ft}dt$$

$$y(t) = \int_{-\infty}^{+\infty} Y(\omega)e^{+j2\pi ft}df$$

(5.0.2)

The Fourier transform pair notation in Eq. (5.0.2) while slightly longer, may actually be slightly more useful to us in this development because its easier to physically relate to frequencies in Hz (rather than radians/sec), and in the digital domain, all frequencies are relative to the sampling frequency. Consider below a short practical example of the Fourier transform for the case $y(t) = \sin(2\pi f_0 t)$.

$$
\begin{aligned}
Y(f) &= \int_{-\infty}^{+\infty} \sin(2\pi f_0 t) e^{-j 2\pi f t} dt \\
&= \int_{-\infty}^{+\infty} \frac{e^{+j 2\pi [f_0 - f]t} - e^{-j 2\pi [f_0 + f]t}}{2j} dt \\
&= \lim_{T \to \infty} \int_{-T/2}^{+T/2} \frac{e^{+j 2\pi [f_0 - f]t}}{2j} dt - \lim_{T \to \infty} \int_{-T/2}^{+T/2} \frac{e^{-j 2\pi [f_0 + f]t}}{2j} dt
\end{aligned}
\tag{5.0.3}
$$

The limit operations are needed to examine the details of the Fourier transform by first evaluating the definite integral. It is straightforward to show the results of the definite integral as seen in Eq. (5.0.4).

$$
Y(f) = \frac{1}{2j} \left\{ \lim_{T \to \infty} \frac{e^{+j\pi[f_0 - f]T}}{j 2\pi[f_0 - f]} - \lim_{T \to \infty} \frac{e^{+j\pi[f_0 - f]T}}{j 2\pi[f_0 - f]} dt \right\} \\
- \frac{1}{2j} \left\{ \lim_{T \to \infty} \frac{e^{+j\pi[f_0 + f]T}}{-j 2\pi[f_0 + f]} - \lim_{T \to \infty} \frac{e^{+j\pi[f_0 + f]T}}{-j 2\pi[f_0 + f]} dt \right\}
\tag{5.0.4}
$$

Combining the complex exponentials yields a more familiar result for analysis in Eq. (5.0.5).

$$
Y(f) = \frac{1}{2j} \left\{ \lim_{T \to \infty} \frac{\sin(\pi[f_0 - f]T)}{\pi[f_0 - f]} - \lim_{T \to \infty} \frac{\sin(\pi[f_0 + f]T)}{\pi[f_0 + f]} \right\}
\tag{5.0.5}
$$

For the cases where f is not equal to $\pm f_0$, $Y(f)$ is zero because the oscillations of the sinusoid in Eq. (5.0.5) integrate to zero in the limit as T approaches infinity. This very important property is the result of *orthogonality* of sinusoids where a residue is generated in the Fourier transform for the corresponding frequency components in $y(t)$. In other words, if $y(t)$ contains one sinusoid only at f_0, $Y(f)$ is zero for all $f \neq f_0$. For the case where $f = +f_0$ and $f = -f_0$, we have an indeterminant condition of the form 0/0 which can be evaluated in the limit as f approaches $\pm f_0$ using L'Hôpital's rule.

$$
\lim_{x \to a} \frac{g(x)}{f(x)} = \lim_{x \to a} \frac{\frac{\partial}{\partial x} g(x)}{\frac{\partial}{\partial x} f(x)}
\tag{5.06}
$$

Taking the partial derivative with respect to f of the numerators and denominators in Eq. (5.0.5) separately and then applying L'Hôpital's rule we find that the magnitude of the "peaks" at $\pm f_0$ is equal to $T/2$.

$$Y(f) = \frac{1}{2j} \lim_{T \to \infty} \left\{ \lim_{f \to f_0} \frac{-\pi T \cos(\pi[f_0 - f]T)}{-\pi} - \lim_{f \to -f_0} \frac{+\pi T \cos(\pi[f_0 + f]T)}{+\pi} \right\}$$

(5.07)

Clearly, the peak at $f = f_0$ has a value of $+T/2j$, or $-jT/2$, and the peak at $f = -f_0$ is $-T/2j$, or $+jT/2$. The imaginary peak values and the skew sign symmetry with frequency are due to the phase of the sine wave. A cosine wave would have two real peaks at $\pm f_0$ with value $+T/2$. A complex sinusoid ($y(t) = e^{j2\pi f 0 t}$) would have a single peak at $f = f_0$ of value $+T$. For real signals $y(t)$, one should always expect the Fourier transform to have a symmetric real part where the value at some positive frequency is the same as that at the "mirror image" negative frequency. The imaginary parts will be opposite in sign at positive and negative frequencies. The real and imaginary components of the Fourier transform of a real signal are said to be Hilbert transform pairs. Figure 1 below shows the frequency response $Y(f)$ for finite T where the amplitude in the figure has been normalized by $2T$.

The width of the peaks are found by noting that the first zero crossing of the sine function in Eq. (5.0.5) occurs when $f - f_0 = 1/T$, giving an approximate width of $1/T$. Therefore, as T approaches infinity, the amplitude of the peak becomes infinite, while the area remains constant at one. This is one way to define the Dirac delta function $\delta(x)$, which is zero for all $x \neq 0$, and infinite, but with unity area at $x = 0$. Therefore, for $y(t) = \sin(2\pi f_0 t)$, $Y(f) = -j\delta(f_0 - f)/2 + j\delta(f_0 + f)/2$, or if the radian notation is used one can simply multiply by 2π to get, $Y(\omega) = -j\pi\delta(\omega_0 - \omega) + j\pi\delta(\omega_0 + \omega)$, as listed in most texts for the sine function. The cosine function is $Y(f) = \delta(f_0 - f)/2 + \delta(f_0 + f)/2$, or if the radian notation is used, $Y(\omega) = \pi\delta(\omega_0 - \omega) + \pi\delta(\omega_0 + \omega)$. For the complex sinusoid $e^{j2\pi f_0 t}$, $Y(f) = \delta(f_0 - f)$, or $Y(\omega) = 2\pi\delta(\omega_0 - \omega)$ which can be verified by inspection using the sifting property of the Dirac delta function. To complete the example for the case of $y(t) = \sin(2\pi f_0 t)$, we simply evaluate the inverse Fourier transform integral.

$$y(t) = \int_{-\infty}^{+\infty} \frac{\delta(f_0 - f) - \delta(f_0 + f)}{2j} e^{+j2\pi f t} df$$

$$= \frac{e^{+j2\pi f_0 t} - e^{-j2\pi f_0 t}}{2j}$$

(5.0.8)

$$= \sin(2\pi f_0 t)$$

Clearly, one can see the symmetry properties of the Fourier transforms of real signals giving a Hilbert transform pair relationship to the symmetric real and skew-symmetric imaginary parts. As T approaches infinity, the resolution of the Fourier transform increases until all frequencies are completely independent, or orthogonal, from each other. As will be seen in the next section, the Discrete Fourier Transform (DFT) has limited resolution due to the finite number of data samples in the digital transform sum.

5.1 SPECTRAL RESOLUTION

We consider a regularly-sampled digital signal (as described in Section 1.2.2) $y[n]$, sampled every T seconds and converted to a signed integer. The sampling frequency $f_s = 1/T$ Hz, or samples per second. The Fourier transform integral in Eq. (5.0.2) becomes the N-point discrete sum

$$Y[f] = \sum_{n=0}^{N-1} y[n] e^{-j\,2\pi fnT} \frac{1}{N} \tag{5.1.1}$$

where $y[n]$ is a digital sample of $y(nT)$, nT is t, and $1/N$ is the equivalent of dt in the integral. Including the factor of $1/N$ here is technically a "N-normalized" discrete Fourier transform (NDFT). Most texts avoid the $1/N$ factor in the forward transform (including it in the inverse transform) since the "height" of Dirac delta function approaches infinity as the time integral gets longer. Therefore, the "height" in the DFT should also increase with increasing N.

However, from a physical signal processing point-of-view, *we prefer here to have the amplitude of the frequency-domain sinusoidal signal independent of N*, the number of points in the transform. This normalization is also applied to the power spectrum in Chapter 6. We must be careful to distinguish the NDFT from the standard DFT to avoid confusion with the literature. Any random noise in $y[n]$ will however be suppressed relative to the amplitude of a sinusoid(s) as the number of points in the NDFT increases. This will also be discussed subsequently in Chapter 6 on spectral density. Consider the NDFT pair where $0 \leq f \leq 0.5/T$ (only analog frequencies less than the Nyquist rate $f_s/2 = 0.5/T$, are present in $y[n]$, which is real).

$$Y[f] = \frac{1}{N} \sum_{n=0}^{N-1} y[n] e^{-j\,2\pi fnT}$$

$$y[n] = \sum_{m=0}^{M-1} Y[(m\Delta f)] e^{+j\,2\pi(m\Delta f)nT} \tag{5.1.2}$$

We have assumed a set of M evenly-spaced frequencies in the inverse NDFT in Eq. (5.1.2) where Δf is f_s/M Hz. However, the resolution Δf and number of frequencies can actually be varied considerably according to the needs of the application, but the underlying resolution available is a function of the integration limits in time or space. In general, the spectral resolution available in a forward DFT can be simply found using the result from the Dirac delta function seen in Figure 1. With N samples spanning a time interval of NT secs in Eq. (5.1.2), the available resolution in the DFT is approximately $1/(NT)$ Hz. Therefore, it is reasonable to set $\Delta f = 1/(NT)$ giving N frequencies between 0 and f_s, or $N = M$ in the DFT pair. Evaluating the DFT at closer-spaced frequencies than Δf will simply result in more frequency points on the "$\sin(x)/x$", or $\mathrm{sinc}(x)$, envelope function for a sinusoidal signal. This envelope is clearly seen in Figure 1 and in Eq. (5.0.5) where the Fourier transform is evaluated over a finite time interval. The only way to increase the spectral resolution (narrowing Δf) is to increase NT, the physical span of the DFT sum for a given f_s. Increasing the number of samples N for a fixed sample interval T (fixed sample rate or bandwidth) simply increases the length of time for the

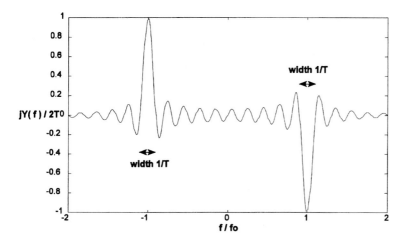

Figure 1   Fourier transform of $\sin(2\pi f_0 t)$ limiting the integration to $\pm T/2$ to evaluate the characteristics of the Dirac delta function.

Fourier transform and allows a finer frequency resolution to be computed. Therefore, one can easily see that a 1-sec forward Fourier transform can yield a 1 Hz resolution in the frequency domain. while a 100 msec transform would have only 10 Hz resolution available, and so on.

Consider the following short proof of the NDFT for an arbitrary $y[n]$, N time samples and N frequency samples giving a $\Delta f = f_s / N$, or $\Delta f\, nT = n/N$, to simplify the algebra and represent a typical discrete Fourier transform operation.

$$y[p] = \sum_{k=0}^{N-1} \left\{ \frac{1}{N} \sum_{n=0}^{N-1} y[n] e^{-j2\pi kn/N} \right\} e^{-j2\pi kp/N} \qquad p = 0, 1, \ldots, N-1 \qquad (5.1.3)$$

The forward DFT in the braces of Eq. (5.1.3) can be seen as $Y(k\Delta f) = Y[k]$. Rearranging the summations provides a more clear presentation of the equality.

$$y[p] = \frac{1}{N} \sum_{n=0}^{N-1} y[n] \left\{ \sum_{k=0}^{N-1} e^{+j2\pi k(p-n)/N} \right\} \qquad p = 0, 1, \ldots, N-1 \qquad (5.1.4)$$

Clearly, the expression in the braces of Eq. (5.1.4) is equal to N when $p = n$. We now argue that the expression in the braces is zero for $p \neq n$. Note that the expression in the braces is of the form of a geometric series.

$$\sum_{k=0}^{N-1} a^k = \frac{1 - a^N}{1 - a} \qquad a = e^{j2\pi(p-n)/N} \qquad (5.1.5)$$

Since $p \neq n$ and $p - n$ goes from $-(N-1)$ to $+(N-1)$, the variable "a" cannot equal 1 and the series is convergent. However, since a^N is $e^{j2\pi(p-n)}$, $a^N = 1$ and the numerator of Eq. (5.1.5) is zero. Therefore, Eq. (5.1.4) is non-zero only for the case where $p = n$, the sum over n reduces to the $p = n$ case, the factors of N and $1/N$ cancel (as they also do for an analogous proof for the DFT), and $y[n] = y[p]$; $p = n$.

We now examine the N-point NDFT resolution for the case of a sine wave. Consider the case where $y[n] = \sin(2\pi f_0 nT) = \sin(2\pi f_0 n/f_s)$ and f_0 is an integer multiple of Δf, or $f_0 = m_0 \Delta f$. We will consider a N-point NDFT and N frequency points $\Delta f = f_s / N$ Hz apart. In other words, the frequency of our sine wave will be exactly at one of the discrete frequencies in the DFT. Since $f_0 = m_0 f_s / N$, we can write our sine wave as a simple function of m_0, n, and N by $y[n] = \sin(2\pi m_0 n / N)$. Equation (5.1.6) shows the N-point NDFT expressed in terms of $Y[m]$, for the mth frequency bin, by again writing the sine function as the sum of two complex exponentials.

$$Y[m] = \frac{1}{N} \sum_{n=0}^{N-1} \sin(2\pi m_0 n/N) e^{-j\pi\frac{mn}{N}}$$

$$= \frac{1}{2jN} \sum_{n=0}^{N-1} e^{+j 2\pi(m_0-m)\frac{n}{N}} - \frac{1}{2jN} \sum_{n=0}^{N-1} e^{-j 2\pi(m_0+m)\frac{n}{N}} \qquad (5.1.6)$$

Applying the finite geometric series formula of Eq. (5.1.5) gives

$$
\begin{aligned}
Y[m] &= \frac{1}{2jN} \left[\frac{1 - e^{+j 2\pi(m_0-m)}}{1 - e^{+j 2\pi(m_0-m)/N}} - \frac{1 - e^{-j 2\pi(m_0+m)}}{1 - e^{-j 2\pi(m_0+m)/N}} \right] \\
&= \frac{e^{+j\pi(m_0-m)\frac{N-1}{N}}}{2jN} \frac{\sin[\pi(m_0-m)]}{\sin[\pi(m_0-m)/N]} \\
&\quad - \frac{e^{-j\pi(m_0-m)\frac{N-1}{N}}}{2jN} \frac{\sin[\pi(m_0+m)]}{\sin[\pi(m_0+m)/N]}
\end{aligned}
\qquad (5.1.7)
$$

As m approaches $\pm m_0$, one can clearly see the indeterminant condition $0/0$ as evaluated using L'Hôpital's Rule in Eq. (5.0.7) above. The result is a peak at $+m_0$ with amplitude $-j/2$, and a peak at $-m_0$ with amplitude $+j/2$. This is consistent with the continuous Fourier transform result in Eqs (5.0.6) and (5.0.7) keeping in mind that we divided our forward DFT by N (as we have defined as the NDFT). Note that the indeterminant condition $0/0$ repeats when $\pi(m_0 - m)/N = k\pi$; $k = 0, \pm 1, \pm 2, ...$, since the numerator sine angle is simply N times the denominator sine angle. The actual frequency of m in Hz is $m\Delta f$, where $\Delta f = f_s/N$, N being the number of frequency points in the NDFT. Therefore, is can be seen that the peak at $+m_0\Delta f$ Hz repeats every f_s Hz up and down the frequency axis to $\pm\infty$, as also does the peak at $-m_0$. Figure 5, page 10, clearly illustrates the frequency aliasing for a cosine wave showing how a wave frequency greater than $f_s/2$ will appear "aliased" as an incorrect frequency within the Nyquist band. If the original continuous waveform $y(t)$ is band-limited to a frequency band less than $f_s/2$, all frequencies observed after performing a NDFT will appear in the correct place. Mathematically, it can be seen that the NDFT (and DFT) *assumes* the waveform within its finite-length buffer repeats periodically outside that buffer for both time and frequency domain signals.

Consider the example of a 16-point NDFT on a unity amplitude sine wave of 25 Hz where the sample rate is 100 Hz. With 16 frequency "bins" in the DFT, we find $\Delta f = 100/16$, or 6.25 Hz. The sampling time interval is 10 msec so the time window of data is 160 msec long, also indicating a possible resolution of 6.25 Hz for the NDFT. Figure 2 shows the NDFT of the sine wave in the Nyquist band (between

± 50 Hz) where the "*" symbols indicate the 16 discrete integer "*m*" values and the curve depicts *Y*[*m*] as a continuous function of *m* showing the resolution limitations for the 160 msec wave. Note that if a 16-point DFT were used instead of the NDFT, the peak magnitudes would be 8, or N/2. The real part of the 16-point NDFT is zero as seen in Figure 3. But, it is more customary to show the magnitude of the NDFT as seen in Figure 4.

The discrete NDFT bin samples in Figure 2, 3, and 4 depicted by the "*" symbols resemble the ideal Dirac delta function (normalized by *N*) in the context of the 16-point NDFT output because the 25 Hz sine wave has exactly 4 wavelengths in the 160 msec long digital input signal. Therefore, the assumptions of periodicity

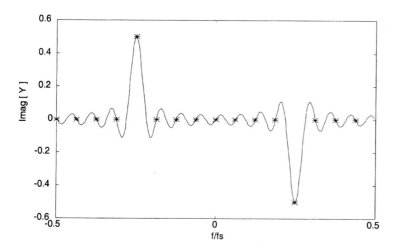

Figure 2   The imaginary part of a 16-point NDFT of a 25 Hz sine wave sampled at 100 Hz showing the discrete NDFT bins (*) and a continuous curve for the available resolution.

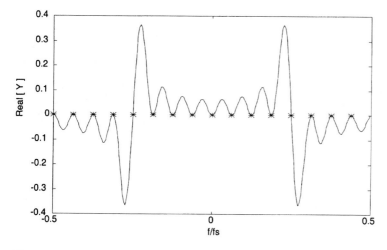

Figure 3   The real part of the 16-point NDFT of the sine wave is zero at the 16 discrete frequency bins.

outside the finite NDFT input buffer are valid. However, if the input sine wave has a frequency of, say, 28.125 Hz, the wave has exactly 4.5 wavelengths in the NDFT input buffer sampled at 100 Hz. For 28.125 Hz, the periodicity assumption is violated because of the half-cycle discontinuity and the resulting magnitude of the NDFT is seen in Figure 5 below.

As Figure 5 clearly shows, when the sine wave frequency does not match up with one of the 16 discrete frequency bins of the NDFT, the spectral energy is "smeared" into all other NDFT bins to some degree. This is called *spectral leakage*. There are three things one can do to eliminate the smeared NDFT (or DFT) resolution. First, one could synchronize the sample frequency to be an integer multiple

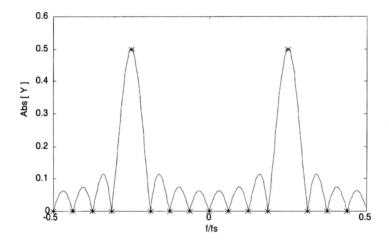

Figure 4   The magnitude of a 16-point NDFT of a 25 Hz sine wave sampled at 100 Hz showing the discrete NDFT nins (*) an a continuous curve for the available resolution.

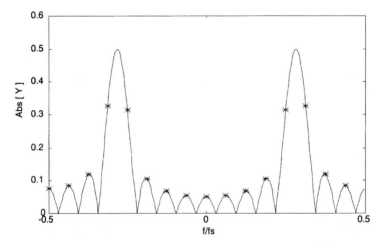

Figure 5   The magnitude of a 16-point NDFT of a 28.125 Hz sine wave sampled at 100 Hz showing the discrete NDFT bins (*) and spectral leakage caused by the mismatched frequency.

of the sine wave frequency, which in the case of 28.125 Hz would require a sample rate of 112.5 Hz, 140.612 Hz, etc. Synchronized Fourier transforms are most often found in "order-tracking" for vibration analysis of rotating equipment such as engines, motor/generators, turbines, fans, etc. Synchronizing the sample rate to the frequencies naturally occurring in the data insures that the vibration fundamental and its harmonics are all well-matched to the DFT bin frequencies giving reliable amplitude levels independent of machinery rotation speed.

The second technique to minimize spectral leakage is to increase the input buffer size, or increase N. Increasing N decreases Δf and the number of output DFT bins. Eventually, with enough increase in resolution, there will be so many bins in the DFT that the sine wave of interest will lie on or very close to one of the discrete bins. For the case of the 16-point NDFT data in Figure 4, doubling the NDFT size to 32 (320 msec input time buffer), narrows the resolution to 3.125 Hz allowing the 28.125 Hz wave to lie exactly on a bin. All the other NDFT bins will be at a zero-crossing of the underlying sinc function giving the appearance of a "normalized Dirac-like" spectrum. Another way to look at doubling the input buffer size is to consider that 28.125 Hz will have exactly 9 full wavelengths in the input buffer, making the periodicity assumptions of the Fourier transform on a finite interval correct. However, if one kept the 16-point input buffer and simply analyzed the NDFT output at many more frequencies closer together than the original 16 NDFT bins, one would simply be computing points on the continuous curves in Figures 2, 3, and 4. These curves represent the underlying resolution available for the 160 msec input buffer.

The third technique is to apply a data envelope window to the DFT or NDFT input. This technique will be discussed in much more detail in Section 5.3. Windowing the data envelope makes the data "appear" more periodic from one input buffer to the next by attenuating the data amplitude near the beginning and end of the input buffer. This of course changes the spectral amplitude as well as causes some spectral leakage to occur even if the input frequency is exactly on one the the discrete DFT frequency bins.

Generally speaking for stationary signals, larger and/or synchronized Fourier transforms are desirable to have a very high resolution, well-defined spectrum. Also, if there is any random broadband noise in the input data, the more NDFT bins one has, the less random noise there will be in each bin since the noise is divided over all the bins. Large NDFT's will tend to focus on periodic signals and suppress random noise which is a very desirable property. However, even if the sines and cosines of the NDFT (or DFT) are computed in advance and stored in memory, the number of complex multiplies in a straightforward N-point DFT is N^2. Therefore, very large broadband DFT's where one can vary the number of frequency bins at will are prohibitively expensive interms of the required computing resources.

5.2 THE FAST FOURIER TRANSFORM

Early on in the history of digital signal processing it was recognized that many of the multiplies in a DFT are actually redundant, making a Fast Fourier Transform (FFT) possible through computer algorithm optimization. We will not present a detailed derivation or discussion of the FFT as there are many signal processing books which do so with eloquence not likely here. It should suffice to say that the FFT is an efficient implementation of the DFT where the number of input samples and output frequency bins are the same, and in general, a power of 2 in length. An N-point FFT where N is a power of 2 (256, 512, 1024, etc.), requires $N \log_2 N$ complex multiplies, as opposed to N^2 multiplies for the exact same N-point output result from the DFT. For a 1024-point transform, that's a reduction from 1,048,576 complex multiplies to 105,240 complex multiplies for the FFT. Clearly, to call the FFT efficient is an understatement for large N!

Both transforms give exactly the same result when the DFT has N equally-spaced output frequency bins over the Nyquist band. In the early days of digital signal processing with vacuum tube flip-flops and small magnetic core memories, the FFT was an absolutely essential algorithm. It is (and likely always will be) the standard way to compute digital Fourier transforms of evenly-spaced data primarily for economical reasons. What is to be emphasized here is that the resolution of the DFT or FFT is always limited by the length of the input buffer, and spectral leakage within the DFT or FFT is caused by a breakdown of the periodicity assumptions in the Fourier transform when a non-integer number of full wavelengths appear in the input buffer. One can use an FFT to evaluate more frequency bins than input samples by "zero-padding" the input buffer to fill out the larger size FFT buffer. This is sometimes a useful alternative to synchronized sampling as a means of finding an accurate peak level even with the underlying sinc function envelope. One can also perform a "zoom-FFT" by bandlimiting the input signal of interest centered at some high center frequency f_c rather than zero Hz, and essentially performing a complex de-modulation down to 0 Hz as described at the end of Section 1.3 and Figure 1.7 before computing the FFT.

Consider briefly the case of an 8-point DFT which we will convert to an 8-point FFT through a process known as time decomposition. First, we split the unnormalized (we'll drop the $1/N$ normalization here for simplicity of presentation) DFT into two $N/2$-point DFT's as seen in Eq. (5.2.1) below.

$$Y[m] = \sum_{n=0}^{\frac{N}{2}-1} y[2n] e^{-j2\pi(2n)m/N} + \sum_{n=0}^{\frac{N}{2}-1} y[2n+1] e^{-j2\pi(2n+1)m/N} \qquad (5.2.1)$$

We can make our notation even more compact by introducing $W_N = e^{-j2\pi/N}$, commonly referred to in most texts on the FFT as the "twiddle factor", or the time/frequency invariant part of the complex exponentials. Equation (5.2.1) is then seen as a combination of 2 $N/2$-point DFT's on the even and odd samples of $y[n]$.

$$Y[m] = \sum_{n=0}^{\frac{N}{2}-1} y[2n] W_N^{2nm} + W_N^m \sum_{n=0}^{\frac{N}{2}-1} y[2n+1] W_N^{2nm} \qquad (5.2.2)$$

Note that by doing two $N/2$-point DFTs rather than one N-point DFT's we've reduced the number of complex multiplies from N^2 to $N^2/4 + N/2$, which in itself is a significant savings. Each of the $N/2$-point DFTs in Eq. (5.2.2) can be split into even and odd samples to give four N/4-point DFTs, as shown in Eq. (5.2.3).

$$
\begin{aligned}
Y[m] = {}& W_N^0 \sum_{n=0}^{\frac{N}{4}-1} y[4n] W_N^{4nm} + W_N^{2m} \sum_{n=0}^{\frac{N}{4}-1} y[4n+2] W_N^{4nm} \\
&+ W_N^m \sum_{n=0}^{\frac{N}{4}-1} y[4n+1] W_N^{4nm} + W_N^{3m} \sum_{n=0}^{\frac{N}{4}-1} y[4n+2] W_N^{4nm}
\end{aligned}
\tag{5.2.3}
$$

For the case $N=8$, Eq. (5.2.3) can be seen as a series of four 2-point DFTs and the time decomposition need not go further. However, with N some power of 2, the time decomposition continues q times where $N = 2^q$ until we have a series of 2-point DFTs on the data. This type of FFT algorithm is known as a radix-2 FFT using time-decomposition. Many other forms of the FFT can be found in the literature including radix 3, 4, 5, etc., as well as frequency decomposition formulations. Radix-2 FFT's are by far the most common used. For blocks of data which are not exactly the length of the particular FFT input buffer, one simply "pads" the input buffer with zeros. The underlying spectral resolution in Hz (wavenumber) is still the inverse of the actual data record in seconds (meters) excluding the zero padded elements.

Figure 6 shows a flow diagram for the 8-point FFT derived above in Eqs (5.2.1)–(5.2.3). The 2-point DFTs are arranged in a convenient order for a recursive algorithm which computes the FFT in-place (the original input data is overwritten during the computation. The 2-point DFTs in the various sections are typically called "butterflies" because of their X-like signal flow patterns in the flow diagram. However, the convenient ordering of the butterflies leads curiously to the outputs being addressed in binary bit-reversed order. This is probably the main reason

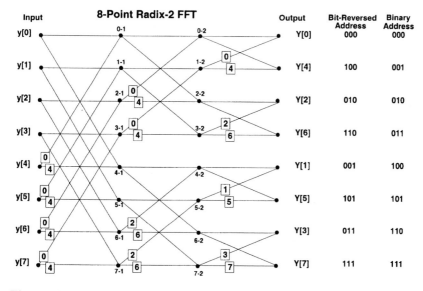

Figure 6 An 8-point radix-2 FFT showing twiddle factors in the square boxes for time decomposition and bit-reversed outputs.

the radix-2 butterfly is so popular. "Unscrambling" of the bit reversed output addresses is actually quite straightforward and most of today's digital signal processing processors (AT&T WE-DSP32C, Texas Instruments. TMS320C30, Motorola 96002, etc.) offer a hardware-based bit-reversed addressing instruction for speedy FFT programs.

The twiddle factor structure in the flow diagram in Figure 6 also has a much more subtle design feature which is not widely understood and is extremely import-ant to algorithm efficiency. Notice that for the input column only $y[4]$ through $y[7]$ are multiplied by twiddle factors while $y[0]$ through $y[3]$ pass straight through to the second column, where again, only half the nodes are multiplied by twiddle factors, and so on. As will be seen, the ordering of the twiddle factors at each node is critical, *because at any one node, the two twiddle factors differ only in sign!* It seems the ingenious algorithm design has also allowed $W_8^0 = -W_8^4$, $W_8^2 = -W_8^6$, $W_8^1 = -W_8^5$, and $W_8^3 = -W_8^7$. Figure 7 graphically shows how the twiddle factors differ in sign.

Obviously, one would do the complex multiply just once at each node, and either add or subtract the result in the computation for the nodes in the next column. For example, $y[4]$ would be multiplied by W_8^0 (which is unity by the way) and saved into complex temporary storage "ctemp1", $y[0]$ is copied into "ctemp2", then is overwritten by $y[0]$+temp1 (seen in node 0–1 of Figure 6). Then, ctemp1 − ctemp2 is seen to be $y[0] + y[4]W_8^4$ and overwrites $y[4]$ in node 4–1. This type of FFT algo-rithm computes the Fourier transform *in-place* using only 2 complex temporary storage locations, meaning that the input data is overwritten and no memory storage space is wasted — a critical design feature in the 1960s and an economical algorithm feature today. To complete the 8-point FFT, one would proceed to compute the unity twiddle factor (W_8^0) butterflies filling out all the nodes in column 1 of Figure 6, then nodes 0–2 through 3–2 in column 2, and finally the outputs $Y[0]$ and $Y[4]$. Then

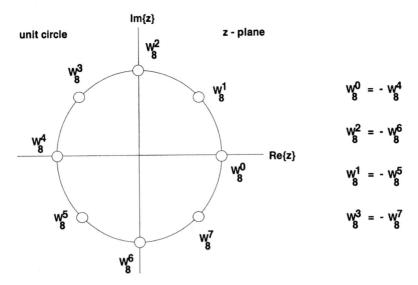

Figure 7 Twiddle factors for the 8-point FFT can be grouped into pairs which differ only in sign to speed computations.

using W_8^2, nodes 4–2 through 7–2 are computed followed by the outputs $Y[2]$ and $Y[6]$. Then W_8^1 can be used to compute the outputs $Y[1]$ and $Y[5]$, followed by W_8^3 to compute the outputs $Y[3]$ and $Y[7]$. A hardware assembly code instruction or a straightforward sorting algorithm in software re-orders the outputs in non bit-reversed address form for use.

It can be seen that for a radix-2 N-point FFT algorithm there will be $\log_2 N$ columns of in-place computations where each column actually requires only $N/2$ complex multiplies and N additions. Since multiplications are generally more intensive than additions, we can focus the computational cost on only multiplications. The radix-2 N-point FFT, where N is a power of 2 ($N = 2^q$) requires $(N/2)\log_2 N$ complex multiplies. Since a significant number of those multiplies are using the twiddle factor W_N^0, which is unity, the required number of complex multiplies is actually less than $(N/2)\log_2 N$. However, a complex multiply actually involves 4 numerical multiplies. Therefore, for complex input and output, the FFT is seen to require $2N\log_2 N$ numerical multiplies. Once again, most FFT applications start with real sampled input data, so the widely accepted computational estimate for the FFT is $N\log_2 N$ multiplies, but the actual number will depend on the specific application and how well the designer has optimized the algorithm. Note that the inverse FFT simply requires the opposite sign to the complex exponential which is most easily achieved by reversing the order of the input data from 0 to $N - 1$ to $N - 1$ to 0. The scale factor of N, or $1/N$ depending on one's definition of the DFT/FFT, is generally handled outside the main FFT operations.

Usually we work with real sampled data as described in Section 1.2, yet most FFT algorithms allow for a fully complex input buffer and provide a complex output buffer. For real input samples, one simply uses zeros for the imaginary component and a significant number of multiplies are wasted unless the user optimizes the FFT algorithm further for real input data. This can be done a number of different ways including packing the even samples into the real part of the input and odd samples into the imaginary part of the input. Another approach is to simply eliminate all unnecessary multiplications involving imaginary components of the input data. Another technique would be to bit-reverse the input data and neglect all computations which lead to the upper half of the in-order output data, and so on. But, perhaps an even more intriguing technique is simultaneously computing the FFT's of two real signals assembled into a single complex array using one complex FFT operation. It was discussed earlier that real input data (imaginary part is zero) yields real frequency domain data with even symmetry (positive and negative frequency components equal), or $Re\{Y[m]\} = Re\{Y[N - m]\}$, $m = 0, 1, \dots, N - 1$. The imaginary part will have skew symmetry, or $Im\{Y[m]\} = - Im\{Y[N - m]\}$, $m = 0, 1, \dots, N - 1$. The amplitudes of the real and imaginary parts are really only a function of the phases of each frequency in the FFT input buffer. If we put our real input data into the imaginary part of the input leaving the real part of the FFT input zero, the opposite is true where $Re\{Y[m]\} = - Re\{Y[N - m]\}$ and $Im\{Y[m]\} = Im\{Y[N - m]\}$.

We can exploit the Hilbert transform pair symmetry by packing two real data channels into the complex FFT input buffer and actually recovering two separate FFTs from the output. For "channel 1" in the real part of our complex input data, and "channel 2" in the imaginary part of our complex input data, the following relationships in Eq. (5.2.4) can be used to completely recover the two separate frequency spectra $Y_1[m]$ and $Y_2[m]$ for positive frequency. Note that the "0 Hz" components are recovered as $Re\{Y_1[0]\} = Re\{Y[0]\}$, $Re\{Y_2[0]\} = Im\{Y[0]\}$, $Im\{Y_1[0]\} = Im\{Y_2[0]\} = 0$.

$$Re\{Y_1[m]\} = \frac{1}{2}\left[\{Re\{Y[m]\} + Re\{Y[N-m]\}\}\right]$$

$$Im\{Y_1[m]\} = \frac{1}{2}\left[\{Im\{Y[m]\} - Im\{Y[N-m]\}\}\right]$$

$$\qquad\qquad\qquad\qquad\qquad\qquad\qquad (5.2.4)$$

$$Re\{Y_2[m]\} = \frac{1}{2}\left[\{Im\{Y[N-m]\} + Im\{Y[m]\}\}\right]$$

$$Im\{Y_2[m]\} = \frac{1}{2}\left[\{Re\{Y[N-m]\} - Re\{Y[m]\}\}\right]$$

5.3 DATA WINDOWING

Controlling the resolution of the Fourier transform using either the NDFT, DFT, or FFT algorithms is an important design consideration when dealing with signal frequencies which may lie in between the discrete frequency bins. The reason why a peak's signal level in the frequency domain is reduced and leakage into other frequency bins increases, when the frequency is between bins, can be seen as the result of a non-periodic waveform in the input buffer. Therefore, it is possible to reduce the leakage effects along with the peak level sensitivity to frequency alignment by forcing the input wave to appear periodic. The most common way this is done is to multiply the input buffer by a raised cosine wave which gradually attenuates the amplitude of the input at either end of the input buffer. This "data window" is known as the Hanning window, after its developer, and has the effect of reducing the signal leakage into adjacent bins when the frequency of the input signal is between bins. It also has the effect of making the peak signal level in the frequency domain less sensitive to the alignment of the input frequency to the frequency bins. However, when the input signal is aligned with a frequency bin and the Hanning window is applied, some spectral leakage, which otherwise would not have occurred, will appear in the bins adjacent to the peak frequency bin. This tradeoff is generally worth the price in most Fourier processing applications. As we will see below, a great deal of artifice has gone into data window design and controlling the amount of spectral leakage. We present a wide range of data windows for the FFT (the NDFT, DFT, and FFT behave the same) and discuss their individual attributes. First, we consider the Hanning window, which is implemented with the following equation.

$$W^{Hn}(n) = \frac{1}{2}\left\{1 - \cos\left(\frac{2\pi n}{N-1}\right)\right\} \quad n = 0, 1, 2, \ldots, N-1 \qquad (5.3.1)$$

The input buffer of the Fourier transform can be seen as the product of the data window times an "infinitely long" waveform of the input signal. The finite sum required by the FFT can be seen as having an input signal which is the product of a rectangular window N samples long and the infinitely long input waveform. A product in the time domain is a convolution in the frequency domain, as explained in detail in Chapter 2, and the sinc-function response of the FFT due to the finite sum can be seen as the convolution of the Fourier transform of the rectangular data window with the Dirac delta function of the infinitely-long sinusoid. The Hanning window can be written as the sum of three terms, each of which has a sinc-like Fourier transform.

$$W^{Hn}(n) = \frac{1}{2} + \frac{1}{4}e^{\frac{+j2\pi n}{N-1}} + \frac{1}{4}e^{\frac{-j2\pi n}{N-1}} \tag{5.3.2}$$

The effect of the three sinc functions convolved with the delta function for the sinusoid is a "fatter", or broader, main lobe of the resulting resolution envelope and lower leakage levels in the side lobes. This is depicted in Figure 8 below which compares the resolution for the rectangular data window to that for the Hanning "raised cosine" data window.

The resolutions shown in Figure 8 are normalized by N (as we defined in the NDFT), but an additional narrowband correction factor of 2.0317 for $N = 64$ (approaches 2.00 for large N) is needed to boost the Hanning window peak level to match the rectangular window peak level. For any given window type, one can "calibrate" a bin aligned sine wave to determine the narrowband correction factor. However, normalizing the integral of the window to match the integral of the rectangular window is a more strict definition of narrowband normalization. Therefore, one simply sums the window function and divides by N, the rectangular window sum, to get the narrowband correction factor. In other words, *the narrowband correction factor is the ratio of the window integral to the integral of a rectangular window of the same length.* The narrowband normalization constant multiplied by the window function is important for signal calibration purposes

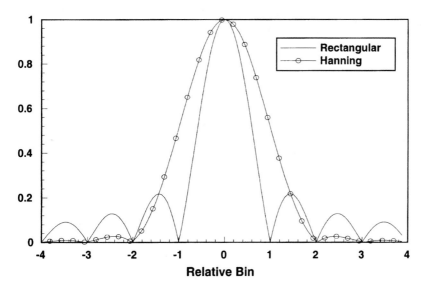

Figure 8 Comparison of the spectral resolution of the rectangular and Hanning data windows where $N = 64$ (a 1024-point FFT of the zero-padded windows is used to show the underlying resolution).

in signal processing systems, as we would like the peak levels of the spectra to match when the peak is bin-aligned for whichever data window is used.

There will be a slightly different correction factor for matching broadband Hanning-windowed signals to the rectangular window levels. Broadband normalization is important for determining the total signal power in the frequency domain by summing all the NDFT bins. Application of a window on the time-domain data before the Fourier transform introduces a small level error in the total spectral power due to the controlled spectral leakage of the window. Since power is measured as a magnitude-squared spectrum, *the broadband correction factor is determined by the square-root of the integral of the window function squared, then divided by N (the integral of the squared rectangular window function).* The broadband correction factor for a $N = 64$ point Hanning window is 1.6459, and 1.6338 for $N = 1024$.

Careful definition of narrowband and broadband normalization constants is consistent with our reasoning for the NDFT, where we would like periodic signal levels to be independent of N, the size of the DFT as well as independent of the type of data window used. By keeping track of time and frequency-domain scale factors we will make applications of frequency domain signal processing techniques to adaptive systems, pattern recognition algorithms, and control applications much more clear. *Narrowband and broadband correction factors are critically important to power spectrum amplitude calibration in the frequency domain.*

There are several other noteworthy data windows which are presented below ranging from highest resolution and side lobe levels to lowest resolution and side lobe levels. For the NDFT these windows need not be normalized by N (as seen in most texts for DFT applications). The calibration factors are presented for $N = 64$ for comparison.

The Welch window is essentially a concave down parabola centered in the middle of the data input buffer. The Welch window requires a narrowband scale factor of 1.5242 to boost its peak level up to the rectangular window's level.

$$W^{Wc}(n) = 1 - \left(\frac{n - \frac{1}{2}(N - 1)}{\frac{1}{2}(N - 1)} \right)^2 \tag{5.3.3}$$

Another simple but very useful data window is the Parzen, or triangle window, described in Eq. (5.3.4), which requires a narrowband scale factor of 2.0323.

$$W^{Wc}(n) = 1 - \left| \frac{n - \frac{1}{2}(N - 1)}{\frac{1}{2}(N - 1)} \right|^2 \tag{5.3.4}$$

Raising the Hanning window a small amount (so the end points are non-zero) produces a significant improvement by narrowing resolution and lowering side lobe leakage as seen in the Hamming window.

$$W^{Hm}(n) \frac{1}{2} \left\{ 1.08 - 0.92 \cos \left(\frac{2\pi n}{N - 1} \right) \right\} \tag{5.3.5}$$

The small change between the Hanning and Hamming windows (small spelling change too) is indicative of the art that has gone into data window design. The

Hamming window requires a narrowband scale factor of 1.8768 for its peak levels to match the rectangular window. One can also define a generic exponential window which can be notified to have almost any spectral width with very smooth low side lobe levels.

$$W^{Ex}(n) = \sqrt{\frac{\pi}{3}} e^{-\frac{1}{2}(n-\frac{N}{2}-1)^2(3.43 \cdot 10^{-k})} \tag{5.3.5}$$

When $k \geq 4$, the exponential window very closely matches the rectangular window and the narrowband scale factor is 1.0347. When $k = 3$, the window very closely matches the Welch window and the narrowband scale factor is 1.5562. However, k also depends on N, so a wide variety of window shapes and resolution can be realized. We show the exponential window below with $N = 64$, $k = 2.6$, and a narrowband scale factor of 2.3229, giving a broader resolution than the other windows but with the lowest side lobe levels.

Figure 9 shows the 6 data windows together in the time domain unnormalized. The narrowband correction scale factors noted above for $N = 64$ and in Table 1 for the windows should be applied (multiply the windows by them) to insure consistent peak or levels for all windows on frequency bin-aligned signals. For FFT sizes not equal to $N = 64$, calculate the narrowband scale factor by the ratio of the window integral to the rectangular window integral and the broadband scale factor by the ratio of the window-squared integral to the integral of the rectangular window. Note that if the input signals are not bin-aligned, or a broadband correction factor is used, their peak levels will differ slightly depending on the particular data window used. The best we can do is to calibrate each data window so the peak levels for each

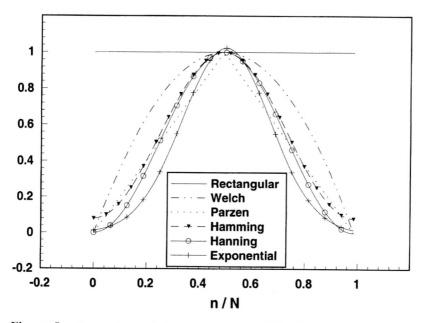

Figure 9 Comparison of the rectangular, Welch, Parzen, Hamming, Hanning, and exponential data windows.

Table 1   Narrowband and Broadband Level Correction Factors for N=64 and N=1024 point windows

Window $N = 64$	Rectangular r	Hanning	Welch	Parzen	Hamming	Exponential $k = 2.6$
Narrowband	1.0000	2.0317	1.5242	2.0323	1.8768	2.3229
Broadband	1.0000	1.6459	1.3801	1.7460	1.5986	1.7890
$N = 1024$						$k = 5.2$
Narrowband	1.0000	2.0020	1.5015	2.0020	1.8534	2.8897
Broadband	1.0000	1.6338	1.3700	1.7329	1.5871	1.6026

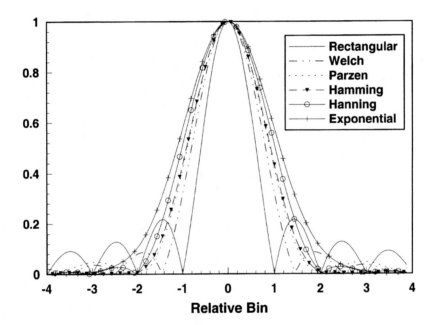

Figure 10   Comparison of the spectral resolution of the rectangular, Welch, Parzen, Hamming, Hanning, and exponential data windows.

window match when the input signal frequency matches the FFT bin. The spectral resolutions shown in Figure 10 include the narrowband correction factors (window calibration) constants named with each window above and in Table 1. Broadband correction factors are applied to insure total signal power is consistent from the (unwindowed) time domain to the frequency domain. Table 1 also presents the broadband corrections for $N = 64$ point windows. All correction factors given will vary slightly with N, so a simple calculation is in order if precise signal level calibrations are needed in the frequency domain.

Comparing the non-rectangular data windows in Figure 9 we see only very slight differences. However, we can make some interesting observations which hold in general for data window applications. First, the broader rectangular and Welch data windows are the ones with better resolution, as seen in Figure 10. Second, the data windows with nearly zero slope at either end (near $n = 0$ and $n = N - 1$) tend to have the lowest side lobe levels, as seen in Figure 10 for the exponential, Hamming and Hanning window's spectral resolution. Its hard to say which data window is the best because it depends on the particular applications emphasis on main lobe resolution and side lobe suppression.

The reason one desires low side lobes is often one has strong and weak signal peaks in the same frequency range. With high leakage, the side lobes of the strong peak can cover up the main lobe of the weak peak. For applications where a wide dynamic range of sinusoidal peak levels are expected, application of a data window is essential. However, other applications require very precise frequency and phase measurement of spectral peaks. In this case, higher resolution is of primary importance and the rectangular window may be best suited. As always, a thorough understanding of the underlying physics of the signal processing application is needed to be sure of an optimal design approach and implementation. Perhaps the most important reason to have a variety of data windows available is for sensor array shading, which is presented in more detail in Sections 7.1, 12.2, and Chapter 13. Array shading using data windows is the reason the spectral leakage is referred to as "side lobes" while the resolution response of the FFT bin is called the "main lobe" These terms have a perfect analogy in array beamforming directivity responses.

5.4 CIRCULAR CONVOLUTION

In this section we briefly discuss some very important and often unappreciated (by amateurs) issues of circular convolution in the proper use of the FFT in real-world signal processing systems. Often one can define a desired system response in the frequency domain using FFTs of various input and output signals or perhaps a particular frequency response function. If the desired system response is found from spectral products and/or quotients, and the system is to be implemented by an inverse FFT to give a FIR digital filter, one has to be very careful about the implications of the finite Fourier integral. The mathematics of the "analog domain" Fourier integral imply that the signals of interest are linear, stationary, and that the finite sum (finite resolution) of the DFT or FFT provides an accurate signal representation (no spectral leakage). Since we can have leakage if the signal frequencies are mis-matched to the FFT bins, the product of two spectral functions leads to the *circular convolution*, rather than linear convolution of the corresponding signals in the time domain giving an erroneous FIR filter from the spectral product. If one has a broadband input signal to a linear system, spectral leakage in the output signal can occur if the system has a sharp (very high Q or low damping) resonance with a narrower bandwidth than the FFT bins. But in general broadband signals will have a circular convolution which very closely matches the linear convolution. One should also note that the inverse FFT of a conjugate multiply in the frequency domain (a cross spectrum) is equivalent to a cross correlation in the time domain, which can also be affected by circular correlation errors when narrowband signals are involved.

Consider the classic optimal filtering problem as depicted in Figure 11. This general filtering problem gives a result known as an Echart filter and will be used

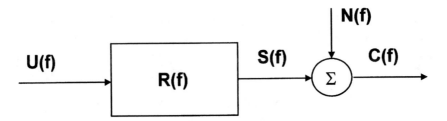

Figure 11 The optimal Weiner filtering problem described in the frequency domain using spectral products to derive a filter which recovers $U(f)$, given $R(f)$, $S(f)$, and $N(f)$.

in Chapters 11 and 12 for optimal signal detection. The least-squares technique will also be seen in a more general form in Chapter 8. The problem in Figure 11 is to recover the system input $U(f)$ given the output $C(f)$ which contains both signal $S(f)$ and $N(f)$ from the known system $R(f)$. With no noise, the answer is almost trivial because $C(f) = S(f) = U(f)R(f)$. Therefore, the "optimal" filter $A(f)$ to recover $U(f)$ from $C(f)$ is simply $A(f) = 1/R(f)$, as seen by $U(f) = C(f)A(f)$. However, in the presence of uncorrelated noise $N(f)$, we have one of the classic adaptive signal processing problems known as the Weiner filtering problem. The beauty of solving the problem in the (analytic) frequency domain is that with infinite spectral resolution and no spectral leakage all frequencies are orthogonal. Therefore, the equations can be solved frequency-by-frequency as a simple algebra equation.

We start by defining our model output $U'(f)$ where $U'(f) = C(f)A'(f)/R(f)$. We wish our model output to match the actual $U(f)$ as close as possible for all frequencies. Therefore, we seek to choose $A'(f)$ such that the total squared spectral error defined in Eq. (5.4.1) is minimized.

$$\int_{-\infty}^{+\infty} E(f)df = \int_{-\infty}^{+\infty} |U'(f) - U(f)|^2 df$$

$$= \int_{-\infty}^{+\infty} \left| \frac{[S(f) + N(f)]A'(f)}{R(f)} - \frac{S(f)}{R(f)} \right|^2 df \qquad (5.4.1)$$

We can let $0 \ge A'(f) \ge 1$ and evaluate the integrand algebraically, dropping the (f) notation for brevity

$$E = \left[|S|^2 + 2Re\{NS\} + |N|^2\right]A'^2 + |S|^2 - 2Re\{SN\}^*A'^*\} - 2A'|S|^2 + A'|N|^2 \qquad (5.4.2)$$

Since the signal S and noise N are assumed uncorrelated, any spectral products involving SN are assumed zero.

$$E = |S|^2\left[A'^2 - 2A' + 1\right] + A'^2|N|^2 \qquad (5.4.3)$$

We can solve Eq. (5.4.3) for an $A'(f)$ which minimizes the squared spectral error $E(f)$ at each frequency by examining the first and second derivatives of E with respect to A'.

$$\frac{\partial E}{\partial A'} = |S|^2(2A' - 2) + 2|N|^2 A' \qquad \frac{\partial E^2}{\partial A'^2} = 2(|S|^2 + |N|^2) \qquad (5.4.4)$$

Clearly, the parabolic structure of the squared spectral error shows a positive second derivative in Eq. (5.4.4), indicating the squared error function is concave up. Therefore, a solution for the first derivative equal to zero is a minimum of the squared error. We will repeatedly use variations of this basic approach to adaptively solving for optimal systems in many places later on in this text. The optimal solution for $A'(f)$ is clearly bounded between zero and unity.

$$A'(f) = \frac{|S|^2}{|S|^2 + |N|^2} \qquad (5.4.5)$$

The optimal filter for recovering $U(f)$ from the noise-corrupted $C(f)$ given that we know $R(f)$ is therefore $H(f) = A'(f)/R(f)$, or the spectral product of $A'(f)$ and $B(f) = 1/R(f)$. For the analog-domain indefinite-integral Fourier transform representation of $A'(f)$ and $B(f)$ there is no mathematical problem with designating the linear convolution of $a'(t)$ and $b(t)$ in the time-domain as the inverse Fourier transform of the spectral product $A'(f)$ and $B(f)$. Note that one could express the cross correlation $R^{ab}(\tau) = E\{a'(t)b(t - \tau)\}$ as the inverse Fourier transform of the spectral conjugate product $A'(f)B'(f)$. These relations are very important for the process of using measured signals from a system of interest to design or optimize performance of signal filters for control or detection of useful information. Therefore it is critical that we recognize the effect of the finite sum in the DFT and FFT on the convolution/correlation relations. *It is only for the case where all narrowband signal frequency components are bin-aligned that the spectral product of two functions results in the equivalent linear correlation or convolution in the time-domain.* For the more common case where leakage is occurring between the discrete spectral bins in the frequency domain, the discrete spectral product results in a *circular convolution* or *circular correlation* (for the conjugate spectral product) in the digital time domain. Unexpected circular convolution or correlation effects in a FIR impulse response from an inverse Fourier transform of the desired frequency response of the FIR filter can be extremely detrimental to the filter's performance in the system.

Consider the following simple example of a system with a short delay modeled in the frequency domain using a single frequency. If the frequency matched one of the DFT bin frequencies (for an exact integer number of wavelengths in the DFT input buffer as seen in the DFT spectra in Figure 4), there would be no problem determining the input-output cross-correlation from the inverse DFT of the spectral product of the input and conjugate of the output signal spectra. However, as seen in Figure 12, the frequency happens to lie exactly inbetween two discrete bins as can be seen by the odd number of half-wavelengths in the DFT input buffer and by the spectral leakage in Figure 5 The true linear correlation is seen in the solid curve in Figure 12 while the inverse DFT of the conjugate spectral product is seen in Figure 12 as the curve denoted with $(-\bigcirc-)$. Clearly, the additional frequencies from the product of the spectral leakage terms leads to significant amplitude and phase errors in the cross correlation.

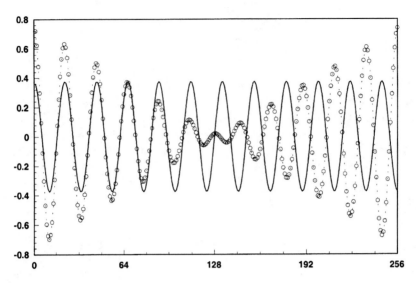

Figure 12   Comparison of frequency-domain estimated circular cross-correlation (– ◯ –) with spectral leakage to actual linear cross-correlation.

One can avoid circular convolution errors only by strictly limiting input and output signals to only have frequency components bin-aligned to the discrete DFT bins or by the following technique for spilling over the circular convolution errors into a range where they can be discarded. Note that only bin-aligned sinusoids will work for the former case. Random noise sequences can be seen to suffer from the leakage problem. To implement the later case, we simply double the number of data points in the DFT or FFT where one of the signals has the later half of its input buffer *zero-padded*, or filled with zeros. Suppose we only really need 128 cross-correlation data points in Figure 12. We would fill one signal buffer with 256 points and compute a 256-point DFT. The other signal buffer would have the most recent 128 points of its 256-point input buffer filled with zeros, the latter 128 points filled with the corresponding input time data, and it 256-point DFT is computed. The zero-padded DFT actually has a bit too much spectral resolution for the input data, but this added frequency sampling of the "sinc" function offers a special trick for the user. When the inverse DFT is computed after the spectral product is computed, all of the circular correlation errors are shifted to the upper half of the system impulse response and the lower half matches the true linear correlation result. Figure 13 shows the results of this operation graphically where the solid curve is again the true linear cross correlation and the (– ◯ –) curve is the result using spectral products.

   Doubling the number of input data points and zero padding the input for one of the spectral products is the accepted method for controlling circular convolution and/or circular cross-correlation errors from spectral products using the DFT or FFT. As seen in Figure 13, the circular correlation errors are completely corrected in the first 128 points, and the latter 128 points are simply discarded. This technique is well-documented in the literature, yet it is still surprising that many system designers fail to embrace its importance. It is likely

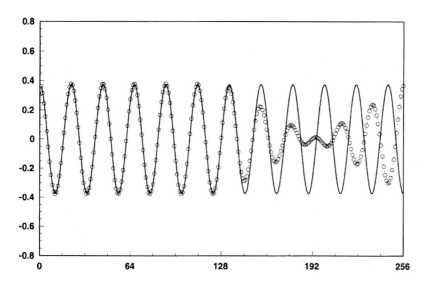

Figure 13 Comparison of the circular-correlation correction (– ○ –) with the actual cross-correlation for the spectral leakage case showing leakage confined to the upper 128 points of the output buffer.

that the widespread use of the analytical Fourier transforms and the corresponding convolution and correlation integral relations make it straightforward to assume that the same relations hold in the digital domain. The fact that these relations do hold but for only a very special case of exact signal frequency bin-alignment for the DFT is the technical point which has ruined many a "robust" adaptive system design! Truly robust adaptive processes need to work with precision for any input and output signals.

5.5   UNEVEN-SAMPLED FOURIER TRANSFORMS

In our discrete summations for the NDFT, DFT, and FFT, we have always assumed regular equally-spaced signal samples. There is a very good reason for this in the use of the trapezoidal rule for integration. However, in real-world signal processing applications, one may be faced with missing or corrupted data samples in the input to an FFT, or irregularly-spaced samples from asynchronous communications between sensor and processor. Perhaps the most common occurrence of uneven samples with spatial Fourier transforms of sensor array systems. As described in wavenumber Fourier transforms, the spatial sampling of a group of sensors may not be regular, requiring special consideration when calculating the Fourier transform. These type of data problems are not at all unusual in real applications of signal processing, and are actually quite common in astrophysics and radio astronomy, where a considerable amount of attention has been given to the subject in the literature.

An area which can frustrate the use of Fourier transforms is the so-called "missing-data" or "unevenly-sampled" data cases. We have generally assumed regular-timed data samples in a continuous numerical stream as this allows simple application of discrete integration to replace the Fourier integral with a discrete summation. The discrete summation approximation can be seen as a finite "trapezoidal rule" summation where the average value of the integrand at two adjacent time sample points is taken as a "height" allowing the area under the integrand curve to be approximated by a rectangle with the width equal to the sampling interval. If the sampling interval is very small and regular, the errors in the approximation are quite small. However, if the sample points are not evenly-spaced, or if various data points are missing from the series of input waveform numbers, the errors in estimating the DFT become very significant. A good example of how unevenly-space input data samples to a Fourier transform can occur in practice can be seen in the application of sensor arrays for spatial Fourier transforms (wavenumber transforms) where the sensors are physically spaced in irregular positions. The irregular spacing can be due to practical limitations of the particular application or it can be purposely done as a simple means of controlling the wavenumber, or spatial (directional), response of the sensor array.

An example of missing or irregular time-sampled data can be seen in global networks of radio astronomy receiver networks. By linking the receiver dishes across the planet, a directional high-gain signal sensitivity response can be obtained. However, due to varying atmospheric conditions and the earth's magnetic field fluctuations, the signal pulses from a distant transmitter such as the Voyager I and II probes do not arrive at each receiver site at precisely the same time. Furthermore, some receiver sites may experience momentary "drop-out" of the received signal due to multipath-induced destructive interference or other technical problems. The central question (among many) is: how should one deal with a section of missing or corrupted data? Should one "clamp" the data at the last known level until a new valid sample comes along? Is it better to simply set the missing or bad data to zero? Should one interpolate or spline fit the missing data to fit the rest of the sequence? A good point to consider is the relationship of the wavelength of interest with respect to the gap in the data sequence. For really long wavelengths, the relatively small irregularity is in general not a problem. However, if the data irregularity is in the order of a quarter wavelength or bigger, clearly a substantial error will occur if the problem is ignored.

In the following examples we will simulate the missing or corrupted input data problem by imposing a Gaussian randomization on the sample interval T making the input data sequence to be actually sampled randomly in time (random spatial samples are the same mathematically). We will then examine the FFT of a low and high frequency with varying degrees of sample time randomization. For small variations in sample period T with respect to the wave period, simply applying the FFT as if there was no variation appears to be the best strategy. However, for relatively high frequencies with a shorter wave period, the variation in sample time is much more significant and the FFT completely breaks down. In the astrophysics literature, a transform developed by N.R. Lomb allows the spectrum to be accurately computed even when significant missing data or sample time variations occur in the input data. The Lomb normalized periodogram, as referred to in the literature, starts out with N samples $y[n]$ $n = 0,1,2,..., N - 1$, sampled at time intervals $t[n]$ $n = 0,1,2,..., N - 1$ where the mean and variance of $y[n]$ are approximated by

$$\bar{y} = \frac{1}{N} \sum_{n=0}^{N-1} y[n] \qquad \sigma^2 = \frac{1}{N} \sum_{n=0}^{N-1} (y[n] - \bar{y})^2 \tag{5.5.1}$$

and used in the normalized periodogram

$$Y^{Lomb}(\omega) = \frac{1}{2\sigma^2} \left\{ \frac{\left[\sum\limits_{n=0}^{N-1} (y[n] - \bar{y}) \cos \omega(t[n] - \tau) \right]^2}{\sum\limits_{n=0}^{N-1} \cos^2 \omega(t[n] - \tau)} + \frac{\left[\sum\limits_{n=0}^{N-1} (y[n] - \bar{y}) \sin \omega(t[n] - \tau) \right]^2}{\sum\limits_{n=0}^{N-1} \sin^2 \omega(t[n] - \tau)} \right\} \tag{5.5.2}$$

where τ is determined by

$$\tau = \frac{1}{2\omega} \tan^{-1} \left\{ \frac{\sum\limits_{n=0}^{N-1} \sin 2\omega t[n]}{\sum\limits_{n=0}^{N-1} \cos 2\omega t[n]} \right\} \tag{5.5.3}$$

The setting of τ in the Lomb transform according to Eq. (5.5.3) not only makes the periodogram independent of time shift, but is actually essential for a least-squared error model of fit of sinusoids to the input data. The model fit is particularly import-ant if the $t[n]$ span is somewhat short. It should be noted that *the Lomb transform is a very slow algorithm* compared to even a DFT since the time shift τ must be estimated for each frequency of the periodogram, which has the equivalent magnitude of the DFT magnitude-squared, normalized by N.

A much more simple transform which is essentially equivalent to the Lomb transform for reasonably large data sets will be referred to here as simply the Uneven Fourier Transform (UFT). The UFT cannot be implemented as an FFT because it requires the precise time sample value for each data value which mathematically requires a unique sample set of complex sinusoids.

$$Y^{UFT}(\omega) = \sum_{n=0}^{N-1} y[n] e^{j\omega t[n]} \tag{5.5.4}$$

We can compare the normalized magnitude squared spectra of Eq. (5.5.4), $|Y^{UFT}(\omega)|^2/N = Y^{lomb}(\omega)$ to examine performance of the two approaches. Consider the following data set of a sinusoid 0.1 times the mean sample frequency ($f/fs = 0.1$) where the sample error standard deviation is 0.5 times the sample interval T as seen in Figure 14 below. The FFT of the signal in Figure 14 is seen in Figure 15.

The FFT of the 0.1fs frequency sine wave in Figure 15 shows a "mirror image" of the spectral peak at 0.9fs on the right side of the Figure. This peak is expected for real data and can be seen as the aliased image of the negative frequency component

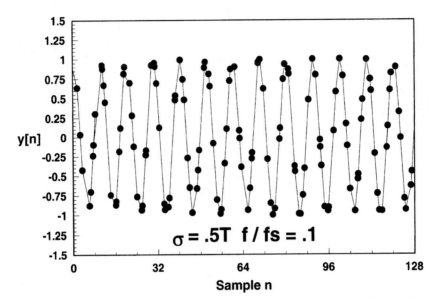

Figure 14   A 0.1*fs* sine wave sampled with a 0.5*T* standard deviation in time interval.

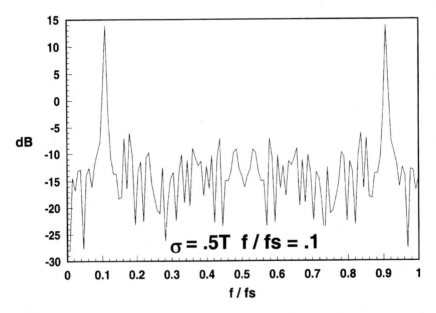

Figure 15   An FFT of the sine wave in Figure 14 clearly shows good results and the "mirror-image" peak at 0.9*fs*.

of the sine wave which would also be seen at −0.1*fs*. The Lomb and Uneven Fourier transforms are seen in Figure 16 and *do not show this "mirror-image"* due to the randomization of the sample times. In other words, the time randomization of the input samples causes some very short sample time intervals, effectively raising

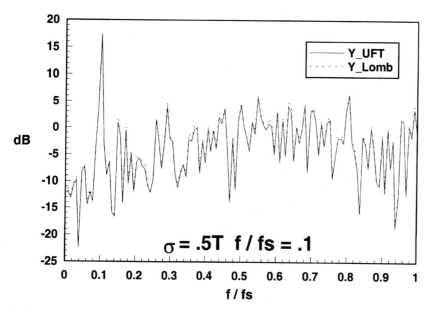

Figure 16 The UFT and Lomb transforms show no mirror image peak at 0.9*fs* but actually have more numerical error than the FFT of the sine wave at 0.1*fs* with a 0.5*T* sampling standard deviation.

the Nyquist rate for the data sequence. However, we must know the precise sample times for each input sample.

Note how the UFT and Lomb transforms are nearly identical. As presented, the two algorithms are not identical and the Lomb approach appears to be more robust on very short data records. However, the UFT is so much more efficient to implement and does not appear to be significantly less accurate than the Lomb algorithm.

Now if we examine a case where the frequency is increased to 0.45*fs* with the same sampling error, the Lomb and UFT transforms outperform the FFT as seen in Figures 17 through 19. Clearly, the results seen in Figure 18 are unacceptable. It can be seen that the sine wave at 0.45*fs* has a wave period nearly equal to twice the sample period *T*. With a standard deviation on the FFT input data samples of 0.5*T*, or approximately a quarter wave period for the sine wave, the integration errors in the FFT are so great as to render the algorithm useless. It is in this relm where the Lomb and UFT algorithms show there value, as seen in Figure 19.

While Figures 14 through 19 are convincing, this next example will be somewhat stunning in its ability to show the usefulness of the UFT or Lomb transforms on extremely erratic-sampled input data. Again, all we require is an accurate sample level and the precise time of the sample. Consider the case where the standard deviation of the data samples is 10*T* (yes, ten sample periods)! Figure 20 shows the input data for the Fourier transform. Even though it looks like a 2-year-old baby's refrigerator artwork, the UFT or Lomb algorithms can easily unscramble the mess and produce a useful Fourier transform! Will also violate the typical even-sampled Nyquist rate by letting the frequency rise to 0.7*fs* just to demonstrate

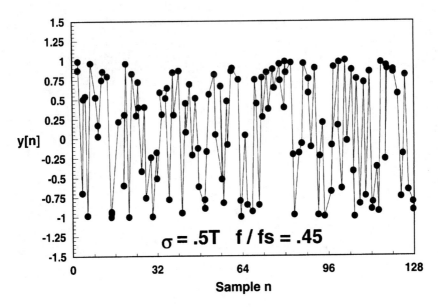

Figure 17 A 0.45*fs* sine wave sampled with a 0.5*T* standard deviation in times interval.

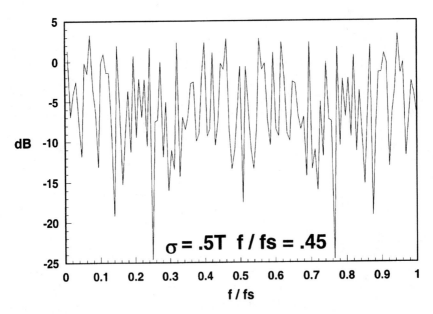

Figure 18 An FFT of the sin wave in Figure 14 clearly shows poor results due to the short wave period at 0.45*fs* as compared to the standard deviation of 0.5*T*.

some positive effects of sample randomization. Figure 21 shows the Fourier transform using the Lomb and UFT algorithms.

Why do the Lomb and UFT algorithms work so well for large randomizations of the input data, yet not so well for the case where a low randomization was given

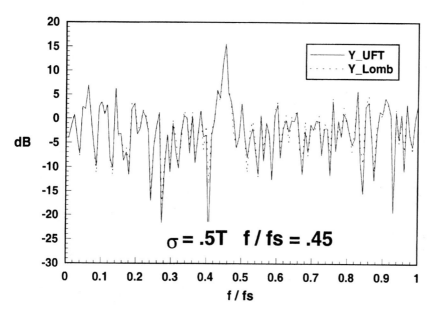

Figure 19 The Lomb and Uneven Fourier transforms perform quite well on randomized-time sampled data even when the sample errors approach a quarter wave period of the signal.

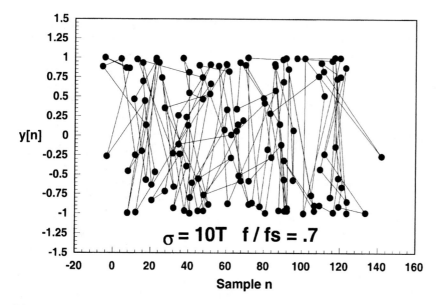

Figure 20 The sine wave input samples for the case where the sampling error standard deviation is 10 samples and the frequency is 0.7*fs* to illustrate randomization effects on the Nyquist rate.

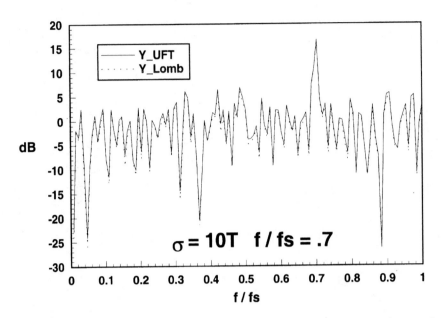

Figure 21   Both the Lomb and UFT algorithms easily provide a Fourier transform from the heavily randomized data in Figure 20.

for random sample times with a standard deviation of $0.5T$? The answer can be seen in the fact that small randomizations of the time samples of $0.5T$ as seen in Figures 14 through 16 are of little consequence to the low frequency wave with frequency $0.1fs$, which has a wave period of $10T$, or 20 times the standard deviation. The random signal levels in the FFT and UFT bins are due to errors in the assumptions in the trapezoidal rule behind the numerical integration carried out by the discrete sum in the DFT. It can be postulated, but will not be proven here, that the regular sample assumptions in the FFT have a better averaging effect on the integration errors than does the UFT and Lomb algorithms. Clearly, the discrete sum in the DFT does not care in what order the products are summed up. *All that really matters is that the correct sample times are used in the Fourier integral.* As the sample time randomizations increase with respect to the wave period, the regular sample assumptions in the FFT simply become a huge source of numerical error. The random noise in the UFT and Lomb algorithms tends to stay at a constant level, depending on the number of data points in the transform. Since the normalized Lomb periodogram is approximately the magnitude-squared of the DFT divided by N, it can be seen that the narrowband signal-to-noise gain the Lomb periodogram is $10 \log_{10} N$, or about 21 dB for a complex sinusoid. Therefore, one would expect a unity amplitude sine wave to be approximately +18 dB in the periodogram and the correspondingly scaled UFT and FFT responses with some small losses due to bin misalignment. It would appear from these simulations that the Lomb Fourier transform periodogram may require more computation without significant improvement in performance when compared to the more simple UFT approach to the random Fourier integral. It should be clear that so long as the precise time and/or sensor position is known for the Fourier input data, an accurate Fourier transform can be computed but at the expense of giving up the efficient FFT algorithm and some added noise in the spectral result.

5.6 SUMMARY, PROBLEMS, AND BIBLIOGRAPHY

The Fourier transform is a fundamental signal processing algorithm which allows time-domain signals to be expressed in the frequency domain as a spectrum of system transfer function. Frequency domain representation of signals is generally more physically intuitive for periodic (sinusoidal) signals. Conversely, impulsive signals such as sonic booms are not very intuitive in the frequency domain, although the impulse response of a physical system is a special case where its Fourier transform is the system frequency response. For discrete signals and systems, the infinite time integration of the Fourier transform is replaced by a finite discrete summation. The result of this practical implementation issue is that the frequency separating resolution of the Discrete Fourier Transform (DFT) is limited by the length of time spanned by the summed discrete signal samples. For example, the maximum resolution in Hz is the inverse of the time span in seconds. For a DFT with 10 Hz bin resolution, the time span of the input data must be at least 100 msec. For 0.1 Hz resolution the time span must be at least 10 sec. The resolution of a particular DFT depends on the discrete sample rate and the number of data points used, which both translate physically into the time span of the input data to the DFT.

The Fast Fourier Transform (FFT) is an engineered DFT such that the number of multiplications and additions are minimized. This is done by clever arranging of the multiplies in the summation so that nothing is repeated. The FFT frequency bins are orthogonal for input frequencies which are exactly aligned with the bin frequencies, and thus, produce no leakage into adjacent FFT bins. The sine and cosine components of the FFT and DFT can be precomputed into tables to minimize redundant computation. Computationally, the FFT is significantly more efficient than the more explicit DFT requiring only $N\log_2 N$ multiplies as compared to the DFT's N^2. For a 1024 point FFT this difference is 10,240 to 1,048,576 or a reduction of 102.4:1. However, in order to use the FFT's marvelous efficiency, the input samples must be regularly-spaced and the number of input samples must equal the total number of FFT bins.

For cases where spectral leakage is unavoidable (nonstationary or uncontrollable input frequencies), one may weight the input data (by multiplying it by a data "window") to keep the spectral leakage approximately the same for both bin-aligned and non-aligned input data frequencies. Window design and use is an approximate art and, like shoes, fast-food, music, and breakfast cereal, there is a wide variety of both old and new designs which do the same basic job slightly differently. We stop short of referring to data windows as pop art, for most applications the Hanning window gives excellent performance. However, one of the most overlooked aspects of the use of data windows is their effect on signal levels in the frequency domain. Using a bin-aligned narrowband reference for the rectangular data window (no window applied), we provide "narrowband correction" scale factors in Table 1) which if multiplied by the window samples, allows bin-aligned narrowband signal levels to be consistent independent of which window is used. The narrowband correction factor is simply N divided by the sum of the window's N samples. For broadband signals the sum of the energy in the time-domain should be consistent with the sum of the spectral energy. The broadband correction factor is found by N divided by the square-root of the sum of the squared window samples. These values can also be found in Table 1 for a wide range of data windows. Both narrowband and broadband signals cannot be simultaneously calibrated in the frequency domain if a data window is used to control spectral leakage, a nonlinear phenomenon.

Another area of Fourier processing of data often overlooked is the effect of circular convolution when spectral products or divisions are inverse Fourier transformed back into the time domain to give impulse responses, cross correlations, or other physical results. If there is spectral leakage in the frequency domain, the effect of the leakage is clearly seen in a plot of the spectral operation but can be far more devastating in the time domain. Spectral products (or divisions) bin by bin are only valid when the spectra accurately represent the data as an orthogonal frequency transform. The assumptions of signal/system linearity and time-invariance carry forward to imply long time records transformed into high resolution Fourier transforms. In the physical world, these assumptions all have limits of applicability which can lead to serious sources of error in frequency domain processing.

Finally, it is possible to transform non-regularly spaced samples into the frequency domain provided the precise sample times are known for each sample. The uneven Fourier transform (UFT) cannot use the efficient form of the FFT, but does offer the benefit of no frequency aliasing when significantly random sample time dithering is present. The UFT may be most useful for spatial Fourier transforms where sensor positions rather than sample times may be randomized for various reasons. The UFT and its more formal brother, the Lomb transform, allow for unevenly sampled data to be transformed to the frequency domain for subsequent analysis and processing.

PROBLEMS

1. A digital signal is sampled 100,000 times per second and contains white noise and two sinusoids. One sinusoid is 20,000 Hz and the other is the same amplitude at 20,072 Hz. Assuming that a discrete Fourier transform can resolve the two sinusoids if at least one bin lies between the two spectral peaks to be resolved, what size DFT (i.e. number of samples) is required to just resolve the two sinusoids? What is the associated integration time in seconds?

2. Given a digital signal processor (DSP) chip capable of 25 million floating-point operations per second (25 MFLOPS), what is the real-time (no data missed) N-point FFT's processing rate assuming N operations are needed to move the real input data into place and N operations are needed to move the $N/2$ complex FFT output away when done if $N = 128$, 1024, 8192? What are the computational times for the FFTs?

3. Given a real time data sequence, the total real spectrum is found by summing the positive and negative frequency bins and the total imaginary spectrum is found by subtracting the negative frequency bin values from the positive frequency bins. This is due to the Hilbert transform pair relationship for the real and imaginary spectra given a real input time data

sequence. Suppose the time data sequence is placed in the imaginary part of the input leaving the real part zero. How does one recover the proper real and imaginary spectra?

4. Suppose we have a data window that's a straight line starting a 1.00 at $n = 1$ and decreasing to 0.00 at $n = N$, the window size. Show that the narrowband correction factor is 2.00 and the broadband correction factor is $3^{1/2}$ or 1.732.

BIBLIOGRAPHY

A. V. Oppenheim, R. W. Schafer Discrete-Time Signal Processing, Englewood Cliffs: Prentice-Hall, 1989.

A. Populus Signal Analysis, New York: McGraw-Hill, 1977.

W. H. Press, B. P. Flannery, S. A. Teukolsky, W. T. Vetterling Numerical Recipies: The Art of Scientific Computing, New York: Cambridge University Press, 1986.

J. J. Shynk "Frequency-domain multirate adaptive filtering," IEEE Signal Processing Magazine, 1, 1992, pp. 1437.

K. Steiglitz An Introduction to Discrete Systems, New York: Wiley, 1974.

S. D. Sterns Digital Signal analysis, Rochelle Park: Hayden, 1975.

REFERENCES

1. J. J. Shynk "Frequency-domain multirate adaptive filtering," *IEEE Signal Processing Magazine*, 1, 1992, pp. 1437.

6

Spectral Density

The spectral density of a signal is a statistical quantity very useful in determining the mean-square value, or power spectral density, for frequency-domain signals. Strictly speaking, the spectral density, as its name implies, is an expected power density per Hz. As the time integral in the forward Fourier transform increases, so does the amplitude of the spectral density while the resolution of the Fourier transform narrows. Therefore, the expected signal power per-Hz in the frequency domain stays the same, matching the expected signal power at that frequency in the time domain. However, many current signal processing texts define the spectral density with some subtle differences which depend on the whether the underlying Fourier transform integral limits are $-T$ to $+T$ or $-T/2$ to $+T/2$, and whether the original time signal is real or complex. Unfortunately for the uninitiated student, both of these subtle parameters can either lead to a "factor of 2" difference in the spectral density definition, or can actually cancel one another deep in the algerbra leading to significant confusion. In the derivation below (which is thankfully consistent with all texts), we will allude to the origins of the potential "factor of 2s" so that the student can easily see the consistency between the many available texts with derivations of spectral density.

We start the derivation of spectral density with the assumption of an analytic time waveform $x(t)$ over the interval $t = -T/2$ to $+T/2$ and zero everywhere else on the range of t. Using the $\pm T/2$ integration interval for the Fourier transform, we define the spectral density as

$$S_X(\omega) = \lim_{T \to \infty} \frac{E\{|X(\omega)|^2\}}{T} \qquad X(\omega) = \int_{-T/2}^{+T/2} x(t)e^{-j\omega t} dt \qquad (6.0.1)$$

Note that if the Fourier transform $X(\omega)$ is defined over the interval $-T$ to $+T$, a factor of $2T$ (rather than T) would be required in the denominator for the spectral density $S_X(\omega)$, as seen in some texts.

A good starting point for the derivation of spectral density is Parseval's theorem for two time functions $g(t)$ and $f(t)$ Fourier-transformable into $G(\omega)$

and $F(\omega)$, respectively.

$$\int\limits_{-\infty}^{+\infty} f(t)g(t)dt = \frac{1}{2\pi} \int\limits_{-\infty}^{+\infty} F(\omega)G(-\omega)d\omega = \int\limits_{-\infty}^{+\infty} F(f)G(-f)df \tag{6.0.2}$$

One can easily see that the only difference between working in radian frequency ω in radians per second and the more familiar physical frequency f in Hz (cycles per second) is the simple factor of $1/2\pi$ from the change of variables. We prefer to use f instead of ω because it will be more straightforward to compare spectral densities in the analog and digital domains. For our real signal $x(t)$ which is nonzero only in the range $-T/2 < t < T/2$, we have

$$\int\limits_{-T/2}^{+T/2} x^2(t)dt = \int\limits_{-\infty}^{+\infty} X(f)X(-f)df \tag{6.0.3}$$

Dividing Eq. (6.0.3) by T and taking the limit as T goes to infinity provides and equation for the expected mean square signal (or signal power), in the time and frequency domains.

$$\lim_{T\to\infty} \frac{1}{T} \int\limits_{-T/2}^{+T/2} x^2(t)dt = \lim_{T\to\infty} \frac{1}{T} \int\limits_{-\infty}^{+\infty} X(f)X(-f)df \tag{6.0.4}$$

For the special (and most common) case where $x(t)$ is a purely real signal (the imaginary component is zero), we note the Hilbert transform pair relationship which allows $X(-f)$ to be seen as the complex conjugate of $X(+f)$. Taking the expected value of both sides of Eq. (6.0.4) gives

$$E\left\{\lim_{T\to\infty} \frac{1}{T} \int\limits_{-T/2}^{+T/2} x^2(t)dt\right\} = \lim_{T\to\infty} \frac{1}{T} \int\limits_{-\infty}^{+\infty} E\{|X(f)|^2\}df \tag{6.0.5}$$

where for stationary signals and random processes, the expected value of the time average of the signal squared is just the mean square signal.

$$\bar{x}^2 = E\{x^2(t)\} = \int\limits_{-\infty}^{+\infty} \lim_{T\to\infty} \frac{E\{|X(f)|^2\}}{T}df = \int\limits_{-\infty}^{+\infty} S_X(f)df \tag{6.0.6}$$

Equation (6.0.6) provides the very important result of the integral of the spectral density over the entire frequency domain is the mean square time-domain signal, or signal power. The integral of the spectral density $S_X(f)$ in Eq. (6.0.6) is referred to as a *two-sided power spectral density (PSD)* because both positive and negative frequencies are integrated. Since $x(t)$ is real, the two-sided PSD is exactly the same as twice the integral of the spectral density over only positive frequencies.

$$\bar{x}^2 = E\{x^2(t)\} = 2 \int\limits_0^{+\infty} S_X(f)df \tag{6.0.7}$$

The integral of the spectral density in Eq. (6.0.7) is known as a *one-sided PSD*. Clearly, if one is not careful about the definition of the time interval ($\pm T$ or $\pm T/2$) for the Fourier transform and the definition of whether a one-sided or two-sided PSD is being used, some confusion can arise. The *power spectrum* estimate, $G_{XX}(f)$, of a real signal, $x(t)$, is defined as twice the spectral density at that frequency as a simple means of including positive and negative frequency power in the estimate.

Consider the following simple example. Let $x(t)$ be equal to a sine wave plus a constant.

$$x(t) = A + B\sin(\omega t) \tag{6.0.8}$$

We can write the mean square signal by inspection as $A^2 + B^2/2$. The Fourier transform of $x(t)$ is

$$X(\omega) = AT\frac{\sin(\omega T/2)}{\omega T/2} + \frac{BT}{2j}\frac{\sin([\omega_1 - \omega]T/2)}{[\omega_1 - \omega]T/2} - \frac{BT}{2j}\frac{\sin([\omega_1 - \omega]T/2)}{[\omega_1 - \omega]T/2} \tag{6.0.9}$$

The spectral density $S_X(\omega)$ is found by taking expected values and noting that the three "$\sin(x)/x$" type functions in Eq. (6.0.9) are orthogonal in the limit as T approaches infinity.

$$S_X(\omega) = \lim_{T\to\infty} \left\{ A^2 T\left|\frac{\sin(\omega T/2)}{\omega T/2}\right|^2 + \frac{B^2}{2}(T/2)\left|\frac{\sin([\omega_1 - \omega]T/2)}{[\omega_1 - \omega]T/2}\right|^2 \right.$$
$$\left. + \frac{B^2}{2}(T/2)\left|\frac{\sin([\omega_1 + \omega]T/2)}{[\omega_1 + \omega]T/2}\right|^2 \right\} \tag{6.0.10}$$

To find the signal power, we simply integrate the two-sided spectral density over positive and negative frequency noting the following important definition of the Dirac delta function.

$$\lim_{T'\to\infty} \int\limits_{-\infty}^{+\infty} T'\left|\frac{\sin(\omega T')}{\omega T'}\right|^2 d\omega = \pi\delta(\omega) = \frac{\delta(f)}{2} \tag{6.0.11}$$

Then, by a straightforward change of variables where $T' = T/2$ and $df = d\omega/2\pi$, then it can be seen that the integral of the two-sided PSD is simply

$$\int\limits_{-\infty}^{+\infty} S_X(f)df = 2A^2\left(\frac{\delta(f)}{2}\right) + \frac{B^2}{2}\left(\frac{\delta(f_1 - f)}{2}\right) + \frac{B^2}{2}\left(\frac{\delta(f_1 + f)}{2}\right) \tag{6.0.12}$$

Since the integral of a Dirac delta function always gives unity area, the signal power is easily verified in Eq. (6.0.12).

We now briefly consider the discrete sampled data case of $x[n] = x(nTs)$, where Ts is the sampling interval in seconds. For an N-point DFT or FFT of $x[n]$, we can compare the digital signal equivalent of Parseval's theorem.

$$\sum_{n=0}^{N-1} x^2[n] = \frac{1}{N} \sum_{m=0}^{N-1} |X[m]|^2 \qquad (6.0.13)$$

The factor of $1/N$ in Eq. (6.0.13) is very interesting. It is required to scale the transform appropriately as is the case for the inverse DFT formula. It essentially takes the place of "df" in the analog-domain inverse Fourier transform. For N equal-spaced frequency samples of $X[m]$, the factor of $1/N$ represents the spectral width of a bin in terms of a fraction of the sample rate. If we consider the mean square digital signal, or expected value of the digital signal power, we have

$$\frac{1}{N} \sum_{n=0}^{N-1} x^2[n] = \frac{1}{N^2} \sum_{m=0}^{N-1} |X[m]|^2 \qquad (6.0.14)$$

the equivalent digital domain spectral density can be seen as the magnitude-squared of the normalized DFT (NDFT) presented in the previous section. The mean-square value (or average power) of the digital signal is the sum of the magnitude-squared of the NDFT bins. But more importantly, the power at some frequency f_k is the sum of the squares of the NDFT bins for $\pm f_k$, or twice the value at f_k for $x[n]$ real. The root mean-square, or rms value, is simply the square root of the power, or more simply, $1/2^{1/2}$ times the magnitude of the NDFT bin. This is why the NDFT is very often convenient for real physical system applications.

Sensor calibration signals are usually specified in rms units at some frequency. Electronic component background noise spectra are usually specified in the standard Volts per square-root Hertz, which is the square-root of the voltage power spectral density. Obviously, one would not want the size of the DFT or FFT to be part of a component noise standard. For intelligent adaptive systems, we generally want to change NDFT size according to some optimization scheme and do not want signal levels in the normalized power spectral density (NPSD) to depend on the number of points in the transform. Throughout the rest of the next section, we will examine the statistics of the data in the NDFT bins and many useful signal functions well-represented by physical applications in the frequency domain. To avoid confusion, we will refer to the mean square values of the NDFT magnitude as the NPSD, while the PSD as defined in the literature is N times the NPSD (i.e. NPSD = PSD/N).

6.1   STATISTICAL MEASURES OF SPECTRAL BINS

One often processes broadband signals in the frequency domain which can be seen as the sum of a large number of sinusoids of randomly distributed amplitudes. There are many naturally occurring random processes which generate broadband signals such as turbulence in the atmosphere, intrinsic atomic vibrations in materials above absolute zero temperature, electron position uncertainty and scattering in conductors, and the least significant bit error in a successive approximation A/D convertor as described in Chapter 1. All physical signals harvested in an A/D and processed in an adaptive signal processing system contain some level of "background" random signals referred to as noise. Even numerical errors in computing can be modeled as random noise, although this kind of noise is (hopefully) quite small. The normalized power spectral density (NPSD) in a given bin can also be seen as a random variable where we would like to know the expected value as well as the likelihood of the bin value falling within a particular range. The central limit theorem states that regardless of the probability distribution of an independent random variable X_n, the sum of a large number of these independent random variables (each of the same probability distribution) gives a random variable Y which tends to have a Gaussian probability distribution.

$$Y = \frac{1}{\sqrt{n}}[X_1 + X_2 + X_3 + \ldots + X_n] \tag{6.1.1}$$

Probability Distributions and Probability Density Functions

A probability distribution $P_Y(y)$ is defined as the probability that the observed random variable Y is less than or equal to the value y. For example, a six-sided die cube has an equal likelihood for showing each of its faces upward when it comes to rest after being rolled vigorously. With sides numbered 1 through 6, the probability of getting a number less than or equal to 6 is 1.0 (or 100%), less than or equal to 3 is 0.5 (50%), equal to 1 is 0.1667 (16.67% or a 1 in 6 chance), etc. A probability distribution has the following functional properties:

(a) $P_Y(y) = Pr(Y \leq y)$

(b) $0 \leq P_Y(y) \leq 1 \qquad -\infty < y < +\infty$

(c) $P_Y(-\infty) = 0 \qquad P_Y(+\infty) = 1$

(d) $\dfrac{dP_Y(y)}{dy} \geq 0$

(e) $Pr(y_1 < Y \leq y_2) = P_Y(y_2) - P_Y(y_1)$

(6.1.2)

It is very convenient to work with the derivative of the probability distribution, or *probability density function (PDF)*. A probability density function $p_Y(y)$ has the following functional properties:

(a) $p_Y(y) = \lim\limits_{\epsilon \to 0} \dfrac{P_Y(y + \epsilon) - P_Y(y)}{\epsilon} = \dfrac{dP_Y(y)}{dy}$

(b) $p_Y(y)dy = Pr(y < Y \leq y + dy)$

(c) $p_Y(y) \geq 0 \qquad -\infty < y < +\infty$

(d) $\displaystyle\int_{-\infty}^{+\infty} p_Y(y)dy = 1$

(6.1.3)

(e) $P_Y(y) = \displaystyle\int_{-\infty}^{y} p_Y(u)du$

(f) $\displaystyle\int_{y_1}^{y_2} p_Y(y)dy = Pr(y_1 < Y \leq y_2)$

For our die example above, the probability distribution is a straight line from 1/6 for a value of 1 to 1.00 for a value less than or equal to 6. The probability density function is the slope of the probability distribution, or simply 1/6. Since the PDF for the die is the same for all values between 1 and 6 and zero elsewhere, it is said to be a *uniform probability distribution*. The expected value, (or average value), of a random variable can be found by computing the 1st moment, which is defined as the mean.

$$m_Y = \bar{Y} = E\{Y\} = \int_{-\infty}^{+\infty} y p_Y(y) dy \tag{6.1.4}$$

The mean-square value is the second moment of the PDF.

$$\bar{Y}^2 = E\{Y^2\} = \int_{-\infty}^{+\infty} y^2 p_Y(y) dy \tag{6.1.5}$$

The variance is defined as the second central moment, or the mean-square value minus the mean value squared. The standard deviation for the random variable, σ_Y, is the square-root of the variance.

$$\sigma_Y^2 = \bar{Y}^2 - m_Y^2 = \int_{-\infty}^{+\infty} (y - m_Y)^2 p_Y(y) dy \tag{6.1.6}$$

For our die example, the mean is 1/6 times the integral from 0 to 6 of y, or 1/12 times y^2 evaluated at 6 and zero, which gives a mean of 3. The mean-square value is 1/6 times the integral from 0 to 6 of y^2, or 1/24 times y^3 evaluated at 6 and 0, which gives a mean-square value of 12. The variance is simply $12 - 9$ or 3 as expected.

Consider the statistics of rolling 100 dice and noting the central limit theorem in Eq. (6.1.1) and the probability distribution for the sum (giving possible numbers between 100 and 600). Clearly, the likelihood of having all 100 dice roll up as 1s or 6s is extremely low compared to the many possibilities for dice sums in the range of 300. This "bell curve" shape to the probability density is well-known to approach a Gaussian PDF in the limit as n approaches infinity. The Gaussian PDF is seen in Eq. (6.1.7) and also in Figure 1 for $m_Y = 0$ and $\sigma_Y = 1$.

$$p_Y(y) = \frac{1}{\sigma_Y \sqrt{2\pi}} e^{\frac{-(y-m_Y)^2}{2\sigma_Y^2}} \tag{6.1.7}$$

For real-world sensor systems where there are nearly an infinite number of random noise sources ranging from molecular vibrations to turbulence, it is not only reasonable, but prudent to assume Gaussian noise statistics. However, when one deals with a low number of samples of Gaussian random data in a signal processing system (say < 1000 samples), the computed mean and variance will not likely exactly match the expected values and a histogram of the observed data arranged to make a digital PDF have a shape significantly different from the expected bell-curve. This problem can be particularly acute where one has scarce data for adaptive algorithm training. There are various intricate tests, such as Student's *t*-test and others, that

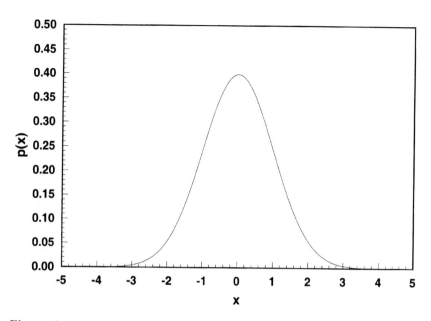

Figure 1 A zero-mean Gaussian (ZMG) probability density function (PDF) with variance equal to unity.

have been developed (but are not presented here) to determine whether one data set is statistically different from another. Results from statistical tests with small data sample sets can be misleading and should only be interpreted in general as a cue, or possible indicator for significant information. The general public very often is presented with statistical results from small sample size clinical trials in the health sciences (human volunteers are rare, expensive, and potentially in harm's way). These statistical sample set tests can also be used in adaptive pattern recognition to insure low bias in small data sets. However, the best advice is to use enormous data set for statistical training of adaptive algorithms wherever possible and to always check the data set distribution parameters to insure the proper assumptions are being applied.

Statistics of the NPSD Bin

Consider the Fourier transform of a real zero-mean Gaussian (ZMG) signal with a time-domain standard deviation σ_t. The Fourier transform (as well as the DFT, FFT, and NDFT defined above) is a linear transform where each transformed frequency component can be seen as the output of a narrowband filter with the ZMG input signal. Since the ZMG signal is spectrally white (it can be represented as the sum of an infinite number of sinusoids with random amplitudes), each bin in a DFT, FFT, or NDFT can be seen as a complex random variable where the real and imaginary components are each ZMG random variables with standard deviation $\sigma_R = \sigma_I = \sigma_f$. If we use the NPSD to compare mean square values in the time and frequency domains (DFT computed using a rectangular data window), we can easily determine σ_f^2 in terms of σ_t^2 using Eq. (6.0.14) as seen in Eq. (6.1.8).

$$\sigma_t^2 = \frac{1}{N} \sum_{n=0}^{N-1} x^2[n] = \sum_{m=0}^{N-1} \frac{X_R[m]^2 + X_I[m]^2}{N^2}$$
$$= N\left(\sigma_R^2 + \sigma_I^2\right) = 2N\sigma_f^2$$

(6.1.8)

Clearly, $\sigma_f^2 = \sigma_t^2/2N$, and the variance of the real and imaginary parts of an N-point NPSD bin, decrease by a factor of $1/2N$ relative to the time-domain variance. If a Hanning or other non-rectangular data window is used in the NDFT calculation, a broadband correction factor (1.6329 for the Hanning window, see Table 5.1 for other window types) must be applied for the broadband variances to match the rectangular window levels. For the PSD, the noise variance in the real and imaginary bins is simply $\frac{1}{2}\sigma_t^2$ while a sinusoid power is amplified by a factor of N^2. If the input data were complex with real and imaginary components each with variance σ_t^2, the NPSD drop in variance is $1/N$ rather than $2/N$ for real signal. It can be seen that the spectrally white input noise variance is equally divided into each of the N discrete Fourier transform bins. Note that the NDFT scaling (and NPSD definition) used allows the mean-square value for a sinusoid in the time domain to be matched to the value in the corresponding NPSD bin. Therefore, the Fourier transform is seen to provide a narrowband signal-to-noise ratio (SNR) enhancement of $10 \log_{10}(N/2)$ dB for real signal input. The larger N is, the greater the SNR enhancement in the power spectrum will be. A more physical interpretation of N is the ratio of the sample rate f_s (in Hz) over the available resolution (in Hz), which is inversely proportional to the total integration time in seconds. Therefore, the SNR enhancement for stationary real sinusoids in white noise is $10 \log_{10}(N/2)$, where N can be derived from the product of the sample rate and the total integration time. A 1024-point PSD or NPSD provides about 27 dB SNR improvement while a 128-point transform gives only 18 dB. PSD measurement is an extraordinary tool for enhancing the SNR and associated observability of stationary sinusoids in white noise whether the normalization is used or not.

To observe an approximation to the expected value of the power spectrum we are clearly interested in the average magnitude-squared value and its statistics for a given Fourier transform bin (with no overlap of input buffers). These statistics for the NPSD will be based on a limited number of spectral averages of the magnitude-squared data in each bin. Let's start by assuming real and imaginary ZMG processes each with variance σ^2. The PDF of the square of a ZMG random variable is found by employing a straightforward change of variables. Deriving the new PDF (for the squared ZMG variable) is accomplished by letting $y = x^2$ be the new random variable. The probability that y is less than some value Y must be the same as the probability that x is between $\pm Y^{1/2}$.

$$P_Y(y \le Y) + P_X(-\sqrt{Y} \le x \le +\sqrt{Y}) = \int_{-\sqrt{Y}}^{+\sqrt{Y}} p(x)dx$$

(6.1.9)

Differentiating Eq. (6.1.9) with respect to Y gives the PDF for the new random variable $y = x^2$.

$$p(Y) = p(x = +\sqrt{Y})\frac{d(+\sqrt{Y})}{dY} - p(x = -\sqrt{Y})\frac{d(-\sqrt{Y})}{dY}$$

$$= \frac{1}{2\sqrt{Y}}\left\{p(x = +\sqrt{Y}) - p(x = -\sqrt{Y})\right\} \qquad Y \geq 0 \tag{6.1.10}$$

Since the Gaussian PDF is symmetric (its an even function), we can write the new PDF simply as

$$p(y) = \frac{1}{\sigma\sqrt{2\pi y}}e^{\left(\frac{-y}{2\sigma^2}\right)} \tag{6.1.11}$$

The PDF in Eq. (6.1.11) is known as a Chi-Square PDF of order $v = 1$ and is denoted here as $p(x^2|v=1)$. The expected value for y is now the variance for x since $E\{y\} = E\{x^2\} = \sigma^2$. The variance for the Chi-Square process is

$$\sigma_y^2 = E\{y^2\} - (E\{y\})^2$$

$$= E\{x^4\} - (E\{x^2\})^2 \tag{6.1.12}$$

$$= 3\sigma^4 - \sigma^4 = 2\sigma^4$$

The 4th moment on the ZMG variable x in Eq. (6.1.12) is conveniently found using the even central moments relationship for Gaussian PDF's (the odd central moments are all zero).

$$E\{(x - \bar{x})^n\} = 1 \cdot 3 \cdot 5 \ldots (n - 1)\sigma^n \tag{6.1.13}$$

We now consider the sum of two squared ZMG variables as is done to compute the magnitude squared in a bin for the PSD. This PDF is a Chi-Square distribution of order $v = 2$. Equation (2.60) gives the general Chi-Square for v degrees of freedom.

$$p(x^2|v) = \frac{(y/\sigma^2)^{v/2-1}}{\sigma^2 2^{v/2}\Gamma(v/2)}e^{-y/2\sigma^2} \qquad y = \sum_{n=1}^{v} x_n^2 \tag{6.1.14}$$

The means and variances for the 2 degree of freedom Chi-Square process simply add up to twice that for the $v = 1$ case. The gamma function $\Gamma(v/2)$ equals $\pi^{1/2}$ for the $v = 1$ case and equals $(M - 1)!$ for $M = 2v$. Since we are interested in the average of pairs of squared ZMG variables, we introduce a PDF for the average of M NPSD bins.

$$p(z|M) = \frac{M(Mz/\sigma^2)^{M-1}}{\sigma^2 2^M(M-1)!}e^{-Mz/2\sigma^2} \qquad z = \frac{1}{M}\sum_{n=1}^{M}(x_{R_n}^2 + x_{I_n}^2) \tag{6.1.15}$$

Figure 2 shows a number of the Chi-Square family PDF plots for an underlying ZMG process with unity variance. The $v = 1$ and $v = 2$ ($M = 1$) cases are classic Chi-Square. However, the $M = 2$, 4, and 32 cases use the averaging PDF in Eq. (6.1.15). Clearly, as one averages more statistical samples, the PDF tends toward a Gaussian distribution as the central limit theorem predicts. Also for the averaging cases ($M > 1$), note how the mean stays the same (2 since $\sigma^2 = 1$) while the variance

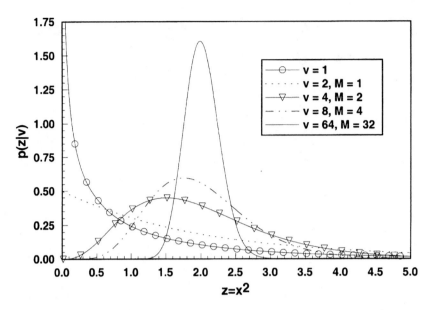

Figure 2 The family of Chi-Square PDFs useful for determining spectral averaging (M > 1) statistics.

decreases with increasing M. In the limit as M approaches infinity, the PDF of Eq. (6.1.15) takes on the appearance of a Dirac delta function.

It is important to note that the $v = 1$ case has a mean of $\sigma^2 = 1$, while all the other means are $2\sigma^2 = 2$. This $2\sigma^2$ mean for the NDFT bin relates to the time-domain variance by a factor of $1/2N$ for real data as mentioned above. It may not at first appear to be true that the mean values are all the same for the spectral averaging cases ($M = 1$, 2, 4, and 32). However, even though the *skewness* (asymmetry) of the spectral averaging PDFs is quite significant for small M as seen in Figure 2, the curves really flatten out as x approaches infinity adding a significant area to the probability integral. The variance for the $v = 1$ case is $2\sigma^4$ while the variances for the averaged cases ($M = 1$, 2, 4, 32) are seen to be $4\sigma^4/M$ which can be clearly seen in Figure 2. Figure 3 shows the probability distribution curves corresponding to the PDFs in Figure 2. Equation (6.1.16) gives the probability of finding a value less than x^2. However, this solution is somewhat cumbersome to evaluate because the approximate convergence of the series depends on the magnitude of x^2.

$$P(x^2|v) = \frac{1}{\Gamma(\frac{v}{2})} \sum_{n=0}^{\infty} \frac{(-1)^n (x^2/2)^{(\frac{v}{2}+n)}}{n!(\frac{v}{2}+n)} \tag{6.1.16}$$

Confidence Intervals For Averaged NPSD Bins

What we are really interested in for spectrally averaged Fourier transforms is the probability of having the value of a particular NPSD bin fall within a certain range of the mean. For example, for the $M = 32$ averages case in Figure 3, the probability of the value in the bin lying below 2.5 is about 0.99. The probability of the value

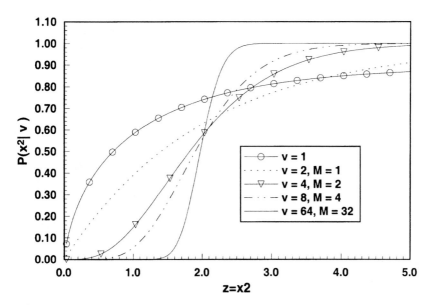

Figure 3   Probability distributions numerically calculated from the PDFs in Figure 2 where $\sigma = 1$.

being less than 1.5 is about 0.02. Therefore, the probability of the averaged value for the bin being between 1.5 and 2.5 is about 97%. With only 4 averages, the probability drops to 0.80–0.30, or about 50%, and with only 1 NPSD bin ($M = 1$), the probability is less than 20%. Clearly, spectral averaging of random data significantly reduces the variance, but not the mean, of the spectral bin data. The only way to reduce the mean of the noise (relative to a sinusoid), is to increase the resolution of the NPSD.

One statistical measure that is often very useful is a *confidence interval* for the spectral bin. A confidence interval is simply a range of levels for the spectral bin and the associated probability of having a value in that range. The 99% confidence interval for the $M = 2$ case could be from 0.04 to 0.49, while for $M = 4$ the interval narrows to 0.08 to 3.7, etc. It can be seen as even more convenient to express the interval in decibels (dB) about the mean value. Figure 4 shows the probability curves on dB value x-axis scale where 0 dB is the mean of $2\sigma^2 = 2.0$. Clearly, using Figure 4 as a graphical aid, one can easily determine the probability for a wide range of dB. For example, the ± 3 dB confidence for the $M = 4$ case is approximately 98%, while for the $M = 1$ case its only about 50%.

Synchronous Time Averaging

We can see that spectral averaging in the frequency domain reduces the variance of the random bin data as the number of complex magnitude-squared averages M increases. This approach is very useful for "cleaning up" spectral data, but it does not allow an increase in narrowband signal-to-noise ratio beyond what the Fourier transform offers. However, if we know the frequency to be detected has period T_p, we can synchronously average the time data such that only frequencies synchronous with $f_p = 1/T_p$ (i.e. f_p, $2f_p$, $3f_p$,...) will remain. All non-coherent fre-

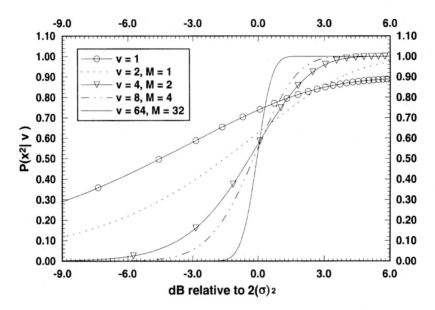

Figure 4   Probability distributions for the Chi-Square average PDFs but with x^2 on a 10 $\log_{10}$ scale relative to the mean at 2.0.

quencies including ZMG random noise will average to zero. This is an extremely effective technique for enhancing known signal detectability. Examples are often found in active sonar and radar, vibrations monitoring of rotating equipment such as bearings in machinery, and measurements of known periodic signals in high noise levels. For the case of rotating equipment, a tachometer signal can provide a trigger to synchronize all the averages such that shaft-rate related vibrations would all be greatly enhanced while other periodic vibrations from other shafts would be suppressed. The averaging process has the effect of reducing the non-coherent signals by a factor of $1/N$ for N averages. Therefore, to increase the detectability by 40 dB, one would need at least 100 coherent time averages of the signal of interest.

Higher-Order Moments

Beyond the mean and mean-square (1st and 2nd general moments) of a probability density function there are several additional moments and moment sequences which are of distinct value to intelligent signal processing systems. In general, these "higher-order" statistics and their corresponding spectra each provide insightful features for statistically describing the data set of interest. We will start by considering the nth general moment of the probability distribution $p(x)$.

$$\overline{(X)^n} = E\{X^n\} = \int\limits_{-\infty}^{+\infty} x^n p(x) dx \tag{6.1.17}$$

The central moment is defined as the moment of the difference of a random variable and its mean as seen in Eq. (6.1.18). The central moments of a random variable are preferred because in many cases a zero mean random variable will result in much simpler mathematical derivations.

$$\overline{(X - \bar{X})^n} = E\{(X - \bar{X})^n\} = \int\limits_{-\infty}^{+\infty} (x - \bar{X})^n p(x)dx \qquad (6.1.18)$$

As noted earlier, the 2nd central moment is the variance σ^2. The 3rd central moment leads one to the *skewness* which is typically defined with the difference between the random variable and its mean normalized by the standard deviation to give a dimensionless quantity.

$$Skew = E\left\{\left(\frac{X - \bar{X}}{\sigma}\right)^3\right\} = \int\limits_{-\infty}^{+\infty} \left(\frac{x - \bar{X}}{\sigma}\right)^3 p(x)dx \qquad (6.1.19)$$

A positive skewness means the distribution tail extends out more in the positive x-direction then in the negative direction. Hence, a positively skewed distribution tends to lean "towards the right", while a negatively skewed PDF leans toward the left. The normalized 4th central moment is is known as the *kurtosis* and is given in Eq. (6.1.20) below.

$$Kurt = E\left\{\left(\frac{X - \bar{X}}{\sigma}\right)^4\right\} - 3 \qquad (6.1.20)$$

The kurtosis is a measure of how "peaky" the PDF is around its mean. A strongly positive kurtosis indicates a PDF with a very sharp peak at the mean and is called (by very few people), leptokurtic. A negative kurtosis indicates the PDF is very flat in the vicinity of the mean and is called (by even fewer people), platykurtic. The measures of skewness and kurtosis are essentially relative to the Gaussian PDF because it has zero skewness and a kurtosis of -1. The mean, variance, skewness and kurtosis form a set of features which provide a reasonable description of the shape of a Gaussian-like uni-modal curve (one which has just one bump). PDFs can be bi-modal (or even more complicated on a bad day) such that other measures of the PDF must be taken for adequate algorithmic description.

The Characteristic Function

Given an analytic function for the PDF or a numerical histogram, the general moments can be calculated through a numerical or analytic integration of Eq. (6.1.17). However, analytic integration of the form in Eq. (6.1.17) can be quite difficult. An alternative analytic method for computing the central moments of a PDF is the *characteristic function* $\phi(u)$. The characteristic function of a random variable X is simply $\phi(u) = E\{e^{juX}\}$.

$$\phi(u) = E\{e^{juX}\} = \int\limits_{-\infty}^{+\infty} p(x)e^{jux}dx \qquad (6.1.21)$$

It can be seen that, except for the positive exponent, the characteristic function is a

Fourier transform of the PDF! Conversely, if we are given a characteristic function, the "negative exponent" inverse transform gives back the PDF.

$$p(x) = \frac{1}{2\pi} \int\limits_{-\infty}^{+\infty} \phi(u) e^{-jux} du \tag{6.1.22}$$

The sign in the exponent is really more of a historical convention than a requirement for the concept to work. To see the "trick" of the characteristic function, simply differentiate Eq. (6.1.21) with respect to u and evaluate at $u = 0$.

$$\left.\frac{d\phi(u)}{du}\right|_{u=0} = j \int\limits_{-\infty}^{+\infty} xp(x)dx = j\bar{X} \tag{6.1.23}$$

As one might expect, the nth general moment is found simply by

$$\overline{X^n} = E\{X^n\} = \frac{1}{j^n} \left.\frac{d^n\phi(u)}{du^n}\right|_{u=0} \tag{6.1.24}$$

Joint characteristic functions can be found by doing 2-dimensional Fourier transforms of the respective joint PDF and the joint nth general moment can be found using the simple formula in Eq. (6.1.25). Its not the first time a problem is much easier solved in the frequency domain.

$$\overline{X^i Y^k} = E\{X^n Y^k\} = \frac{1}{j^{i+k}} \left.\frac{\partial^{i+k}\phi_{XY}(u, v)}{\partial u^i \partial v^k}\right|_{u=v=0} \tag{6.1.25}$$

Cumulants and Polyspectra

The general moments described above are really just the 0th lag of a moment time sequence defined by

$$m_n^x(\tau_1, \tau_2, \ldots \tau_{n-1}) = E\{x(k)x(k + \tau_1) \ldots x(k + \tau_{n-1})\} \tag{6.1.26}$$

Therefore, one can see that $m_1^x = E\{x\}$, or just the mean while $m_2^x(0)$ is the mean square value for the random variable $x(k)$. The sequence $m_2^x(\tau)$ is defined as the autocorrelation of $x(k)$. The *2nd order cumulant* is seen as the *covariance sequence* since the mean is subtracted from the mean-square.

$$c_2^x(\tau_1) = m_2^x(\tau_1) - (m_2^x)^2 \tag{6.1.27}$$

The 0th lag of the covariance sequence is simply the variance. The 3rd order cumulant is

$$c_3^x(\tau_1, \tau_2) = m_3^x(\tau_1, \tau_2) - m_1^x[m_2^x(\tau_1) + m_2^x(\tau_2) + m_2^x(\tau_1 - \tau_2)] + 2(m_1^x)^3 \tag{6.1.28}$$

The (0,0) lag of the 3rd order cumulant is actually the "unnormalized" skewness, or simply the skewness times the standard deviation cubed. As one might quess, the 4th order cumulant (0,0,0) lag is the unnormalized kurtosis. The 4th order cumulant can be written more compactly if we can assume a zero mean $m_1^x = 0$.

$$
\begin{aligned}
c_4^x(\tau_1, \tau_2, \tau_3) = & \, m_4^x(\tau_1, \tau_2, \tau_3) - m_2^x(\tau_1)m_2^x(\tau_1)m_2^x(\tau_3 - \tau_2) \\
& - m_2^x(\tau_2)m_2^x(\tau_3 - \tau_1) - m_2^x(\tau_3)m_2^x(\tau_2 - \tau_1)
\end{aligned}
\tag{6.1.29}
$$

These cumulants are very useful for examining the just how "Gaussian" a random noise process is, as well as, for discriminating linear processes from nonlinear processes. To see this we will examine the spectra of the cumulants known as the *power spectrum* for the 2nd cumulant, *bispectrum* for the 3rd cumulant, *trispectrum* for the 4th cumulant, and so on. While the investigation of *polyspectra* is a relatively new area in signal processing, the result of the power spectrum being the Fourier transform of the 2nd cumulant is well-known as the Weiner–Khintchine theorem. Consider the bispectrum which is a 2-dimensional Fourier transform of the 3rd cumulant.

$$
C_3^x(\omega_1, \omega_2) = \sum_{\tau_1=-\infty}^{+\infty} \sum_{\tau_2=-\infty}^{+\infty} c_3^x(\tau_1, \tau_2)e^{-j(\omega_1\tau_1+\omega_2\tau_2)}
\tag{6.1.30}
$$

$$
|\omega_1|, |\omega_2|, |\omega_1 + \omega_2| \leq \pi
$$

It can be shown that the bispectrum in the first octant of the ω_1, ω_2 plane ($\omega_2 > 0$, $\omega_1 \geq \omega_2$), bounded by $\omega_1 + \omega_2 \leq \pi$ to insure no aliasing, is actually all that is needed due to a high degress of symmetry. The trispectrum is a 3-dimensional Fourier transform of the 4th cumulant.

$$
C_4^x(\omega_1, \omega_2, \omega_3) = \sum_{\tau_1=-\infty}^{+\infty} \sum_{\tau_2=-\infty}^{+\infty} \sum_{\tau_3=-\infty}^{+\infty} c_4^x(\tau_1, \tau_2, \tau_3)e^{-j(\omega_1\tau_1+\omega_2\tau_2+\omega_3\tau_3)}
\tag{6.1.31}
$$

$$
|\omega_1|, |\omega_2|, |\omega_3|, |\omega_1 + \omega_2 + \omega_3| \leq \pi
$$

The trispectrum covers a 3-dimensional volume in ω and is reported to have 96 symmetry regions. Don't try to implement Eqs (6.1.30) and (6.1.31) as written on a computer. There is a far more computationally efficient way to compute polyspectra without any penalty given the N-point DFT of $x(k)$, $X(\omega)$. While the power spectrum (or power spectral density PSD) is well known to be $C_4^x(\omega) = X(\omega)X^*(\omega)/N$, the bispectrum and trispectrum can be computed as

$$
\begin{aligned}
C_3^x(\omega_1, \omega_2) &= \frac{1}{N} X(\omega_1)X(\omega_2)X^*(\omega_1 + \omega_2) \\
C_4^x(\omega_1, \omega_2, \omega_3) &= \frac{1}{N} X(\omega_1)X(\omega_2)X(\omega_3)X^*(\omega_1 + \omega_2 + \omega_3)
\end{aligned}
\tag{6.1.32}
$$

which is a rather trivial calculation compared to Eqs (6.1.28) through (6.1.31). For purposes of polyspectral analysis, the factor of $1/N$ is not really critical and is included here for compatibility with power spectra of periodic signals.

Let take an example of a pair of sinusoids filtered through a nonlinear process and use the bispectrum to analyze which frequencies are the result of the nonlinearity. The bispectrum allows frequency components which are phase-coupled to coherently integrate, making them easy to identify. Let $x(k)$ be sampled at $f_s = 256$

Hz and

$$x(k) = A_1 \cos\left(2\pi\frac{f_1}{f_s}k + \theta_1\right) + A_2 \cos\left(2\pi\frac{f_2}{f_s}k + \theta_2\right) + w(k) \qquad (6.1.33)$$

where $A_1 = 12, f_1 = 10, \theta_1 = 45°$, $A_2 = 10, f_2 = 24, \theta_1 = 170°$, and $w(k)$ is a ZMG noise process with standard deviation 0.01. Our nonlinear process for this example is

$$z(k) = x(k) + \epsilon\, x(k)^2 \qquad (6.1.34)$$

where $\epsilon = 0.10$, or a 10% quadratic distortion of $x(k)$. The nonlinearity should produce a harmonic for each of the two frequencies (20 and 48 Hz) and sum and difference tones at 14 and 34 Hz. Figure 5 clearly shows the original PSD of $x(k)$ (dotted curve) and its two frequencies and the PSD of $z(k)$ (solid curve), showing 6 frequencies. The 10% distortion places the harmonics about 10 dB down from the original levels. But since the distorted signal is squared, the levels are not quite that far down.

Harmonic generation is generally known in audio engineering community as *Harmonic Distortion*, while the sum and difference frequency generation is known as *Intermodulation Distortion*. It is interesting to note that if f_1 and f_2 were harmonically related (say 10 and 20 Hz), the sum and difference tones would also be harmonics of the 10 Hz fundamental and no "intermodulated" tones would be readily seen. For most music and speech signals, the harmonic content is already very rich, so a little harmonic distortion is acceptable to most people. However, in some forms of modern rock music (such as "heavy metal" and "grunge"), more harmonics are desirable for funky artistic reasons and are usually produced using

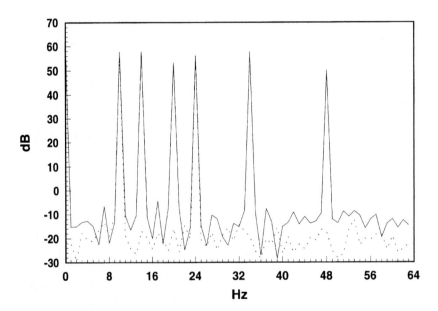

Figure 5 Comparison of a two sinusoid signal (dotted curve) and the addition of a 10% quadratic distortion (solid curve).

overdriven vacuum tube amplifiers. Out of the 6 frequencies in Figure 5, only two are phase coupled via the nonlinearity to a respective fundamental. Figure 6 shows the bispectrum where bright areas on the plot indicate frequency pairs of high coherence.

Clearly, there is perfect symmetry about the $f_1 = f_2$ line in Figure 6 suggesting that all one needs is the lower triangular portion of the bispectrum to have all the pertinent information. The bright spots at 20 and 48 Hz indicate that these two frequencies are phase coupled to their respective fundamentals due to the distortion. The symmetry spot at $\langle 48, 20 \rangle$ and verticle and horizontal lines, are due to the fact that the same nonlinearity produced both harmonics. The diagonal spots appear to be due to a spectral leakage effect.

We now test the bispectrum of the same set of 6 frequencies, except only the 24 Hz sinusoid is passed through the nonlinearity, where the other tones are simply superimposed. The bispectrum for the single sinusoid nonlinearity is seen in Figure 7. Comparing Figures 6 and 7 one can clearly see the relationship between a fundamental and the generated harmonic from the nonlinearity. The difference tones at 14 and 34 Hz should be seen in the tri-spectrum. The bi-spectrum, or 3rd cumulent provides a measure of the skewness of the PDF. Since the nonlinear effect increases

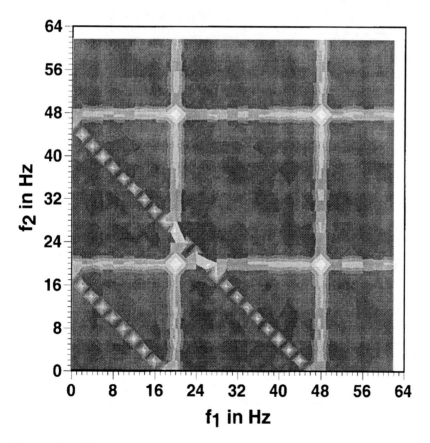

Figure 6 Bispectrum of the pair of sinusoids in white noise passed through the 10% quadratic nonlinearity.

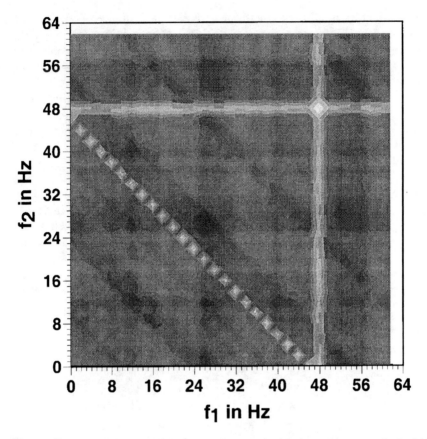

Figure 7 The bispectrum of the same 6 frequencies in Figure 5 except only the 24 Hz passes through the nonlinearity.

with signal amplitude, one can see that a PDF signal representation will be skewed because of the nonlinearity.

6.2 TRANSFER FUNCTIONS AND SPECTRAL COHERENCE

Perhaps the most common method to measure the frequency response of a transducer or electrical/mechanical system is to compute the transfer function from the simultaneous ratio of the output signal spectrum to the input signal spectrum. For modal analysis of structures, the classic approach is to strike the structure with an instrumented hammer (it has an embedded force-measuring accelerometer), and record the input force impulse signal simultaneously with the surface acceleration response output signal(s) at various points of interest on the structure. The ratio of the Fourier transforms of an output signal of interest to the force impulse input provides a measure of the mechanical force-to-acceleration transfer function between the two particular points. Noting that velocity is the integral of acceleration, one can divide the acceleration spectrum by "$j\omega$" to obtain a velocity spectrum. The ratio of velocity over force is the mechanical mobility (the inverse of mechanical impedance). An even older approach to measuring transfer functions involves the use of a swept sinusoid as the input signal and an rms level for the output magnitude and phase meter level for system phase. The advantage of the swept sinusoid is a high controllable signal to noise ratio and the ability to synchronous average the signals in the time domain to increase the signal-to-noise ratio. The disadvantage is, the sweep rate should be very slow in the regions of sharp peaks and dips in the response function and tracking filters must be used to eliminate harmonic distortion. More recently, random noise is used as the input and the steady-state response is estimated using spectral averaging.

Once the net propagation time delay is separated from the transfer function to give a minimum phase response, the frequencies where the phase is 0 or π (imaginary part is zero) indicate the system modal resonances and anti-resonances, depending on how the system is defined. For mechanical input impedance (force over velocity) at the hammer input force location, a spectral peak indicates that a large input force is needed to have a significant structural velocity response at the output location on the structure. A spectral "dip" in the impedance response means that a very small force at that frequency will provide a large velocity response at the output location. It can therefore be seen that the frequencies of the peaks of the mechanical mobility transfer function are the structural resonances and the dips are the anti-resonances. Recalling from system partial fraction expansions in Chapter 2, the frequencies of resonance are the system modes (poles) while the phases between modes determine the frequencies of the spectral dips, or zeroes (anti-resonances). Measurement of the mechanical system's modal response is essential to design optimization for vibration isolation (where mobility zeros are desired at force input locations and frequencies) or vibration communication optimization for transduction systems. For example, one needs a flat (no peaks or dips) response for good broadband transduction, but using a structural resonance as a narrowband "mechanical amplifier" is also often done to improve transducer efficiency and sensitivity.

It is anticipated that adaptive signal processing systems will be used well into the 21st century for monitoring structural modes for changes which may indicate a likely failure due to structural fatigue or damage, as well as to build in a sensor-health "self-awareness", or sentient capability for the monitoring system. It is therefore imperative that we establish the foundations of transfer function measurement especially in the area of measurement errors and error characterization. Consider the case of a general input spectrum $X(f)$, system response $H(f)$, and coherent output spectrum $Y(f)$ as seen in Figure 8.

Clearly, $H(f) = Y(f)/X(f)$, but for several important reasons, we need to express $H(f)$ in terms of expected values of the ratio of the short time Fourier transforms $X_n(f,T)$ and $Y_n(f,T)$ integrated over the time interval T where each block "n" will be averaged.

$$X_n(f, T) = \int_0^T x_n(t)e^{-j2\pi ft}dt \qquad (6.2.1)$$

The data $x_n(t)$ in Eq. (6.2.1) can be seen as being sliced up into blocks T sec long which will then be Fourier transformed and subsequently averaged. This approach for spectral averaging is based on the property of *ergodic random processes*, where the expected values can be estimated from a finite number of averages. This requires that the signals be stationary. For example, the blows of the modal analysis hammer will in general not be exactly reproducible. However, the average of, say 32, blows of the hammer should provide the same average force impulse input as the average of thousands of hammer blows if the random variability (human mechanics, background noise, initial conditions, etc.) is ergodic. The same is true for the PSD Gaussian random noise where each DFT bin is a complex Gaussian random variable. However, after averaging the NDFT bins over a finite period of time (say $n = 32$ blocks or $32T$ sec), the measured level in the bin will tend toward the expected mean as seen in Figure 9 for the cases of $M = 1$ and $M = 32$ and the autospectrum.

Given the resolution and frequency span in Figure 9, it can be seen that the underlying NDFT is 1024-point. For a unity variance zero-mean Gaussian (ZMG) real time-domain signal, we expect the normalized power spectrum to have mean $2\sigma^2 = 2/N$, or approximately 0.002. The bin variance of the $M = 32$ power spectrum is approximately $1/32$ of the $M = 1$ case. For the transfer function measurement, $H_{xy}(f)$, we can exploit the ergodic random processes by estimating our transfer func-

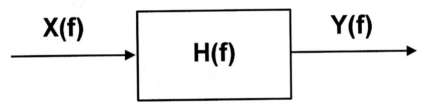

Figure 8   A general transfer function $H(f)$ to be measured using the input spectrum $X(f)$ and output spectrum $Y(f)$.

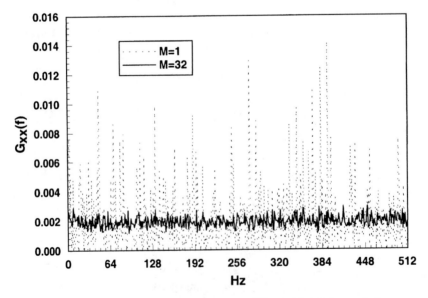

Figure 9   Comparison of 1 and 32 averages of the 1-sided 1 Hz resolution power spectrum of ZMG process of unity variance.

tion as the expected value of the ratio of Fourier transforms.

$$H_{xy}(f) = E\left\{\frac{Y_n(f, T)}{X_n(f, T)}\right\}$$ (6.2.2)

The expected value in Eq. (6.2.2) is the average of the ratio of the output spectrum over the input spectrum. It is not the ratio of the average output spectrum over the average input spectrum which does not preserve the input–output phase response. There is however, an unnecessary computational burden with Eq. (6.2.2) in that complex divides are very computationally demanding. However, if we pre--multiply top and bottom by $2X_n^*(f, T)$ we can define the transfer function as the ratio of the input–output cross spectrum $G_{xy}(f)$ over the input autospectrum (referred to earlier as the 1-sided power spectrum) $G_{xx}(f)$.

$$H_{xy}(f) = E\left\{\frac{2X_n^*(f, T)Y_n(f, T)}{2X_n^*(f, T)X_n(f, T)}\right\} = \frac{E\{2X_n^*(f, T)Y_n(f, T)\}}{E\{2X_n^*(f, T)X_n(f, T)\}} = \frac{G_{xy}(f)}{G_{xx}(f)}$$ (6.2.3)

Recall that the 2-sided power spectrum $S_X(f)$ for real $x(t)$ is defined as

$$S_X(f) = \lim_{T \to \infty}\left\{\frac{|X(f, T)|^2}{T}\right\} + \lim_{T \to \infty}\left\{\frac{|X(-f, T)|^2}{T}\right\}$$ (6.2.4)

and the 1-sided power spectrum is $G_{xx}(f)$, or the autospectrum of $x(t)$.

$$
\begin{aligned}
G_{xx}(f) &= 2 \lim_{T \to \infty}\left\{\frac{|X(f, T)|^2}{T}\right\} \quad f > 0 \\
&= \lim_{T \to \infty}\left\{\frac{|X(f, T)|^2}{T}\right\} \quad f = 0 \\
&= 0 \quad f < 0
\end{aligned}
$$ (6.2.5)

The autospectrum gets its name from the fact that a conjugate multiply $X^*(f)X(f)$ in the frequency domain is an auto-correlation, not convolution, in the time domain. The autospectrum, (or power spectrum), $G_{xx}(f)$, is the Fourier transform of the autocorrelation of $x(t)$ in the time domain.

$$G_{xx}(f) = 2 \int_{-\infty}^{+\infty} R^x(\tau)e^{-j2\pi f\tau}d\tau \quad f > 0$$ (6.2.6)

$$R^x(\tau) = E\{x(t)x(t + \tau)\}$$

Likewise, the input–output cross spectrum $G_{xy}(f)$ is the Fourier transform of the cross correlation of x and y in the time domain.

$$G_{xx}(f) = 2 \int_{-\infty}^{+\infty} R^{xy}(\tau)e^{-j2\pi f\tau}d\tau \quad f > 0$$ (6.2.7)

$$R^{xy}(\tau) = E\{x(t)x(t + \tau)\}$$

Equations (6.2.3), (6.2.6), and (6.2.7) allow for a computationally efficient methodology for computing the transfer function $H_{xy}(f)$. One simply maintains a running average of the cross spectrum $G_{xy}(f)$ and autospectrum $G_{xx}(f)$. After the averaging process is complete, the two are divided only once, bin by bin, to give the transfer function estimate $H_{xy}(f)$.

However, due to numerical errors which can be caused by spectral leakage, environmental noise, or even the noise from the least-significant bit of the A/D convertors, we will introduce a measurement parameter known as an *ordinary coherence function* for characterization of transfer function accuracy.

$$\gamma_{xy}^2(f) = \frac{|G_{xy}(f)|^2}{G_{xx}(f)G_{yy}(f)} \tag{6.2.8}$$

The coherence function will be unity if no extraneous noise is detected leading to errors in the transfer function measurement. The exception to this is when only 1 average is computed giving a coherence estimate which is algebraically unity. For a reasonably large number of averages, the cross spectrum will differ slightly from the input and output autospectra, if only from the noise in the A/D system. However, the system impulse response is longer than the FFT buffers, some response from the previous input buffer will still be reverberating into the time frame of the subsequent buffer. Since the input is ZMG noise in most cases for system identification, the residual reverberation is uncorrelated with the current input, and as such, appears as noise to the transfer function measurement. A simple coherence measurement will identify this effect. Correction requires that the FFT buffers be at least as long as the system impulse response, giving corresponding frequency-domain resolution higher than the system peaks require. Significant differences will be seen at frequencies where an estimation problem occurs. We will use the coherence function along with the number of non-overlapping spectral averages to estimate error bounds for the magnitude and phase of the measured transfer function.

Consider the example of a simple digital filter with two resonances at 60.5 and 374.5 Hz, and an antiresonance at 240.5 Hz, where the sample rate is 1024 Hz and the input is ZMG with unity variance. Figure 10 shows the measured transfer function (dotted line), true transfer function (solid line), and measured coherence for a single buffer in the spectral average ($M = 1$).

Obviously, with only one pair of input and output buffers, the transfer function estimate is rather feeble due primarily to spectral leakage and non-uniform input noise. The coherence with only one average is algebraically unity since

$$\gamma_{xy}^2(f) = \frac{|G_{xy}(f)|^2}{G_{xx}(f)G_{yy}(f)} \approx \frac{X_1^*(f)Y_1(f)Y_1^*(f)X_1(f)}{X_1^*(f)X_1(f)Y_1^*(f)Y_1(f)} = 1 \tag{6.2.9}$$

Figure 11 gives the estimated result with only 4 averages. Considerable improvement can be seen in the frequency ranges where the output signal level is high. With only 4 averages, the cross spectrum and auto spectrum estimates begin to differ slightly where the output signal levels are high, and differ significantly where the spectral leakage dominates around the frequency range of the 240.5 Hz dip. The

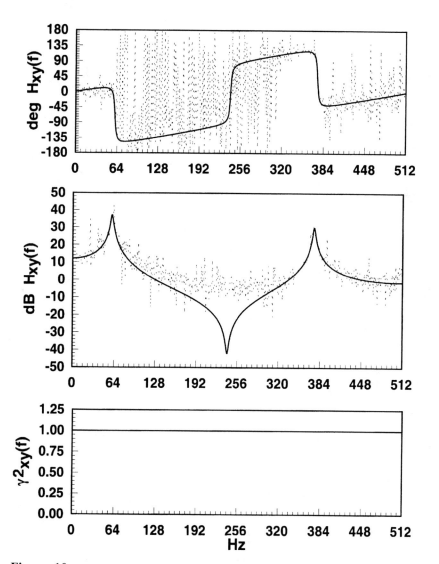

Figure 10 Magnitude and phase of the true (solid line) and measured (dotted line) transfer function and coherence $\gamma_{xy}^2(f)$.

expected values can be approximated by sums.

$$\gamma_{xy}^2(f) = \frac{|G_{xy}(f)|^2}{G_{xx}(f)G_{yy}(f)} \approx \frac{\left|\sum_{n=1}^{4} X_n^*(f)Y_n(f)\right|^2}{\left[\sum_{n=1}^{4} X_n^*(f)X_n(f)\right]\left[\sum_{n=1}^{4} Y_n^*(f)Y_n(f)\right]} \tag{6.2.10}$$

Figure 12 clearly shows how application of a Hanning data window significantly reduces the spectral leakage allowing the transfer function estimate to be accurately measured in the frequency range of the dip at 240 Hz. Note that the

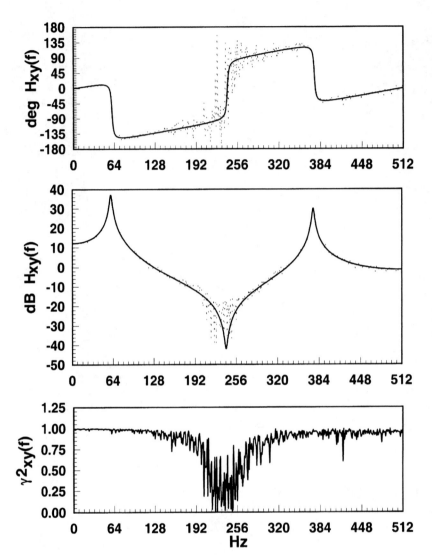

Figure 11   Transfer function and coherence measurement with 4 averages show the effect of spectral leakage in the dip region.

coherence drops slightly at the peak and dip frequencies due to the rapid system phase change there and the loss of resolution from the Hanning window. Clearly, the utility of a data window for transfer function measurements is seen in the improvement in the dip area and throughout the frequency response of the measured transfer function.

We now consider the real-world effects of uncorrelated input and output noise leaking into our sensors and causing errors in our transfer function measurement. Sources of uncorrelated noise are, at the very least, the random error in the least significant bit of the A/D process, but typically involve things like electromagnetic noise in amplifiers and intrinsic noise in transducers. A block

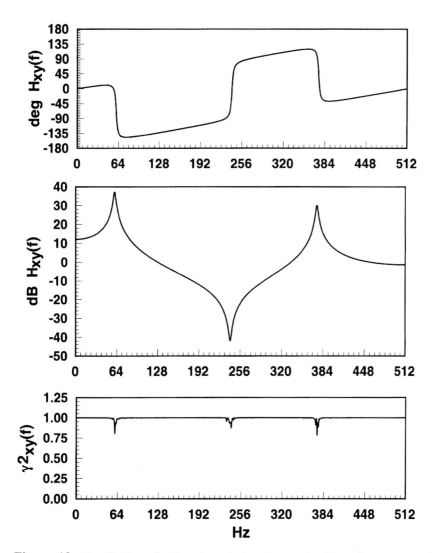

Figure 12 Application of a Hanning window gives a significant improvement in the dip region by reducing spectral leakage although coherence is reduced at the peak and dip regions.

diagram of the transfer function method including measurement noise is seen in Figure 13.

Consider the effect of the measurement noise on the estimated transfer function. Substituting $U(f) + Nx(f)$ for $X(f)$ and $V(f) + Ny(f)$ for $Y(f)$ in Eq. (6.2.3) we have

$$H_{xy}(f) = \frac{E\{[U(f) + Nx(f)]^*[V(f) + Ny(f)]\}}{E\{[U(f) + Nx(f)]^*[U(f) + Nx(f)]\}} \tag{6.2.11}$$

We simplify the result if we can assume that the measurement noises $Nx(f)$ and $Ny(f)$ are uncorrelated with each other and uncorrelated with the input signal $U(f)$

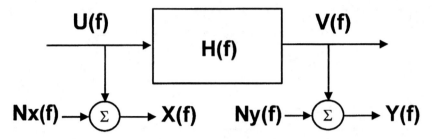

Figure 13   The transfer function $H(f)$ measured using the input $X(f)$ and output $Y(f)$ which contain measurement noise.

and the output signal $V(f)$.

$$H_{xy}(f) = \frac{G_{UV}(f)}{G_{UU}(f) + G_{NxNx}(f)} \tag{6.2.12}$$

Equation (6.2.12) clearly shows a bias whenever the input SNR is low which causes the transfer function amplitude to be low. An overall random error is seen in Eq. (6.2.11) when the output SNR is low. The ordinary coherence function also shows a random effect which reduces the coherence to < 1 with uncorrelated input and output measurement noise.

$$\gamma_{xy}^2(f) = \frac{|G_{UV}(f)|^2}{[G_{UU}(f) + G_{NxNx}(f)][G_{VV}(f) + G_{NyNy}(f)]} \tag{6.2.13}$$

Equation (6.2.13) clearly shows a breakdown in coherence whenever either the input, output, or both SNRs become small. Figure 14 below shows the effect of low input SNR (both $U(f)$ and $Nx(f)$ are ZMG signals with unity variance) on the measured transfer function and coherence. The 0 dB input SNR gives about a 3 dB bias to the transfer function magnitude, and an average coherence of about 0.5. The variances of the magnitude, phase, and coherence will be reduced if the number of averages in increased, but not the bias offsets.

   While the low input SNR in Figure 14 is arguably a worse case and can easily be avoided with good data acquisition practice, it serves us to show the bias effect on the transfer function magnitude. If the output measurement noise is ZMG, we should only expect problems in frequency ranges where the output of $H_{xy}(f)$ is low. Figure 15 shows the effect of unity variance ZMG output noise only for 4 averages when a Hanning window is used on both input and output data.

   Clearly, one can easily see the loss of coherence and corresponding transfer function errors in the frequency region where the output of our system $H_{xy}(f)$ is low compared to the output measurement noise with unity variance. Since the output noise for our example is uncorrelated with the excitation input $U(f)$, it is possible to partially circumvent the low output SNR by doing a large number of averages. Figure 16 below shows the result for 32 averages which shows a mild improvement for the regions around 250 Hz and 500 Hz, and an overall reduced spectral variance.

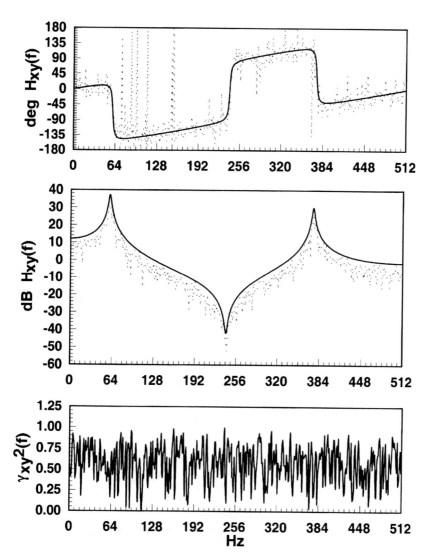

Figure 14 Transfer function magnitude, phase, and coherence for 4 averages, Hanning window, and 0 dB input SNR ($G_{UU}(f)$ and $G_{N_X N_X}(f)$ both have unity variance ZMG noise).

The situation depicted in Figure 16 is more typical of a real-world measurement scenario where the dynamic range of the system being measured exceeds the available SNR of the measurement system and environment. The errors in the transfer function magnitude and phase are due primarily to the breakdown in coherence as will be seen below. While one could arduously average the cross and input autospectra for quite some time to reduce the errors in the low output frequency ranges, the best approach is usually to eliminate the source of the coherence losses.

The variance of the transfer function error is usually modeled by considering the case of output measurement noise (since the input measurement noise case is more easily controlled), and considering the output SNR's impact on the real

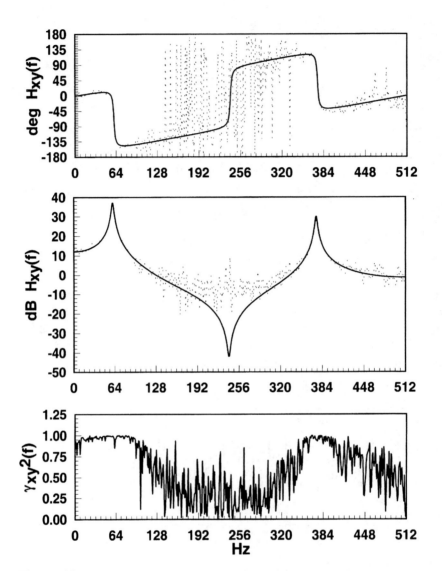

Figure 15   Transfer function magnitude, phase, and coherence for an output measurement ZMG noise with unity variance using 4 spectral averages and a Hanning window.

and imaginary parts of the transfer function (1) using the diagram depicted in Figure 17 below.

The circle in Figure 17 represents the standard deviation for the transfer function measurement which is a function of the number of averages n_d, the inverse of the output SNR, and the true amplitude of the transfer function represented by the "H" in the figure. Recall that for output noise only, the transfer function can be written as an expected value.

$$E\{H_{xy}(f)\} = H(f)\left(1 + E\left\{\frac{Ny(f)}{V(f)}\right\}\right)$$ (6.2.14)

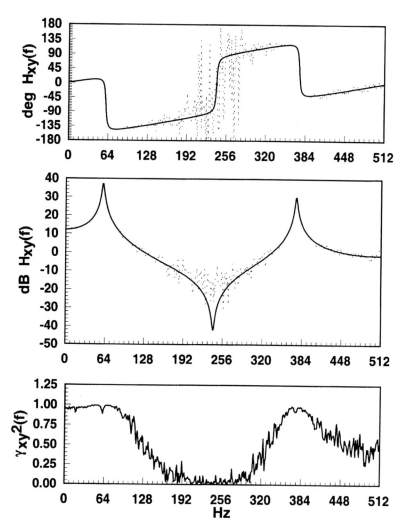

Figure 16 Transfer function magnitude, phase, and coherence for an output measurement ZMG noise with unity variance using 32 spectral averages and a Hanning window.

The expected value $E\{Ny(f)/V(f)\}$ is zero because for a ZMG input $U(f)$ into the system, which is uncorrelated with the output noise $Ny(f)$, the real and imaginary bins of both $Ny(f)$ and $V(f)$ are ZMG processes. This centers the error circle around the actual $H(f)$ in Figure 17 where the radius of the circle is the standard deviation for the measurement. The mean magnitude-squared estimate for the transfer function is

$$\{|H_{xy}(f)|^2\} = |H(f)|^2\left(1 + E\left\{\frac{Ny(f)}{V(f)}\right\} + E\left\{\frac{Ny^*(f)}{V^*(f)}\right\} + E\left\{\frac{Ny^*(f)Ny(f)}{V^*(f)V(f)}\right\}\right)$$

$$= |H(f)|^2\left(1 + \frac{G_{NyNy}(f)}{G_{VV}(f)}\right)$$

$$(6.2.15)$$

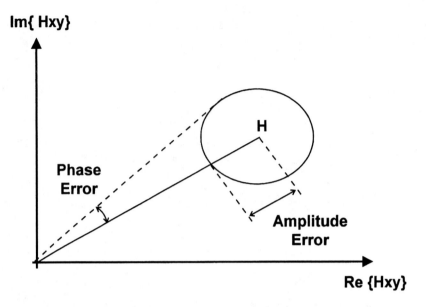

Figure 17   Phasor diagram showing the magnitude and phase errors of the transfer function due to the output SNR.

The variance of the amplitude error for $H_{xy}(f)$ is therefore

$$\sigma^2_{H_{xy}} = E\{|H_{xy}|^2\} - |H(f)|^2 = \frac{1}{2}\left(1 + \frac{G_{N_yN_y}(f)}{G_{VV}(f)}\right)|H_{xy}(f)|^2 \frac{1}{n_d} \qquad (6.2.16)$$

where the factor of $1/(2n_d)$ takes into account the estimation error for the division of two autosprectra averaged over n_d buffers. Clearly, $G_{N_yN_y}(f)/G_{VV}(f)$ is the magnitude-squared of the noise-to-signal ratio (NSR). We can write the NSR in terms of ordinary coherence by observing

$$G_{VV}(f) = |H_{xy}(f)|^2 G_{xx}(f) = \frac{|G_{xy}(f)|^2}{G_{xx}(f)G_{yy}(f)}G_{yy}(f) = \gamma^2_{xy}(f)G_{yy}(f) \qquad (6.2.17)$$

and

$$G_{N_yN_y}(f) = G_{yy}(f) - G_{VV}(f) = [1 - \gamma^2_{xy}(f)]G_{yy}(f) \qquad (6.2.18)$$

allowing the magnitude-squared of the NSR to be written as

$$\frac{G_{N_yN_y}(f)}{G_{VV}(f)} = \frac{G_{N_yN_y}(f)}{G_{yy}(f)}\frac{G_{yy}(f)}{G_{VV}(f)} = \frac{[1 - \gamma^2_{xy}(f)]}{\gamma^2_{xy}(f)} \qquad (6.2.19)$$

We can also write the ordinary coherence function in terms of the output SNR magnitude-squared

$$\gamma^2_{xy}(f) = \frac{|SNR(f)|^2}{1 + |SNR(f)|^2} \qquad (6.2.20)$$

The transfer function amplitude error standard deviation (the circle radius in Figure 17), assuming no input measurement noise is therefore

$$\sigma_{h_{xy}} = \sqrt{\frac{1}{2n_d} \frac{[1 - \gamma_{xy}^2(f)]}{\gamma_{xy}^2(f)}} |H_{xy}(f)| \tag{6.2.21}$$

The transfer function phase error standard deviation (angle error in Figure 17), is found by an arctangent.

$$\sigma_{\theta_{xy}} = \tan^{-1} \left\{ \sqrt{\frac{1}{2n_d} \frac{[1 - \gamma_{xy}^2(f)]}{\gamma_{xy}^2(f)}} \right\} \tag{6.2.22}$$

Clearly, as the coherence drops, the accuracy expected in magnitude and phase also decreases. Spectral averaging improves the variance of the coherence, but does not restore coherence. As a consequence, compensating poor coherence in the transfer function input–output signals with increased spectral averaging has only a limited benefit. For example, where the coherence is below 0.1, the model estimated magnitude standard deviation is about 0.375 times the transfer function magnitude, or about −20 to −40 dB as is confirmed in Figure 16. The phase error standard deviation is only about 20 degrees and approaches 90 degrees as the coherence approaches zero. This too is confirmed in Figure 16 although some of the more significant phase jumps are due to wrapping of the phase angle to fit in the graph.

For systems where the impulse response is effectively longer than the FFT buffers, a loss of coherence occurs because some of the transfer function output is due to input signal before the current input buffer, and therefore is incoherent due to the random nature of the noise. It can be seen in these cases that the spectral resolution of the FFTs is inadequate to precisely model the sharp system resonances. However, a more problematic feature is that the measured magnitude and phase of a randomly-excited transfer function will be inconsistent when the system impulse response is longer than the FFT buffers. If the FFT buffers cannot be increased, one solution is to use a chirped sinusoid as input, where the frequency sweep covers the entire measurement frequency range within a single FFT input buffer. Input–output coherence will be very high allowing a consistent measured transfer function, even if the spectral resolution of the FFT is less than optimal. The swept-sine technique also requires a tracking filter to insure that only the fundamental of the sinusoid is used for the transfer function measurement. However, tracking harmonics of the input fundamental is the preferred technique to measuring harmonic distortion. Again, the bandwidth of the moving tracking filter determines the amount of time delay measurable for a particular frequency. Accurate measurements of real systems can require very slow frequency sweeps to observe all the transfer function dynamics.

For applications where input and output noise are a serious problem, time-synchronous averaging of a periodic input–output signal can be used to increase SNR before the FFTs are computed. The time buffers are synchronized so that the input buffer recording starts at exactly the same place on the input signal waveform with each repetition of the input excitation. The simultaneously triggered and recorded input–output buffers may then be averaged in the time domain greatly enhancing the input signal and its response in the output signal relative to outside non-synchronous noise waveforms which average to zero. This technique is very common in active sonar, radar, and ultrasonic imaging to suppress clutter and noise. However, the time-bandwidth relationship for linear time-invariant system transfer functions still requires that the time buffer length must be long enough to allow the desired frequency resolution (i.e. $\Delta f = 1/T$). If the system represents a medium (such as in sonar) one must carefully consider time-bandwidth of any Doppler frequency shifts from moving objects or changing system dynamics. Optimal detection of non-stationary systems may require a wide-band processing technique using wavelet analysis, rather than narrowband Fourier analysis.

6.3   INTENSITY FIELD THEORY

There are many useful applications of sensor technology in mapping the power flow from a field source, be it electrical, magnetic, acoustic, or vibrational. Given the spatial power flow density, or *intensity* which has units of Watts per meter-squared, one can determine the total radiated power from a source by scanning and summing the intensity vectors over a closed surface. This is true even for field region where there are many interfering sources so long as the surface used for the integration encloses only the power source of interest. Recalling Gauss's law, one obtains the total charge in a volume either by integrating the charge density over the volume or, by integrating the flux density over the surface enclosing the volume. Since it is generally much easier to scan a sensor system over a surface than throughout a volume, the flux field over the surface, and the sensor technologies necessary to measure it are of considerable interest. In electromagnetic theory, the power flux is known as the *Poynting Vector*, or $\mathbf{S} = \mathbf{E} \times \mathbf{H}$. The cross product of the electric field vector $\mathbf{E}$ and the magnetic field vector $\mathbf{H}$ has units of Watts per meter-squared. A substantial amount of attention has been paid in the literature to *acoustic* intensity where the product of the acoustic pressure and particle velocity vector provides the power flux vector which, can be used with great utility in noise control engineering. The same is true the use of vibrational intensity in structural acoustics, however the structural intensity technique is somewhat difficult to measure in practice due to the many interfering vibrational components (shear, torsional, compressional waves, etc.). The Poynting vector in electromagnetic theory is of considerable interest in antenna design. Our presentation of intensity will be most detailed for acoustic intensity but, will also show continuity to structural and electromagnetic intensity field measurements.

Point Sources and Plane Waves

When the size of the radiating source is much smaller than the radiating wavelength, it is said to be a *point source*. Point sources have the distinct characteristic of radiating power equally in all directions giving a spherical radiation pattern. At great distances with respect to wavelength from the source, the spherically-spreading wave in an area occupied by sensors can be thought of as nearly a plane wave. Intensity calculation for a plane wave is greatly simplified since the product of the potential field (pressure, force, electric fields) and flux field (particle velocity, mechanical velocity, magnetic fields) gives the intensity magnitude. The vector direction of intensity depends only on the particle velocity for acoustic waves in fluids (gasses and liquids), but depends on the vector cross-product for vibrational waves in solid and electromagnetic waves. One can get an idea of the phase error for a plane wave assumption given a measurement region approximately the size of a wavelength at a distance R from the source as seen in Figure 18.

The phase error is due to the distance error $R - R\cos\theta$ between the chord of the plane wave and the spherical wave. The phase error $\Delta\phi$ is simply

$$\Delta\phi = 2\pi\frac{R}{\lambda}\left[1 - \cos\left(\tan^{-1}\left\{\frac{\lambda}{2R}\right\}\right)\right] \qquad (6.3.1)$$

where λ is the wavelength. A plot of this phase error in degrees is seen in Figure 19 which clearly shows that for the measurement plane to have less than 1 degree phase error, the distance to the source should be more than about 100 wavelengths. For

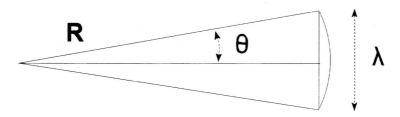

Figure 18 Over a wavelength-sized patch a distance R from the source, a plane-wave assumption introduces a phase error.

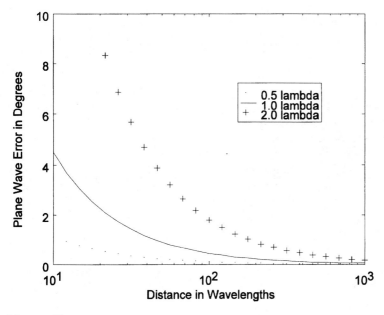

Figure 19 Phase error due to plane wave assumption as a function of distance from source and measurement aperture.

sound, the wavelength of 1 kHz is approximately 0.34 m and one would need to be over 30 m away from the source for the plane wave assumption to be accurate. Even at 40 kHz where many ultrasonic sonars for industrial robotics operate the plane wave region only exists beyond about a meter. The phase error scales with the size of the measurement area shown in Figure 19 in terms of wavelength. Clearly, the plane wave assumption is met for small apertures far from the source. Larger sensor array sizes must be farther away then smaller arrays for the plane wave assumption to be accurate. For multiple point sources, the radiation pattern is no longer spherical but the distance/aperture relationship for the plane wave assumption still holds. This issue of plane wave propagation will be revisited again in Sections 12.3 and 12.4 for field reconstruction and propagation modeling techniques.

The point of discussing the plane wave assumption is to make clear the fact that one simply cannot say that the power is proportional to the square of the pressure, force, electric fields, etc., unless the measurement is captured with a small array many wavelengths from the source. This is a very important distinction between field theory and circuit theory where the wave propagation geometry is much less than a wavelength. Since we need to understand the field in more detail to see the signal processing application of field intensity measurement, we develop the wave equation for acoustic waves as this is the most straightforward approach compared to electric or vibration fields.

Acoustic Field Theory

To understand acoustic intensity one must first review the relationship between acoustic pressure and particle velocity. Derivation of the acoustic wave equation is fairly straightforward and highly useful in gaining insight into acoustic intensity and the physical information which can be extracted. The starting point for the wave equation derivation is usually the *Equation of State*, which simply put, states that the change in pressure relative to density is a constant. In general, the pressure is a monic, but nonlinear function of fluid density (gasses will also be referred to here as "fluids"). However, the static pressure of the atmosphere at sea level is about 100,000 Pascals where 1 Pa = 1 Nt/m^2. An rms acoustic signal of 1 Pa is about 94 dB, or about 100 times louder than normal speech, or about as loud as an obnoxious old lawnmower. So even for fairly loud sounds, the ratio of acoustic pressures to the static pressure is about 1:10^5 making the linearization assumption quite valid. However, for sound levels above 155 dB linear acoustics assumptions are generally considered invalid and linear wave propagation is replaced by shock wave propagation theory. We can derive the equation of state for an ideal gas using Boyle's law

$$pv^{\gamma} = RT \tag{6.3.2}$$

where v is the volume of 1 mole of fluid, R is 8310 J/kmole $^{\circ}$K, and T is the temperature in degrees Kelvin. The thermodynamics are assumed to be *adiabatic*, meaning that no heat flow occurs as a result of the acoustic pressure wave. This is reasonable for low heat conduction media such as air and the relatively long wavelengths and rapid pressure/temperature changes of audio band acoustics.

Taking the natural logarithm of Eq. (6.3.2) we have

$$\gamma \ln v + \ln p = \ln(RT) \tag{6.3.3}$$

We examine the differential of the pressure with respect to volume by computing

$$\gamma \frac{\partial v}{v} + \frac{\partial p}{p} = 0 \tag{6.3.4}$$

Since the volume v of one mole of fluid is the molecular mass M divided by the density ρ, $dv/d\rho = -M/\rho^2$ and we can simplify the derivative of pressure with respect to volume as seen in Eq. (6.3.5).

$$\frac{dp}{dv} = -\frac{\gamma p}{v} = -\frac{\rho^2}{M}\frac{dp}{d\rho} \tag{6.3.5}$$

The *Linearized Acoustic Equation of State* is simply

$$\frac{dp}{d\rho} = \frac{\gamma p}{\rho} = c^2 \tag{6.3.6}$$

The parameter c^2 has the units of meters-squared per second-squared, so c is taken as the speed of sound in the fluid. For small perturbations $p = RT/v$, we can write the speed of sound in terms of the fluid's physical parameters.

$$c = \sqrt{\frac{\gamma RT}{M}} \tag{6.3.7}$$

At a balmy temperature of 20°C (293°K) we find nitrogen ($M = 28$) to have a speed of 349 m/sec, and oxygen ($M = 32$) to have a speed of 326 m/sec. Assuming the atmosphere is about 18% oxygen and almost all the rest nitrogen, we should expect the speed of sound in the atmosphere at 20°C to be around 345 m/sec, which is indeed the case. Note how the speed of sound in an ideal gas does not depend on static pressure since the density also changes with pressure. The main contributors to the acoustic wave speed (which can be measured with a little signal processing), are temperature, mass flow along the direction of sound propagation, and fluid properties. Therefore, for intelligent sensor applications where temperature, fluid chemistry, and mass flow are properties of interest, measurement of sound speed can be an extremely important non-invasive observation technique. However, for extremely intense sound waves from explosions, sonic booms, or devices such as sirens where the sound pressure level exceeds approximately 155 dB, many other loss mechanisms become important and the "shock wave" speed significantly exceeds the linear wave speed of sound given by Eq. (6.3.7).

The second step in deriving the acoustic wave equation is to apply Newton's second law ($f = ma$) to the fluid in what has become known as *Euler's Equation*. Consider a cylindrical "slug" of fluid with circular cross-sectional area S and length dx. The force acting on the slug from the left is $Sp(x)$ while the net force acting on the slug from the right is $Sp(x + dx)$. With the positive x-direction pointing to the right, the net force on the slug resulting in an acceleration du/dt to the right of the mass $Sdx \, \rho$ is $S[p(x) - p(x+dx)]$ or

$$Sdx\frac{p(x) - p(x+dx)}{dx} = -Sdx\frac{dp}{dx} = Sdx\rho\frac{du}{dt} \tag{6.3.8}$$

which simplifies to

$$\frac{dp}{dx} = -\rho\left(\frac{du}{dt} + u\frac{du}{dx}\right) \tag{6.3.9}$$

The term du/dt in Eq. (6.3.9) is simply the acoustic particle acceleration while u du/dx is known as the convective acceleration due to flow in and out of the cylinder of fluid. When either the flow u is zero, or the spatial rate of change of flow du/dx is zero, *Euler's Equation* is reduced to

$$\frac{dp}{dx} = -\rho\frac{du}{dt} \tag{6.3.10}$$

The Continuity Equation expresses the effect of the spatial rate of change of velocity to the rate of change of density. Consider again our slug of fluid where we have a "snapshot" dt sec long where the matter flowing into the volume from the left is $S\rho(x) \, u(x) \, dt$ and the matter flowing out to the right is $S\rho(x+dx) \, u(x+dx)$ dt. If the matter entering our cylindrical volume is greater than the matter leaving, the mass of fluid will increase in our volume by $\partial\rho Sdx$.

$$-Sdx\frac{\partial(\rho u)}{\partial x}dt = -Sdx\left[\rho_0\frac{\partial u}{\partial x} + \frac{\partial\rho}{\partial x}u_0\right]dt \tag{6.3.11}$$

The change in density due to sound relative to the static density ρ_0 is assumed small in Eq. (6.3.11) allowing a linearization in Eq. (6.3.12)

$$\partial\rho_0 Sdx \approx -Sdx\frac{\partial u}{\partial x}\rho_0 \, dt \tag{6.3.12}$$

which is rearranged to give the standard form of the continuity equation.

$$\frac{\partial u}{\partial x} = -\frac{1}{\rho_0}\frac{\partial\rho}{\partial t} = \frac{1}{\rho_0 c^2}\frac{\partial p}{\partial t} \tag{6.3.13}$$

The Acoustic Wave Equation is derived by time differentiating the continuity equation in (6.3.13), taking a spatial derivitive of Euler's equation in (6.3.10), and equating the terms $(d^2u/dx)dt$.

$$\frac{\partial^2 p}{\partial x^2} = \frac{1}{c^2}\frac{\partial^2 p}{\partial t^2} \tag{6.3.14}$$

Acoustic Intensity

For a plane wave, the solution of the one-dimensional wave equation in cartesian coordinates is well-known to be of the form

$$p(x, t) = C_1 e^{j(\omega t - kx)} + C_2 e^{j(\omega t + kx)} \tag{6.3.15}$$

where for positive moving time t (the causal world) the first term in Eq. (6.3.15) is an outgoing wave moving in the positive x-direction and the second term is incoming. The parameter k is radian frequency over wave speed ω/c, which can also be expressed as an inverse wavelength $2\pi/\lambda$. For an outgoing plane wave in free space many wavelengths from the source, we can write the velocity using Euler's equation.

$$u(x, t) = \frac{-1}{\rho} \int \frac{\partial p(x, t)}{\partial x} dt = \frac{1}{\rho c} p(x, t) \tag{6.3.16}$$

Since the pressure and velocity are in phase and only differ by a constant, one can say under these strict *far-field* conditions that the time-averaged acoustic intensity is

$$\langle I(x) \rangle_t = \langle p(x, t) u^*(x, t) \rangle_t = \frac{1}{2\rho c} |p(x, t)|^2 \tag{6.3.17}$$

The specific acoustic impedance p/u of a plane wave is simply ρc. However, for cases where the intensity sensor is closer to the source, the pressure and velocity are not in phase and accurate intensity measurement requires both a velocity and a pressure measurement.

$$p(r, t) = \frac{A}{r} e^{j(\omega t - kr)} \tag{6.3.18}$$

Consider a finite-sized spherical source with radius r_0 pulsating with velocity $u_0 e^{j\omega t}$ to produce a spherically-spreading sinusoidal wave in an infinite fluid medium. From the LaPlacian in spherical coordinates it is well-known that the solution of the wave equation for an outgoing wave in the radial direction r iswhere A is some amplitude constant jet to be determined. Again, using Euler's equation the particle velocity is found.

$$u(r, t) = \frac{A}{r} \frac{1}{\rho c} \left(\frac{1}{jkr} + 1 \right) e^{j(\omega t - kr)} \tag{6.3.19}$$

At the surface of the pulsating sphere $r = r_0$, the velocity is $u_0 e^{j\omega t}$ and it can be shown that the amplitude A is complex.

$$A = \frac{jkr_0^2 u_0 \rho c}{1 + jkr_0} e^{jkr_0} \tag{6.3.20}$$

The time-averaged acoustic intensity is simply the product of the pressure and velocity conjugate replacing the complex exponentials by their expected value of $1/2$ for real waveforms. If the amplitude A is given as an rms value, the factor of $1/2$ is dropped.

$$\langle I(r) \rangle_t = \frac{|A|^2}{r^2} \frac{1}{2\rho c} \left(1 + \frac{j}{kr} \right) \tag{6.3.21}$$

Equation (6.3.21) shows that for distances many wavelengths from the source ($kr \gg 1$), the intensity is approximately the pressure amplitude squared divided by $2\rho c$. However, in regions close to the source ($kr \ll 1$), the intensity is quite different and more complicated. This complication is particularly important when one is attempting to scan a surface enclosing the source(s) to determine the radiated power. One must compute the total field intensity and use the real part to determine the total radiated power by a source. Figure 20 shows the magnitude and phase of the intensity as a function of source distance. It can be seen that inside a few tens of wavelengths we have an acoustic *nearfield* with non-plane wave propagation, and beyond an acoustic *farfield* where one has mainly plane wave propagation. The consequence of the non-zero phase of the intensity in the nearfield is stored energy or *reactive intensity* in the field immediately around the source. The propagating part of the field is known as the *active intensity* and contains the real part of the full intensity field.

   Figure 21 plots the radiation impedance for the source and shows a "mass load" in the nearfield consistent with the reactive intensity as well as the spherical wavefront curvature near the source. Interestingly, the range where the nearfield and farfield meet, say around 20 wavelengths, does not depend on the size of the source so long as the source is radiating as a monopole (pulsating sphere). Real-world noise sources such as internal combustion engines or industrial ventilation systems have many "monopole" sources with varying amplitudes and phases all radiating at the same frequencies to make a very complex radiation pattern. Scanning the farfield radiated pressure (100s of wavelengths away) to integrate $p^2/2\rho c$ for the total radiated power is technically correct but very impractical. Scanning closer to the noise source and computing the full complex acoustic intensity provides tremendous insight into the physical noise mechanisms (including machinery health) as well as allowing the real part to be integrated for the total radiated power.

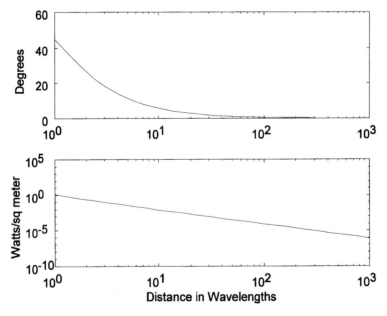

Figure 20   Intensity magnitude and phase as a function of distance for a $r_0 = \lambda$ source pulsating with a velocity of 10 cm/sec.

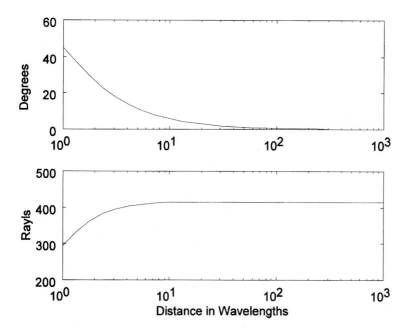

Figure 21 Radiation impedance of a spherical wave as a function of distance from the source.

A great deal of consistency can be seen in comparing Figures 19, 20, and 21 where the acoustic nearfield and acoustic farfield can be seen to meet in the range of 10 to 100 wavelengths from the source. When the wavefront has a spherical curvature the intensity indicates energy storage in the field and the wave impedance clearly shows a "mass-like" component. Note how the phase of the intensity field is also the phase of the wave impedance. If a sensor design places one conveniently in the farfield, great simplifications can be made. But, if nearfield measurements must be made or are desirable, techniques such as intensity are invaluable. The conservation of energy corollary governing all systems is seen "verbally" in Eq. (6.3.22).

$$\begin{pmatrix} Power \\ From \\ Sources \end{pmatrix} + \begin{pmatrix} Power \\ Out\ Of \\ Surface \end{pmatrix} + \begin{pmatrix} Rate\ Of \\ Energy\ Storage \\ In\ Field \end{pmatrix} = 0 \qquad (6.3.22)$$

If the radiation is steady-state and the medium passive, the rate of energy storage eventually settles to zero and the energy radiating out from the surface enclosing the sources is equal to the radiated energy of the sources. This is also well-known in field theory as Gauss's Law. Intelligent adaptive sensor systems can exploit the physical aspects of the waves they are measuring to determine the accuracy and/or validity of the measurement being made.

Structural Intensity

Structural acoustics is a very complicated discipline associated with the generation and propagation of force waves in solids, beams, shells, etc. A complete treatment of structural intensity is well beyond the scope of this and many other books. However, from the perspective of intelligent sensor and control system design, it is important for us to contrast the differences between structural and fluid waves towards the goal of understanding the sensor system signal processing requirements. There are three main types of waves in solids: compressional, or "p-waves" where the force and propagation directions are aligned together; transverse shear, or "s-waves" where a moment force and shear force are orthogonal to the propagation direction; and torsional shear, where a twisting motion has a shear force aligned with the rotation angle θ and is orthogonal to the propagation of the wave. There are also numerous composite forces such as combinations of p, s, and torsional waves in shells and other complex structural elements.

There are some simplified structural elements often referred to in textbooks which tend to have one dominant type of wave. For example, "rod" vibration refers to a long stiff element with compressional waves propagating end-to-end like in the valves of a standard internal combustion engine. "String" vibration refers to mass-tension transverse vibration such as the strings of a guitar or the cables of a suspension bridge. "Shaft" vibration refers to torsional twisting of a drive shaft connecting gears, motors, etc. Rod, string, and shaft vibrations all have a second order wave equation much like the acoustic wave equation where the wave speed is constant for all frequencies and the solution is in the form of sines and cosines. "Beam" vibration refers to a long element with transverse bending shear vibrations such as a cantilever. "Plate" vibration refers to bending vibrations in two dimensions. Beams and plates have both shear and moment forces requiring a 4th-order wave equation, giving a wave speed which increases with increasing frequency, and a solution in the form of sines, cosines, and hyperbolic sines and hyperbolic cosines. "Shell" vibration is even more complex where the bending s-waves are coupled into compressional p-waves.

These "textbook" structural models can be analytically solved (or numerically using finite element software) to give natural frequencies, or modes, of vibration. The modes can be seen as the "energy storage" part of the consevation of energy corollary in Eq. (6.3.22) making intensity measurement essential to measurement of the vibration strength of the internal sources. When dealing with fluid-loaded structures such as pipes and vessels, strong coupling between the "supersonic" vibration modes (vibrations with wave speeds faster than the fluid wave speed) will transmit most of their energy to the fluid while the non-radiating "subsonic" remain in the structure dominating the remaining vibrations there. One can observe the net radiated power using a combination of modal filtering and structural intensity integration. Real-world structures are far more complex than the simplified "textbook" structural elements due to effect of holes, bolts, welds, ribs, etc. These discontinuities in structural impedance tend to scatter modes. For example, a low frequency bending mode which is subsonic (non-radiating) can spillover its stored energy into a supersonic mode at the discontinuity in the structure. Not surprisingly, required structural elements such as ribs, bolts, rivets, and weld seams tend to be radiated noise "hot spots" on the structure. The modal scattering at discontinuities also contributes to high dynamic stress making the hot spots susceptible to fatigue cracks, corrosion, and eventual structural failure. The vibration and intensity response of the structure will change as things like fatigue and corrosion reduce the stiffness of the structural elements. Therefore, in theory at least, an intelligent sensor monitoring system can detect signs of impending structural failure from the vibrational signature in time to save lives and huge sums of money.

P-waves in solids are much the same as acoustic waves except they can have three components in three dimensions. Each component has the force and propagation direction on the same axis. In Cartesian coordinates, each p-wave component is orthogonal with respect to the other components. For the direction unit vectors a_x, a_y, and a_z, the p-wave time averaged intensity for a sinusoidal p-wave is simply

$$\langle \vec{I}^p(\omega) \rangle_t = \frac{1}{2} \left[F_x^p u_x^{p*} a_x + F_y^p u_y^{p*} a_y + F_z^p u_z^{p*} a_z \right] \tag{6.3.23}$$

where the factor of $1/2$ can be dropped if rms values are used for the compressional forces and velocity conjugates. Generally, p-waves in solids are extremely fast compared to s-waves and most acoustic waves. This is because solids are generally quite stiff (hard) allowing the compressional force to be transmitted to great depths into the solid amost instantly. Torsional intensity is found in much the same manner except the shear force velocity in the θ direction produce an intensity vector in the orthogonal direction.

Bending waves on lossless beams, or s-waves, there are two kinds of restoring forces, shear and moment, and two kinds of inertial forces, transverse and rotary mass acceleration. Combining all forces (neglecting damping), we have a fourth-order wave equation

$$EI \frac{\partial^4 y}{\partial x^4} + \rho S \frac{\partial^2 y}{\partial t^2} = f(x, t) \tag{6.3.24}$$

where E is Young's modulus, I is the moment of inertia, ρ is the mass density, and S is the cross-sectional area. The transverse vibration is in the y direction and the wave propagation is along the beam in the x direction. To compute the intensity, one needs to measure the shear force $F^s = EI(\partial^3 y/\partial x^3)$, moment force $M^s = EI(\partial^2 y/\partial x^2)$, transverse velocity $u^s = (\partial y/\partial t)$, and rotary velocity $\Omega^s = (\partial^2 y/\partial x \partial y)$. Again, assuming peak values for force and velocity and a sinusoidal s-wave, the bending wave time-averaged intensity is

$$\langle I^s(\omega) \rangle_t = \frac{1}{2}(F^s u^{s*} + M^s \Omega^{s*}) = \frac{EI}{2} \left[\frac{\partial^3 y}{\partial x^3} \left(\frac{\partial y}{\partial t} \right)^* + \frac{\partial^2 y}{\partial x^2} \left(\frac{\partial^2 y}{\partial x \partial t} \right)^* \right] \tag{6.3.25}$$

where * indicates a complex conjugate of the time varying part of the bending wave only. The spatial derivatives needed for the shear and moment forces are generally measured assuming EI and using a small array of accelerometers and finite difference approximations, where the displacement is found by integrating the acceleration twice with respect to time (multiply by $-1/\omega^2$). However, other techniques exist such as using strain gauges or lasers to measure the bending action. Bending wave intensity, while quite useful for separating propagating power from the stored energy in standing wave fields, is generally quite difficult to measure accurately due to transducer response errors and the mixture of various waves always present in structures. Intensity measurements in plates and shells are even more tedious, but likely to be an important advanced sensor technique to be developed in the future.

Electromagnetic Intensity

Electromagnetic intensity is quite similar to acoustic intensity except that the power flux is defined as the curl of the electric and magnetic fields, $\mathbf{S} = \mathbf{E} \times \mathbf{M}$. While the acoustic pressure is a scalar, the electric field is a full vector in three dimensions as well as the magnetic field. In simplest terms, if the electric field is pointing upwards and the magnetic field is pointing to the right, the power flux is straight ahead. Measurement of the electric field simply requires emplacement of unshielded conductors in the field and measuring the voltage differences relative to a ground point. The $\mathbf{E}$ field is expressed in Volts per meter (V/m) and can be decomposed into orthogonal a_x, a_y, and a_z components by arranging three antennae separated along Cartesian axes a finite distance with the ground point at the origin. The *dynamic* magnetic field $\mathbf{H}$ can be measured using a simple solenoid coil of wire on each axis of interest. Note that static magnetic fields require a flux-gate magnetometer which measures a static offset in the coils hysteresis curve to determine the net static field. Flux-gate magnetometers are quite commonly used as electronic compasses. Three simple coils arranged orthogonally can be used to produce a voltage proportional to the time derivative of the magnetic flux Φ along each of three orthogonal axes. If the number of turns in the coil is N, the electromotive force, or emf voltage will be $v = -N\,d\Phi/dt$, where $\Phi = \mu \mathbf{H} S$, S being the cross-sectional area and μ the permeability of the medium inside the coil (μ for air is about $4\pi \times 10^{-7}$ Henrys per meter). Unlike a electric field sensor, the coil will not be sensitive to static magnetic fields. For Cartesian coordinates, the curl of the electromagnetic field can be written as a determinant.

$$\mathbf{E} \times \mathbf{H} = \begin{vmatrix} a_x & a_y & a_z \\ E_x & E_y & E_z \\ H_x & H_y & H_z \end{vmatrix} \tag{6.3.26}$$

which can be written out in detail for the time-averaged intensity assuming a sinusoidal wave.

$$\langle \mathbf{E} \times \mathbf{H} \rangle_t = \frac{a_x}{2}\left(E_y H_z^* - E_z H_y^*\right) + \frac{a_y}{2}\left(E_x H_z^* - E_z H_x^*\right)$$
$$+ \frac{a_x}{2}\left(E_x H_y^* - E_y H_x^*\right) \tag{6.3.27}$$

The electromagnetic intensity measurement provides the power flux in Watts per meter-squared at the measurement point, and if scanned, the closed surface integral can provide the net power radiation in Watts. The real part of the intensity can be used to avoid the nearfield effects demonstrated for the acoustic case which still hold geometrically for electromagnetic waves. Clearly, the nearfield for electromagnetic waves can extend for enormous distances depending on frequency. The speed of light is 3×10^8 m/s giving a 100 MHz signal a wavelength of 3 m. For a modern computer or other digital consumer electronic device, electromagnetic radiation can interfere with radio communications, aircraft navigation signals, and the operation of digital electronics. For hand-held cellular telephones, there can be potential health problems with the transmitter being in very close proximity to the head and brain, although these problems are extremely difficult to get objective research data due to the varying reaction of animal tissues to electromagnetic fields. However, for most frequencies of concern, a simple field voltage reading is not sufficient to produce an accurate measure of net power radiation in the farfield.

6.4   INTENSITY DISPLAY AND MEASUREMENT TECHNIQUES

The measurement of intensity can involve separate measurement of the potential and kinetic field components with specialized sensors, as long as sensor technologies are available which provide good field component separation. In acoustics, velocity measurements in air are particularly difficult. Typically one indirectly measures the acoustic particle velocity by estimating the gradient of the acoustic pressure from a finite difference approximation using two pressure measurements. The gradient estimation and intensity computation when done in the frequency domain can be approximated using a simple cross-spectrum measurement, which is why this technique is presented here. While we will concentrate on the derivation of the acoustic intensity using cross-spectral techniques here, one could also apply the technique to estimating the Poynting vector using only electric field sensors. We also present the structural intensity technique for basic compressional and shear waves in beams. Structural intensity measurement in shells and complicated structures are the subject of current acoustics research and are beyond the scope of this book.

We begin by deriving acoustic intensity and showing its relation to the cross spectrum when only pressure sensors are used. An example is provided using an acoustic dipole to show the complete field reconstruction possible from an intensity scan. Following the well-established theory of acoustic intensity in air, we briefly show the intensity technique in the frequency domain for simple vibrations in beams. Finally we will very briefly explore the Poynting vector field measurement and why intensity field measurements are important to adaptive signal processing.

Graphical Display of the Acoustic Dipole

A dipole is defined as a source which is composed of two distinct closely-spaced sources, or "monopoles", of differing phases. If the phases are identical and the spacing much less than a wavelength, the two identical sources can be seen to couple together into a single monopole. For simplicity, we consider the two sources to be of equal strength, opposite phase, and of small size with respect to wavelength ($r_0 \ll \lambda$, therefore $kr_0 \ll 1$). The pressure amplitude factor A in Eq. (6.3.20) can be simplified to $A = jk\rho c Q/4\pi$, where $Q = 4\pi r_0^2 u_0$ is the source strength, or volume velocity in m^3/sec. Figure 21 shows the pressure response in Pascals for two opposite phased 0.1 m diameter spherical sources separated by 1 m and radiating 171.5 Hz with 1 mm peak surface displacement. The sources are located at $y = 0$ and $x = \pm 0.5$ m and the net pressure field is found from summing the individual point source fields.

The classic dipole "figure-8" directivity pattern can be seen in Figure 22 by observing the zero pressure line along the $x = 0$ axis where one expects the pressures from the two opposite-phased sources to sum to zero. For the assumed sound speed of 343 m/sec, the sources are separated by exactly $1/2 \lambda$ so no cancellation occurs along the $y = 0$ axis. Figure 23 shows the pressure response at 343 Hz giving "4-leaf clover" directivity pattern.

The velocity field of a dipole is also interesting. Using Eq. (6.3.19), we can calculate the velocity field of each source as a function of distance. In the $x - y$

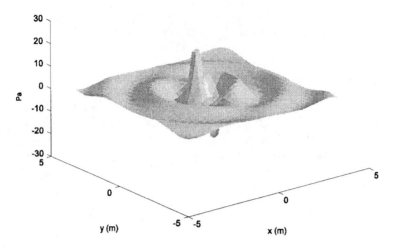

Figure 22 Sound pressure field for a 171.5 Hz opposite-phased acoustic dipole separated by 1 m on the $y = 0$ axis.

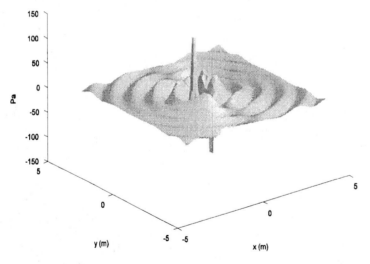

Figure 23 At 343 Hz, a 4-lobe directivity pattern appears due to the additional cancellation along the $y = 0$ axis.

plane of the dipole, the velocity field is decomposed into x and y components for each source and then summed according to the phase between the sources. The resulting field is displayed using a vector field plot as seen in Figure 24 for 171.5 Hz. For each of the measured field points, the direction of the vectors indicates the particle velocity direction, and the length of the vectors indicates the relative magnitude of the particle velocity. For a 1 mm peak displacement of the 10 cm diameter spherical sources at 171.5 Hz, the surface velocity is 1.078 m/sec and the Q is 0.0338 m^3/sec. The vector field in Figure 24 is scaled for the best display and, like the pressure field in Figure 23, represents a "snapshot" of the velocity field at one time instant.

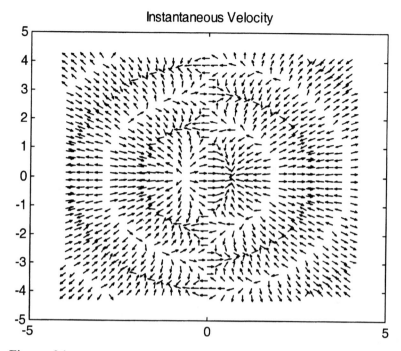

Figure 24 Instantaneous particle velocity of the 171.5 Hz dipole clearly showing elliptical motion along the null axis of $x = 0$.

One can compute, point by point, the instantaneous intensity $I(x,y,t) = p(x,y,t) u(x,y,t)$. However, the direction of real power flow is not always the same as the velocity because both pressure and velocity are complex, and because the product of a negative velocity and negative pressure yields a positive instantaneous intensity. In Figure 25 we purposely change the sign of the source intensity field for the $\langle +0.5, 0 \rangle$ source to show the details of the dipole field. Figure 26 shows another representation of the field using time-averaged superposition.

Neither Figure 25 nor Figure 26 are technically correct representations of the actual intensity field! However, they can be seen as quite useful in describing the power flow between the two sources. It is expected that the instantaneous intensity field in Figure 25 would be zero where either the pressure or velocity is zero, but this can be misleading since the waves are propagating outward from the sources. The time-averaged intensity is found for sinusoidal excitation using $\langle I(x,y) \rangle_t = \frac{1}{2} p(x,y,t) u^*(x,y,t)$, where the real part represents the active intensity. Computing the time-averaged intensity for each source in the dipole and then superpositioning them in opposite phase yields the vector plot in Figure 26, which is physically intuitive since one can visualize the arrows oscillating back and forth with the phases of the two sources. However, Figure 26, like Figure 25, is misleading since it is neither a true time-average field nor does power actually flow into the source at $x = \langle +0.5, 0 \rangle$. To graphically depict the dipole intensity field, we first need a model for the total pressure and velocity fields in the radial direction for the dipole. It is straightforward to show that for a separation distance d on the y axis symmetric about the origin the

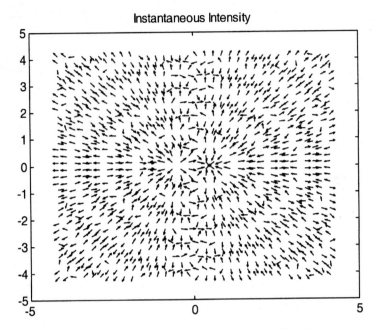

Figure 25 Instantaneous intensity for the 171.5 Hz dipole (vector lengths are on a logarithmic scale to show detail).

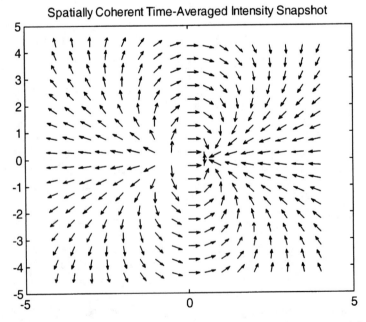

Figure 26 Superposition of the opposite-phased time-averaged intensities of the two sources in the 171.5 Hz dipole.

total pressure field is

$$p(r, \theta, t) = j\frac{k\rho c Q e^{j(\omega t - kr)}}{4\pi r}\left[e^{+j\frac{kd}{2}\cos\theta} - e^{-j\frac{kd}{2}\cos\theta}\right] \tag{6.4.1}$$

where $r = (x^2 + y^2)^{1/2}$ and θ is measured counterclockwise from the positive x-axis. The radial component of velocity for the complete dipole field is

$$u(r, \theta, t) = j\frac{kQe^{j(\omega t - kr)}}{4\pi r}\left(1 - j\frac{i}{kr}\right)\left[e^{+j\frac{kd}{2}\cos\theta} - e^{+j\frac{kd}{2}\cos\theta}\right] \tag{6.4.2}$$

Finally, the time averaged intensity for the dipole field is found using $\langle I(r, \theta)\rangle_t = \frac{1}{2}$ $p(r, \theta, t)\, u^*(r, \theta, t)$ as seen in Eq. (6.4.3). We would drop the factor of $1/2$ for rms values of $p(r, \theta, t)$ and $u(r, \theta, t)$ rather than peak values.

$$\langle I(r, \theta)\rangle_t = \frac{k^2 c Q^2}{8\pi^2 r}\left(1 - j\frac{i}{kr}\right)\sin^2\left(\frac{kd}{2}\cos\theta\right) \tag{6.4.3}$$

A vector plot of the correct time-averaged dipole intensity along the radial direction at 171.5 Hz is seen in Figure 27. We do not include the circumferential direction intensity component since its contribution to radiated power on a spherical surface is zero. Figure 27 correctly shows radiated power from both sources and a null axis along $x = 0$ corresponding to the zero pressure axis seen in Figure 22 and elliptical particle velocities in Figure 24.

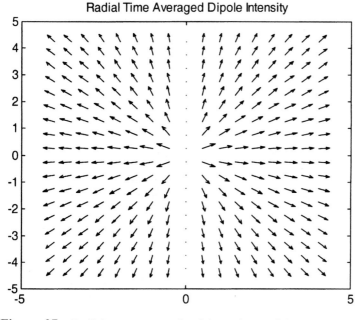

Figure 27 Radial component only of the real part of the acoustic intensity of the 171.5 Hz dipole.

It is important to understand the graphical display techniques in order to obtain correct physical insight into the radiated fields being measured. The vector plots in Figures 24 through 27 are typically generated from very large data sets which require significant resources to gather engineering-wise. Even though one is well inside the nearfield, the net radiated power can be estimated by integrating the intensity over a surface enclosing the sources of interest. Any other noise sources outside the enclosing scan surface will not contribute to the radiated power measurements. This surface is conveniently chosen to be a constant coordinate surface such as a sphere (for radial intensity measurements) or a rectangular box (for x-y-z intensity component measurements). The intensity field itself can be displayed creatively (such as in Figure 25 and 27) to gain insight into the physical processes at work. Intensity field measurements can also be used as a diagnostic and preventative maintenance tool to detect changes in systems and structures due to damage.

Calculation of Acoustic Intensity From Normalized Spectral Density

Given spectral measurements of the field at the points of interest, the time-averaged intensity can be calculated directly from the measured spectra. However, we need to be careful to handle spectral density properly. Recall that twice the bin magnitude for the normalized discrete Fourier transform (NDFT, or DFT divided by the size of the transform N) with real data input gives the peak amplitude at the corresponding bin frequency. Therefore, the square root of 2 times the NDFT bin amplitude gives a rms value at the corresponding bin frequency. When the necessary field components are measurable directly, the calibrated rms bin values can be used directly in the field calculations resulting in vector displays such as that in Figure 27. When spatial definitives are required to compute field components only indirectly observable (say via pressure or structural acceleration sensors only), some interesting spectral calculations can be done. For the acoustic intensity case, two pressure sensors signals $p_1(\omega)$ and $p_2(\omega)$ are separated by a distance Δr, where $p_2(\omega)$ is in the more positive position. From Euler's equation given here in Eq. (6.3.10), the velocity can be estimated at a position inbetween the two sensors from the pressure by using a spatial derivative finite difference approximation.

$$u(\omega) = \frac{-1}{\rho} \int \frac{\partial p(\omega)}{\partial r} dt \approx j \frac{p_2(\omega) - p_1(\omega)}{\omega \rho \Delta r} \tag{6.4.4}$$

The pressure sensors need to be close together for the finite difference approximation to be accurate, but if they are too close, the finite difference will be dominated by residual noise since the sensor signals will be nearly identical. Somewhere between about $1/16$ and $1/4$ of a wavelength spacing can be seen as a near optimum range. The pressure used in the intensity calculation is estimated at the center of the sensors by a simple average. The rms intensity is simply

$$
\begin{aligned}
I(\omega) &= \frac{1}{2}\Big\langle p(\omega)u(\omega)^*\Big\rangle_t = \frac{1}{2}\Big\langle p(\omega)^*u(\omega)\Big\rangle_t \\
&= \frac{-j}{4\omega\rho\Delta r}\Big\langle \big(p_1(\omega) + p_2(\omega)\big)\big(\big(p_2(\omega)^* - p_1(\omega)^*\big)\big)\Big\rangle_t \\
&= \frac{-j}{4\omega\rho\Delta r}\Big\langle \big(p_1(\omega)p_2(\omega)^* + |p_2(\omega)|^2 - |p_1(\omega)|^2 - p_2(\omega)p_1(\omega)^*\big)\Big\rangle_t
\end{aligned}
\tag{6.4.5}
$$

Noting that both $p_1(\omega)$ and $p_2(\omega)$ have real and imaginary components, i.e. $p_2(\omega) = p_2^R(\omega) + j\,p_2^I(\omega)$, the real part of the intensity, or *active intensity*, is

$$I^R(\omega) = \frac{1}{2\omega\rho\Delta r}\langle p_1^I(\omega)p_2^R(\omega) - p_1^R(\omega)p_2^I(\omega)\rangle_t \qquad (6.4.6)$$

which represents the propagating power in the field in the units of Watts per meter-squared (W/m^2). The active intensity can be very conveniently calculated using the negative of the imaginary part of the cross-spectrum $G^{12}(\omega) = \langle p_1^*(\omega)p_2(\omega)\rangle_t$

$$I^R(\omega) \approx \frac{-Im\{G^{12}(\omega)\}}{2\omega\rho\Delta r} = \frac{\langle p_2^R(\omega)p_1^I(\omega) - p_1^R(\omega)p_2^I(\omega)\rangle_t}{2\omega\rho\Delta r} \qquad (6.4.7)$$

The imaginary part of the intensity, or *reactive intensity* depicts the stored energy in the field and is given in Eq. (6.4.8).

$$I^I(\omega) = \frac{1}{4\omega\rho\Delta r}\langle |p_1(\omega)|^2 - p_2(\omega)|^2\rangle_t \qquad (6.4.8)$$

The reactive intensity provides a measure of the stored energy in the field, such as the standing wave field in a duct or the nearfield of a dipole. Since the energy stored in the field eventually is coupled to the radiating part of the field, the transient response of a noise source can be significantly affected by the reactive intensity. For example, a loudspeaker in a rigid-walled enclosure with a small opening radiating sound to a free space would reproduce a steady-state sinusoid well, but be very poor for speech intelligibility. The standing-waves inside the enclosure would store a significant part of the loudspeaker sound and re-radiate it as the modes decay causing words to be reverberated over seconds of time. The nearfields of active sources for sonar and radar can also have their impulse responses affected by the stored energy in the reactive intensity field. The term nearfield generally refers to any component of the field in the vicinity of the source which does not propagate to the far-field. This could be the evanescent part of the velocity field as well as the "circulating" components of the pressure field which propagate around the source rather than away from it. One can consider the nearfield as an energy storage mechanism which can effect the farfield transient response of a transmitter source.

Calculation of Structural Intensity for Compressional and Bending Waves

Calculation of structural intensity most commonly uses finite-differences to estimate the force and velocity from spatial acceleration measurements and known physical parameters of the structure material, such as Young's modulus E and the moment of inertia I. Structural vibrations are enormously complex due to the presence of compressional, shear, and torsional vibrations in three dimensions. At discontinuities (edges, holes, joints, welds, etc.) each kind of wave can couple energy into any of the others. The finite size of a structure leads to spatial modes of vibration, which with low damping are dominated by one frequency each, and again can couple to any other mode at an impedance discontinuity. Measurement of structural vibration power is not only important from a noise/vibration control point of view, but also from a fatigue monitoring and prognosis view. Controlling the dissipation of fatigue causing vibrations in a structure will have important economical implications for every thing from vehicles to high performance industrial equipment. The technical issues with structural intensity measurement are analogous to the acoustic case in that one wants the sensors as close together as possible for accurate finite difference approximations, yet far enough apart to yield high signal-to-noise ratios in the finite difference approximations. For simplification one generally assumes plane waves. Separation of, say, shear and compressional waves involves complicated sensor arrays for wavenumber separation through the use of wavenumber filtering and "beamforming", discussed later in this text. We note that newer sensor technologies in the area of laser processing and micro-machined strain-gauges offer the potential to eliminate the finite difference approximations, and the associated systematic errors.

The type of sensor preferred for a particular vibration measurement depends on the frequency range of interest, wave speed, and effect of the sensor mass on the vibration field. At very low frequencies (say below the audio range of around 20 Hz), displacement sensing is preferred because velocity is quite small and acceleration even smaller (they scale as $j\omega$ and $-\omega^2$, respectively). Typical very low frequency/static displacement sensors would be capacitive or magnetic proximity transducers, strain gauges, or optical sensors. At medium to low frequencies, velocity sensors such as geophones are preferred since a moving coil in a magnetic field can be made to have very high voltage sensitivity to a velocity excitation in the range of a few Hz to several hundred Hz. Above a few hundred Hz, accelerometers are preferred due to their high sensitivities and low cost. An accelerometer is essentially a strain gauge with a known "proof" mass attached to one side. The known mass produces a traceable stress on the accelerometer material when accelerated along the common axis of the accelerometer material and mass. The accelerometer material is usually a piezoelectric material such as natural quartz or the man-made lead-zirconate-titanate (PZT) ceramic. These materials produce a small voltage potential when mechanical stress in applied. In recent years, a new class of accelerometer has become widely available making use of micro-machined strain gauge technologies. Micromachining (accomplished chemically rather than with micro-milling) allows very robust and low cost manufacture of sensors with precise properties and integrated signal conditioning electronics. For light weight structures, the mass of an attached sensor is very problematic due to its effect of the vibration field. For light weight structural vibration measurement, non-contact laser vibrometers are the most common sensors.

Fundamental to all intensity measurement is the need to simultaneously measure force and velocity at a field position. The preferred approach is to employ a sensor which directly measures force and another sensor which directly measures velocity. In heavy structures, velocity is easily observed directly using a geophone, by a simple time integration (low pass filter with 6 dB/octave rolloff) of acceleration, or by a time differentiation (high pass filter with 6 dB/octave rollup) of measured displacement. Force measurement can be done with a direct strain gauge measurement which is usually a thin metal film attached under tension so that its electrical resistance changes with the resulting strain in the structure. However, an easier approach to force measurement is to use the known material properties of the structure to relate the spatial difference in, say, acceleration, to the force. For example, Young's modulus E for a particular material is known (measured a priori) and has units of Nt/m^2 (often listed as pounds per square-inch — psi) or the ratio of longitudal stress (Nt/m^2) over longitudal strain (change in length over initial length). Given two accelerometers spaced by Δ meters, the difference in acceleration spectra, $A_2(\omega) - A_1(\omega)$, where point 2 is further along the positive x-axis than point 1, can be multiplied by $-1/(\Delta\omega^2)$ to give the longitudal strain spectrum. The structure cross-sectional area S times Young's modulus times the longitudal strain spectrum gives the compressional force spectrum, or p-force wave.

$$F^p(\omega) = \frac{ES}{\Delta\omega^2}\{A_1(\omega) - A_2(\omega)\} \tag{6.4.9}$$

The compressional force spectrum in Eq. (6.4.9) represents the force between the two accelerometers oriented along the structure (usually referred to as a rod for this type of structural vibration). The velocity is simply the average of the two estimated velocities from the acceleration signals.

$$u^p(\omega) = \frac{1}{2j\omega}\{A_1(\omega) + A_2(\omega)\} \tag{6.4.10}$$

The rms compressional wave intensity in a rod is therefore

$$I^p(\omega) = \frac{ES}{4\Delta j\omega^3}\{|A_1|^2 - A_2A_1^* + A_2^*A_1 - |A_2|^2\} \tag{6.4.11}$$

where the factor of $1/2$ can be dropped if rms calibrated acceleration spectra are used (multiply by 2). Equation (6.4.11) provides a one-dimensional intensity p-wave which can be integrated into the 3-dimensional compressional time-averaged intensity for solids in Eq. (6.3.23)

The largest source of error in Eq. (6.4.11) comes from the acceleration differences in Eq. (6.4.9), which can be near zero for long wavelenths and short accelerometer separations. The compressional wave speed can be quite fast and is calculated by the square-root of Young's modulus divided by density. For a given steel, if $E = 50 \times 10^{10}$ and density $\rho = 500$ kg/m³, the compressional wave speed is 31,623 m/sec, or nearly 100 times the speed of sound in air. Therefore, the accelerometers need to be significantly separated to provide good signal to noise ratio (SNR) for the intensity measurement.

Bending wave intensity follows from Eq. (6.3.25) where the various components of shear are calculated using finite difference approximations for the required spatial derivatives. The time-averaged bending wave intensity in a structural beam is seen to be

$$\langle I^s(\omega)\rangle_t = \frac{1}{2}(F^s u^{s*} + M^s\Omega^{s*}) = \frac{EI}{2}\left[\frac{\partial^3 y}{\partial x^3}\left(\frac{\partial y}{\partial t}\right)^* + \frac{\partial^2 y}{\partial x^2}\left(\frac{\partial^2 y}{\partial x\partial t}\right)^*\right] \tag{6.4.12}$$

where again, the factor of $1/2$ may be dropped if rms spectra are used. Typically one would use accelerometers to measure the beam response to bending waves where the displacement $y(x)$ is $-1/\omega^2$ times the acceleration spectra. One could estimate the spatial derivatives using a 5-element linear array of accelerometers each spaced a distance Δ meters apart, to give symmetric spatial derivative estimates about the middle accelerometer, $A_3(\omega)$. Consider the accelerometers to be spaced such that

$A_5(\omega)$ is more positive on the x-axis than $A_1(\omega)$, etc. Using standard finite difference approximations for the spatial derivatives, the bending (shear wave) rms intensity is

$$I^s(\omega) = j\frac{EI}{4\omega^3\Delta^3}\left\{(A_5 - 2A_4 + 2A_2 - A_1)A_3^* - (A_4 - 2A_3 + A_1)(A_4^* - A_2^*)\right\}$$

$$(6.4.13)$$

where the accelerations $A_i(\omega)$ are written as A_i, $i = 1,2,3,4,5$; to save space. Clearly, the bending wave intensity error is significant by finite difference approximation errors which happen if the sensors are widely spaced (Δ must always be less than $\lambda/2$), and by SNR if Δ is too small. The bending wave wavelength for a beam with moment of inertia I, density ρ, cross-sectional area S, Young's modulus E, and frequency ω, is seen in Eq. (6.4.14).

$$\lambda = \frac{2\pi^4\sqrt{EI}}{\sqrt[4]{\rho S\omega^2}} = \frac{2\pi}{\sqrt{\omega}} \cdot \left(\frac{EI}{es}\right)^{1/4}$$

$$(6.4.14)$$

Clearly, the bending wavelength is much slower than the compressional wave. But even more interesting, is that the bending wave speed can be seen as a function of frequency to be slower at low frequencies than at higher frequencies. This is because a given beam of fixed dimensions (moment of inertia) is stiffer to shorter wavelengths than it is to longer wavelengths. While measurement of vibrational power flow (bending or compressional) is of itself valuable, changes in vibrational power flow and wavelength due to material changes could be of paramount importance for monitoring structural integrity and prediction of material fatigue.

Calculation of The Poynting Vector

Electromagnetic sensors can rather easily detect the electric field with a simple voltage probe and the alternating magnetic field with a simple wire coil. The static magnetic field requires a device known as a flux-gate magnetometer, where filtering a measured hysteresis response of a coil provides an indirect measurement of the static magnetic field. For strong magnetic fields in very close proximity to magnetic sources, a Hall effect transistor can be used. We are interested here in the process of measuring electrical power flow in fields using the intensity technique, or Poynting vector. While one could use finite difference approximations with either electric or magnetic field sensors, it is much simpler to directly observe the field using a directional voltage probe and search coil. The electric field is measured in volts per meter and can be done using a pair of probes, each probe exposing a small unshielded detector to the field at a known position. Probe voltage differences along each axis of a cartesian coordinate system normalized by the separation distances provide the 3-dimensional electric field vector, expressed as a spectrum where each bin has three directional components. This information alone is sufficient to resolve the direction of wave propagation with surprising accuracy. The time derivative of the magnetic field can be indirectly estimated from the curl of the electric field as seen in Eq. (6.3.27), allowing the magnetic field to be estimated from spatial derivatives of the electric field. However, the presence of a solenoid coil is not problematic for the field measurement (back emf from currents in the coil may affect the field to be measured), a direct magnetic field measurement can be made

$$H(\omega) = j\frac{\zeta(\omega)}{NA\mu\omega} \tag{6.4.15}$$

where the solenoid has N circular loops with average cross sectional area A. The parameter $\zeta(\omega)$ in Eq. (6.4.15) is the voltage produced by the coil due to Faraday's law. The simple and effective use of coils to measure magnetic field is preferable due to the ability to detect very weak fields, while finite-difference approximations to the electric field would likely produce very low SNR. Also, for a given level of magnetic field strength, one can see from Eq. (6.4.15) that a large coil with a large number of turns could produce substantial voltage for easy detection by a sensor system. However, too many turns in the coil will produce too much inductance for easy detection of high frequencies. Direct measurement of electric and magnetic fields allows one to compute the electromagnetic intensity using Eq. (6.3.27) with spectral products in the frequency domain.

6.5 SUMMARY, PROBLEMS, AND BIBLIOGRAPHY

Most applications of the Fourier transform are for the purposes of steady-state sinusoidal signal detection and analysis, for which, the Fourier transform provides an enhancement relative to random noise. This is because of the orthogonality of sinusoids of different frequencies in the frequency domain. Provided that the time domain signal recording is long enough, conversion to the frequency domain allows the levels of different frequencies to be easily resolved as peaks in the magnitude spectrum output. Peaks in the time domain are best detected in the time domain because the broad spectral energy would be spread over the entire spectrum in the frequency domain. For non-stationary frequencies, the length of time for the Fourier transform should be optimized so that the signal is approximately stationary (within the transforms frequency resolution). For most signal detection applications, we are interested in the statistics of the Fourier spectrum bin mean and standard deviation, given that the spectrum is estimated by averaging many Fourier transforms together (one assumes an ergodic input signal where each averaged buffer has the same underlying statistics). If the input signal is zero-mean Gaussian (ZMG) noise, both the real and imaginary bins are ZMG processes. The magnitude-squared of the spectral bin is found by summing the real part squared plus the imaginary part squared and results in a 2nd-order chi-square probability density function (pdf). For a time domain ZMG signal with σ_t^2 variance, a N-point normalized Fourier transform produces a magnitude-squared frequency bin with mean σ_t^2/N and variance σ_t^4/N^2. As one averages M Fourier magnitude-squared spectra, the mean for each bin stays the same, but the variance decreases as $\sigma_t^4/(MN^2)$ as M increases. The pdf for the averaged spectral bin is chi-square of order $2M$. Given the underlying statistics of averaged spectra, one can assign a confidence interval where the bin value is say, 90% of the time. Examination of higher-order spectral statistics is also very useful in extracting more information about the spectrum, such as nonlinearities as seen in the bispectrum.

The transfer function of a linear time-invariant system can be measured by estimating the ratio of the time-averaged input–output cross-spectrum divided by the input autospectrum. Again, the statistics of the spectral bins plays a crucial role in determining the precision of measured system magnitude and phase. The spectral coherence is a very useful function in determining the precision of transfer function measurements as it is very sensitive to showing weak SNR as well as interfering signals in the input–output signal path. Using the observed coherence and the known number of averages, one can estimate a variance for the transfer function magnitude and phase responses. This error modeling is important because it allows one to determine the required number of spectral averages to achieve a desired precision in the estimated transfer function.

However, if the transfer function has very sharp modal features (spectral peaks and/or dips in the frequency response), the input–output buffers may have to be rather long for consistent ergodic averages. This is because the buffer must be long enough to measure the complete system response to the input. With random input signals in each buffer, some of the "reverberation" from previous input buffers are not correlated with the current input, giving transfer function results which do not seem to improve with more averaging. This requires switching to a broadband periodic input signal such as an impulse or sinusoidal chirp, which is exactly reproduced in each input buffer. The magnitude response for a synchronous periodic input will quickly approach the spectral resolution limit define by the buffer length, but some phase errors may still be present due to the "reverberation" from previous buffers. Synchronous input signals also allow one to average in the time domain, or average the real and imaginary spectral bins separately, before computing the transfer function and magnitude and phase. This "time-synchronous averaging" virtually eliminates all interfering signals leaving only the input and output to be processed and is an extremely effective technique to maximize SNR on a difficult measurement.

While the details to which we have explored acoustic, structural, and electromagnetic intensity here are somewhat involved, we will make use of these developments later in the text. The intensity techniques all generally involve spectral products, cross-spectra, or various forms of finite difference approximations using measured spectra. The underlying statistics of the spectral measurements translate directly into the precision of the various power flow measurements. Detailed measurements of the full field through active and reactive intensity scans can provide valuable information for controlling radiated power, optimizing sensor system performance, or for diagnostic/prognostic purposes in evaluating system integrity. Clearly, the most sophisticated adaptive sensor and control systems can benefit from the inclusion of the physics of full field measurements.

PROBLEMS

1. Broadband electrical noise is known to have white spectral characteristics, zero mean, and a variance of 2.37 v^2. A spectrum analyzer (rectangular window) is used to average the magnitude squared of 15 FFTs each 1024 points in size. What is the mean and standard deviation of on of the spectral magnitude-squared bins? If the sample rate is 44.1 kHz, whats the noise level in $\mu v^2 / Hz$?

2. Suppose a Hanning data window was used by mistake for the data in question 1. What would the measured spectral density be for the noise?

3. A transfer function of a transducer is known to have a very flat response. Would you expect a reasonable measurement of the transfer function using a rectangular window?

4. A transfer function is measured with poor results in a frequency range where the coherence is only 0.75. How many spectral averages are needed to bring the standard deviation of the transfer magnitude to with 5% of the actual value?

5. Given the measured intensity in 3-dimensions over at random points on a closed surface around a source of interest, how does one compute the total radiated power? If the source is outside the closed surface, what is the measured power?

BIBLIOGRAPHY

W. S. Burdic Underwater Acoustic System Analysis, 2nd ed., Chapters 9 and 13, Englewood Cliffs: Prentice-Hall, 1984.

G. W. Elko "Frequency Domain Estimation of Complex Acoustic Intensity and Acoustic Energy Density", PhD dissertation, The Pennsylvania State University, University Park, PA, 1984.

F. J. Fahy Sound Intensity, New York: Elsevier Science, 1989.

W. H. Press, B. P. Flannery, S. A. Teukolsky, W. T. Vetterling Numerical Recipes: The Art of Scientific Computing, New York: Cambridge University Press, 1986.

F. W. Sears, M. W. Zamansky, H. D. Young University Physics, Reading: Addison-Wesley, 1977.

REFERENCE

1. J. S. Bendat, A. G. Piersol Engineering Applications of Correlation and Spectral Analysis, New York: Wiley, 1993.

7

Wavenumber Transforms

The wavenumber k is physically the radian frequency divided by wave speed (ω/c), giving it dimensions of radians per meter. Wavenumbers can be seen as a measure of wavelength relative to 2π. Actually, k also equals $2\pi/\lambda$ and can be decomposed into k_x, k_y, and k_z components to describe the wave for 3-dimensional spaces. The wavenumber k is the central parameter of *spatial signal processing* which is widely used in radar, sonar, astronomy, and in digital medical imaging. Just as the phase response in the frequency domain relates to time delay in the time domain, the phase response in the wavenumber domain relates to the spatial position of the waves. The wavenumber transform is a Fourier transform, where time is replaced by space. Wavenumber processing allows spatial and even directional information to be extracted from waves sampled with an array of sensors at precisely known spatial positions. The array of sensors could be a linear array of hydrophones towed behind a submarine for long range surveillance, a line of geophones used for oil exploration, or a two-dimensional phased-array radar such as the one used in the Patriot missile defense system. The most common two-dimensional signal data is video which can benefit enormously from wavenumber processing in the frequency domain. Wavenumber processing can be used to enhance imagery from telescopes, digitized photographs, and video where focus or camera steadiness is poor.

We begin our analysis of wavenumber transforms by defining a plane wave moving with speed c m/sec in the positive r direction as being of the form $e^{j(\omega t - kr)}$. With respect to the origin and time t, the wave is outgoing because the phase of the wave further out from the origin corresponds to a time in the past. As t increases a constant phase point on the wave propagates outward. If $r = (x^2 + y^2 + z^2)^{1/2}$ represents a distance relative to the origin, the respective wavenumber components k_x, k_y, and k_z in the x, y, and z directions can be used to determine the direction of wave propagation. Recall that to measure temporal frequency, a signal is measured at one field point and sampled at a series of known times allowing a time Fourier transform on the time series to produce a frequency spectrum. To measure the wavenumber spectrum, the waveform is sampled at one time "snapshot" over a series of known field positions allowing a spatial Fourier transform to be applied, producing a wavenumber spectrum. Since the wavenumber is $2\pi/\lambda$, long wavelengths correspond to small wavenumbers and short wavelength correspond to high wavenumber.

Consider the pinhole camera in Figure 1. On any given spot on the outside of the camera box, the light scattered from the field produces a "brightness" wavenumber spectrum. Another way to look at the physics is that light rays from every object in the field can be found on any given spot on the outside of the camera box. By making a pin hole, the camera box allows reconstruction of the object field as an upside-down image on the opposite side of the pinhole.

One can describe the magic of the pinhole producing an image inside the camera as a 2-dimension inverse wavenumber transform where the wavenumber spectrum at the pinhole spot is $I(k_x, k_y)$.

$$i(x, y) = \int_{-\infty}^{+\infty} \int_{-\infty}^{+\infty} I(k_x, k_y)e^{j(k_x x + k_y y)}dx\, dy \qquad (7.0.1)$$

The ray geometry which allows reconstruction of the image is represented by the wavenumber components k_x and k_y for the image coordinates x and y. Small wavenumbers near zero correspond to light with slowly varying brightness (long wavelengths) across the pinhole side of the camera while the large wavenumbers correspond to sharply varying brightness such as edges from shadows.

Some of the earliest cameras used a simple pinhole and extremely long exposures on the photographic plate due to the low light levels. Lens-type cameras are far more efficient at gathering and focusing light. The shape and index of refraction for the lens bends the light rays creating a focal point and an inverted image. The focal point is geometrically analogous to the pinhole, but light is highly concentrated there by the lens. This allows much faster exposure times. The amount of light is controlled by the aperture which is also referred to as the entrance pupil in most optics texts. A border usually forming a rectangular boundary around the photographic film is the exit pupil. Dilating or constricting the aperture does

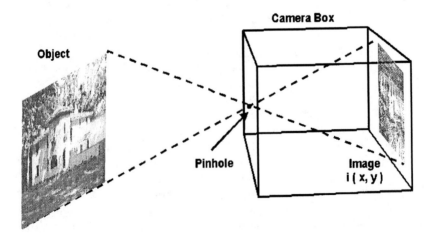

Figure 1 A pinhole camera produces a clear upside down image due to the coherence of the light rays through the small aperture, but not much light gets to the image plane.

not effect the exit pupil boundary, but rather just impacts the brightness of the image.

As can be seen in Figure 2, the aperture can have an effect on focus of objects near and far. Note that for the far object parallel rays enter the lens and are refracted depending on the angle of incidence. A near object produces a focal point closer than a far object, therefore far objects in the background can appear out of focus when a wide aperture is used. Narrowing the aperture will tend to allow only the nearly parallel rays into the camera thus improving the focus of both near and far objects. The entire field of view is in focus for the pinhole camera. Focus can be seen as the situation where the wavenumbers are coherently "filtered" by the camera to faithfully reproduce a sharply-defined image. The "fuzzyness" of out-of-focus objects is really due to "leakage" of wavenumber components into other areas of the image surrounding the "true" intended spot. The physics to be noted here is that "focus" for a camera system is definable as a wavenumber filter and that out-of-focus objects can (in theory) be digitally recovered if the lens system wavenumber filtering response is known. The transfer function of the lens system is quite complicated and will change depending on aperture, focus, and also for objects near and far.

Before the corrective optics were installed in the orbiting Hubble Space Telescope, the focus of initial images was controlled using Fourier image processing. We will examine this technique in detail below. In short, the "fuzzyness" in a astronomical image can be dealt with by picking a distant star in the field of view. Under ideal conditions, this distant star might be detected in only one pixel. However,

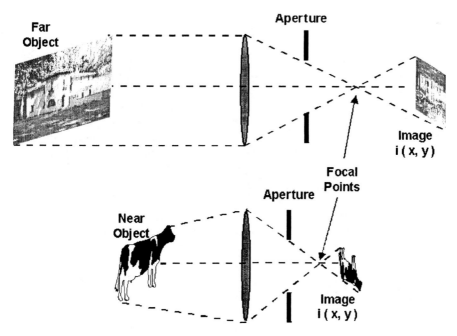

Figure 2   A lens system increases the amount of light for the photograph, but also introduces a depth of focus field.

the reality in the early days of Hubble was that the distant star's light would smear over all the in a given area of the image in what photographers sometimes call "circles of confusion" of poor focus. From a signal processing point of view, the circles of confusion can be seen as a spatial impulse response of a wavenumber filter. Performing a 2-dimensional Fourier transform of the spatial impulse response, or *point spread function* gives the wavenumber frequency response. The optimal sharpening filter for the distant star is found from inverting the wavenumber frequency response for the fuzzy distant star and multiplying the entire image's wavenumber frequency response by this corrective wavenumber filter. The inverse 2-dimensional Fourier transform of the corrected wavenumber frequency response gives an image where the formerly fuzzy distant star appears sharp along with the rest of the image (assuming the fuzzyness was homogeneously distributed). The technique is analogous to measuring and inverting the frequency response of a microphone to calculate the optimal FIR filter via inverse 1-dimensional Fourier transform which will filter the microphone signals such that the microphone's frequency response appears perfectly uniform. Astronomers have been applying these very powerful image restoration techniques for some time to deal with atmospheric distortions, ray multipaths, and wind buffeting vibrations of ground-based telescopes.

7.1 SPATIAL FILTERING AND BEAMFORMING

A very common coherent wavenumber system is the parabolic reflector used widely in radar, solar heating, and for acoustic sensors. Consider an axi-symetric paraboloid around the y-axis with the equation $y = ar^2 + b$, where r is a radial measure from the y-axis. To find the focal point, or focus, one simply notes where the slope of the paraboloid is unity. Since the angle of incidence equals the angle of the reflected wave, the rays parallel to the y-axis reflect horizontally and intersect the y-axis at the focus (see the lightly dotted line in Figure 3). This unity slope ring on the parabola will be at $r = 1/(2a)$ giving a focus height of $y = 1/(4a) + b$. For rays parallel to the y-axis, all reflected rays will have exactly the same path length from a distant source to the focus. Physically, this has the effect of integrating the rays over the dish cross-section of diameter d into the sensor located at the focus (1). This gives a wavelength λ dependent gain of $1 + 2d/\lambda$, which will be presented in more detail in Section 12.2. For high frequencies where the gain is quite large, the dish provides a very efficient directional antenna which allows waves from a particular direction to be received with high SNR. At low frequencies, the directional effect and gain are quite small. This gain defines a "beam" where the parabolic reflector can be aimed towards a source of interest for a very high gain communication channel.

The "beam" created by the parabolic reflector is the result of coherent wavenumber filtering. All waves along the "look direction" of the beam add coherently at the focus F in Figure 3. Waves from other directions will appear as well at the focus, but much weaker relative to the look direction beam. For example, some waves will scatter in all directions at the reflector's edges but the area for this scattering is extremely small compared to the area of the reflector as a whole which produces the coherent reflections for the look direction beam.

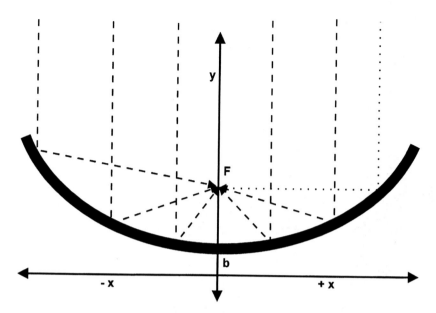

Figure 3 Parabolic reflectors focus substantial wave energy from one direction (one range of wavenumbers) onto the sensor.

Parabolic reflectors are unique in that the amplification and beam geometry are identical for all frequencies provided the reflector can give a near perfect reflection of the wave. Therefore, the material of the parabolic reflector plays an important role in defining the frequency range of useful operation. For example, a mirror finish is required for visible light frequencies, a conductive wire mesh will suffice for radar frequencies, and a solid dense material such as plastic will provide an atmospheric acoustic reflection. For underwater applications, one might provide an air voulme in the shape of a parabolic dish to achieve a high reflection. The parabolic shape is somewhat problematic to manufacture, and often a spherical shape is used as an approximation. The parabolic reflector can be very useful as a movable "search beam" to scan a volume or plane for wave emitting sources or reflections from an adjacent active transmitter allowing the locations of the reflecting sources to be determined. As will be discussed below, applications to direction-finding and location of sources or wave scatterers is a major part of wavenumber filtering.

Moving of the parabolic dish mechanically is sometimes inconvenient and slow, so many modern radar systems electronically "steer" the search beam. These operations are generally referred to as "array processing" because an array of sensors is used to gather the raw wavenumber data. The amount of computation is generally alarming for electronically steered beams, but the operations are well-suited for parallel processing architectures. Current microprocessors provide a very economical means to steer multiple search beams electronically as well as design beams which also have complete cancellation of waves from particular directions.

Consider a "line array" of acoustic sensors (radio frequency antennae could also be used) as seen in Figure 4. where 3 distant sources "A", "B", and "C" over

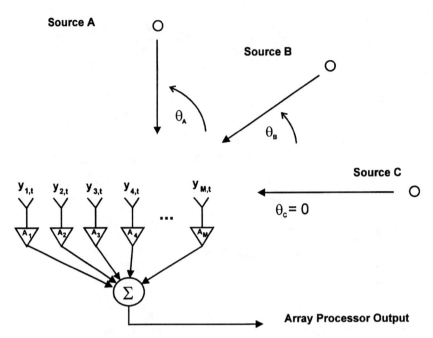

Figure 4 A line array of sensors can determine the direction of arrival of the plane waves from distant sources.

100 wavelengths away provide plane waves. The outputs of each of the array elements are filtered to control relative amplitude, time delay, and phase response and combined to give the line array the desired spatial response.

If the array processing filters are all set to unity gain and zero phase for all frequencies and the sensor frequency responses are all identical in the array, the processor output for source A will be coherently summed while the waves from sources B and C will be incoherently summed. This is because the plane wave from source A has the same phase across the array while the phases of the waves from sources B and C will have a varying phase depending on the wavelength and the angle of incidence with respect to the array axis. A simple diagram for the first two array elements in Figure 5 illustrates the relative phase across the array from a distance source's plane waves. Obviously, when θ is 90°, the distance R to the source is the same, and thus, the plane wave phase is the same across the array and the summed array output (with unity weights) enhances the array response in the $\theta = 90°$ direction. The phase of the wave at the first element $y_{1,t}$ relative to the phase of the source is simply kR, where k is the wavenumber. Remembering that the wavenumber has units of radians per meter, k can be written as either $2\pi/\lambda$, or $2\pi f/c = \omega/c$, where λ is wavelength, $2\pi f = \omega$ is the radian frequency, and c is the speed of the wave in meters per second (m/sec). With all the array weights $A_i = 1$; $i = 1,2,...,M$, the summed array output s_t is simply

$$s_t = P_0 e^{j(\omega t - kR)} \sum_{m=1}^{M} A_m e^{+jk(m-1)d\cos\theta} \tag{7.1.1}$$

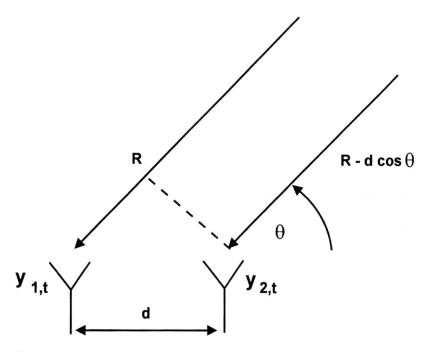

Figure 5 A distant source's plane wave will have a varying phase across the array depending on the angle of incidence.

where R is the distance from the source and d is the array element spacing. Note the opposite sign of the ωt and the kR in the exponent. As time increases, the distance R must increase for a given phase point on the plane wave. When $\theta = 90°$ s_t is M times louder than if a single receiver was used to detect the incoming plane wave. The mathematical structure of Eq. (7.1.1) is identical to that of a Fourier transform. For other angles of arrival, the array output level depends on the ratio of d/λ and the resulting relative phases of the array elements determined by the exponent $j2\pi(m-1)\cos\theta/\lambda$ in Eq. (7.1.1). When d/λ is quite small (low frequencies and small array spacings) the spatial phase changes across the array will be small and the array output will be nearly the same for any angle of arrival.

The net phase from the wave propagation time delay is really not of concern. The relative phases across the array are of interest because this information determines the direction of arrival within the half-plane above the line array. Given that we can measure frequency and array element positions with great accuracy, a reasonable assumption for the speed of sound or even wavelength for a given plane wave frequency is all that is actually needed to turn the measured relative phase data into source bearing information.

Clearly, the phase difference, $\Delta\phi_{21}$, found by subtracting the measured phase of sensor 1 from sensor 2 is simply $kd\cos\theta$. Therefore, the bearing to the distant source can be measured from the phase difference by

$$\theta = \cos^{-1}\left(\frac{\Delta\phi_{21}}{kd}\right) \tag{7.1.2}$$

The simple formula in Eq. (7.1.2) is sometimes referred to as a "direction cosine" or "phase interferometric" bearing estimation algorithm. Early aircraft navigation systems essentially used this direction-finding approach with a dipole antenna to determine the direction to a radio transmitter beacon. This technique is extremely simple and effective for many applications, but it is neither beamforming nor array processing, just a direct measurement of the angle of arrival of a single plane wave frequency. If there are multiple angles of arrival for the same frequency then the direction cosine technique will not work. However, the relationship between measured spatial phase, the estimated wavenumber, and the angle of arrival provides the necessary physics for the development of direction-finding sensor systems.

Returning now to the array output equation in (7.1.1), we see that the array output depends on the chosen number of array elements M, weights A_i, source angle, and the ratio of element spacing to wavelength. Figure 6 displays the response of a 16-element line array with unity weights for ratios of element spacing to wavelength (d/λ) of 0.05, 0.10, and 0.50. As the frequency increases towards a wavelength of $2d$, the beam at 90° becomes highly focused. If the element spacing is greater than half a wavelength their will be multiple angles of arrival where the array response will have high gain at a focused angle. These additional beams are called "grating lobes" after light diffraction gratings which separate light into its component colors using a periodic series of finely-spaced slits. The same physics are at work for the operation of the line array beamforming algorithm and the optical spectroscopy device using diffraction gratings to allow detection of the various color intensities in light. Figure 7 shows the beam response as the frequency is increased to give d/λ ratios of 0.85, 1.00, and 1.50. In general, the array element spacing should be at most a half wave-

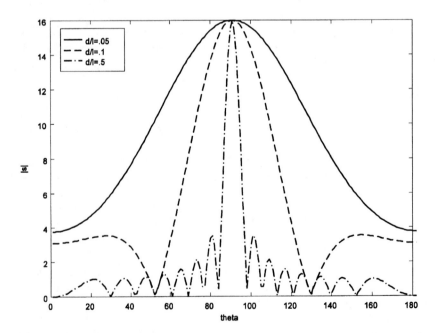

Figure 6   16 element line array output magnitude for ratios of element spacing to wavelength (d/λ) of 0.05, 0.1, and 0.5.

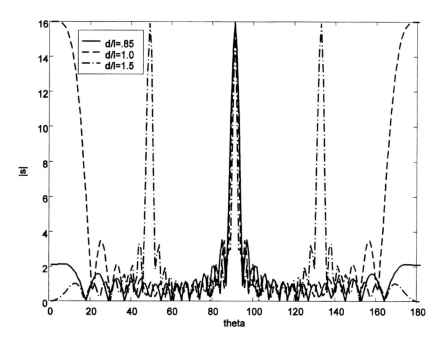

Figure 7 Beam response of a 16-element line array at high frequencies where the ratio d/λ is 0.85, 1.0, and 1.5.

length to insure no grating lobes for any angle of arrival. The presence of grating lobes can be seen as aliasing in spatial sampling. Adding more elements to the array while keeping the same aperture will eliminate the grating lobes. Also, the narrowness of the main look direction beam depends on the total array aperture Md. It can be shown that for large arrays ($M \gg 16$), the beamwidth in degrees is approximately $180\lambda/Md\pi$, so when the array aperture is many wavelengths in size, the resulting beam will be narrow and highly desirable for detecting plane waves from specific directions.

One of the more interesting aspects of the weights A_m in Figure 4 and Eq. (7.1.1) is that the magnitudes of the weights can be "shaded" using the spectral data windows described in Chapter 5 to suppress the side lobe leakage. The expense of this is a slight widening of the main lobe, but this effect is relatively minor compared to the advantage of having a single main beam with smooth response. Figure 8 compares a Hanning window to a rectangular window for the weights A_m. Recall that the Hanning window is $A_m = \frac{1}{2}[1 - \cos(2\pi m/M)]$ for an M-point data window; i.e. $m = 1, 2, ..., M$. For large M, the narrowband data normalization factor is 2, allowing "on-the-bin" spectral peaks to have the same amplitude whether a rectangular or Hanning window is used. For small arrays, we have to be a little more careful and integrate the window and normalize it to M, the integral of the rectangular window. For the $M = 16$ Hanning window, the integral is 8.5 making the narrowband normalization factor $16/8.5$ or 1.8823. As M becomes large the normalization factor approaches 2.0 as expected.

The phase of the array weights A_m can also be varied to "steer" the main lobe in a desired look direction other than 90°. Electronic beam steering is one of the main

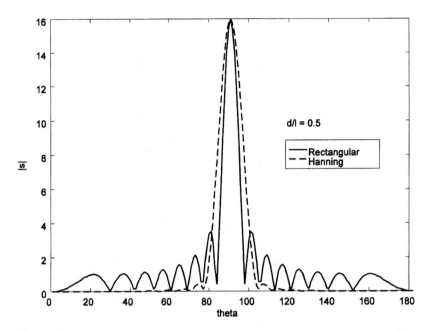

Figure 8 Use of data windows to "shade" the array element outputs at either end greatly reduces sidelobe leakage.

advantages of the array processing technique because it avoids the requirement for mechanical steering of devices such as the parabolic dish to move the beam. Furthermore, given the signals from each array element, parallel processing can be used to build several simultaneous beams looking in different directions. A snapshot of array data can be recorded and then scanned in all directions of interest to search for sources of plane waves. To build a steering weight vector one simply subtracts the appropriate phase from each array element to make a plane wave from the desired look direction θ_d, sum coherently in phase. The steering vector including the Hanning window weights is therefore

$$A_m = \frac{A_0}{2}\left[1 - \cos\left(\frac{2\pi m}{M}\right)\right]e^{-j2\pi\frac{d(m-1)}{\lambda}\cos\theta_d} \tag{7.1.3}$$

where A_0 is the narrowband normalization scale factor for the Hanning window. Recalling Eq. (7.1.1), the application of the weights in Eq. (7.1.3) would allow a plane wave from a direction θ_d to pass to the array output in phase coherency giving an output M times that of. a single sensor. Figure 9 shows the beam response for steering angles of 90°, 60°, and 30° with a Hanning window and a spacing to aperture ratio of 0.5.

Some rather important physics are revealed in Figure 9. As the beam is steered away from the broadside direction ($\theta_d = 90°$) towards the endfire directions ($\theta_d = 0°$ or 180°), the beam get wider because the effective aperture of the line array decreases towards the endfire directions. This decrease in effective aperture can be expressed approximately as $Md \sin \theta_d$ for steering angles near 90°. But, as one steers the beam near the endfire directions, the beam response depends mainly

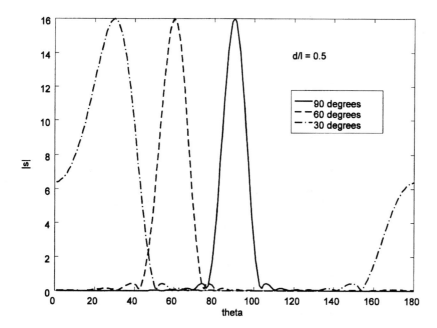

Figure 9 Adjusting the phase of the array weights allows the main look direction beam to be steered to a desired direction.

on the spacing to wavelength ratio, as will be demonstrated shortly. Since the line array is 1-dimensional, its beam patterns are symmetric about the axis of the array. Figure 10 shows the complete 360° response of the line array for the three steering angles given in Figure 9. Note the appearance of a "backward" beam near 180° for the beam steered to 30°. This "forward–backward" array gain is often overlooked in poor array processing designs, but tends to go away at lower frequencies as seen in Figure 11 for several frequencies steered to 0°.

The way one makes use of the steered beams in an electronic scanning system is to record a finite block of time data from the array elements and scan the data block in all directions of interest. The directions where the steered beam has a high output level correspond to the directions of distant sources, provided the beam pattern is acceptable. Figure 12 shows a scanned output for the 16-element line array where the 4 sources at 45°, 85°, 105°, and 145° with levels of 10, 20, 25, and 15, respectively. The frequency to spacing ratio is 0.4 and a Hanning window is used to shade the array and suppress the sidelobe leakage.

Figure 12 clearly shows 4 distinct peaks with angles of arrival and levels consistent with the sources (the array output magnitude is divided by 16). However, at lower frequencies the resolution of the scanning beam is not sufficient to separate the individual sources. Limited resolution is due to the finite aperture Md which must be populated with sufficient elements so that a reasonable high frequency limit is determined by the intra-element spacing to wavelength ratio. Obviously, one must also limit the number of elements to keep computational complexity and cost within reason. It is extremely important to understand the fundamental physics governing array size, resolution, element spacing, and upper frequency limits.

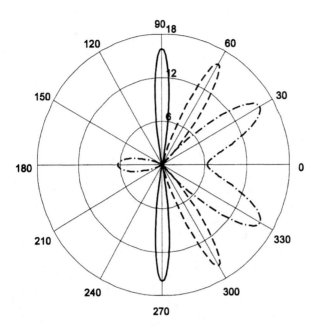

Figure 10   The complete 360° response of a line array shows the symmetry of the beam patterns around the array axis for steering angles of 90(__), 60(- -), (_._) degrees.

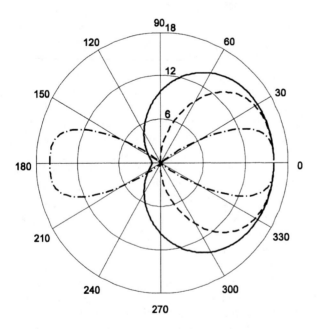

Figure 11   Complete beam response for steering angles of 0° at frequencies of $d/\lambda = 0.05$ (solid curve), 0.1 (dashed curve), and 0.5 (dash-dot curve) showing forward–backward gain problems.

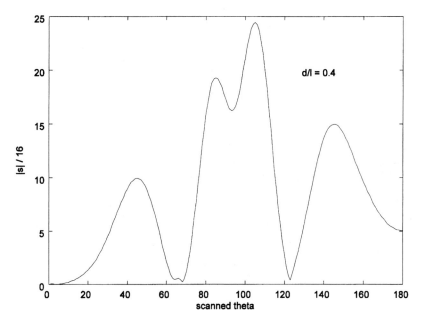

Figure 12 A scanned 16-element line array output clearly showing 4 distinct sources radiating plane waves.

While a 1-dimensional line array can determine a plane wave angle of arrival within a half-plane space, a 2-dimensional planar array can determine the azimuth and elevation angles of arrival in a half space. We define the angle ψ as the angle from the normal to the plane of the 2-dimensional array. If the array plane is horizontal, a plane wave from the direction $\psi = 0$ would be coming from a direction straight up, or normal to the array plane. Waves from $\psi = 90°$ would be coming from a source on the horizon with an azimuthal direction defined by θ, as defined earlier for the 1-dimensional line array. The elevation angle of the incoming plane wave is very important because it effects the measured wavelength in the array plane by making it appear longer as the elevation angle approaches zero. Figure 13 graphically depicts a 16 element planar array arranged in a 4 by 4 equally-spaced sensor matrix.

Since the effective wave speed in the x–y plane of the 2-dimensional array is $c_{xy} = c / \sin\psi$, where c is the wave speed along the direction of propagation, we define the wavenumber k_{xy} for the x–y plane as $2\pi f / c_{xy}$, or more explicitly, $2\pi f \sin\psi / c$. Using the sensor layout in Figure 13 where the spacing along the x-axis d is the same as the spacing along the y-axis, we can define a beam steering weight $A_{m,n}$, by simply subtracting the appropriate phase for a given sensor to make the array output phase coherent for the desired look direction in θ and ψ.

$$A_{m,n}(f) = A_{m,n}^w e^{-j\left[\frac{2\pi f d \sin\phi}{c}\{(m-1)\cos\theta + (n-1)\sin\theta\}\right]} \tag{7.1.4}$$

The frequency independent parameter $A_{m,n}^w$ in Eq. (7.1.4) represents the spatial data window for controlling side-lobe leakage. The 2-dimensional Hanning window is

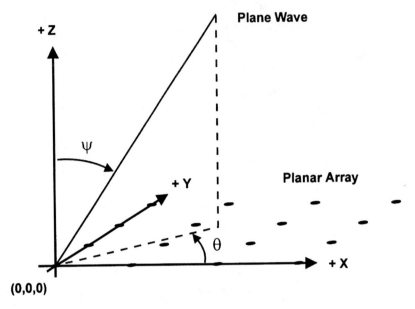

Figure 13 A 2-dimensional planar array showing the azimuthal angle θ and elevation angle Ψ relative to the cartesian coordinate system.

defined as

$$A_{m,n}^{w} = \frac{A_0}{4}\left[1 - \cos\left(\frac{2\pi m}{M}\right)\right]\left[1 - \cos\left(\frac{2\pi n}{N}\right)\right] \qquad (7.1.5)$$

where A_0 is the narrowband normalization factor to equalize the array output peak levels with what one would have using a rectangular window where all $A_{m,n}^{w}$ equal unity. For a desired beam look direction defined by θ_d and ψ_d, one calculates the array beam steering weights $A_{m,n}(f)$ for each frequency of interest. A convenient way to process the array data is to compute the Fourier spectra for each array element and multiply each Fourier frequency bin by the corresponding frequency steering weight in a vector dot product to produce the array output. One can then produce an array output spectrum for each desired look direction to scan for potential source detections. The steering vectors for each frequency and direction are independent of the signals from the array elements. Therefore, one typically "precomputes" a set of useful steering vectors and methodically scans a block of Fourier spectra from the array elements for directions and frequencies of high signal levels. Many of these applications can be vectorized as well as executed in parallel by multiple signal processing systems.

7.2 IMAGE ENHANCEMENT TECHNIQUES

One of the more astute signal processing techniques makes use of two dimensional Fourier transforms of images to control focus, reduce noise, and enhance features. The first widespread use of digital image enhancement is found among the world's astronomers, where the ability to remove noise from imagery allows observations of new celestial bodies as well as increased information. Obviously, satellite imagery of Earth, has for some time now, been fastidiously examined for state surveillance purposes. The details of particular image enhancement techniques and how well they work for satellite surveillance are probably some of the tightest held state technology secrets. However, we can examine the basic processing operations through the use of simple examples which hopefully will solidify the concepts of wavenumber response and filtering. As presented in Section 4.3, image processing techniques can be applied to any two, or even higher, dimensional data to extract useful information.

The point spread function (PSF) for an imaging system is analogous to the impulse response of a filter in the time domain. The input "impulse" to a lens system can be simply a black dot on a white page. If the lens system is perfectly focused, a sharp but not perfect dot will appear on the focal plane. The dot cannot be perfect because the index of refraction for a glass or plastic lens is not exactly constant as a function of light wavelength. Blue light will refract slightly different from red, and so on. Our nearly perfect dot on the focal plane will upon close examination have a "rainbow halo" around it from chromatic aberrations as well as other lens system imperfections. The two-dimensional Fourier transform (2DFFT) of the PSF gives the wavenumber response of the camera. If our dot were perfect, the 2DFFT would be constant for all wavenumbers. With the small halo around the dot, the 2DFFT would reveal a gradual attenuation for the higher wavenumbers (shorter wavelengths) indicating a small loss of sharpness in the camera's imaging capability. As the lens is moved slightly out-of-focus, the high wavenumber attenuation increases. Focus can be restored, or at least enhanced, by normalizing the 2DFFT of the image by the 2DFFT of the PSF for the out-of-focus lens producing the image. This is analogous to inverse filtering the output of some system to obtain a signal nearly unchanged by that system. For example, one could amplify the bass frequency range of music before it passes through a loudspeaker with weak bass response to reproduce the music faithfully. However, for imagery, one's ability to restore focus is quite limited due to limited signal-to-noise ratios and that fact that the PSF is attenuated rapidly into the noise when spread circularly in two dimensions.

Astronomers can use distant stars in their images to try to measure the PSF, which includes the light refracting and scattering affect of the atmosphere as well as the telescope. The "twinkling" of stars actually happens due to atmospheric turbulence and scattering from dust. The PSF for a distant star can sometimes look like a cluster of stars which fluctuate in position and intensity. Inverting the 2DFFT for the distant star section of the image, one can use the inverse filter to clarify the entire image. The technique usually works well when the same refractive multipath seen in the distant star section of the image applies throughout the image, which is reasonable physically since the angle of the telescope's field of view is extremely small. Recently, real-time feedback controllers have been used to bend flexible mirrors using a "point source" PSF criteria on a distant star to effectively keep the image sharp while integrating.

Consider the house picture shown earlier in the book, but this time clipped into a 256 by 256 pixel image where each pixel is an 8-bit gray scale in Figure 14. The house image reveals some noise as well as pixelation. Figure 15 shows a 2DFFT of the house where a log base 10 amplitude scale is used to make the details more visible. The 2DFFT reveals some very interesting image characteristics. The

Figure 14 The house picture shown earlier clipped to be 256×256 8-bit gray scale pixels.

Figure 15 A 2DFFT of the house image in Figure 14 shown on a $\log_{10}$ scale to make more of the higher wavenumber details visible.

equation for computing the 2-dimensional Fourier transform of an image $i(x, y)$ is

$$I(k_x, k_y) = \int\limits_{-\infty}^{+\infty} \int\limits_{-\infty}^{+\infty} i(x, y)e^{j(k_x x + k_y y)}dx\,dy \tag{7.2.1}$$

The equation for the inverse 2-dimensional Fourier transform is seen in Eq. (7.0.1).

One can easily see some dominant horizontal bands near $k_y = 0$ (across the middle of Figure 15) indicating large horizontal bands of light and dark across the original image in Figure 14 (the dark tile roof, white stucco, and dark ground). Horizontally, the arched windows and many different vertical band widths give a "sinusoidal" pattern in the region along $k_x = 0$ (the vertical band in the middle of Figure 15). The phase information in the 2DFFT places the various bright/dark regions on the correct spot on the original image. One can see a fair amount of noise throughout the 2DFFT as well as the original image, which we will show can be suppressed without much loss of visual information.

In the 2-dimensional wavenumber domain, one can do filtering to control sharpness just like one can filter audio signals to control "brightness" or "treble". The high wavenumber components which correspond to the sharp edges in the image are found in the outskirts of the 2DFFT in Figure 15. We can suppress the sharp edges by constructing a 2-dimensional wavenumber filter for the M by M discrete wavenumber spectrum.

$$w^{lp}(m_x, m_y) = \left[1 - \cos\left(\frac{2\pi m_x}{M}\right)\right]^{12}\left[1 - \cos\left(\frac{2\pi m_y}{M}\right)\right]^{12} \tag{7.2.2}$$

For an N by N image, the discrete 2-dimensional Fourier transform is

$$I(m_x, m_y) = \sum_{n_x=1}^{N}\sum_{n_y=1}^{N} i(n_x, n_y)e^{-j2\pi\left(\frac{m_x n_x}{N} + \frac{m_y n_y}{N}\right)} \tag{7.2.3}$$

Multiplying the wavenumber transform in Eq. (7.2.3) by the low-pass filter in Eq. (7.2.2) one obtains a "low-pass filtered" wavenumber transform of the house image as seen in Figure 16 on a log scale. This is not a matrix multiply, but rather a matrix "dot product", where each wavenumber is attenuated according to the filter function in Eq. (7.2.2). The filter function is rather steep in its "roll-off" so that the effect on focus is rather obvious.

Given the filtered M by M wavenumber spectrum (M is usually equal to N) I^{lp} $(m_x, m_y) = I\,(m_x, m_y)\,w^{lp}(m_x, m_y)$, the low pass filtered N by N image $i^{lp}(n_x, n_y)$ can be computed using a 2-dimensional inverse Fourier transform.

$$i^{lp}(n_x, n_y) = \sum_{m_x=1}^{M}\sum_{m_y=1}^{M} I^{lp}(m_x, m_y)e^{-j2\pi\left(\frac{m_x n_x}{M} + \frac{m_y n_y}{M}\right)} \tag{7.2.4}$$

The result of the low pass filtering can be seen in Figure 17.

What happened to all the sharp edges in Figure 17? We can find out by examining what was removed from the original image using a high-pass filter somewhat more gradual than the low pass filter. The high-pass filter is unity at the highest

Figure 16 Low pass filtering of the 2D wavenumber transform suppresses all but the long wavelengths represented near the center of the spectrum.

Figure 17 Low pass filtering of the wavenumber spectrum has the effect of "softening" or "defocusing" the image by removal of the short wavelengths.

wavenumbers and declines to zero for zero wavenumber (at the center) as $1/k$. The inverse Fourier transform of the removed high frequency components is seen in Figure 18.

From an engineering point of view, one must ask how much visual information is actually in a given image, how can one determine what information is important, and how one can manage visual information to optimize cost. A reasonable approach is to threshold detect the dominant wavenumbers and attempt to reconstruct the image using information compression based on the strongest wavenumber components. But, the practical situation turns out to be even a bit more complicated than that since important textures and features can be lost by discarding all but the most dominant wavenumbers. The short answer is that the optimal image data compression generally depends on the type of image data being compresses. The dominant wavenumber approach would work best on images with periodic patterns. Simple thresholding in the image domain would work well for line art, or images with large areas on one solid shade. One of the dominant strategies in MPEG and MPEG2 image/movie compression algorithms is run length encoding, where large areas of constant shade can be reduced to only a few bytes of data without loss of any information. For our case, we can eliminate nearly half the image data by focusing on the strong vertical and horizontal energy as seen in the filtered wavenumber response in Figure 19 on a log scale. The result is seen in Figure 20 showing very little loss of visual information. Overall the image is much softer than the original, the sharpness along the dominant horizontal and vertical edges is not lost. This can be clearly seen along the arches of the windows where strong horizontal and vertical edges are seen next to softer curves.

Figure 18 High-frequency components of the image found by high-pass wavenumber filtering and inverse Fourier transforming the wavenumber spectrum.

Figure 19 Filtered wavenumber response to suppress unnecessary noise and sharpness along diagonal directions while keeping useful high frequency information along the vertical and horizontal.

Figure 20 The resulting image from the wavenumber filtering in Figure 19 which reduces the necessary image data by about half without significant loss of visual information, but notice the horizontal and vertical matting (canvas effect) from the wavenumber filtering.

Clearly, many interesting and useful visual effects can be implemented using wavenumber filtering on image data. Even a brief survey of all the popular techniques is beyond the scope of this book. However, wavenumber processing can be used for many applications, even on non-visual images made up from 2 or more dimensional data. Our attempt here is to explain the physics and signal processing of the wavenumber transform with some useful examples of image data.

7.3 COMPUTER-AIDED TOMOGRAPHY

Computer-aided tomography, or "CAT-Scans" are becoming well known to the general public through its popular use in medical diagnosis. While the medical CAT-Scan is synonymous with an X-ray generated "slice" of the human body, without actually cutting any tissue, it also has many industrial uses for inspection of pipes and vessels, structures, and in the manufacture of high technology materials and chemicals. Webster's definition of tomography reads simply as "roentgenography of a selected plane in the body". After looking up "roentgenography", we find that it is named after Wilhelm Konrad Roentgen 1845–1923, the German physicist who discovered X-rays around the turn of the last century in 1895. X-rays have extremely short wavelengths (approximately 10^{-10} m) and are produced when electrons accelerated through a potential difference of perhaps 1 kV to 1 MV strike a metal target. The electrons in the inner atomic electron shells jump into higher energy states in the surrounding electron shells and produce the X-rays when they return to their equilibrium state back in the original shell. This electron-to-photon "pumping" also occurs between the outermost electron shells at much lower voltages producing only visible light. The obvious utility of X-rays is their ability to penetrate many solid objects. Continued animal exposure to high levels of X-rays has been shown to cause radiation poisoning and some cancers. But, modern engineering has nearly perfected low-level X-ray systems for safe periodic use by the medical community.

An X-ray image is typically created using high resolution photographic film placed on the opposite side of the subject illuminated by the X-ray source. The extremely short wavelength of X-rays means that plane wave radiation is achieved easily within a short distance of the source. The X-ray image is composed of 3-dimensional translucent visual information on a 2-dimensional medium. Examination of X-ray images requires a specially trained medical doctor called a *radiologist* to interpret the many subtle variations due to flesh, organs, injury, or pathology. It was not until 1917 when Radon (2) published what is now referred to as the Radon transform, that the concept of combining multiple 2-dimensional images to produce a volumetric cross-slice was available. It took another 50 years of computing and signal processing technology advancement before G.N. Hounsfield at EMI Ltd in England and A.M. Cormack at Tufts University in the United States developed a practical mathematical approach and electronic hardware making the CAT scan a real piece of technology in 1972. In 1979, Hounsfield and Cormack shared the Nobel prize for their enormous contribution to humanity and science.

A straightforward walk-through tour of the Radon transform and its application to the CAT scan follows. The reader should refer to an outstanding book by J.C. Russ (3) for more details on the reconstruction of images. We start by selecting some artwork representative of a CAT scan of a human brain as seen in Figure 21. The top part of the drawing is the area behind the forehead while the bottom part shows the cerebellum and brain stem.

Figure 22 shows a sketch of a single scan along an image angle of $+40°$ above the horizontal. The scan line could be simply a line sampled from a 2-dimensional X-ray image or simply an array of photovoltaic cells sensitive to X-ray energy.

The bright and dark areas along the detection array are the result of the integration of the X-ray absorption along the path from the source through the brain to the detectors. If the X-ray propagation were along the horizontal from left to right, the spatial Fourier transform of the detector array output corresponds exactly to a $k_x = 0$, k_y 2-dimensional Fourier transform where the result would be graphed along the k_y vertical axis passing through the $k_x = 0$ origin. At the $40°$ angle in Figure 22, the wavenumber-domain line would be along $130°$ (the normal to this line is $40°$), as seen in Figure 23.

For an $N \times N$ Fourier transform, we can write

$$I(m'_x, m'_y) = \sum_{n'_x=1}^{N} \sum_{n'_y=1}^{N} i(n'_x, n'_y)e^{-j\frac{2\pi}{N}(m'_x n'_x + m'_y n'_y)} \qquad (7.3.1)$$

where $m'_x = m_x\cos\theta + m_y\sin\theta$, $m'_x = m_y\cos\theta\ m_x\sin\theta$, $n'_x = n_x\cos\theta + n_y\sin\theta$, and $n'_y = n_y\cos\theta - n_x\sin\theta$. One simply rotates the x'-axis to align with the X-ray propagation path so that the rotated y'-axis aligns with the sensor array. The Fourier transform of the y' data gives the wavenumber spectra for m'_y with $m'_x = 0$. With the integration of the absorption along the propagation path of the X-rays being done physically,

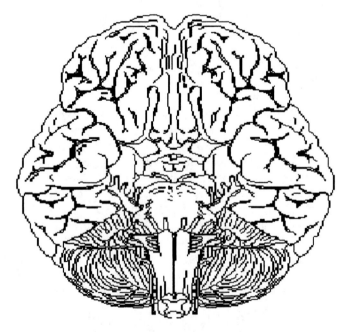

Figure 21 Cross-section drawing of the human brain used for tomographic reconstruction demonstration.

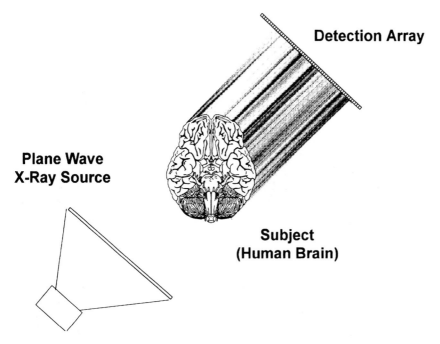

Figure 22 Layout of a single scan line produced by X-rays along a +40° angle with respect to the horizontal.

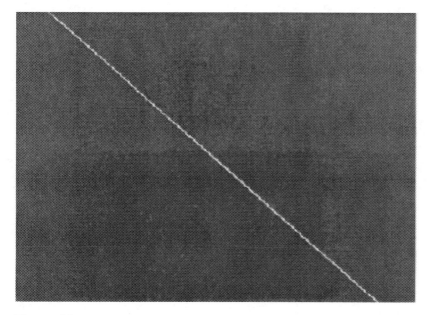

Figure 23 The spatial Fourier transform line corresponding integration of the brain cross-section along a 40° angle with respect to the horizontal.

rather than mathematically, only one of the summations in Eq. (7.3.1) is necessary (m_x' is zero). The inverse Fourier transform of the data in Figure 23 is seen in Figure 24. Equation (7.3.1) is known as a discrete Radon transform. For the simple case where $\theta = 0$, we are given a response where the only variation is along the vertical y-axis in the spatial image domain. The corresponding wavenumber response only has data along the vertical y-axis where $k_x = 0$. For $k_x = 0$, we have an infinite wavelength ($k = 2/\lambda$), hence the straight integration along the x-axis.

Increasing the number of scan directions will begin to add additional image information which is combined in the wavenumber domain and then inverse Fourier transformed to present a reconstruction of the original image. Figure 25 shows the wavenumber response for 8-scans, each at 22.5° spacing. The corresponding inverse Fourier transform is seen in Figure 26. Only the basic structures of the brain are visible and many lines are present as artifacts of the 65k pixel image being represented by only 256 times 8, or 2048 pixels in wavenumber space. Figures 27 and 28 show the wavenumber response and image reconstruction when using 32 scans. Figure 29 and 30 show the wavenumber response and image reconstruction when using 128 scans. A typical high resolution CAT scan would use several hundred scan angles over a range from 0° to 180°. Its not necessary to re-scan between 180° and 360°, and even if one did, the chances of misregistration between the upper and lower scans are great in many applications using human or animal subjects.

It is truly amazing that a "slice" through the cross-section of the subject can be reconstructed from outside one-dimensional scans. The utility of being able to extract this kind of detailed information has been invaluable to the lives of the millions of people who have benefitted by the medical CAT scan. However, what is even more fascinating is that the Fourier transforms can be completely eliminated

Figure 24   Inverse Fourier transform of the single scan wavenumber response in Figure 23.

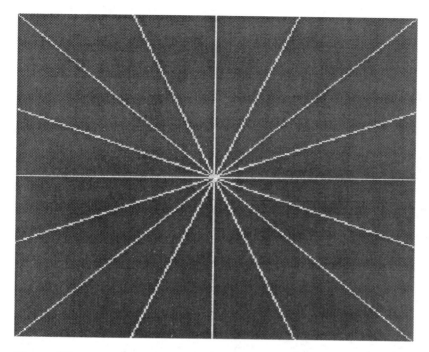

Figure 25 8 scans as seen in the wavenumber domain for the brain image.

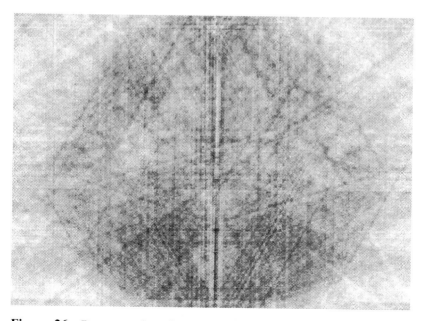

Figure 26 Reconstruction of the Fourier wavenumber data for the 8 scans.

Figure 27   32 scans of the brain cross-section in wavenumber space.

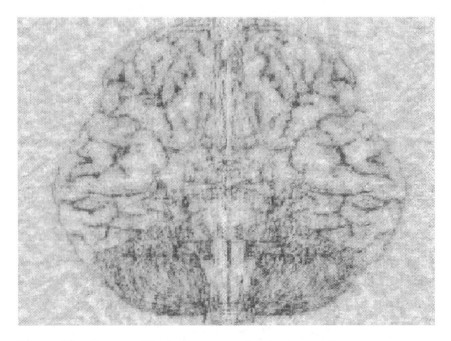

Figure 28   32 scans of the brain cross-section in wavenumber space.

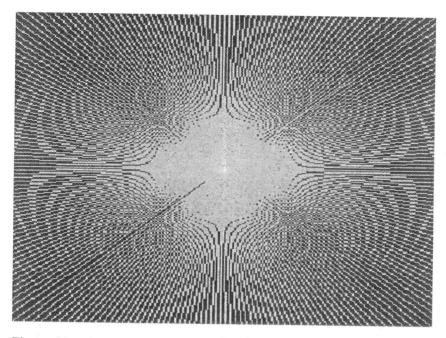

Figure 29 128 scan wavenumber response for the brain image.

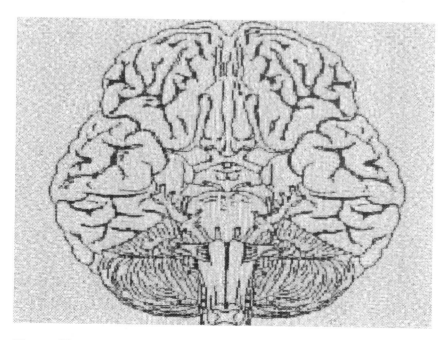

Figure 30 Corresponding image reconstruction from the 128 scan data seen in Figure 29.

by using the filtered backprojection technique. The largest computational burden is not all the individual scans, but the inverse 2-dimensional Fourier transform. For a 256×256 pixel crude image, each 256-point FFT requires only 2048 multiply operations, but the 256×256 inverse Fourier transform to reconstruct the image requires over 4 million multiplies. A typical medical CAT scan image will have millions of pixels (over 1024×1024) requiring hundreds of millions of multiplies for the 2-dimensional inverse FFT, as well as consist of dozens of "slices" of the patient to be processed.

The filtered backprojection technique is extremely simple, thereby making CAT-scan equipment even more affordable and robust. In back propagation, one simply "stacks" or adds up all the scan data (as seen in Figure 24) on a single image without doing any Fourier transforms. Adding all this scan data up pixel by pixel gives a sharp, yet foggy, image as seen in Figure 31 for 128 scans. The effect is due to the fact that the pixels in the center of the image are "summed" and "resummed" with essentially the same low wavenumber scan information with each added scan. The high wavenumber data near the center of the image tends to sum to zero for the different scan angles. In the wavenumber domain, this means that the low wavenumbers near the origin are being overemphasized relative to the high wavenumbers which correspond to the sharp edges in the image. A simple high-pass wavenumber filter corrects the fogginess as seen in Figure 32 yielding a clear image without a single Fourier transform. The wavenumber filtering can be accomplished very simply in the time domain using an FIR filter on the raw scan data. The filtered backprojecttion technique also allows filtering to enhance various image features to help medical diagnosis. Newer techniques such as the backpropagation technique even allow for propagation multipath and wave

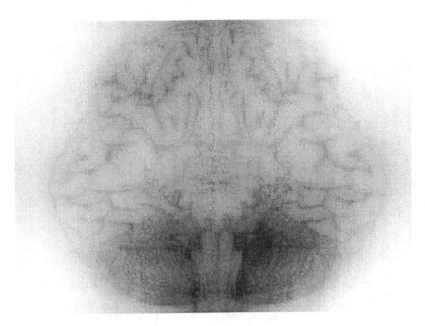

Figure 31 Unfiltered backprojection reconstruction of the brain cross-section from 128 scans.

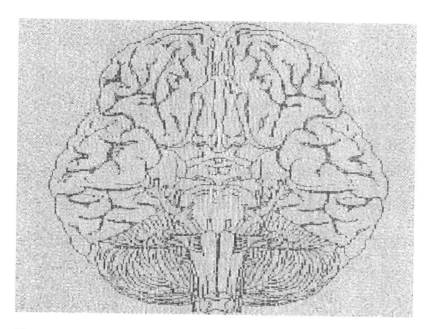

Figure 32 High-pass filtered backprojection reconstruction of the 128 scan image data.

scattering effects to be (at least in theory) be controlled for tomographic applications in ultrasonic imaging.

7.4 SUMMARY, PROBLEMS, AND BIBLIOGRAPHY

Waves can be represented in either time or space as a Fourier series of sinusoids which can be filtered in the time-frequency or spatial-wavenumber domain depending on the chosen wave representation. We have shown that for sensor arrays spaced fractions of a wavelength apart (up to half of a wavelength), spatial filtering can be used to produce a single array output having very high spatial gain in a desired look-direction. The look direction "beam" can be electronically steered using signal processing to give a similar search beam to that for a parabolic reflector which would be mechanically steered. Using parallel signal processing, one could have several simultaneous beams independently steerable, where each beam produces a high-gain output on a signal from a particular angle of arrival. Later in the book, Chapters 12 and 13 will discuss in detail adaptive beamforming techniques for sonar and radar which optimization of the search beam by steering beam nulls in the directions of unwanted signals.

Examination of images from optical sensors as wavenumber systems requires consideration of a whole different set of physics. Visible light wavelengths range around 1×10^{-6} m, or 1 μm, and frequencies around 1×10^{14} Hz, making spatial filtering using electronic beamforming impossible with today's technology. However, when optical wave information is projected on a screen using either a pinhole aperture or lens system, we can decompose the 2-dimensional image data into spatial waves using Fourier's theorem. The image element samples, or pixels, form a 2-dimensional set of data for Fourier series analysis. We can control the sharpness of the image by filtering the wavenumbers in an analogous manner to the way one can control the treble in an audio signal by filtering frequencies. The short wavelengths in an audio time-domain signal correspond to the high frequencies while the short wavelengths in image data correspond to the high wavenumbers making up the sharp edges in the picture. If the pixels have enough bits of resolution, an out-of-focus image can be partially recovered by inverting the "point spread function" for the lens in its out-of-focus position. The success of focus recovery is sensitive to the number of bits of pixel brightness resolution because the out-of-focus lens spreads the light information out in 2-dimensions into adjacent pixels, which attenuates the brightness substantially in the surrounding pixels. We can also use the 2-dimensional Fourier transform to detect the strongest wavenumber components, rejecting weaker (and noisier) wavenumbers, and thereby compress the total image data while also reducing noise.

For scanning systems such as X-rays, the massive 3-dimensional data on an X-ray image can be resolved into a synthetic thin "slice" cross-wise by combining many rays from a wide range on angles. The technique of computer-aided tomography solves an important problem in X-ray diagnosis by simplifying the amount of information on the translucent X-ray image into a more organized virtual slice. Radon recognized the need and solved the problem mathematically in the early part of the 20th century. But, it was not until the arrival of computers and the genius of Cormack and Hounsfield in the early 1970s that the CAT-Scan became practical. The understanding of the Radon transform and wavenumber domain signal processing allows an even more simple approach to be used called the filtered backpropagation tomographic reconstruction. Filtered backprojection produces a sharp image without the computational demands of Fourier transforms by filtering each scan to better balance the wavenumbers in the final image. The CAT-Scan is considered one of the premier technological achievements of the 20th century because of the technical complexities which had to be solved and the enormous positive impact it has achieved for the millions of people who have benefitted from its use in medical diagnosis.

PROBLEMS

1. For an arbitrary 3-element planer (but not linear) sensor array with sensor spacings 1–2 and 1–3 each less than $1/2$ wavelength, derive a general formula to determine the angle-of arrival of a plane wave given the phases of the three sensors at a particular frequency.

2. A linear array of 16 1-m spaced hydrophones receives a 50 Hz acoustic signal from a distant ship screw. If the angle of arrival is $80°$ relative to the array axis (the line of sensors), what is the steering vector which best detects the ship's signal? Assume a sound speed of 1530 m/sec.

3. Consider a circular array of 16 evenly-spaced microphones with diameter 2 m (sound speed in air is 345 m/sec). Derive an equation for the array response as a function of k and θ when the array is steered to θ'.

4. Show that the sharpness operator W^s in Eq. (2.1.14) is a high pass filter by examining the operator's response in the wavenumber domain.

BIBLIOGRAPHY

W. S. Burdic Underwater Acoustic System Analysis, Englewood Cliffs: Prentice-Hall, 1991.
Oswego and Miller, adaptive arrays.
S. J. Orfandis Optimum Signal Processing, 2nd ed., NewYork: McGraw-Hill, 1988.

REFERENCES

1. R. E. Colin Antennas and Radio Wave Propagation, McGraw-Hill: New York, 1985.
2. J. Radon Über die Bestimmung von Funktionen durch ihre Integralwerte längs gewisser
    Mannigfaltkeiten, Berlin Sächsische Akad. Wissen., Vol. 29, pp. 262–279, 1917.
3. J. C. Russ The Image Processing Handbook, Boca Raton: CRC Press, 1994.

Part III

Adaptive System Identification and Filtering

Adaptive signal processing is a fundamental technique for intelligent sensor and control systems which uses the computing resources to optimize the digital system parameters as well as process the digital signals. We generally use microprocessors today as a stable, flexible, and robust way to filter sensor signals precisely for information detection, pattern recognition, and even closed-loop control of physical systems. The digital signal processing engine allows for precise design and consistent filter frequency response in a wide range of environments with little possibility for response drift characteristic of analog electronics in varying temperatures, etc. However, the microprocessor is also a computer which while filtering digital signals can also simultaneously execute algorithms for the analysis of the input–output signal waveforms and update the digital filter coefficients to maintain a desired filter performance. An adaptive filter computes the optimal filter coefficients based on an analytic cost minimization function and adapts the filter continuously to maintain the desired optimal response. This self-correction feature requires some sort of simple cost function to be minimized as a means of deciding what adjustments must be made to the filter. In general, a quadratic cost function is sought since it only has one point of zero slope, which is either a maximum or minimum. For linear time-invariant filters, the cost function is usually expressed as the square of the error between the adaptive filter output and the desired output. Since the filter output is a linear (multidimensional) function of the filter coefficients, the squared-error is a positive quadratic function of the filter coefficients. The minimum of the squared-error cost function is known as a least squared-error solution.

In Chapter 8, we present a very concise development of a least squared-error solution for an FIR filter basis function using simple matrix algebra. A projection operator is also presented as a more general framework which will be referred to in the development of adaptive lattice filter structures latter in Chapter 9. The recursive adaptive algorithms in Chapters 9 and 10 allow the implementation of real-time adaptive processing where the filter coefficients are updated along with the filter computations for the output digital signal. All adaptive filtering algorithms presented are linked together through a common recursive update formula. The most simple and robust form of the recursive update is the least mean-square (LMS) error adaptive filter algorithm. We present some important convergence properties of the LMS algorithm and contrast the convergence speed of the LMS to other more complicated but faster adaptive algorithms. Chapter 10 details a wide range of adaptive filtering applications including Kalman filtering for state vector adaptive updates and frequency-domain adaptive filtering. Recursive system identification is also explored using the adaptive filter to model an "unknown" system. The issues of mapping between the digital and analog domains are again revisited from Chapter 2 for physical system modeling.

8

Linear Least-Squared Error Modeling

We owe the method of least-squares to the genius of Carl Friedrich Gauss (1777–1855), who at the age of only 24, made the first widely accepted application of least-squared error modeling to astronomy in the prediction of the orbit of the asteroid Ceres from only a few position measurements before it was lost from view (1). This was an amazing calculation even by today's standards. When the asteroid reappeared months later, it was very close to the position Gauss predicted it would be in, a fact which stunned the astronomical community around the world. Gauss made many other even more significant contributions to astronomy and mathematics. But, without much doubt, least-squared error modeling is one of the most important algorithms to the art and science of engineering, and will likely remain so well into the 21st century.

8.1 BLOCK LEAST-SQUARES

We begin my considering the problem of adaptively identifying a linear time-invariant causal system by processing only the input and output signals. Figure

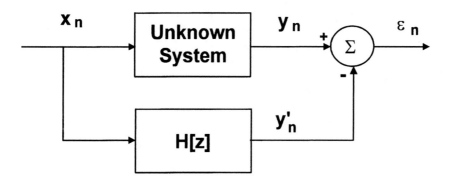

Figure 1   Block diagram depicting adaptive system identification using least-squared error system modeling.

1 depicts the process with a block diagram sketch showing the "unknown" system, its digital input $x[n] = x_n$, its digital output y_n, our digital FIR filter model $H[z]$, and the model output y'_n. The difference between the unknown system output y_n and the model output y'_n gives an error signal ϵ_n which is a linear function of the model filter coefficients. If we can find the model filter coefficients which give an error signal of zero, we can say that our model exactly matches the unknown system's response to the given input signal x_n. If x_n is spectrally white (an impulse, zero-mean Gaussian noise, sinusoidal chirp, etc.), or even nearly white (some signal energy at all frequencies), then our model's response should match the unknown system. However, if the unknown system is not well represented by an FIR filter, or if our model has fewer coefficients, the error signal cannot possibly be made exactly zero, but can only be minimized. The *least-squared error* solution represents the best possible match for our model to the unknown system given the constraints of the chosen model filter structure (FIR) and number of coefficients.

Real-world adaptive system identification often suffers from incoherent noise interference as described in Section 6.2 for frequency domain transfer functions. In general, only in the sterile world of computer simulation will the error signal actually converge to zero. The issue of incoherent noise interference will be addressed in Section 10.3 on Weiner filtering applications of adaptive system identification. The FIR filter model $H[z]$ has $M + 1$ coefficients as depicted in Eq. (8.1.1).

$$H[z] = h_0 + h_1 z^{-1} + h_2 z^{-2} + \ldots + h_M z^{-M} \tag{8.1.1}$$

We define a *basis function* for the least-squared error model ϕ_n as a 1 by $M + 1$ row vector

$$\phi_n = [x_n \ x_{n-1} \ x_{n-2} \ldots x_{n-M}] \tag{8.1.2}$$

and a model impulse response $M + 1$ by 1 column vector as

$$H = [h_0 \ h_1 \ h_2 \ldots h_M]^T \tag{8.1.3}$$

The FIR filter model output y'_n can now be written as a simple vector dot product.

$$y'_n = \phi_n H \tag{8.1.4}$$

For this application of FIR system identification, the basis function is a vector of the sampled digital inputs from sample n back to sample $n - M$. Other forms of basis functions are possible such as a power series, function series (exponents, sinusoids, etc.), or other functions defined by the structure of the model which one wishes to obtain a least-squared error fit to the actual unknown system of interest. Least-squares modeling using other forms of basis function will be addressed in Section 8.3. The model error is simply the unknown system output minus the modeled output signal.

$$\varepsilon_n = y_n - y'_n = y_n - \phi_n H \tag{8.1.5}$$

The least-squared error solution is computed over an observation window starting at sample n and ranging backwards to sample $n - N + 1$, or a total of N samples. The error signal over the observation window is written as a N by 1 column vector.

$$
\begin{bmatrix} \varepsilon_n \\ \varepsilon_{n-1} \\ \varepsilon_{n-2} \\ \vdots \\ \varepsilon_{n-N+1} \end{bmatrix} = \begin{bmatrix} y_n \\ y_{n-1} \\ y_{n-2} \\ \vdots \\ y_{n-N+1} \end{bmatrix} - \begin{bmatrix} x_n & x_{n-1} & x_{n-2} & \cdots & x_{n-M} \\ x_{n-1} & x_{n-2} & x_{n-3} & \cdots & x_{n-M-1} \\ x_{n-2} & x_{n-3} & x_{n-4} & \cdots & x_{n-M-2} \\ \vdots & \vdots & \vdots & \cdots & \vdots \\ x_{n-N+1} & x_{n-N} & x_{n-N-1} & \cdots & x_{n-N+1-M} \end{bmatrix} \begin{bmatrix} h_0 \\ h_1 \\ h_2 \\ \vdots \\ h_M \end{bmatrix}
$$

$$(8.1.6)$$

Eq. (8.1.6) is written more compactly in matrix form

$$\bar{\varepsilon} = \bar{y} - \bar{X}H \tag{8.1.7}$$

where the rows of $\bar{X}$ are the basis functions for the individual input–output signal samples.

We now write a compact matrix expression for the sum of the squared error over the observation window by computing a complex inner product for the error signal vector (superscript H denotes Hermitian transpose–transpose and complex conjugate).

$$
\begin{aligned}
\bar{\varepsilon}^H \bar{\varepsilon} &= (\bar{y} - \bar{X}H)^H (\bar{y} - \bar{X}H) \\
&= \bar{y}^H \bar{y} - \bar{y}^H \bar{X}H - H^H \bar{X}^H \bar{y} + H^H \bar{X}^H \bar{X}H
\end{aligned}
\tag{8.1.8}
$$

If H is a scalar ($M = 0$), the sum of the squared error in Eq. (8.1.8) is clearly a quadratic function of the FIR filter coefficient h_0. If $M = 1$, one could visualize an bowl-shaped error "surface" which is a function of h_0 and h_1. For $M > 1$, the error surface is multi-dimensional and not very practical to visualize. However, from a mathematical point of view, Eq. (8.1.8) is quadratic with respect to the coefficients in H. The desirable aspect of a quadratic matrix equation is that there is only one extremum where the slope of the error surface is zero. The value of H where the slope of the error surface is zero represents a minimum of the cost function in Eq. (8.1.8) provided that the second derivatives with respect to H is positive definite, indicating that the error surface is concave up. If the second derivative with respect to H is not positive definite, the Eq. (8.1.8) represents a "profit" function, rather than a cost function, which is maximized at the value of H where the error slope is zero.

Calculating the derivative of a matrix equation is straightforward with the exception that we must include the components of the derivative with respect to H and H^H in a single matrix-dimensioned result. To accomplish this, we simply compute a partial derivative with respect to H (treating H^H as a constant) and adding this result to the Hermitian transpose of the derivative with respect to H^H (treating H as a constant). Note that this approach of summing the two partial derivatives provides a consistent result with the scalar case.

$$
\begin{aligned}
\frac{\partial \bar{\varepsilon}^H \bar{\varepsilon}}{\partial H} + \left\{ \frac{\partial \bar{\varepsilon}^H \bar{\varepsilon}}{\partial H^H} \right\}^H &= -\bar{y}^H \bar{X} + H^H \bar{X}^H \bar{X} + \left\{ -\bar{X}^H \bar{y} + \bar{X}^H \bar{X}H \right\}^H \\
&= -2\bar{y}^H \bar{X} + 2H^H \bar{X}^H \bar{X}
\end{aligned}
\tag{8.1.9}
$$

The solution for the value of H^H which gives a zero error surface slope in Eq. (8.1.9) is

$$H^H = \bar{y}^H \bar{X} (\bar{X}^H \bar{X})^{-1} \tag{8.1.10}$$

The Hermitian transpose of (8.1.10) gives the value of H for zero slope.

$$H = (\bar{X}^H \bar{X})^{-1} \bar{X}^H \bar{y} \tag{8.1.11}$$

Given that we now have a value for H where the error surface is flat, the second derivative of the squared error is calculated to verify that the surface is concave up making the solution for H in Eq. (8.1.11) a least-squared error solution.

$$\frac{\partial^2 \bar{\varepsilon}^H \bar{\varepsilon}}{\partial H^2} = 2 \bar{X}^H \bar{X} = 2 \sum_{k=0}^{N-1} \phi_{n-k}^H \phi_{n-k} \tag{8.1.12}$$

Closer examination of the complex inner product of the basis function reveals that the second derivative is simply the autocorrelation matrix times a scalar.

$$2 \sum_{k=0}^{N-1} \phi_{n-k}^H \phi_{n-k} \approx 2N \begin{bmatrix} R_0^x & R_1^x & R_2^x & \cdots & R_M^x \\ R_{-1}^x & R_0^x & R_1^x & \cdots & R_{M-1}^x \\ \cdot & \cdot & & & \cdot \\ \cdot & & \cdot & & \cdot \\ R_{-M}^x & R_{-M+1}^x & \cdots & R_{-1}^x & R_0^x \end{bmatrix} \tag{8.1.13}$$

where $R_j^x = E\{x_n^* x_{n-k}\}$ is the jth autocorrelation lag of x_n. If x_n is spectrally white, the matrix in Eq. (8.1.13) is diagonal and positive. If x_n is nearly white, then the matrix is approximately diagonal (or can be made diagonal by Gaussian elimination or QR decomposition). In either case, the autocorrelation matrix is positive definite even for complex signals, the error surface is concave up, and the solution in Eq. (8.1.11) is the least-squared error solution for the system identification problem. As the observation window grows large, the least-squared error solution can be seen to asymptotically approach

$$\lim_{N \to \infty} H = \begin{bmatrix} R_0^x & R_1^x & R_2^x & \cdots & R_M^x \\ R_{-1}^x & R_0^x & R_1^x & \cdots & R_{M-1}^x \\ \cdot & \cdot & R_0^x & \cdot & \cdot \\ R_{-M}^x & R_{-M+1}^x & \cdots & R_{-1}^x & R_0^x \end{bmatrix}^{-1} \begin{bmatrix} R_0^{xy} \\ R_{-1}^{xy} \\ \vdots \\ R_{-M}^{xy} \end{bmatrix} \tag{8.1.14}$$

where $R_j^{xy} = E\{x_n^* y_{n-j}\}$ is the cross correlation of the input and output data for the unknown system and the scale factor N divides out of the result.

Equation (8.1.14) has an exact analogy in the frequency domain where the frequency response of the filter $H[z]$ can be defined as the cross spectrum of the input–output signals divided by the autospectrum of the input signal as described in Section 6.2. This should come as no surprise, since the same information (the input and output signals) are used in both the frequency domain and adaptive system identification cases. It makes no difference to the least-squares solution, so long as the Fourier transform resolution is comparable to the number of coefficients chosen for the filter model (i.e. M-point FFTs should be used). However, as we will see in

Chapters 9 and 10, the adaptive filter approach can be made to work very efficiently as well as can be used to track nonstationary systems.

Figure 2 presents an example where a 2-parameter FIR filter ($M = 1$) with $h_0 = -1.5$ and $h_1 = -2.5$ showing the squared-error surface for a range of model parameters. The least-squared error solution can be seen with the coordinates of H matching the actual "unknown" system where the squared-error is minimum.

The term "block least-squares" is used to depict the idea that a least-squared error solution is calculated on a block of input and output data defined by the N samples of the observation window. Obviously, the larger the observation window, the better the model results will be assuming some level of noise interference is unavoidable. However, if the unknown system being modeled is not a linear function of the filter coefficients H, the error surface depicted in Figure 2 would have more than one minima. Nonlinear system identification is beyond the scope of this book. In addition, it is not clear from the above development what effect the chosen model order M has on the results of the least-squared error fit. If the unknown system is linear and FIR in structure, the squared error will decline as model order is increased approaching the correct model order. Choosing a model order higher than the unknown FIR system will not reduce the squared error further (the higher-order coefficients are computed at or near zero). If the unknown system is linear and IIR in structure, the squared error will continue to improve as model order is increased.

8.2 PROJECTION-BASED LEAST-SQUARES

In this section we present the least-squared error solution for modeling a linear time-invariant system in the most general mathematical terms. The reasons for this development approach may become of more interest later when we develop the

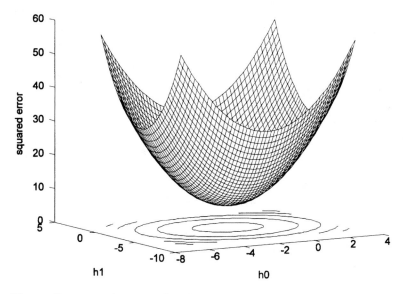

Figure 2 Squared-error surface for 2-parameter FIR system with $h_0 = -1.5$ and $h_1 = -2.5$ showing the minimum error for the correct model.

fast-converging adaptive least-squares lattice filter and Schur recursions. The systems presented in this book can all be modeled as a weighted sum of linearly independent functions. For example, Chapter 2 showed how any pole-zero filter could be expressed as a weighted sum of resonances, defined by conjugate pole pairs. For a stable causal system with real input and output signals, the conjugate pole pairs each constitute a subsystem with impulse response representable by a simple damped sinusoid, or mode. The total system response to any set of initial conditions and excitation signal can be completely described by the proper weighted sum of modes. The system response can be seen as a finite subset of the infinite number of possibilities of signals.

An abstract Hilbert space $\hat{H}$ is an infinite-dimensional linear inner product space (2). A linear time-invariant system can be seen as a linear manifold spanned by the subspace $\hat{G}$. The difference between a Hilbert space and a subspace is that the Hilbert space has an infinite number of linearly independent elements and the subspace has only a finite number of linearly independent elements. Casting our linear time-invariant system into a subspace allows us to write the least-squared error signal as an orthogonal projection of the system output signal onto the subspace. Why do we care? The subspace can be expanded using mathematically-defined orthogonal projections, which represent by definition, the minimum distance between the old subspace and the new expanded subspace. For new observations added to the signal subspace, the orthogonal expansion represents the least-squared error between the model output prediction and the unknown system output. The mathematical equations for this orthogonal expression give exactly the same least-squared error solution derived in the previous section.

The signal subspace can also be expanded in model order using orthogonal projections, allowing a least-squared error $M + 1$st system model to be derived from the Mth order model, and so on. Given a new observation, one would calculate the $m = 0$ solution, then calculate the $m = 1$ solution using the $m = 0$ information, and so on up to the $m = M$ order model. Since the response of linear time-invariant systems can be represented by a weighted sum of linearly independent functions, it is straightforward to consider the orthogonal-projection order updates a well matched framework for identification of linear systems. The projection operator framework allows the subspace to be decomposed in time and order as an effective means to achieve very rapid convergence to a least-squared error solution using very few observations. The block least-squares approach will obtain exactly the same result for a particular model order, but only that model order. The projection operator framework actually allows the model order to be evaluated and updated along with the model parameters. Under the conditions of noiseless signals and too high a chosen model order, the required matrix inverse in block least-squares is ill-conditioned due to linear dependence. Updating the model order in a projection operator framework allows one to avoid over determining the model order and the associated linear dependence which prohibits the matrix inversion in the least-squares solution. The projection operator framework can also be seen as a more general representation of the Eigenvalue problem and singular value decomposition. The recursive update of both the block least-squares and projection operator framework will be left to Chapter 9.

Without getting deep into the details of linear operators and Hilbert space theory, we can simply state that $\hat{G}$ is a subspace of the infinite Hilbert space $\hat{H}$.

Given a new observation y_n in the vector $\bar{y}$ but not in the subspace spanned by $\hat{G}$, we define a vector g as the projection of $\bar{y}$ onto the subspace $\hat{G}$, and f as a vector orthogonal to the subspace $\hat{G}$ which connects g to $\bar{y}$. One can think of the projection of a 3-dimensional vector onto a 2-dimensional plane as the "shadow" cast by the vector onto the plane. Simply put, $\bar{y} = g + f$. One can think of g as the component of $\bar{y}$ predictable in the subspace $\hat{G}$, and f is the least-squared error (shortest distance between the new observation and the prediction as defined by an orthogonal vector) of the prediction for $\bar{y}$. The prediction error is $\bar{\varepsilon} = f = \bar{y} - g$. Substituting the least-squared error solution for our model H, we can examine the implications of casting our solution in a projection operator framework.

$$
\begin{aligned}
\bar{\varepsilon} &= \bar{y} - \bar{X}H \\
&= \bar{y} - \bar{X}(\bar{X}^H \bar{X})^{-1} \bar{X}^H \bar{y}
\end{aligned}
\tag{8.2.1}
$$

The projection of $\bar{y}$ onto the subspace $\hat{G}$, denoted as the vector g above is simply

$$
P_X \bar{y} = \bar{X}(\bar{X}^H \bar{X})^{-1} \bar{X}^H \bar{y}
\tag{8.2.2}
$$

where P_X is a projection operator for the subspace $\hat{G}$ spanned by the elements of $\hat{H}$. The projection operator outlined in Eqs (8.2.1)–(8.2.2) has the interesting properties of being *bounded, having unity norm,* and being *self-adjoint.* Consider the square of a projection operator.

$$
\begin{aligned}
P_X^2 &= \bar{X}(\bar{X}^H \bar{X})^{-1} \bar{X}^H \bar{X}(\bar{X}^H \bar{X})^{-1} \bar{X}^H \\
&= \bar{X}(\bar{X}^H \bar{X})^{-1} \bar{X}^H = P_X
\end{aligned}
\tag{8.2.3}
$$

The projection operator orthogonal to the subspace $\hat{G}$ is simply

$$
I - P_X = I - \bar{X}(\bar{X}^H \bar{X})^{-1} \bar{X}^H
\tag{8.2.4}
$$

thus, it is easy to show that $(I - P_X)P_X = 0$, where I is the identity matrix (all elements zero except the main diagonals which are all unity). Therefore, the error vector can be written as the orthogonal projection of the observations to the subspace as seen in Figure 3 and Eq. (8.2.5).

$$
\bar{\varepsilon} = (I - P_X)\bar{y}
\tag{8.2.5}
$$

This still seems like a roundabout way to re-invent least-squares, but consider updating the subspace spanned by the elements of $\bar{X}$ to the space spanned by $\bar{X} + \bar{S}$. We can update the subspace by again applying an orthogonal projection.

$$
\bar{X} + \bar{S} = \bar{X} + \bar{S}(I - P_X)
\tag{8.2.6}
$$

The projection operator for the updated subspace is

$$
\begin{aligned}
P_{\{X+S\}} &= P_X + P_{\{S(I-P_X)\}} \\
&= P_X + (I - P_X)\bar{S}\left[\bar{S}^H(I - P_X)\bar{S}\right]^{-1} \bar{S}^H(I - P_X)
\end{aligned}
\tag{8.2.7}
$$

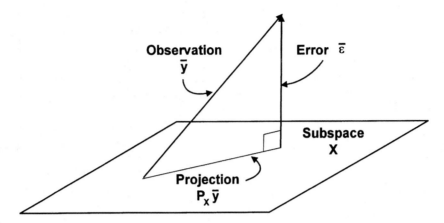

Figure 3 Graphical depiction of projection-based least-squared error prediction showing the error vector as an orthogonal projection.

The orthogonal projection operator to the updated subspace $S + X$ is simply

$$I - P_{\{X+S\}} = I - P_X - (I - P_X)\bar{S}\left[\bar{S}^H(I - P_X)\bar{S}\right]^{-1}\bar{S}^H(I - P_X) \qquad (8.2.8)$$

Equation (8.2.8) can be used to generate a wide range of recursive updates for least-squared error algorithms. The recursion in Eq. (8.2.8) is remarkably close to the recursion derived from the matrix inversion lemma presented in Chapter 9. But, this difference is due only to the different applications of matrix inversion and orthogonal projection and will be revisited in more detail in Chapter 9. However, it is enlightening to observe the mathematical uniformity of least-squares recursions. Casting an error signal as an orthogonal projection by definition guarantees the minimum error as the shortest distance between a plane and a point in space is a line normal to the plane and intersecting the point.

8.3 GENERAL BASIS SYSTEM IDENTIFICATION

Consider a more general framework for least-squared error system modeling. We have shown a straightforward application for system identification using a digital FIR filter model and the input–output signals. The basis function for the FIR filter model was a digital filter's tapped delay line, or input sequence

$$[x_n \; x_{n-1} \; x_{n-2} \ldots x_{n-M}] \qquad (8.3.1)$$

where x_n is the input signal to the filter. Gauss's technique, developed nearly two centuries ago, was used for many things long before digital filters even existed. By applying several other basis functions, we will see the great power of least-squared error system modeling, even for non-linear systems. The reader should understand that any basis function can be used to produce a linear least-squared error model. But the choice of basis function(s) will determine the optimality of the the least-squared error model.

Consider a very simple task of fitting a polynomial curve to four pairs of observations $\bar{y} = [1.1\ 6.0\ 8.0\ 26.5]$ at the ordinates $x = [1\ 2\ 3\ 4]$. For a linear curve fit, the basis function is simply $\phi_n = [x_n\ 1]$ and the subspace X is spanned by the elements $\bar{X}$ of defined by

$$\bar{X} = \begin{bmatrix} 1 & 1 \\ 2 & 1 \\ 3 & 1 \\ 4 & 1 \end{bmatrix} \tag{8.3.2}$$

The least-squared error solution for a linear (straight line) fit is

$$H = (\bar{X}^H \bar{X})^{-1} \bar{X}^H \bar{y}$$

$$\begin{bmatrix} 7.82 \\ -9.15 \end{bmatrix} = \begin{bmatrix} 0.2 & -0.5 \\ -0.5 & 1.5 \end{bmatrix} \begin{bmatrix} 1 & 2 & 3 & 4 \\ 1 & 1 & 1 & 1 \end{bmatrix} \begin{bmatrix} 1.1 \\ 6 \\ 8 \\ 26.5 \end{bmatrix} \tag{8.3.3}$$

If we were to write to write the equation of the line in slope-intercept form, $y = 7.82x - 9.15$, where the slope is 7.82 and the intercept of the y-axis is -9.15. The projection operator for the linear fit is symmetric about both diagonals, but the main diagonal does not dominate the magnitudes indicating a rather poor fit as seen in the error vector.

$$P_X = \begin{bmatrix} 0.7 & 0.4 & 0.1 & -0.2 \\ 0.4 & 0.3 & 0.2 & 0.1 \\ 0.1 & 0.2 & 0.3 & 0.4 \\ -0.2 & 0.1 & 0.4 & 0.7 \end{bmatrix} \quad \bar{\varepsilon} = (I - P_X)\bar{y} = \begin{bmatrix} 2.43 \\ -0.49 \\ -6.31 \\ 4.37 \end{bmatrix} \tag{8.3.4}$$

The mean-squared error for the linear fit in Eqs (8.3.2) through (8.3.4) is 16.26.

Fitting a quadratic function to the data simply involves expanding the basis function as $\phi_n = [x_n^2\ x_n\ 1]$.

$$H = (\bar{X}^H \bar{X})^{-1} \bar{X}^H \bar{y}$$

$$\begin{bmatrix} 3.40 \\ -9.18 \\ 7.85 \end{bmatrix} = \begin{bmatrix} 0.25 & -1.25 & 1.25 \\ -1.25 & 6.45 & -6.75 \\ 1.25 & -6.75 & 7.75 \end{bmatrix} \begin{bmatrix} 1 & 4 & 9 & 16 \\ 1 & 2 & 3 & 4 \\ 1 & 1 & 1 & 1 \end{bmatrix} \begin{bmatrix} 1.1 \\ 6 \\ 8 \\ 26.5 \end{bmatrix} \tag{8.3.5}$$

The projection operator and error vector for the quadratic fit are

$$P_X = \begin{bmatrix} 0.95 & 0.15 & -0.15 & -0.05 \\ 0.15 & 0.55 & 0.45 & -0.15 \\ -0.15 & 0.45 & 0.55 & 0.15 \\ 0.05 & -0.15 & 0.15 & 0.95 \end{bmatrix} \quad \bar{\varepsilon} = (I - P_X)\bar{y} = \begin{bmatrix} -0.97 \\ 2.91 \\ -2.91 \\ 0.99 \end{bmatrix} \tag{8.3.6}$$

The mean-squared error for the quadratic fit is 4.71, significantly less than the linear fit. Clearly, the projection operator for the quadratic fit, while still symmetric about both diagonals, is dominated by the main diagonal element magnitudes, indicating a better overall fit for the data. Fitting a cubic basic function to the data we get the model parameters $H = [3.233 - 20.85\ 44.82 - 26.10]$, P_X is the identity matrix, and the error is essentially zero. Figure 4 shows the linear, quadratic, and cubic modeling results graphically.

It is expected that the least-squared error fit for the cubic basis function would be zero (within numerical error limits — the result in this case was 1×10^{-23}). This is because there are only 4 data observations and four elements to the basis function, or four equations and four unknowns. When the data is noisy (as is the case here), having too few observations can cause a misrepresentation of the data. In this case the data is really from a quadratic function, but with added noise. If we had many more data points it would be clear that the quadratic model really provides the best overall fit, especially when comparing the cubic model for the 4-observation data to a more densely-populated distribution of observations. In any modeling problem with noisy data it is extremely important to overdetermine the model by using many more observations than the basis model order. This is why distribution measures such as the Student's t-test, F-test, etc., are used extensively in clinical trials where only small numbers of subjects are available. In most adaptive signal processing models, large data sets are readily available, and this should be exploited wherever possible.

Consider the highly nonlinear frequency-loudness response of the human ear as an example. Audiologists and noise control engineers often use a weighted dB value to describe sound in terms of human perception. According to Newby (3), the A weighted curve describes a dB correction factor for the human ear for sound levels below 55 dB relative to 20 μPa, the B curve for levels between 55 and 85 dB,

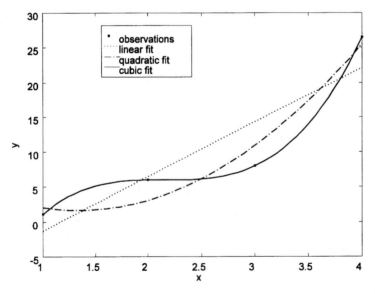

Figure 4 Linear, quadratic, and cubic least-squared error models fitting 4 observation points.

and the C curve for levels over 85 dB. The A and C curves are predominantly used. Most hearing conservation legislation uses the A curve since occupational hearing impairment occurs in the middle frequency range where the A curve is most sensitive. For loud sounds, the brain restricts the levels entering the ear via tendons on the bones of the middle ear and neural control of the mechanical response of the cochlea in the organ of Corti. This action is analogous to the action of the iris in the eye closing down in response to bright light. In the age of signal processing tools such as spectrum analyzers and Matlab, it is useful to have audioband curves to quickly convert a narrowband sound spectrum to an A weighted level. To obtain a straightforward algorithm for curve generation, the method of least-squared error is employed to fit a curve accurately through a series of well-published dB-frequency coordinates. The ear is an absolutely remarkable intelligent sensor system which uses adaptive response to optimize hearing in a wide range of environments. It is very useful from a sensor technology point-of-view, to analyze the physics and physiology of the ear in the context of the least-squared error modeling of its frequency-loudness response to acoustic signals.

Mechanics of the Human Ear

The human ear is a remarkable transducer, which by either evolution or divine design, has mechanisms to adapt to protect itself while providing us detection of a wide range of sounds. If we define 0 dB, or 20 μPa, as the "minimum" audible sound detectable above the internal noise from breathing and blood flow for an average ear, and 130 dB as about the loudest sound tolerable, our ears have typically a 10^6 dynamic pressure range. This is greater than most microphones and certainly greater than most recording systems capable of the ear's frequency response from 20 Hz to 20 kHz. We know that a large part of human hearing occurs in the brain, and that speech is processed differently from other sounds. We also know that the brain controls the sensitivity of the ear through the auditory nerve. Figure 5 shows the anatomy of the right ear looking from front to rear.

Like the retina of the eye, overstimulation of the vibration-sensing "hair cells" in the cochlea can certainly lead to pain and even permanent neural cell damage. The brain actually has two mechanisms to control the stimulation levels in the cochlea. First, the bones of the middle ear (mallus, incus, and stapes) can be restricted from motion by the *tendon of Stapedious muscle* as seen in Figure 5 in much the same way the iris in the eye restricts light. As this muscle tightens, the amplitude of the vibration diminishes and the frequency response of the middle ear actually flattens out so that low, middle, and high frequencies have nearly the same sensitivity. For very low-level sounds, the tendon relaxes allowing the stapes to move more freely. The relaxed state tends to significantly enhance hearing sensitivity in the middle speech band (from 300 to 6000 Hz approximately). Hence the A weighting reflects a significant drop in sensitivity at very low and very high frequencies. The second way the brain suppresses overstimulation in the cochlea involves the response of the hair cells directly. It is not known whether the neural response for vibration control affects the frequency response of the ear, or whether the tendon of Stapedious in conjunction with the neural response together cause the change in the ear's frequency response as a function of loudness level. However, the average frequency responses of healthy ears have been objectively measured for quite some

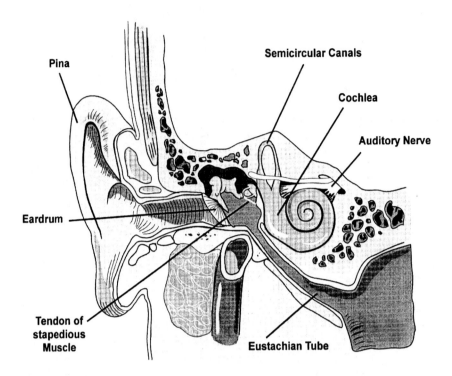

Figure 5 Anatomy of the right human ear looking from front to rear showing the tendons used by the brain to control the loudness of the sound reaching the inner ear cochlea and auditory nerve.

time. The "loudness" button on most high fidelity music playback systems inverts the ear's frequency response in conjunction with the "volume" control so that music can have a rich sounding bass response at low listening levels. Figure 6 shows the relative frequency correction weightings for the A, B, and C curves.

It is useful to have the ability to precisely generate the A, B, or C weighting curves for the human ear's response for any desired frequency in the audible range. This capability allows one to easily convert a power spectrum in dB re 20 μPa of some arbitrary frequency resolution directly to an A, B, or C-weighted dB reading. The American National Standards Institute (ANSI) standard for sound level meter specifications provides tables of relative dB weightings as a function of frequency for the human ear (4). Additional tables are given specifying the accuracy required in ± dB vs. frequency for a sound level meter (SLM) to be considered Type 0 (most accurate), Type 1 (roughly ±1 dB in the 50 Hz to 4000 Hz range), or Type 2 (economical accuracy). The ANSI standard does not specify A, B, and C curve equations, just the Tables representing the accepted "normal" human response.

Least-Squares Curve Fitting

To generate a model for a continuous curve to be used to map narrowband FFT spectra calibrated in Pascals to A, B, or C weighted sound pressure levels, the least-squared error technique is employed. We begin with 7 decibel observations

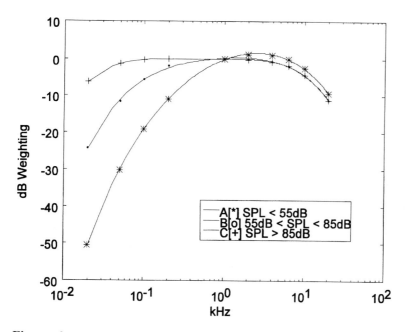

Figure 6 A, B, and C, weighting curves for modeling the human ear's relative frequency response to sound at various levels of loudness.

Table 1 dB Correction Factors for A, B, and C Weightings

Freq Hz	19.95	50.12	100	199.5	1000	1995	3981	6310	10000	20000
A	−50.5	−30.2	−19.1	−10.9	0	1.2	+1.0	−0.1	−3.0	−9.3
B	−24.2	−11.6	−5.6	−2	0	−0.1	−0.7	−1.9	−4.3	−11.1
C	−6.2	−1.3	−0.3	0	0	−0.2	−0.8	−2.0	−4.0	−11.2

at 7 frequencies which are simply read from existing A, B, and C curves in the literature given in Table 1. These data observations can be easily seen in Figure 6 as the symbols plotted on the corresponding curves. As noted earlier, the B curve is generally only used in audiology, but is useful to illustrate the sensitivity changes in the ear between 55 dB and 85 dB.

To improve the conditioning of the curve-fitting problem, we use the base-10 logarithm of the frequency in kHz in the basis function and a curve model of the form

$$dB_H^x(f) = \sum_{m=1}^{M} H_m^x f_\ell^{m-1} \tag{8.3.7}$$

where f_ℓ is the log base 10 of the frequency in kHz, the superscript "x" refers to either A, B, or C weighting, and the subscript "H" means the dB value is predicted using the weights H_m. The use of the base 10 logarithm of the frequency in kHz may seem unnecessary, but it helps considerably in conditioning the matrices in the least-squares fitting problem over the 9 octave frequency range from 20 Hz to 20 kHz. The error between our model in Eq. (8.2.7) and the ANSI table data is

$$\mathscr{E}(f) = dB_T^x(f) - dB_H^x(f) = dB_T^x(f) - \left[1 f_\ell f_\ell^2 \ldots f_\ell^{M-1}\right] \begin{bmatrix} H_1^x \\ H_2^x \\ \vdots \\ H_M^x \end{bmatrix} \tag{8.3.8}$$

where the subscript "T" means that the dB data is from the ANSI tables. We want our model, defined by the weights H_m and basis function $[1\ f_\ell\ f_\ell^2 \ldots f_\ell^{M-1}]$, to provide minimum error over the frequency range of interest. Therefore, we define Eq. (8.3.8) in matrix form for a range of N frequencies.

$$\begin{bmatrix} \mathscr{E}(f_1) \\ \mathscr{E}(f_2) \\ \vdots \\ \mathscr{E}(f_N) \end{bmatrix} = \begin{bmatrix} dB_T^x(f_1) \\ dB_T^x(f_2) \\ \vdots \\ dB_T^x(f_N) \end{bmatrix} - \begin{bmatrix} 1 & f_{\ell,1} & f_{\ell,1}^2 & \cdots & f_{\ell,1}^{M-1} \\ 1 & f_{\ell,2} & f_{\ell,2}^2 & \cdots & f_{\ell,2}^{M-1} \\ & & \vdots & & \\ 1 & f_{\ell,N} & f_{\ell,N}^2 & \cdots & f_{\ell,N}^{M-1} \end{bmatrix} \begin{bmatrix} H_1^x \\ H_2^x \\ \vdots \\ H_M^x \end{bmatrix} \tag{8.3.9}$$

Equation (8.3.9) is simply written in compact matrix form as

$$\bar{E} = \bar{D} - \bar{F}\bar{H} \tag{8.3.10}$$

Therefore, the least-squared error solution for the weights which best fit a curve through the ANSI table dB values is seen in Eq. (8.3.11). The resulting 5th-order model coefficients for the A, B, and C-weighted curves in Figure 6 are given in Table 2.

$$\bar{H} = (\bar{F}'\bar{F})^{-1}\bar{F}'\bar{D} \tag{8.3.11}$$

Pole-Zero Filter Models

However, historically, the functions for the A, B, and C curves are given in terms of pole-zero transfer functions, which can be implemented as an analog circuit directly in the SLM. The method for generating the curves is wisely not part of the ANSI standard, since there is no closed form model defined from physics. One such pole-zero model is given in Appendix C of ANSI S1.4-1983 (with the disclaimer

Table 2   5th Order Least-Squares Fit Coefficients for A, B, and C Curves

m	H^A	H^B	H^C
1	−0.1940747	+0.1807204	−0.07478983
2	+8.387643	+1.257416	+0.3047574
3	−9.616735	−3.32772	−0.2513878
4	−0.2017488	−0.7022932	−2.416345
5	−1.111944	−1.945072	−2.006099

"for informational purposes only"). The C-weighting curve in dB is defined as

$$dB^C(f) = 10 \log_{10}\left(\frac{K_1 f^4}{(f^2 + f_1^2)^2 (f^2 + f_4^2)^2}\right)$$

(8.3.12)

where K_1 is 2.242881×10^{16}, $f_1 = 20.598997$, $f_4 = 12194.22$, and f is frequency in Hz. The B-weighting is defined as

$$dB^B(f) = 10 \log_{10}\left(\frac{K_2 f^2}{f^2 + f_5^2}\right) + dB^C(f)$$

(8.3.13)

where K_2 is 1.025119 and f_5 is 158.48932. The A-weighting curve is

$$dB^A(f) = 10 \log_{10}\left(\frac{K_3 f^4}{(f^2 + f_2^2)^2 (f^2 + f_3^2)}\right) + dB^C(f)$$

(8.3.14)

where K_3 is 1.562339, f_2 is 107.65265, and f_3 is 737.86223. The curves generated by Eqs (8.3.12)–(8.3.13) are virtually identical to those generated using least-squares. A comparison between the least-squares fit and the "ANSI" pole-zero models is seen in Figure 7.

Since the highest required precision for a SLM is Type 0 where the highest precision in any frequency range is ± 0.7 dB, the variations in Figure 7 for the least-squares fit are within a Type 0 specification. The European Norm EN60651 (IEC651) has slightly different formulation for a pole-zero model for the A, B, and C curves, but the listed numbers tabulated are identical to the ANSI curve

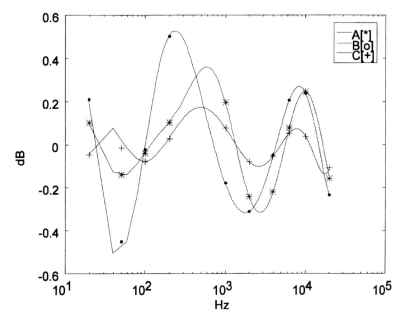

Figure 7 dB error between the least-squares fit and the pole-zero model given "for informational purposes only" in Appendix C of ANSI S1.4-1983.

responses. Using either curve definition, one can access the A, B, and C weightings for conversion of narrowband spectra to broadband weighted readings for modeling the human ear's response to sound. Accuracy of the least-squared error model can be improved by using more frequency samples in between the samples given in Table 1. In general, the least-squares fit will best match the model *at the basis function sample points*. In between these input values, the model can be significantly off the expected trend most notably when the number of input samples in the basis function N is close to the model order M. N must be greater than M for a least-squared error solution to exist, but $N \gg M$ for a very accurate model to be found from potentially noisy data.

As our final example basis function in this section, we consider the Fourier transform cast in the form of a least-squares error modeling problem. The basis function is now a Fourier series of complex exponentials of chosen particular frequencies and our N observations $\bar{y} = [y_n \; y_{n-1} \; \cdots \; y_{n-N+1}]$ are of some time series of interest.

$$\bar{\varepsilon} = \bar{y} - \begin{bmatrix} e^{j\omega_1 nT} & e^{j\omega_2 nT} & \cdots & e^{j\omega_M nT} \\ e^{j\omega_1 (n-1)T} & e^{j\omega_2 (n-1)T} & \cdots & e^{j\omega_M (n-1)T} \\ \vdots & \vdots & \cdots & \vdots \\ e^{j\omega_1 (n-N+1)T} & e^{j\omega_2 (n-N+1)T} & \cdots & e^{j\omega_M (n-N+1)T} \end{bmatrix} \begin{bmatrix} A_1 \\ A_2 \\ \vdots \\ A_M \end{bmatrix} \qquad 8.3.15)$$

The least-squared error solution for the chosen basis functions (choice of ω^s and M) is simply

$$A_m = \frac{1}{N} \sum_{n'=n}^{n'=n-N+1} y_n' e^{-j\omega_m n' T} \qquad (8.3.16)$$

which is consistent with the normalized discrete Fourier transform presented in Chapter 5, Eq. (5.1.2). The solution is greatly simplified by the orthogonality of the complex sinusoids (provided that the span of N samples constitute an integer number of wavelengths in the observation window). Orthogonality of the basis functions simplifies the matrix inverse $(\bar{X}^H \bar{X})^{-1}$ by making it essentially a scalar N^{-1} times the identity matrix I. Complex sinusoids are also orthonormal when the factor of N is removed by scaling. Non-orthogonal basis functions can always be used in a least-squared error model, but the matrix inverse must be completely calculated.

If the basis function is a spectrum of the input to an unknown system $X(\omega)$, and the observations are a spectrum of the output signal from an unknown system $Y(\omega)$, then the model is the frequency response of the unknown system $H(\omega)$. It is useful to compare the least-squared error spectrum solution to the transfer function measurement previously developed in Section 6.2. The transfer function solution using least squares and time-domain input–output data is seen in Eq. (8.3.17) for $X(\omega)$

white.

$$
\begin{bmatrix} h_0 \\ h_1 \\ \vdots \\ h_{M-1} \end{bmatrix} = \begin{bmatrix} R_0^x & R_1^x & \cdots & R_{M-1}^x \\ R_{-1}^x & R_0^x & \cdots & R_{M-2}^x \\ \vdots & & \ddots & \vdots \\ R_{-M+1}^x & R_{-M+2}^x & \cdots & R_0^x \end{bmatrix}^{-1} \begin{bmatrix} R_0^{xy} \\ R_{-1}^{xy} \\ \vdots \\ R_{-M+1}^{xy} \end{bmatrix} = \frac{1}{R_0^x} \cdot \begin{bmatrix} R_0^{xy} \\ R_{-1}^{xy} \\ \vdots \\ R_{-M+1}^{xy} \end{bmatrix}
$$

$$(8.3.17)$$

Equation (8.3.17) is the same result presented in Eq. (8.1.14) in the limit as N approaches infinity. Since $X(\omega)$ is spectrally white, the basis functions are all "digital" Dirac delta functions making the matrix inverse in Eq. (8.3.17) a simple scalar inverse. Taking Fourier transforms of both sides of Eq. (8.3.17) we have the transfer function expression from Section 6.2.

$$
H(\omega) = \frac{G^{xy}(\omega)}{G^{xx}(\omega)}
\tag{8.3.18}
$$

The projection operator is the identity matrix when the input signal is orthogonal for the observation window N. For large N and reasonably white $X(\omega)$ the characteristics of Eqs (8.3.17)–(8.3.18) generally hold as stated. However, for non-orthogonal N (spectral leakage cases), the frequency domain solution for the frequency response $H(\omega)$ in Eq. (8.3.18) may have significant error compared to the time domain solution including the full matrix inverse. The reason for this is that in the time domain solution, the model is free to move its zeros anywhere to best match the dominant peaks in the spectral response. But with a frequency domain basis function, the specific frequencies are pre-chosen and can result in a bias error in the model. Obviously, for very large N and high model orders, the two solutions are essentially identical.

8.4 SUMMARY, PROBLEMS, AND BIBLIOGRAPHY

The manual computations required in Gauss's day severely limited the applications of least-squares to only a few well-defined problems. It is humbling, to say the least, that an individual at the turn of the 19th century could manually make the numerical calculations necessary to predict the position of an asteroid 9 months in advance. Remarking once that the least-squares technique would be useful on a much wider scale if a suitable machine could be built to automate some of the calculations, Gauss made probably the greatest understatement of the last two centuries! The advent of the digital computer has made application of least-squared error modeling as pervasive in business law, sociology, psychology, medicine, and politics as it is in engineering and the hard sciences.

The least-squared error problem can be cast into a projection operator inner product space which is very useful for derivation of fast recursive adaptive processing algorithms. The observation data for a particular problem can be seen as a subspace where the system of interest is a linear manifold, the response of which can be modeled as a weighted linear combination of orthogonal eigenvectors. The subspace spanned by the observations of the system input signal form the basis of a projection operator. The orthogonal projection of the observed system output signal vector onto the input signal subspace determines the least-squared error solution. The value of the projection operator framework is that the subspace can be expanded using orthogonal projections for model order as well as observation window size. The orthogonal decomposition of the subspace allows for very efficient (fast converging) recursive adaptive algorithms to be developed such as the least-squares lattice adaptive algorithm.

We have shown a very straightforward application of least-squared error modeling on an FIR filter where the basis function is a simple delay line of filter input samples. However, a much wider range of basis functions can be used, including nonlinear functions. What is critical is that the error response, the difference between the actual and predicted outputs, be a linear function of the model coefficients. The linear error model allows a quadratic squared error surface which is minimized through the choice of model coefficients which gives a zero gradient on the error surface. The fact that the zero-gradient solution is a minimum of the error surface is verified by examination of the second derivative for a positive definite condition.

PROBLEMS

1. Given 4 temperatures and times, find β and T_0 for the model $T(t) = T_0 e^{-t/\beta}$. $T(60) = 63.8$, $T(120) = 19.2$, $T(180) = 5.8$, and $T(240) = 1.7$. Assume t is in seconds and T is $°C$.

2. Do basis functions have to be orthogonal for the least-squared error technique to work?

3. Does the underlying error model have to be linear for the least-squared error technique to work?

4. A stock market index has the following values for Monday through Thursday: {1257, 1189, 1205, 1200}.

   (a) Using a 4-day moving average linear model fit, what do you predict the index will be by Friday close?

   (b) Using a 5-day period sinusoidal basis function, what do you expect Friday's index to be?

5. Show that the optimal mean-squared error can be written as $\bar{\varepsilon}^H \bar{\varepsilon} = \frac{1}{N} \bar{y}^H (I - P_X) \bar{y}$.

6. Show that the projection operator is self-adjoint and has unity norm.

7. Show that as the projection operator P_X approaches the identity matrix, the mean square error must go to zero.

8. Show that $P_X \bar{y} = \bar{X} H_0$, where H_0 is the least-squared error set of coefficients.

9. You buy a stock at $105/share. Mondays closing price is $110. Tuesday's closing price drops to $100. Wednesday's closing price is back up to $108. But then Thursday's closing price only drops to $105. Based on a linear model, should you sell on Friday morning?

BIBLIOGRAPHY

M. Bellanger Adaptive Digital Filters and Signal Analysis, Marcel-Dekker: New York, 1988.

G. E. P. Box, Jenkins Time Series Analysis: Forecasting and Control, San Francisco: Holden-Day, 1970.

T. Kailath Linear Systems, Englewood Cliffs: Prentice-Hall, 1980.

S. J. Orfandis Optimum Signal Processing, 2nd Ed., New York: McGraw-Hill, 1988.

B. Widrow, S. D. Sterns Adaptive Signal Processing, Englewood Cliffs: Prentice-Hall, 1985.

REFERENCES

1. C. B. Boyer A History of Mathematics, 2nd Ed., New York: Wiley, 1991, pp. 496–508.

2. N. I. Akhiezer, I. M. Glazman Theory of Linear Operators in Hilbert Space, New York: Dover, 1993.

3. Hayes A. Newby Audiology, 4th Ed., Englewood Cliffs: Prentice-Hall, 1979.

4. ANSI SI.4-1983, "Specification for Sound Level Meters", ASA Catalog No. 47-1983, American Institute of Physics, 335 East 45th St, New York, NY 10017, (516) 349–7800.

9

Recursive Least-Squares Techniques

The powerful technique of fitting a linear system model to the input–output response data with least-squared error can be made even more useful by developing a recursive form with limited signal data memory to adapt with nonstationary systems and signals. One could successfully implement the block least-squares in Section 8.1 on a sliding record of N input–output signal samples. However, even with today's inexpensive, speedy, and nimble computing resources, a sliding-block approach is ill-advised. It can be seen that the previous $N-1$ input–output samples and their correlations are simply being recomputed over and over again with each new sample as time marches on. By writing a recursive matrix equation update for the input data autocorrelations and input–output data cross correlations we can simply add on the necessary terms for the current input–output signal data to the previous correlation estimates, thereby saving a very significant amount of redundant computations. Furthermore, the most precise and efficient adaptive modeling algorithms will generally require the least amount of overall computation. Every arithmetic operation is a potential source of numerical error, so the fewer redundant computations the better. Later we will show simplifications to the recursive least-squares algorithm which require very few operations but converge more slowly. However, these simplified algorithms can actually require more operations over many more iterations to reach a least-squared error solution and may actually not produce as precise a result. This is particularly true for non-stationary signals and systems.

To make the recursive least-squares algorithm adaptive, we define an exponentially decaying memory window, which weights the most recent data the strongest and slowly "forgets" older signal data which is less interesting and likely from an "out-of-date" system relative to the current model. Exponential memory weighting is very simple and only requires the previous correlation estimate. However, a number of other shapes of recursive data memory windows can be found (1), but are rarely called for. Consider a simple integrator for a physical parameter w_t for the wind speed measured by a cup-vane type anemometer.

$$\bar{W}_t = \alpha \bar{W}_{t-1} + \beta w_t; \qquad \alpha = \frac{N-1}{N} \qquad \beta = \frac{1}{N} \tag{9.0.1}$$

The parameter N in Eq. (9.0.1) is the exact length of a linear memory window in terms of number of wind samples. If the linear memory window starts at $N = 1$, $\alpha = 0$ and $\beta = 1$, at $N = 2$ $\alpha = \frac{1}{2}$ and $\beta = \frac{1}{2}$, at $N = 3$, $\alpha = \frac{2}{3}$ and $\beta = \frac{1}{3}$, $N = 4$ $\alpha = \frac{1}{4}$ and $\beta = \frac{3}{4}$, and so on. For any given linear data window length N, the computed α and β provide an *unbiased* estimate of the mean wind speed. However, if at sample 100 we fix $\alpha = 0.99$ and $\beta = 0.01$, we have essentially created an exponentially- weighted data memory window effectively 100 samples long. In other words, a wind speed sample 100 samples old is discounted by $1/e$ in the current estimate of the mean wind speed. The "forgetting factor" of 0.99 on the old data can be seen as a low-pass moving average filter. The wind speed data is made more useful by averaging the short-term turbulence while still providing dynamic wind speeds in the changing environment. If the anemometer is sampled at a 1 Hz rate, the estimated mean applies for the previous 2 minutes (approximately), but is still adaptive enough to provide measurements of weather fronts and other events. The basic integrator example in Eq. (9.0.1) is fundamental in its simplicity and importance. All sensor information has a value, time and/or spatial context and extent, as well as a measurement confidence. To make the most use of raw sensor information signals in an intelligent signal processing system, one must optimize the information confidence as well as context which is typically done through various forms of integration and filtering.

9.1 THE RLS ALGORITHM AND MATRIX INVERSION LEMMA

The recursive least-squares (RLS) algorithm simply applies the recursive mean estimation in Eq. (9.0.1) to the autocorrelation and crosscorrelation data outlined in Section 8.1 for the block least-squares algorithm. Using the formulation given in Eqs (8.1.1)–(8.1.7), a recursive estimate for the input autocorrelation matrix data is

$$(\bar{X}^H \bar{X})_{n+1} = \alpha(\bar{X}^H \bar{X})_n + \phi_{n+1}^H \phi_{n+1} \tag{9.1.1}$$

where we do not need a β term on the right since

$$\lim_{n \to N} E\{(\bar{X}^H \bar{X})_n\} = N \lim_{n \to N} E\{\phi_n^H \phi_n\} \tag{9.1.2}$$

where N is the size of the data memory window. A similar relation is expressed for the cross correlation of the input data x_t and output data y_t.

$$\lim_{n \to N} E\{(\bar{X}^H \bar{y})\} = N \lim_{n \to N} E\{\phi_n^H y_n\} \tag{9.1.3}$$

A recursive update for the cross correlation is simply

$$(\bar{X}^H \bar{y})_{n+1} = \alpha(\bar{X}^H \bar{y})_n + \phi_{n+1}^H y_{n+1} \tag{9.1.4}$$

However, the optimal filter H requires the inverse of Eq. (9.1.1) times (9.1.4). Therefore, what we really need is a recursive matrix inverse algorithm.

The matrix inversion lemma provides a means to recursively compute a matrix inverse when the matrix itself is recursively updated in the form of "$A_{\text{new}} = A_{\text{old}} +$

BCD". The matrix inversion lemma states

$$(A + BCD)^{-1} = A^{-1} - A^{-1}B(C^{-1} + DA^{-1}B)^{-1}DA^{-1} \qquad (9.1.5)$$

where A, C and $DA^{-1}B$, and BCD are all invertible matrices or invertible matrix products. It is straightforward to prove the lemma in Eq. (9.1.5) by simply multiplying both sides by $(A + BCD)$. Rewriting Eq. (9.1.1) in the $A + BCD$ form, we have

$$\alpha^{-1}(\bar{X}^H \bar{X})_{n+1} = (\bar{X}^H \bar{X})_n + \phi_{n+1}^H \alpha^{-1} \phi_{n+1} \qquad (9.1.6)$$

Taking the inverse and applying the matrix inversion lemma gives

$$\alpha(\bar{X}^H \bar{X})_{n+1}^{-1} = (\bar{X}^H \bar{X})_n^{-1} - \frac{(\bar{X}^H \bar{X})_n^{-1} \phi_{n+1}^H \phi_{n+1}(\bar{X}^H \bar{X})_n^{-1}}{\alpha + \phi_{n+1}(\bar{X}^H \bar{X})_n^{-1} \phi_{n+1}^H} \qquad (9.1.7)$$

which is more compactly written in terms of a Kalman gain vector K_{n+1} in Eq. (9.1.8).

$$(\bar{X}^H \bar{X})_{n+1}^{-1} = \alpha^{-1}\big[I - K_{n+1}\phi_{n+1}\big](\bar{X}^H \bar{X})_n^{-1} \qquad (9.1.8)$$

The Kalman gain vector K_{n+1} has significance well beyond a notational convenience. Equations (9.1.7) and (9.1.9) show that the denominator term in the Kalman gain is a simple scalar when the basis function vector ϕ_{n+1} is a row vector.

$$K_{n+1} = \frac{(\bar{X}^H \bar{X})_n^{-1} \phi_{n+1}^H}{\alpha + \phi_{n+1}(\bar{X}^H \bar{X})_n^{-1} \phi_{n+1}^H} \qquad (9.1.9)$$

Clearly, inversion of a scalar is much less a computational burden than inversion of a matrix. When x_n is stationary, one can seen that the Kalman gain simply decreases with the size of the data memory window N as would be expected for an unbiased estimate. However, if the statistics (autocorrelation) of the most recent data is different from the autocorrelation matrix, the Kalman gain will automatically increase causing the recursion to "quickly forget" the old outdated data. One of the more fascinating aspects of recursive adaptive algorithms is the mathematical ability to maintain least-squared error for nonstationary input–output data. Several approximations to Eq. (9.1.9) will be shown later which still converge to the same result for stationary data, but more slowly due to the approximations.

Our task at the moment is to derive a recursion for the optimal (least-squared error) filter H_{n+1}, given the previous filter estimate H_n and the most recent data. Combining Eqs (9.1.8) and (9.1.4) in recursive form gives

$$H_{n+1} = \big[\alpha^{-1}(\bar{X}^H \bar{X})_n^{-1} - \alpha^{-1}K_{n+1}\phi_{n+1}(\bar{X}^H \bar{X})_n^{-1}\big]\big[\alpha(\bar{X}^H \bar{y})_n + \phi_{n+1}^H y_{n+1}\big] \qquad (9.1.10)$$

which can be shown to reduce to the result in Eq. (9.1.11).

$$H_{n+1} = H_n + K_{n+1}\big(y_{n+1} - y'_{n+1}\big) \qquad (9.1.11)$$

The prediction of y_{n+1} (denoted as y'_{n+1}) is derived from $\phi_{n+1}H_n$, and in some texts is denoted as $y_{n+1|n}$, which literally means "the prediction of y_{n+1} given a model last updated at time n." Again, the significance of the Kalman gain vector can be seen intuitively in Eq. (9.1.11). The filter does not change if the prediction error is zero. But, there is always some residual error, if not due to extraneous noise or model error then to the approximation of the least significant bit in the analog-to-digital convertors in a real system. The impact of the residual noise on the filter coefficients is determined by the Kalman gain K_{n+1}, which is based on the data memory window length and the match between the most recent basis vector statistics and the long-term average in the correlation data. It is humanistic to note that the algorithm has to make an error in order to learn. The memory window as well as variations on the Kalman gain in the form of approximations determine the speed of convergence to the least-squared error solution.

The RLS Algorithm is summarized in Table 1 at step "$n+1$" given new input sample x_{n+1} and output sample y_{n+1}, and the previous estimates for the optimal filter H_n and inverse autocorrelation matrix $P_n = (\bar{X}^H \bar{X})_n^{-1}$

Approximations to RLS can offer significant computational savings, but at the expense of slower convergence to the least-squared error solution for the optimal filter. So long as the convergence is faster than the underlying model changes being tracked by the adaptive algorithm through its input and output signals, one can expect the same optimal solution. However, as one shortens the data memory window, the adaptive algorithm becomes more reactive to the input–output data resulting in more noise in the filter coefficient estimates. These design tradeoffs are conveniently exploited to produce adaptive filtering systems well-matched to the application of interest. The largest reduction in complexity comes from eliminating P_n from the algorithm. This is widely known as the projection algorithm.

$$K_{n+1}^{PA} = \frac{\gamma \phi_{n+1}^H}{\alpha + \phi_{n+1}\phi_{n+1}^H}; \qquad 0 < \gamma < 2, \quad 0 < \alpha < 1 \qquad (9.1.13)$$

It will be shown later in the section on the convergence properties of the LMS algorithm why γ must be less than 2 for stable convergence. Also, when α is positive K_{n+1} remains stable even if the input data becomes zero for a time.

Table 1 The Recursive Least Squares (RLS) Algorithm

Description	Equation
Basis function containing x_{n+1}	$\phi_{n+1} = [x_{n+1}x_nx_{n-1} \ldots x_{n-M+1}]$
Inverse autocorrelation matrix; output prediction	$P_n = (\bar{X}^H \bar{X})_n^{-1}; \quad y'_{n+1} = \phi_{n+1}H_n$
Kalman gain using RLS and exponential memory window N samples long. $\alpha = (N-1)/N$	$K_{n+1} = \dfrac{P_n\phi_{n+1}^H}{\alpha + \phi_{n+1}P_n\phi_{n+1}^H}$
Update for autocorrelation matrix inverse	$P_{n+1} = \alpha^{-1}[I - K_{n+1}\phi_{n+1}]P_n$
Optimal filter update using prediction error	$H_{n+1} = H_n + K_{n+1}[y_{n+1} - y'_{n+1}]$

Another useful approximation comes from knowledge that the input data will always have some noise (making $\phi\phi^H$ nonzero) allowing the elimination of α which makes the memory window essentially of length M. This is known as the stochastic approximation to RLS. Of course, choosing a value of $\gamma < 2$ effectively increases the data memory window as desired.

$$K_{n+1}^{SA} = \frac{\gamma\phi_{n+1}^H}{\phi_{n+1}\phi_{n+1}^H}; \qquad 0 < \gamma < 2 \tag{9.1.14}$$

By far the most popular approximation to RLS is the least-mean squared error, or LMS algorithm defined in Eq. (9.1.15).

$$K_{n+1}^{LMS} = 2\mu\phi_{n+1}^H; \qquad \mu \leq E\{\phi_{n+1}\phi_{n+1}^H\}^{-1} \tag{9.1.15}$$

The LMS algorithm is extremely simple and robust. Since the adaptive step size μ is determined by the inverse of the upper bound of the model order $M+1$ times the expected mean-square of the input data, it is often referred to as the normalized LMS algorithm as presented here. The beauty of the LMS adaptive filtering algorithm is that only one equation is needed as seen in Eq. (9.1.16), and no divides are needed except to estimate μ. In the early days of fixed-point embedded adaptive filters, the omission of a division operation was a key advantage. Even with today's very powerful DSP processors, the simple LMS adaptive filter implementation is important because of the need for processing wider bandwidth signals with faster sample rates. However, the slowdown in convergence can be a problem for large model-order filters. Still, an additional factor $\mu_{rel} < 1$ seen in Eq. (9.1.16) is generally needed in the LMS algorithm as a margin of safety to insure stability. A small μ_{rel} has the effect of an increased memory window approximately as $N = 1/\mu_{rel}$, which is often beneficial in reducing noise in the parameter estimates at the expense of even slow convergence. We will see in the next Section that the LMS algorithm convergence depends on the eigenvalues of the input data.

$$\boldsymbol{H}_{n+1} = \boldsymbol{H}_n + 2\mu_{max}\mu_{rel}\phi_{n+1}^H\left(y_{n+1} - y'_{n+1}\right) \tag{9.1.16}$$

The parameter $\mu_{max} = 1/\{\sigma_x^2\}$ and σ_x^2 is the variance of the input data ϕ_{n+1} and μ_{rel} is the step size component which creates an effective "data memory window" with an exponential forgetting property of approximate length $N = 1/\mu_{rel}$. It can be seen that the shortest possible memory window length for a stable LMS algorithm is the filter length $M+1$ and occurs if $\mu_{rel} = 1/(M+1)$ such that $\mu_{max}\mu_{rel} < 1/\{\phi_{n+1}\phi_{n+1}^H\}$.

9.2 LMS CONVERGENCE PROPERTIES

The least-mean squared error adaptive filter algorithm is a staple technique in the realm of signal processing where one is tasked with adaptive modeling of signals or systems. In general, modeling of a system (system of linear dynamical responses) requires both the input signal exciting the system and the system's response in the form of the corresponding output signal. A signal model is obtainable by assuming the signal is the output of a system driven by zero-mean Gaussian (ZMG), or spectrally white, random noise. The signal model is computed by "whitening" the available output signal using an adaptive filter. The inverse of the converged filter can then be used to generate a model of the signal of interest from a white noise input, called the *innovation*. We will first discuss the dynamical response of the LMS algorithm by comparing it to the more complicated RLS algorithm for system modeling, also known as Wiener filtering, where both input and output signals are available to the "unknown system", as seen in Figure 1.

System modeling using adaptive system identification from the input–output signals is a powerful technique made easy with the LMS or RLS algorithms. Clearly, when the input–output response (or frequency response) of the filter model $H[z]$ in Figure 1 closely matches the response of the unknown system, the matching error $\epsilon_n = y_n - y'_n$ will be quite small. The error signal is therefore used to drive the adaptive algorithm symbolized by the system box with the angled arrow across it. This symbolism has origins in the standard electrical circuitry symbols for variable capacitance, variable inductance, or the widely used variable resistance (potentiometer). It's a reasonable symbolic analogy except that when the error signal approaches zero, the output of $H[z]$ is not zero, but rather $H[z]$ stops adapting and should in theory be closely matched to the unknown system.

The LMS algorithm offers significant savings in computational complexity over the RLS algorithm for adaptive filtering. Examination of the convergence properties of the LMS algorithm in detail will show where the effects of the simplification of the algorithm are most significant. We will show both mathematically and through numerical experiment that the performance penalty for the LMS simplification can be seen not only for the modeling of more complex systems, but also for system input excitation signals with a wide eigenvalue spread. System identification with a ZMG "white noise" input excitation and the LMS algorithm will converge nearly identically to the more complex RLS algorithm, except for systems with large model orders (requiring a smaller, slower, step size in the LMS algorithm). When the input signal has sinusoids at the low and/or high frequency extremes of the Nyquist band, the resulting wide eigenvalue spread of the input signal correlation matrix results in very slow convergence of the LMS algorithm.

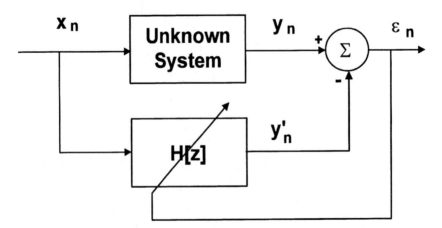

Figure 1 Block diagram depicting adaptive system identification using an adaptive filter with input x_n, output y_n, predicted output y'_n and error signal ε_n.

Consider the LMS coefficient update in Eq. (9.1.16) where we set $\mu_{rel} = 1$ for the moment and write the predicted output signal y'_{n+1} as $\phi_{n+1}H_n$.

$$H_{n+1} = H_n + 2\mu\phi_{n+1}^H(y_{n+1} - \phi_{n+1}H_n) \tag{9.2.1}$$

Rearranging and taking expected values, we have

$$E\{H_{n+1}\} = \left[I - 2\mu E\{\phi_{n+1}^H\phi_{n+1}\}\right]E\{H_n\} + 2\mu E\{\phi_{n+1}^H y_{n+1}\} \tag{9.2.2}$$

If we define $H_{opt} = E\{(\phi^H\phi)^{-1}\phi^H y\}$ and $R^x = E\{\phi^H\phi\}$ as seen in Eq. (9.2.3)

$$E\{H_{n+1}\} = \left[I - 2\mu R^x\right]E\{H_n\} + 2\mu R^x H_{opt} \tag{9.2.3}$$

we can write the LMS recursion as a coefficient vector error as given in Eq. (9.2.4).

$$
\begin{aligned}
E\{H_{n+1}\} - H_{opt} &= [I - 2\mu R^x][E\{H_n\} - H_{opt}] \\
&= [I - 2\mu R^x]^2[E\{H_{n-1}\} - H_{opt}] \\
&\;\;\vdots \qquad\quad \vdots \qquad\quad \vdots \\
&= [I - 2\mu R^x]^{n+1}[E\{H_0\} - H_{opt}]
\end{aligned}
\tag{9.2.4}
$$

We note that R^x, the input signal autocorrelation matrix (see Eqs (8.1.13) and (8.1.14)) is positive definite and invertible. In a real system, there will always be some residual white noise due to the successive approximation error in the analog-to-digital convertors for the least significant bit. The autocorrelation matrix can be diagonalized to give the signal eigenvalues in the form $D = Q^H R^x Q$, where the columns of Q have the eigenvectors which correspond to the eigenvalues λ_k, $k = 0, 1, 2, ..., M$ on the main diagonal of D. Since the eigenvectors are orthonormal, $Q^H Q = I$ and Eq. (9.2.4) becomes

$$E\{H_{n+1}\} - H_{opt} = Q\begin{bmatrix} (1-2\mu\lambda_0)^{n+1} & 0 & 0 & \ldots & 0 \\ 0 & (1-2\mu\lambda_1)^{n+1} & 0 & \ldots & 0 \\ \vdots & \vdots & \vdots & \vdots & \ldots & \vdots \\ 0 & \ldots & 0 & 0 & (1-2\mu\lambda_M)^{n+1} \end{bmatrix} Q^H[E\{H_0\} - H_{opt}] \tag{9.2.5}$$

We can now define an upper limit for μ, the LMS step size to insure a stable adaptive recursion where the coefficient vector error is guaranteed to converge to some small stochastic value. The estimated eigenvalues will range from a minimum value λ_{min} to some maximum value λ_{max}. For each coefficient to converge to the optimum value $|1 - 2\mu\lambda_k| < 1$. Therefore $0 < \mu < 1/(2|\lambda_{max}|)$. Setting μ to the maximum allowable step size for the fastest possible convergence of the LMS algorithm, one can expect the kth coefficient of H to converge proportional to

$$h_{n+1,k} - h_{opt,k} \propto e^{-\frac{\lambda_k}{\lambda_{max}}n} \tag{9.2.6}$$

Therefore the coefficient components associated with the maximum eigenvalue (signal component with the maximum amplitude) will converge the fastest while the components due to the smaller eigenvalues will converge much more slowly. For input signals consisting of many sinusoids or harmonics with wide separation in power levels, the time-domain LMS algorithm performs very poorly.

Determining λ_{max} in real time in order to set μ for the fastest possible stable convergence performance of the LMS algorithm requires a great deal of computation, defeating the advantages of using the LMS algorithm. If the input signals are completely known and stationary one could solve for μ once (in theory at least), set up the LMS algorithm, and enjoy the fastest possible LMS performance. However, in practice, one usually does not know the input signal statistics *a priori* requiring some efficient method to make μ adaptive. We note that for the white noise input signal case, the eigenvalues will all be identical to the signal variance, making the sum of the eigenvalues equal to the signal power times the adaptive filter model order. For sinusoidal inputs, the eigenvalues (and eigenvalue spread) depend on the amplitudes and frequencies of the input sinusoids. However, if one kept the amplitudes the same and varied the frequencies, the sum of the eigenvalues stays the same while the eigenvalue spread changes as seen in the below examples. Therefore, we can be guaranteed that λ_{max} is always less than the adaptive filter model order times the input signal variance. Adaptive tracking of the input signal power using an exponential memory data window as described in Eq. (9.0.1) provides a real-time input signal power estimate which can be used to very simply compute μ for reasonable LMS performance. The *Normalized LMS Algorithm* is the name most commonly used to describe the use of an input signal power tracking μ in the adaptive filter.

Consider a system identification problem where the input signal consists of unity variance ZMG white noise plus a 25 Hz sinusoid of amplitude 5 where the sample rate is 1024 samples/sec. To further illustrate the parameter tracking abilities of the LMS and RLS adaptive filter algorithms, we twice change the parameters of the so-called unknown system to test the adaptive algorithms' ability to model these changes using only the input and output signals from the system. The "unknown system" simply consists of a pair of complementary zeros on the unit circle which start at ± 100 Hz, then at iteration 166 immediately changes to ± 256 Hz, then again at iteration (time sample) 333 immediately changes to ± 400 Hz. Since the zeroes are on the unit circle, only h_1 changes in the unknown system as seen in Figure 2.

Figure 2 shows poor convergence performance of the LMS algorithm relative to the more complicated RLS algorithm in the beginning and end of the trial, where the unknown system's zero is away from the center of the Nyquist band. In the center region, the two algorithms appear comparable, even though the input signal autocorrelation matrix has a fairly wide eigenvalue spread of -11.2 and 38.2. Note that the sum of the eigenvalues is 27, or $2(1+12.5)$ which is the model order times the noise variance plus the sinusoid power $(25/2)$. The parameter μ_{rel} in Eq. (9.1.16) was set to 0.05 for the simulation which translates to an equivalent exponential memory window of about 20 samples ($\alpha = 0.95$ in the RLS algorithm). Note that the time constant for the RLS algorithm (and LMS algorithm in the middle range) in response to the step changes is also about 20 samples. The error signal response corresponding to the trial in Figure 2 is seen in Figure 3. Separating the LMS step size into a factor μ set adaptively to inversely track the input signal power, and μ_{rel} to control the effective data memory window length (approximately equal to $1/\mu_{rel}$) for the algorithm is

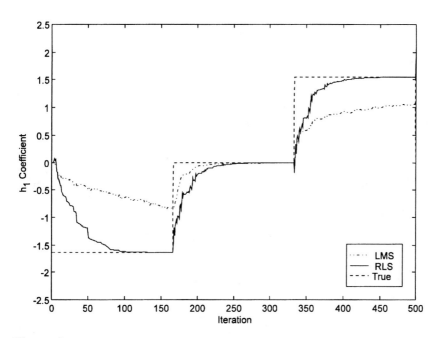

Figure 2 LMS and RLS h_1 coefficient tracking for an input signal x_n consisting of unity variance white Gaussian noise plus a 25 Hz sinusoid of amplitude 5 with a sampling rate f_s of 1024 Hz.

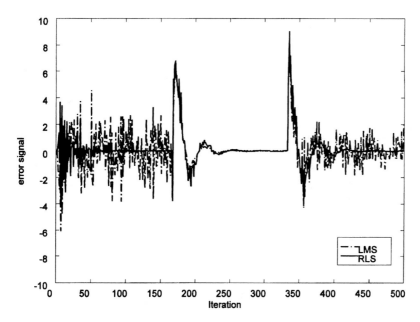

Figure 3 Error signal responses for the LMS and RLS algorithm for the trial seen in Figure 2.

very useful. However, the LMS convergence properties will only be comparable to the RLS algorithm for cases where the input signal is white noise or has frequencies in the center of the Nyquist band.

If we move the 25 Hz sinusoid to 256 Hz in the input signal we get the far superior coefficient tracking seen in Figure 4 which shows essentially no difference between the RLS and LMS algorithms. The eigenvalues are both 13.5 for the ZMG white noise only input signal autocorrelation matrix. Note that the sum of the eigenvalues is still 27 as was the case for the 25 Hz input. The factors μ_{rel} and α are set to 0.05 and 0.95 as before. The exponential response of the LMS and RLS coefficient convergence is completely evident in Figure 4. Yet, the LMS is slightly slower than the RLS algorithm for this particular simulation. Further analysis is left to the student with no social life. Clearly, the simplicity of the LMS algorithm and the nearly equivalent performance make it highly advantageous to use LMS over RLS is practical applications where the input signal is white. The error signal for this case is seen in Figure 5.

Signal modeling using adaptive signal-whitening filters is another basic adaptive filtering operation. The big difference between system modeling and signal modeling is that the input to the unknown system generating the available output signal is not available. One assumes a ZMG unity variance white noise signal as the *innovation* for the available output signal. The signal can be thought to be generated by a digital filter with the signal innovation as input. The frequencies of the signal are thus the result of poles in the generating digital filter very near the unit circle. The relative phases of the frequencies are determined by the zeros of the digital filter. Therefore, our signal model allows the parameterization of

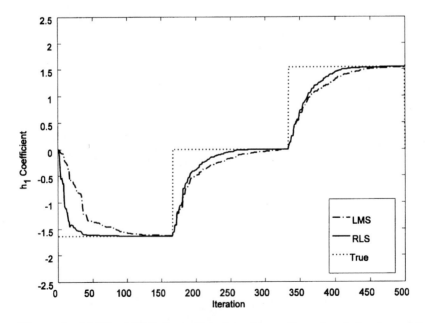

Figure 4   LMS and RLS H_1 coefficient tracking with an input signal consisting of unity variance white Gaussian noise and a 256 Hz sinusoid sampled at 1024 Hz.

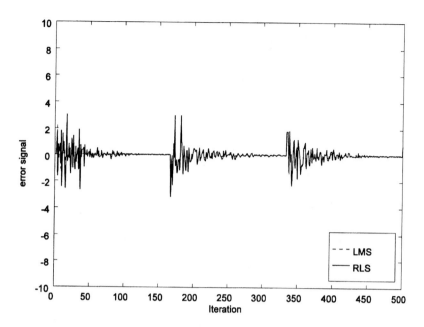

Figure 5 Error signals for the LMS and RLS coefficient tracking trial seen in Figure 4.

the signal in terms of poles and zeros of a digital filter with the white noise innovation. By adaptively filtering the signal to remove all the spectral peaks and dips (whitening the signal to reproduce the ZMG white noise innovation as the error signal), one can recover the modeled signal parameters in the form of the poles and/or zeros of the whitening filter. The converged adaptive whitening filter models the inverse of the signal generation filter. In other words, inverting the converged adaptive whitening filter provides the digital generation filter parameters which are seen as the underlying model of the signal of interest. Figure 6 depicts the signal flow and processing for signal modeling using adaptive whitening filters.

The block diagram in Figure 6 shows a typical whitening filter arrangement where, rather than a full blown ARMA signal model (see Section 3.2), a simpler AR signal model is used. The whitening filter is then a simple MA filter and the resulting error signal will be approximately (statistically) $b_0 w_n$ if the correct model order M is chosen and the whitening filter is converged to the least-squared error solution. Chapter 10 will formulate the ARMA whitening and other filter forms. Note the sign change and the exclusion of the most recent input y_n to the linear prediction signal y'_n. For the AR signal model, the prediction is made solely from past signal outputs (the input signal innovation is not available).

Development of an LMS whitening filter also provides an opportunity to illustrate an alternative heuristic approach to the development of the LMS algorithm. We start with the linear prediction model.

$$y'_n = -a_1 y_{n-1} - \ldots - a_M y_{n-M}$$

(9.2.7)

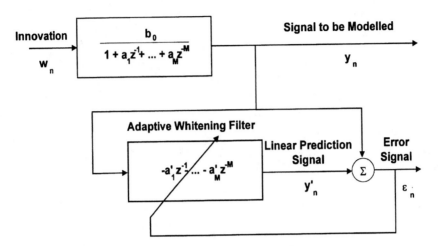

Figure 6 Block diagram showing adaptive whitening filtering to recover signal parameters in terms of a digital generating filter.

In the RLS whitening filter, one simply makes $\phi_n = [-y_{n-1} - y_{n-2} \ldots - y_{n-M}]$ and omits the 0th coefficient (which is not used in whitening filters) from the coefficient vector. But in the LMS algorithm the sign must be changed on the coefficient update to allow $a_{k,n} = a_{k,n-1} - 2\mu\mu_{rel} y_{n-k} \epsilon_n$ rather than the plus sign used in Eq. (9.1.16). One can see the need for the sign change by looking at the gradient of the error, $\epsilon_n = y_n - y_n'$, which is now positive with respect to the whitening filter coefficients. To achieve a gradient descent algorithm, one must *step-wise adapt the LMS coefficient weights in the opposite direction of the error gradient.* Whitening filters have a positive error gradient with respect to the coefficients, while a system identification LMS application has a negative error gradient. Therefore, an LMS whitening filter has a negative sign to the coefficient update while a system identification LMS coefficient update has a positive sign. Setting this sign incorrectly leads to very rapid divergence of the LMS algorithm. Note that the model orders for the system identification and whitening filter examples are both two, even though the system identification filter has three coefficients while the whitening filter only has two. The third whitening filter coefficient is fixed to unity by convention for an AR process.

A numerical example of a whitening filter is seen in Figure 7 where a 50 Hz sinusoid (1024 Hz sample rate) of amplitude 1.0 is mixed with ZMG white noise of standard deviation 0.01 (40 dB SNR). To speed up convergence, $\mu_{rel} = 0.2$ and $\alpha = 0.8$, giving an approximate data memory window of only 5 data samples. The signal autocorrelation matrix eigenvalues are -0.4532 and 1.4534 giving a slow convergence of the whitening filter. The sum of the eigenvalues gives the model order times the signal power as expected. Figure 8 shows the error signal response for the whitening filter.

Simply moving the sinusoid frequency up to 200 Hz (near the midpoint of the Nyquist band) makes the eigenvalue spread of the signal autocorrelation matrix closer at 0.1632 and 0.8370. A much faster convergence is seen in Figure 9 for the LMS algorithm. The convergence of the RLS and LMS filters are identical

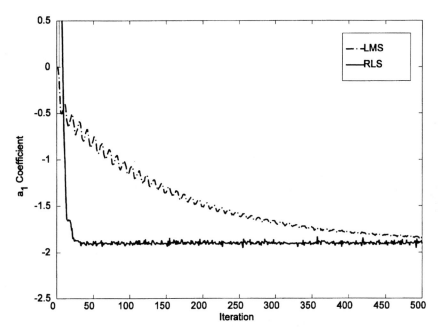

Figure 7   Whitening filter a_1 coefficient response for a 50 Hz sinusoid in white noise for the LMS and RLS algorithms.

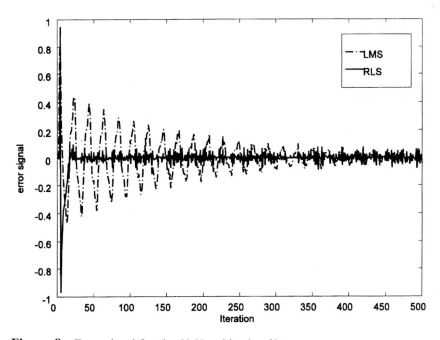

Figure 8   Error signal for the 50 Hz whitening filter.

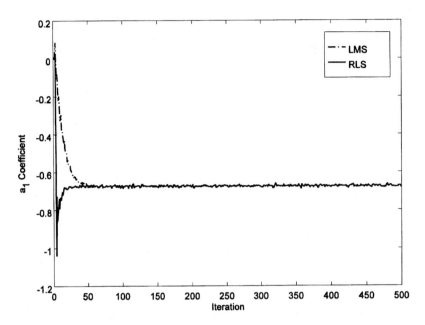

Figure 9 Whitening filter parameter a_1 for a 200 Hz signal near the center of the Nyquist band for the LMS and RLS filters.

for the sinusoid at 256 Hz. LMS whitening performance is best for signal frequencies in the center of the Nyquist band.

The convergence properties of the LMS and RLS algorithms have been presented for two basic adaptive filtering tasks: system identification and signal modeling. The LMS algorithm is much simpler than the RLS algorithm and has nearly identical convergence performance when the input signal autocorrelation matrix for the adaptive filter have narrow eigenvalue spreads (ZMG white noise or sinusoids in the middle of the Nyquist band). When the signal frequency range of interest resides at low or high frequencies in the Nyquist band, the LMS performance becomes quite slow, limiting applications to very stationary signal processes where slow convergence is not a serious drawback. The reason the RLS algorithm performs gracefully regardless of input eigenvalue spread is that the Kalman gain vector is optimized for each coefficient. In the more simplified LMS algorithm, the upper limit for a single step size μ is determined by the maximum eigenvalue and this "global" step size limits the convergence performance of the algorithm. Generally, one approximates the maximum step size for μ by the inverse of the adaptive filter model order times the variance of the input signal as described for the normalized LMS algorithm.

9.3 LATTICE AND SCHUR TECHNIQUES

Lattice filters are of interest because they offer the fast convergence properties of the RLS algorithm with a significant reduction in computational complexity for large model order adaptive filter applications. The lattice filter structure contains a series of nearly independent stages, or Sections, where the filter coefficients are called partial correlation, or PARCOR, coefficients. One stage is needed for each model order increment. Each stage passes two signals: a forward error signal is passed directly; and a backward error signal is passed through a time delay of one sample. The cross correlation of the forward and delayed backward error signals are calculated in the PARCOR coefficients. The PARCOR coefficients times their respective forward and backward error signals are used to subtract any measurable cross correlations from the error signals before they exit the stage. The structure of successive lattice stages, each removing the correlation between forward and backward error signals, is unique in that adding an $M+1$st stage has no effect on any of the lower-order stage PARCOR coefficients. Increasing the model order for an RLS or LMS filter requires that all the adaptive filter coefficients change. While the PARCOR coefficients (also called Schur coefficients) and the LMS or RLS filter coefficients are not the same, the equivalent FIR filter coefficients for the lattice are computed by using a Levinson recursion. Forward and backward prediction error, PARCOR coefficients, and the Levinson recursion are new concepts which are introduced below.

The "forward" prediction error is the error signal defined in the previous Sections. It represents the error in making a future prediction for our signal y_n using an Mth order linear combination of past samples.

$$\varepsilon_n = y_n - y'_n = y_n + a_{M,1}y_{n-1} + a_{M,2}y_{n-2} + \ldots + a_{M,M}y_{n-M} \tag{9.3.1}$$

The predicted signal in Eq. (9.3.1) is simply

$$y'_n = -a_{M,1}y_{n-1} + a_{M,2}y_{n-2} - \ldots - a_{M,M}y_{n-M} \tag{9.3.2}$$

The backward prediction error represents a prediction backward in time $M+1$ samples using only a linear combination of the available M samples from time n to $n - M + 1$.

$$r_{n-1} = a^r_{M,M}y_{n-1} + a^r_{M,M-1}y_{n-2} + \ldots + a^r_{M,1}y_{n-M} + y_{n-M-1} \tag{9.3.3}$$

Clearly, the backward prediction error given in Eq. (9.3.3) is based on the prediction backward in time of

$$y'_{n-M-1} = -a^r_{M,M}y_{n-1} - a^r_{M,M-1}y_{n-2} - \ldots - a^r_{M,1}y_{n-M} \tag{9.3.4}$$

where y'_{n-m-1} is a linear prediction of the data $M+1$ samples in the past. Why bother? We had the sample y_{n-M-1} available to us just one sample ago.

The reason for backward and forward prediction can be seen in the symmetry for linear prediction for forward and backward time steps. Consider a simple sinusoid as y_n. Given a 2nd-order linear predictor for y_n two past samples, i.e. $y'_n = -a_{2,1}y_{n-1} - a_{2,2}y_{n-2}$, one can show that $a_{2,1}$ is approximately $-2\cos(2\pi f_0/f_s)$ and $a_{2,2}$ is approximately unity. The approximation is to insure that the magnitude of the poles is slightly less than unity to insure a stable AR filter for generating our modeled estimate of y_n. Sinusoids have exactly the same shape for positive and negative time. Therefore, one could predict y'_{n-3} using the linear predictor coefficients in opposite order, i.e. $y'_{n-3} = -a_{2,2}y_{n-1} - a_{2,1}y_{n-2}$. Note that the highest indexed coefficient is always the farthest from the predicted sample. An important aspect of the forward–backward prediction symmetry is that for a stationary signal the backward prediction coefficients are always equal to the forward predictor coefficients in reverse order. For complex signals the backward coefficients are the complex conjugate of the forward coefficients in reverse order. Therefore, one of the reasons adaptive lattice filters have superior convergence performance over transversal LMS filters is that for non-stationary signals (or during convergence of the algorithm) the forward–backward prediction symmetry is not the same, and the algorithm can use twice as many predictions and errors to "learn" and converge. A sketch of an adaptive lattice filter and the equivalent LMS transversal filter is seen in Figure 10.

Figure 10 shows that the lattice filter is clearly much more complicated than the LMS filter. The complexity is actually even worse due to the need for several divide operations in each lattice stage. The LMS algorithm can be executed with simple multiplies and adds (and fewer of them) compared to the lattice. However, if the model order for the whitening filter task is chosen higher than the expected model order needed, the lattice PARCOR coefficients for the extraneous stages will be approximately zero. The lattice filter allows the model order of the whitening filter to be estimated from the lattice parameters when the lattice model order is over determined. The fast convergence properties and the ability to add on stages without affecting lower-order stages makes the lattice filter a reasonable choice for high performance adaptive filtering without the extremely high computational requirements of the RLS algorithm.

To derive the PARCOR coefficients from the signal data, we need to examine the relationship between the signal autocorrelation and the prediction error cross correlation. We begin with the prediction error in Eq. (9.3.1) and proceed to derive

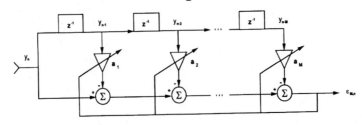

Figure 10   Mth order recursive least squares lattice and FIR transversal LMS whitening filters for the signal y_n.

the respective correlations.

$$\varepsilon_{M,n} = y_n + \sum_{i=1}^{M} a_{M,i} y_{n-i} \tag{9.3.5}$$

The subscripts M,n mean that the error at time n is from a whitening filter model order M. For the linear prediction coefficients, the subscripts M,i refers to the ith coefficient of an Mth order whitening filter model. Multiplying both sides of (9.3.5) by y_{n-j} and taking expected values (averaging) yields

$$
\begin{aligned}
E\{\varepsilon_{M,n}y_{n-j}\} &= E\{y_n y_{n-j}\} + \sum_{i=1}^{E} a_{M,i} E\{y_{n-i}\,y_{n-j}\} \\
&= R_j^y + \sum_{i=1}^{M} a_{M,i} R_{j-1}^y = \begin{cases} 0 & j>0 \\ R_M^\varepsilon & j=0 \end{cases}
\end{aligned} \tag{9.3.6}
$$

where R_M^ε is the Mth order forward prediction error variance. The reason the forward prediction error cross correlation with the signal y_n is zero for $j > 0$ is that $\varepsilon_{M,n}$ is uncorrelated with y_{n-j} as well as $\varepsilon_{M,n-j}$ since the current signal innovation can only appear in past predictions. Following a similar approach for the backward prediction error in Eq. (9.3.3).

$$
\begin{aligned}
E\{r_{M,n-1}y_{n-j}\} &= E\{y_{n-M-1}y_{n-j}\} + \sum_{i=1}^{E} a_{M,M+1}^r E\{y_{n-i}y_{n-j}\} \\
&= R_{j-M-1}^y + \sum_{i=1}^{M} a_{M,M+1-i}^r R_{j-1}^y = \begin{cases} 0 & j>0 \\ R_M^r & j=M+1 \end{cases}
\end{aligned} \tag{9.3.7}
$$

As with the forward prediction error, the current backward prediction error is uncorrelated with future backward prediction errors. We now may write a matrix equation depicting the forward and backward error correlations and their relationship to the signal autocorrelation matrix.

$$
\begin{bmatrix} 1 & a_{M,1} \cdots a_{M,M-1} & a_{M,M} \\ a_{M,M}^r a_{M,M-1}^r \cdots a_{M,1}^r & 1 \end{bmatrix}
\begin{bmatrix} R_0^y & R_1^y & & \cdots & R_M^y \\ R_{-1}^y & R_0^y & R_1^y & \cdots & R_{M-1}^y \\ & \ddots & \ddots & \ddots & \\ R_{-M}^y & \cdots & & R_{-1}^y & R_0^y \end{bmatrix} \tag{9.3.8}
$$

$$
= \begin{bmatrix} R_M^\varepsilon & 0 & 0 & \cdots & 0 \\ 0 & \cdots & 0 & 0 & R_M^r \end{bmatrix}
$$

The signal autocorrelation matrix in Eq. (9.3.8) has a symmetric Toeplitz structure where all the diagonal elements are equal on a given diagonal. For a complex signal, the positive and negative lags of the autocorrelation are complex conjugates. This can also be seen in the covariance matrix in Eq. (8.1.13). If we neglect the backward error components, Eq. (9.3.8) is also seen as the Yule–Walker equation, named after two British astronomers who successfully predicted periodic sunspot activity using an autoregressive filter model around the beginning of the 20th century. Note that Eq. (9.3.8) is completely solvable by simply assuming a white prediction error for the given signal autocorrelations. However, our motivation here is to examine the algebraic relationship between successive model orders as this will lead to an algorithm for generating the $p+1$st order whitening filter from the pth model.

Suppose we make a trial solution for the $M+1$st order whitening filter coefficients by simply letting $a_{M+1,k}=a_{M,k}$ and let $a_{M+1,M+1}=0$. As seen in Eq. (9.3.9) this obviously leads to a non-white prediction error where the autocorrelation of the error is no longer zero for the non-zero time lags.

$$
\begin{bmatrix} 1\ a_{M,1}\dots a_{M,M-1}a_{M,M}\ 0 \\ 0\ a^r_{M,M}a^r_{M,M-1}\dots a^r_{M,1}\ 1 \end{bmatrix}
\begin{bmatrix} R^y_0 & R^y_1 & & \cdots & R^y_{M+1} \\ R^y_{-1} & R^y_0 & R^y_1 & \cdots & R^y_M \\ \ddots & & \ddots & \ddots & \\ R^y_{-M-1} & \cdots & & R^y_{-1} & R^y_0 \end{bmatrix}
$$

$$
= \begin{bmatrix} R^\varepsilon_M & 0 & 0 & \cdots & \Delta^\varepsilon_{M+1} \\ \Delta^r_{M+1} & \cdots & 0 & 0 & R^r_M \end{bmatrix}
\tag{9.3.9}
$$

However, we can make the $M+1$st order prediction error white through the following multiplication.

$$
\begin{bmatrix} 1 & -\Delta^\varepsilon_{M+1}/R^r_M \\ -\Delta^r_{M+1}/R^\varepsilon_M & 1 \end{bmatrix}
\begin{bmatrix} 1\ a_{M,1}\dots a_{M,M-1}a_{M,M}\ 0 \\ 0\ a^r_{M,M}a^r_{M,M-1}\dots a^r_{M,1}\ 1 \end{bmatrix}
$$

$$
\times \begin{bmatrix} R^y_0 & R^y_1 & & \cdots & R^y_{M+1} \\ R^y_{-1} & R^y_0 & R^y_1 & \cdots & R^y_M \\ \ddots & & \ddots & \ddots & \\ R^y_{-M-1} & \cdots & & R^y_{-1} & R^y_0 \end{bmatrix}
\tag{9.3.10}
$$

$$
= \begin{bmatrix} 1 & -\Delta^\varepsilon_{M+1}/R^r_M \\ -\Delta^r_{M+1}/R^\varepsilon_M & 1 \end{bmatrix}
\begin{bmatrix} R^\varepsilon_M & 0 & 0 & \cdots & \Delta^\varepsilon_{M+1} \\ \Delta^r_{M+1} & \cdots & 0 & 0 & R^r_M \end{bmatrix}
$$

$$
= \begin{bmatrix} R^\varepsilon_{M+1} & 0 & 0 & \cdots & 0 \\ 0 & \cdots & 0 & 0 & R^r_{M+1} \end{bmatrix}
$$

From Eq. (9.3.10) we have a Levinson recursion for computing the $M+1$st model order whitening filter from the Mth model coefficients and error signal correlations.

$$
\begin{bmatrix} 1 & -\Delta^{\varepsilon}_{M+1}/R^r_M \\ -\Delta^r_{M+1}/R^{\varepsilon}_M & 1 \end{bmatrix} \begin{bmatrix} 1 & a_{M,1} \ldots a_{M,M-1} a_{M,M} & 0 \\ 0 & a^r_{M,M} a^r_{M,M-1} \ldots a^r_{M,1} & 1 \end{bmatrix}
$$
$$
= \begin{bmatrix} 1 & a_{M+1,1} \ldots a_{M+1,M} a_{M+1,M+1} \\ a^r_{M+1,M+1} a^r_{M+1,M} \ldots a^r_{M+,1} & 1 \end{bmatrix}
\tag{9.3.11}
$$

The error signal cross correlations are seen to be

$$
\begin{aligned}
\Delta^{\varepsilon}_{M+1} &= R^y_{M+1} + \sum_{i=1}^{M} a_{M,i} R^y_{M-i-1} \\
&= R^y_{M+1} + \sum_{i=1}^{M} a_{M,i} R^y_{M-i-1} \\
&= \Delta^{r^H}_{M+1}
\end{aligned}
\tag{9.3.12}
$$

where for the complex data case $\Delta^{\varepsilon}_{M+1}$ equals the complex conjugate of $\Delta^{\varepsilon}_{M+1}$, depicted by the H symbol for Hermitian transpose in Eq. (9.3.12). Even though we are deriving a scalar lattice algorithm, carrying the proper matrix notation will prove useful for reference by latter Sections of the text. To more clearly see the forward and backward error signal cross correlation, we simply write the expression in terms of expected values.

$$
\begin{aligned}
\Delta_{M+1} &= [1 \; a_{M,1} \ldots a_{M,M} \; 0] E\{[y_n \ldots y_{n-M-1}]^H [y_n \ldots y_{n-m-1}]\} \\
&\times [0 \; a^r_{M,M} \ldots a^r_{M,1} \; 1]^H = E\{\varepsilon_{M,n} r_{M,n-1}\}
\end{aligned}
\tag{9.3.13}
$$

The PARCOR coefficients are defined as

$$
K^{\varepsilon}_{M+1} = \Delta^H_{M+1}/R^{\varepsilon}_M \quad K^r_{M+1} = \Delta_{M+1}/R^r_M
\tag{9.3.14}
$$

We also note that from Eq. (9.3.10) the updates for the $M+1$st forward and backward prediction error variances are, respectively

$$
R^{\varepsilon}_{M+1} = R^{\varepsilon}_M - K^r_{M+1} \Delta^H_{M+1}
\tag{9.3.15}
$$

and

$$
R^r_{M+1} = R^r_M - K^{\varepsilon}_{M+1} \Delta_{M+1}
\tag{9.3.16}
$$

It follows from Eqs (9.3.13)–(9.3.16) that the forward and backward error signal updates are simply

$$
\varepsilon_{M+1,n} = \varepsilon_{M,n} - K^r_{M+1} r_{M,n-1}
\tag{9.3.17}
$$

and

$$
r_{M+1,n} = r_{M,n-1} - K^{\varepsilon}_{M+1} \varepsilon_{M,n}
\tag{9.3.18}
$$

The structure of a lattice stage is determined by the forward and backward error signal order updates given in Eqs (9.3.17) and (9.3.18). Starting with the 1st stage, we let $\varepsilon_{0,n} = r_{0,n} = y_n$ and estimate the cross correlation Δ_1, between the forward error $\varepsilon_{0,n}$, and the delayed backward error signal $r_{0,n-1}$. The 0th stage error signal variances are equal to R_0^y, the mean square of the signal y_n. The PARCOR coefficients are then calculated using (9.3.14) and the error signals and error signal variance are updated using Eqs (9.3.15)–(9.3.18). Additional stages are processed until the desired model order is achieved, or until the cross correlation between the forward and backward error signals becomes essentially zero.

The PARCOR coefficients are used to calculate the forward and backward linear predictors when desired. The process starts off with $-K_1^r = a_{1,1}$ and $-K_1^\varepsilon = a_{1,1}^r$. Using the Levinson recursion in Eq. (9.3.11) we find $a_{2,2} = -K_2^r$ and $a_{2,1} = a_{1,1} - K_2^r a_{1,1}^1$, and so on. A more physical way to see the Levinson recursion is to consider making a copy of the adaptive lattice at some time instant allowing the PARCOR coefficients to be "frozen" in time with all error signals zero. Imputing a simple unit delta function into the frozen lattice allows the forward linear prediction coefficients to be read in succession from the forward error signal output of the stage corresponding to the desired model order. This view makes sense because the transversal FIR filter generated by the Levinson recursion and the lattice filter must have the same impulse response. One of the novel features of the lattice filter is that using either the Levinson recursion or frozen impulse response technique, the linear prediction coefficients for all model orders can be easily obtained along with the corresponding prediction error variances directly from the lattice parameters. Figure 11 shows the detailed algorithm structure of the $p+1$st lattice filter stage. The lattice structure with its forward and backward prediction allows an adaptive Gram–Schmidt orthogonalization of the linear prediction filter. The backward error signals are orthogonal between stages in the lattice allowing each stage to adapt as rapidly as possible independent of the other lattice stages.

The orthogonality of the stages in the lattice structure is seen as the key feature which makes adaptive lattice filters so attractive for fast convergence without the computational complexity of the recursive least-squares algorithm for transversal

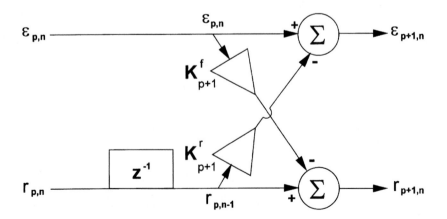

Figure 11 Detailed structure of an adaptive lattice filter stage.

FIR-type filters. The error signal variances and cross-correlations can be estimated using simple expected values such as that suggested in Eq. (9.0.1) for the exponentially weighted data memory window. Since the PARCOR coefficients are the error signal cross correlation divided by the variance, a common bias in the cross correlation and variance estimates is not a problem. For a exponentially decaying data window of effective length N, the cross correlation is estimated using

$$\Delta_{p+1,n} = \alpha\Delta_{p+1,n-1} + \varepsilon_{p,n}r_{p,n-1} \tag{9.3.19}$$

where N is effectively $1/(1-\alpha)$ and $\Delta_{p+1,n} \cong N\,E\{\varepsilon_{p,n}\,r_{p,n-1}\}$. The forward error signal variance is

$$R^{\varepsilon}_{p,n} = \alpha R^{\varepsilon}_{p,n-1} + \varepsilon_{p,n}\varepsilon_{p,n} \tag{9.3.20}$$

and the backward error signal variance is

$$R^{r}_{p,n-1} = \alpha R^{r}_{p,n-2} + r_{p,n-1}r_{p,n-1} \tag{9.3.21}$$

The PARCOR coefficients are calculated using the expressions in Eq. (9.3.22).

$$K^{\varepsilon}_{p+1,n} = \Delta^{H}_{p+1,n}/R^{\varepsilon}_{p,n} \quad K^{r}_{p+1,n} = \Delta_{p+1,n}/R^{r}_{p,n-1} \tag{9.3.22}$$

We show the lattice cross correlation, variances, and PARCOR coefficients as time dependant in Eqs (9.3.19)–(9.3.22) to illustrate the recursive operation of the lattice algorithm. For stationary data and time periods long after the start-up of the lattice Eqs (9.3.19)–(9.3.22) are optimum. However, for nonstationary signals and/or during the initial start up of the lattice, some additional equations are needed to optimize the lattice algorithm. Consider that the delay operators in the backward error of the lattice do not allow the residual for the data into the pth stage until $p+1$ time samples have been processed. The transient effects on the PARCOR coefficients will persist, slowing the algorithm convergence. However, if we use the orthogonal decomposition of the Hilbert space spanned by the data outlined in Section 8.2, we can make the lattice recursions least-squared error even during the transients from start up or nonstationary data. While somewhat complicated, subspace decomposition gives the lattice algorithm an effective convergence time on the order of the number of lattice stages on start up and N for tracking nonstationary data.

9.4 THE ADAPTIVE LEAST-SQUARES LATTICE ALGORITHM

Projection Operator Subspace Decomposition can also be used to derive the complete least squares lattice algorithm including the optimal time updates for the error signal variances and cross correlations. As shown in Figure 11, the PARCOR coefficients may be computed any number of ways including even an LMS-type gradient descent algorithm. One of the interesting characteristics about adaptive processing is that even if the algorithm is not optimal, the result of the learning processing still gives reasonable results, they're just not the best result possible. We now make use of the projection operator orthogonal decomposition given in Section 8.2 and in particular in Eqs (8.2.6)–(8.2.8) to optimize the lattice equations for the fastest possible convergence to the least-squared error PARCOR coefficients.

Application of orthogonal subspace decomposition to the development of fast adaptive filtering can be attributed to a PhD dissertation by Martin Morf (2). The lattice structure was well-known in network theory, but its unique processing properties for adaptive digital filtering generally were not well appreciated until the 1970s. By the early 1980s a number of publications on lattice filters (also called ladder or even "wave" filter structures) appeared, one of the more notable by Friedlander (3). A slightly different notation is used below from what has previously been presented in Section 8.2 to facilitate the decomposition matrix equations. We start by constructing a vector of signal samples defined by

$$y_{n-N:n-1} = [y_{n-N} \ldots y_{n-2}y_{n-1}] \tag{9.4.1}$$

For a time block of N samples and a whitening filter model order p, we define a p-row, N-column data matrix.

$$Y_{p,N} = \begin{bmatrix} y_{n-N\,:\,n-1} \\ y_{n-N-1\,:\,n-2} \\ \vdots \\ y_{n-p-N+1\,:\,n-p} \end{bmatrix} = \begin{bmatrix} y_{n-N} & \cdots & y_{n-2} & y_{n-1} \\ y_{n-N-1} & \cdots & y_{n-3} & y_{n-2} \\ & & \vdots & \\ y_{n-p-N+1} & \cdots & y_{n-p-1} & y_{n-p} \end{bmatrix} \tag{9.4.2}$$

The linear prediction error from time $n - N+1$ to time n can now be written as seen in Eq. (9.4.3)

$$\varepsilon_{p,n-N+1\,:\,n} = Y_{n-N+1\,:\,n} + [a_{p,1}a_{p,2} \ldots a_{p,p}] Y_{p,N} \tag{9.4.3}$$

where $a_{p,2}$ is the second linear prediction coefficient in a pth order whitening filter. The least squared error linear prediction coefficients can be written as

$$[a_{p,1}a_{p,2} \ldots a_{p,p}] = y_{n-N+1\,:\,n} Y_{p,N}^{H} \left(Y_{p,N} Y_{p,N}^{H} \right)^{-1} \tag{9.4.5}$$

which is essentially the transpose form of Eq. (8.1.11). Equation (9.4.3) is expressed in the form of a orthogonal projection

$$\varepsilon_{p,n-N+1\,:\,n} = y_{n-N+1\,:\,n}\left(I - P_{Y_{p,N}} \right) \tag{9.4.5}$$

where $P_{Yp,N} = Y_{p,N}^{H}(Y_{p,N}Y_{p,N}^{H})^{-1}Y_{p,N}$ is the projection operator. Equations (9.4.1)–(9.4.5) can be compared to Eqs (8.2.1)–(8.2.5) to see that the effect of the change in variable definitions is really only a simple transpose. We can "pick out" the prediction error at time n by adding a post-multiplication of the form

$$\varepsilon_{p,n} = Y_{n-N+1\,:\,n}\left(I - P_{Y_{p,N}} \right)\pi^{H} \tag{9.4.6}$$

where $\pi = [\,0\ 0\ \ldots\ 0\ 1\,]$. Shifting the data vector $y_{n-N+1:n}$ $p+1$ samples to the right yields a data vector suitable for computing the backward prediction error

$$r_{p,n-1} = y_{n-N+1\,:\,n}^{p+1}\left(I - P_{Y_{p,N}} \right)\pi^{H} \tag{9.4.7}$$

where

$$y_{n-N+1\,:\,n}^{p+1} = y_{n-N-p\,:\,n-p-1} \tag{9.4.8}$$

We have gone to some lengths to present the projection operator back in Section 8.2 and here to cast the lattice forward and backward prediction errors into an orthogonal projection framework. After presenting the rest of the lattice variables in the context of the projection operator framework, we will use the orthogonal decomposition in Eq. (8.2.8) to generate the least-squared error updates. The error signal variances for the order p are given by post-multiplying Eqs (9.4.6)–(9.4.7) by their respective data vectors.

$$R_{p,n}^{\varepsilon} = y_{n-N+1:n}(I - P_{Y_{p,N}})y_{n-N+1:n}^{H} \tag{9.4.9}$$

$$R_{p,n-1}^{r} = y_{n-N+1:n}^{p+1}(I - P_{Y_{p,N}})y_{n-N+1:n}^{p+1^{H}} \tag{9.4.10}$$

The error signal cross correlation is simply

$$\Delta_{p+1,n} = y_{n-N+1:n}(I - P_{Y_{p,N}})y_{n-N+1:n}^{p+1^{H}} \tag{9.4.11}$$

However, a new very important variable arises out of the orthogonal projection framework called the likelihood variable. It is unique to the least-squares lattice algorithm and is responsible for making the convergence as fast as possible. The likelihood variable $\gamma_{p-1,n-1}$ can be seen as a measure of how well recent data statistically matches the older data. For example, as the prediction error tends to zero, the main diagonal of the projection operator tends toward unity, as seen in the example described in Eqs (8.3.1)–(8.3.6). Pre-multiplying and post-multiplying the orthogonal projection operator by the π vector gives $1 - \gamma_{p-1,n-1}$.

$$1 - \gamma_{p-1,n-1} = \pi(I - P_{Y_{p,N}})\pi^{H} \tag{9.4.12}$$

If the projection operator is nearly an identity matrix, the likelihood variable $\gamma_{p-1,n-1}$ approaches zero (the parameter $1 - \gamma_{p-1,n-1}$ will tend toward unity) for each model order, indicating the data-model fit is nearly perfect. As will be seen for the update equations below, this makes the lattice adapt very fast for the near-perfect data. If the data is noisy or from a new distribution, $1 - \gamma_{p-1,n-1}$ approaches zero ($\gamma_{p-1,n-1}$ approaches unity) and the lattice will weight the recent data much more heavily, thereby "forgetting" the older data which no longer fits the current distribution. This "intelligent memory control" is independent of the data memory window, which also affects the convergence rate. The ingenious part of the likelihood variable is that it naturally arises from the update equation as an independent adaptive gain control for optimizing the nonstationary data performance in each lattice stage.

Consider the update equation for the Hilbert space orthogonal decomposition back in Section 8.2. We now pre- and post-multiply by arbitrary vectors V and W^{H}, respectively, where the existing subspace spanned by the rows of $Y_{p,N}$ is being updated to include the space spanned by the rows of the vector S.

$$V(I - P_{\{Y_{p,N}+S\}})W^{H} = V(I - P_{Y_{p,N}})W^{H}$$
$$- V(I - P_{Y_{p,N}})S[S^{H}(I - P_{Y_{p,N}})S]^{-1}S^{H}(I - P_{Y_{p,N}})W^{H} \tag{9.4.13}$$

If we add a row vector on the bottom of $Y_{p,N}$ we have an order update to the subspace spanned by the rows of $Y_{p,N}$.

$$Y_{p+1,N} = \begin{bmatrix} Y_{p,n} \\ y_{n-N+1:n}^{p+1} \end{bmatrix} \tag{9.4.14}$$

Therefore, choosing $S = y_{n-N+1:n}^{p+1}$ allows the following order updates when $V = y_{n-N+1:n}$ and $W = \pi$,

$$\begin{aligned}
\varepsilon_{p+1,n} &= \varepsilon_{p,n} - \Delta_{p+1,n}[R_{p,n-1}^r]^{-1} r_{p,n-1} \\
&= \varepsilon_{p,n} - K_{p+1,n}^r r_{p,n-1}
\end{aligned} \tag{9.4.15}$$

and when $V = y_{n-N+1:n}$ and $W = y_{n-N+1:n}$,

$$\begin{aligned}
R_{p+1,n}^\varepsilon &= R_{p,n}^\varepsilon - \Delta_{p+1,n}[R_{p,n-1}^r]^{-1} \Delta_{p+1,n}^H \\
&= R_{p,n}^\varepsilon - K_{p+1,n}^r \Delta_{p+1,n}^H
\end{aligned} \tag{9.4.15}$$

Setting $V = W = \pi$ yields one of several possible expressions for the likelihood variable.

$$1 - \gamma_{p,n-1} = 1 - \gamma_{p-1,n-1} - r_{p,n-1}^H [R_{p,n-1}^r]^{-1} r_{p,n-1} \tag{9.4.17}$$

The likelihood variable gets its name from the fact that the probability density function for the backward error is simply

$$p(r_{p,n-1}) = \frac{1}{\sqrt{2\pi R_{p,n-1}^r}} e^{-\frac{1}{2} r_{p,n-1}^H [R_{p,n-1}^r]^{-1} r_{p,n-1}} = p(y_n) \tag{9.4.18}$$

where the probability density for the backward error signal is the same as the probability density for the data because they both span the same subspace. When $1 - \gamma_{p,n-1}$ is nearly unity ($\gamma_{p,n-1}$ is small), the exponent in (9.4.18) is nearly zero, making the probability density function a very narrow Gaussian function centered at zero on the ordinate defined by the backward error signal for the pth lattice stage. If the lattice input signal y_n changes level or spectral density distribution, both the forward and backward error signal probability density functions become wide, indicating that the error signals have suddenly grown in variance. The likelihood variable detects the error signal change before the variance increase shows up in the error signal variance parameters, instantly driving $1 - \gamma_{p-1,n-1}$ towards zero and resulting in rapid updates of the lattice estimates for error variance and cross correlation. This will be most evident in the time-update recursions shown below, but it actually occurs whether one uses the time update forms, or the time and order, or order update forms for the forward and backward error signal variances.

Consider a time-and-order subspace update which amounts to adding a row to the top of $Y_{p,N}$ which increases the model order and time window by one.

$$Y_{p+1,N+1} = \begin{bmatrix} y_{n-n:n} \\ 0 \; Y_{p,N} \end{bmatrix} = \begin{bmatrix} y_{n-N} & y_{n-N-1} & \cdots & y_{n-1} & y_n \\ 0 & y_{n-N} & \cdots & y_{n-2} & y_{n-1} \\ 0 & y_{n-N-1} & \cdots & y_{n-3} & y_{n-2} \\ & & \vdots & & \\ 0 & y_{n-p-N+1} & \cdots & y_{n-p-1} & y_{n-p} \end{bmatrix} \quad (9.4.19)$$

Therefore, for the subspace spanned by the rows of $Y_{p,N}$ we can choose $S = y_{n-N+1:n}$ allowing the following order updates when $V = y^{p+1}_{n-N+1:n}$ and $W = \pi$,

$$\begin{aligned} r_{p+1,n} &= r_{p,n-1} - \Delta^H_{p+1,n}[R^\varepsilon_{p,n}]^{-1}\varepsilon_{p,n} \\ &= r_{p,n-1} - K^\varepsilon_{p+1,n}\varepsilon_{p,n} \end{aligned} \quad (9.4.20)$$

and when $V = y^{p+1}_{n-N+1:n}$ and $W = y^{p+1}_{n-N+1:n}$,

$$\begin{aligned} R^r_{p+1,n} &= R^r_{p,n-1} - \Delta^H_{p+1,n}[R^\varepsilon_{p,n-1}]^{-1}\Delta_{p+1,n} \\ &= R^r_{p,n-1} - K^\varepsilon_{p+1,n}\Delta_{p+1,n} \end{aligned} \quad (9.4.21)$$

Setting $V = W = \pi$ yields another one of the several possible expressions for the likelihood variable.

$$1 - \gamma_{p,n} = 1 - \gamma_{p-1,n-1} - \varepsilon^H_{p,n}[R^\varepsilon_{p,n}]^{-1}\varepsilon_{p,n} \quad (9.4.22)$$

Finally, we present the time update form of the equations. A total of 9 equations are possible for the lattice, but only 6 update equations are needed for the algorithm. The error signal cross correlation can only be calculated using a time update. The importance and function of the likelihood variable on the lattice equation will become most apparent in the time update equations.

$$\begin{aligned} Y_{p,N+1} &= \begin{bmatrix} y_{n-N:n} \\ Y_{n-N-1:n-1} \\ \vdots \\ Y_{n-p-N+1:n-p+1} \end{bmatrix} = \begin{bmatrix} y_{n-N} & \cdots & y_{n-2} \; y_{n-1} \; y_n \\ y_{n-N-1} & \cdots & y_{n-3} \; y_{n-2} \; y_{n-1} \\ & \vdots & \\ y_{n-p-N+1} & \cdots & y_{n-p-1} \; y_{n-p} \; y_{n-p+1} \end{bmatrix} \\ &= \begin{bmatrix} & & y_n \\ & & y_{n-1} \\ Y_{p,N} & & \vdots \\ & & y_{n-p+1} \end{bmatrix} \end{aligned}$$

$$(9.4.23)$$

It's not clear from our equations how one augments the signal subspace to represent the time update. However, we can say that the projection operator $P_{Y_{p,N+1}}$ is $N+1$ by $N+1$ rather than N by N in size. So, again for the subpace spanned by the rows of $Y_{p,N}$ a time update to the projection operator can be written as

$$
P_{\{Y_{p,N}+\pi\}} = P_{\{Y_{p,N-1}\}} + P_\pi
$$

$$
= \begin{bmatrix} P_{Y_{p,N-1}} & 0 \\ & \vdots \\ 0 & \cdots & 0 \end{bmatrix} + \pi^H (\pi^H \pi)^{-1} \pi
\tag{9.2.24}
$$

$$
= \begin{bmatrix} P_{Y_{p,N-1}} & 0 \\ & \vdots \\ 0 & \cdots & I \end{bmatrix}
$$

Equation (9.4.24) can be seen to state that the projection operator for the sub-space augmented by π is actually the previous time iteration projection operator! However, recalling from Section 8.3, as the data are fit more perfectly to the model, the projection operator tends toward an identity matrix. By defining the time update to be a perfect projection (although backwards in time) we are guaranteed a least-squared error time update. The orthogonal projection operator for the time-updated space is

$$
(I - P_{\{Y_{p,N}+\pi\}}) = \begin{bmatrix} (I - P_{Y_{p,N-1}}) & 0 \\ & \vdots \\ 0 & \cdots & 0 \end{bmatrix}
\tag{9.4.25}
$$

which means $V(I - P_{\{Yp,N + \pi\}})W^H$ in the decomposition equation actually corresponds to the old parameter projection, and the term $V(I - P_{Yp,N})W^H$ corresponds to the new time updated parameter. This may seem confusing, but the time updates to the least squares lattice are by design orthogonal to the error and therefore optimum. Choosing $S = \pi$, $V = y_{n-N+1:n}$ and $W = y_{n-N+1:n}^{p+1}$ we have the time update for the cross correlation coefficient. A forgetting factor α has been added to facilitate an exponentially decaying data memory window. Note the plus sign which results from the rearrangement of the update to place the new parameter on the left-hand side.

$$
\Delta_{p+1,n} = \alpha \Delta_{p+1,n-1} + \varepsilon_{p,n}[1 - \gamma_{p-1,n-1}]^{-1} r_{p,n-1}^H
\tag{9.4.26}
$$

Choosing $S = \pi$ and $V = W = y_{n-N+1:n}$ gives a time update for the forward error signal variance.

$$
R_{p,n}^\varepsilon = \alpha R_{p,n-1}^\varepsilon + \varepsilon_{p,n}[1 - \gamma_{p-1,n-1}]^{-1} \varepsilon_{p,n}^H
\tag{9.4.27}
$$

Choosing $S = \pi$ and $V = W = y_{n-N+1:n}^{p+1}$ gives a time update for the backward error signal variance.

$$
R_{p,n-1}^r = \alpha R_{p,n-2}^r + r_{p,n-1}[1 - \gamma_{p-1,n-1}]^{-1} r_{p,n-1}^H
\tag{9.4.28}
$$

Comparing equations to (9.3.26)–(9.3.28) we immediately see the role of the likelihood variable. If the recent data does not fit the existing model, the likelihood variable (and corresponding main diagonal element for the projection operator) will be small, causing the lattice updates to quickly "forget" the old outdated data and pay attention to the new data. Eqs (9.4.17) and (9.4.22) can be combined to give a pure time update for the likelihood variable which is also very illustrative of its operation.

$$(1 - \gamma_{p,n}) = \alpha(1 - \gamma_{p,n-1}) + r_{p,n-1}^H [R_{p,n-1}^r]^{-1} r_{p,n-1} - e_{p,n}^H [R_{p,n}^\varepsilon]^{-1} \varepsilon_{p,n} \qquad (9.4.29)$$

Clearly, if either the forward or backward error signals start to change, $1 - \gamma_{p,n}$ will tend towards zero giving rise to very fast adaptive convergence and tracking of signal parameter changes. Figure 12 compares the whitening filter performance of the least-squares lattice to the RLS and LMS algorithms for the 50 Hz signal case seen previously in Figure 7.

The amazingly fast convergence of the lattice can be attributed in part to the initialization of the error signal variances to a small value (1×10^{-5}). Many presentations of adaptive filter algorithms in the literature performance comparisons which can be very misleading due to the way the algorithms are initialized. If we initialize the signal covariance matrix P^{-1} of the RLS algorithm to be an identity matrix times the input signal power, and initialize all of the forward and backward error signal variances in the lattice to the value of the input signal power, a more representative comparison is seen in Figure 13.

Figure 13 shows a much more representative comparison of the RLS and Lattice algorithms. The lattice filter still converges slightly faster due to the likelihood variable and its ability to optimize each stage independently. The effective memory window is only 5 samples, and one can see from Figures 12 and 13 that the convergence time is approximately the filter length (3 samples) plus

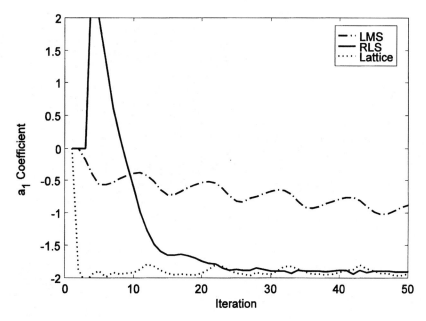

Figure 12 Comparison of LMS, RLS and Lattice for the case of the 50 Hz Whitening filter corresponding to Figure 7.

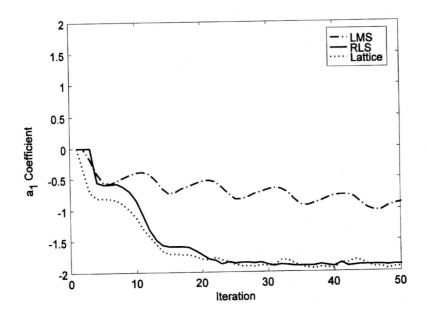

Figure 13 Comparison of the LMS, RLS, and Lattice where both the RLS and Lattice algorithms have identical initial signal covariances equal to the input signal power.

the memory window length, but also depends on the initialization of the algorithms. Initialization of the lattice error signal variances to some very small value makes any change in error signals rapidly drive the convergence by the presence of the likelihood variable in the updates, making the lattice converge astonishingly fast in time intervals on the order of the filter length. Initialization of the lattice error variances at higher levels makes the start-up transients in the error signals seem less significant and the parameters converge at a rate determined mainly by the memory window length.

Table 2 summarizes the least-squares lattice implementation. There are 9 possible update equations where only 6 are actually needed. Naturally, there is some flexibility available to the programmer as to which equation forms to choose based on storage requirements and convenience. The equation sequence in Table 1 is just one possible implementation. Note that the error signal variances cannot be initialized to zero due to the required divisions in the algorithm. Divisions are much more computationally expensive than multiplications in signal processing engines. However, the availability of fast floating-point processors makes the lattice filter an attractive choice for high performance adaptive filtering in real-time.

The linear prediction coefficients for all model orders up to the lattice whitening filter model order M can be calculated from an updated set of PARCOR coefficients using the Levinson recursion in Table 3. Another way to visualize how the Levinson recursion works is to consider the impulse response of the lattice. As the impulse makes its way down each stage of the lattice, the forward error signal at the Mth stage yields the Mth-order linear prediction coefficients. It can be very useful for the student to calculate this sequence manually for a small lattice of say three stages and compare the results to the Levinson recursion. Checking

Table 2 The Least-Squares Lattice Algorithm

Description	Equation
Initialization at time $n = 0$ for $p = 0, 1, 2, \ldots, M$	$R_{p,0}^{\varepsilon} = R_{p,0}^{r} = E\{y_n y_n^H\} \quad 1 - \gamma_{-1,-1} = 1$
Input error signals at 1st stage	$\varepsilon_{0,n} = r_{0,n} = y_n$
Lattice stage update sequence $\alpha = (N-1)/N$	$for \quad p = 0, 1, 2, \ldots, M$
Time update for error signal cross correlation	$\Delta_{p+1,n} = \alpha\Delta_{p+1,n-1} + \varepsilon_{p,n} r_{p,n-1}^H / (1 - \gamma_{p-1,n-1})$
Time update of forward prediction error variance	$R_{p,n}^{\varepsilon} = \alpha R_{p,n-1}^{\varepsilon} + \varepsilon_{p,n}\varepsilon_{p,n}^H / (1 - \gamma_{p-1,n-1})$
Time update of backward prediction error variance	$R_{p,n-1}^{r} = \alpha R_{p,n-2} + r_{p,n-1} r_{p,n-1}^H / (1 - \gamma_{p-1,n-1})$
Order update of likelihood variable	$(1 - \gamma_{p,n-1}) = (1 - \gamma_{p-1,n-1}) - r_{p,n-1}^H R_{p,n-1}^{-r} r_{p,n-1}$
Forward PARCOR coefficient	$K_{p+1,n}^{\varepsilon} = \Delta_{p+1,n}^H R_{p,n}^{-\varepsilon}$
Backward PARCOR coefficient	$K_{p+1,n}^{r} = \Delta_{p+1,n} R_{p,n-1}^{-r}$
Order update of forward error signal	$\varepsilon_{p+1,n} = \varepsilon_{p,n} - K_{p+1,n}^{r} r_{p,n-1}$
Time and order update of backward error signal	$r_{p+1,n} = r_{p,n-1} - K_{p+1,n}^{\varepsilon} \varepsilon_{p,n}$

Table 3 The Levinson Recursion

Description	Equation
Lowest forward and backward prediction coefficients	$a_{p,0} = a_{p,0}^{r} = 1 \quad for \quad p = 0, 1, 2, \ldots, M$
Highest forward and backward prediction coefficients	$a_{p+1p+1} = K_{p+1}^{r} \quad a_{p+1,p+1}^{r} = -K_{p+1}^{\varepsilon}$
Linear prediction coefficient recursion	$\left. \begin{array}{l} a_{p+1,j} = a_{p,j} - K_{p+1}^{r} a_{p,p-j+1}^{r} \\ a_{p+1,p-j+1}^{r} = a_{p,p-j+1}^{r} - K_{p+1}^{\varepsilon} a_{p,j} \end{array} \right\} j = 1, 2, \ldots, p$

the impulse response is also a useful software debugging tool for lattice Levinson recursion algorithms.

Consider the case where both the input and output are known and we use a lattice filter for system identification rather that just signal modeling. The system identification application of adaptive filter is generally known as Wiener filtering. It is well-timed to introduce the lattice Wiener filter here, as the changes necessary from the whitening filter arrangement just presented are very illustrative. Since we have both x_n and y_n and wish to fit an FIR filter model to the unknown linear system relating x_n and y_n, recall that x_n is the input to the LMS Wiener filter in Section 9.2. The same is true for the Wiener lattice as seen in Figure 14. To see what to do with the output y_n we have to revisit the backward error signal definition

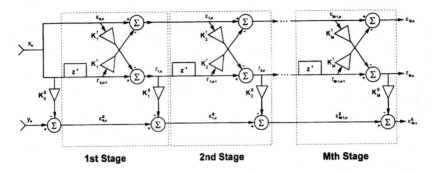

Figure 14 The Wiener lattice filter structure for adaptive system identification given both the input x_n and out y_n.

in Eq. (9.3.4), replacing y_n (used in the whitening filter problem) with x_n.

$$
\begin{bmatrix}
r_{0,n} \\
r_{1,n} \\
\vdots \\
r_{M-1,n} \\
r_{M,n}
\end{bmatrix}
=
\begin{bmatrix}
1 & 0 & \cdots & & 0 \\
a_{1,1}^r & 1 & 0 & \cdots & 0 \\
\vdots & & \ddots & & \vdots \\
a_{M-1,M-1}^r & & \cdots & 1 & 0 \\
a_{M,M}^r & a_{M,M-1}^r & \cdots & a_{M,1}^r & 1
\end{bmatrix}
\begin{bmatrix}
x_n \\
x_{n-1} \\
\vdots \\
x_{n-M+1} \\
x_{n-M}
\end{bmatrix}
= \bar{r}_{0,n:M,n} = L\phi_n^T
\quad (9.4.30)
$$

The lower triangular matrix L in Eq. (9.4.30) also illustrates some interesting properties of the lattice structure. Post-multiplying both sides by $\bar{r}_{0,n:M,n}^H$ and taking expected values leads to $R_M^r = LR_M^x L^H$ where and $R_M^r = diag\{R_0^r R_1^r \ldots R_M^r\}$ and $R_M^x = E\{\phi_n^H \phi_n\}$, the covariance matrix for the input data signal. The importance of this structure is seen when one considers that the covariance matrix inverse is simply

$$
[R_M^x]^{-1} = L^H [R_M^r]^{-1} L
\quad (9.4.31)
$$

where the inverse of the diagonal matrix R_M^r is trivial compared to the inverse of a fully populated covariance matrix. The lower triangular backward error predictor matrix inverse L^{-1} and inverse Hermitian transpose L^{-H} are the well-known LU Cholesky factors of the covariance matrix. The orthogonality of the backward error signals in the lattice makes the Cholesky factorization possible. The forward error signals are not guaranteed to be orthogonal.

Also recall from Eqs (8.1.4) and (9.4.30) that the predicted output is $y_n' = \phi_n H = H^T L^{-1} \bar{r}_{0,n:M,n}^T$. We can therefore define a PARCOR coefficient vector $K_{0:M}^g = [K_0^g K_1^g \ldots K_M^g]$ which represents the cross correlation between the backward error signals and the output signal y_n as seen in Figure 14.

The prediction error for the output signal is simply

$$
\varepsilon_{p,n}^g = y_n - K_p^g r_{p,n} \quad p = 1, \ldots, M
\quad (9.4.32)
$$

Since the backward error signals for every stage are orthogonal to each other, we can write an error recursion similar to the forward and backward error recursions in the lattice filter.

$$\varepsilon_{p,n}^g = \varepsilon_{p-1,n}^g - K_p^g r_{p,n} \qquad p = 0, 1, \ldots, M; \quad \varepsilon_{-1,n} = y_n \tag{9.4.33}$$

The PARCOR coefficients for the output data are found in an analogous manner to the other lattice PARCOR coefficients. The output error signal signal cross correlation is

$$\Delta_{p,n}^g = \alpha \Delta_{p,n-1}^g + \varepsilon_{p-1,n}^g r_{p,n}/(1 - \gamma_{p-1,n}) \qquad p = 0, 1, \ldots, M; \quad \varepsilon_{-1,n}^g = y_n \tag{9.4.34}$$

and $K_{p,n}^g = \Delta_{p,n}^g / R_{p,n}^r = \Delta_{p,n}^g R_{p,n}^{-r}$. The updates of the additional equations for the Wiener lattice are very intuitive as can be seen in Figure 14. To recover H one simply computes $H = L^H K_{0:M}^g$ in addition to the Levinson recursion in Table 3 for the whitening filter problem. The transpose of the lower-triangular backward predictors times the output PARCOR vector reduces to

$$h_p = K_p^g + \sum_{i=p+1}^{M} a_{i,i-p}^{rH} K_i^g \quad p = 0, 1, 2, \ldots, M \tag{9.4.35}$$

which is a very straightforward recursion using the backward error prediction coefficients and lattice output signal PARCOR coefficients.

Figure 15 shows the Wiener lattice performance compared to the standard RLS and LMS algorithms for the exact same data case as presented in Figure 2 where the input signal is unity variance white noise plus a 25 Hz sinusoid of amplitude 5. The Wiener lattice and RLS algorithms have clearly nearly identical parameter tracking

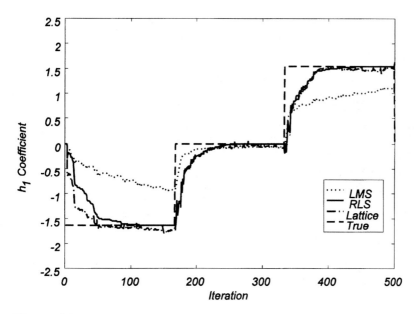

Figure 15 Comparison of LMS, standard RLS, and Wiener lattice performance on the nonstationary system identification previously seen in Figure 2.

performance for this case. However, the lattice results in Figure 15 do indicate some additional numerical noise is present.

While generally not seen for a whitening filter application, the additional divides in the Wiener lattice can be seen as a source of numerically-generated noise due to roundoff error. It is fortunate that this particular example happens to show an excellent example of numerical error processing noise in an adaptive algorithm. We can suppress the noise somewhat by eliminating some of the division operations. Believe it or not, even double precision floating-point calculations are susceptible to roundoff error. Implementing an adaptive filter on a signal processing chip requires careful attention, because double precision floating-point is not usually available without resorting to very slow software-based double precision calculations. Roundoff error is a major concern for any fixed-point signal processing implementation. Figure 16 shows the same comparison but with the use of the double/direct Weiner lattice algorithm developed by Orfanidis (4), which demonstrates superior numerical performance compared to the conventional lattice algorithm.

The Double/Direct lattice gets its name from the use of double error signal updates (one set updated with the previous PARCOR coefficients and one updated the conventional way) and a direct PARCOR update equation. The Double/Direct lattice eliminates the divisions involving the likelihood variable by making error signal predictions *a priori* to the PARCOR coefficient updates, and combining them with *a posteriori* error signals (as computed in the conventional lattice) in the updates for the PARCOR coefficients. An important formulation is the direct PARCOR update, rather than the computation of a cross correlation involving a divide, and then the PARCOR coefficient involving a second divide. The a priori forward

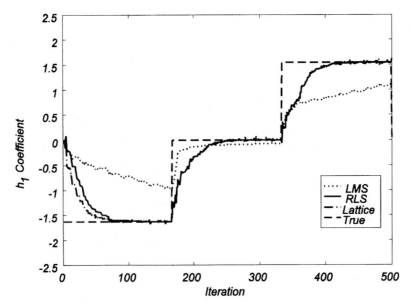

Figure 16 Comparison of the Double/Direct Wiener lattice to the RLS and LMS Wiener filtering algorithms.

and backward error signals are simply

$$\varepsilon_{p,n}^+ = \frac{\varepsilon_{p,n}}{(1-\gamma_{p-1,n-1})} \quad \varepsilon_{p,n}^{+g} = \frac{\varepsilon_{p,n}^g}{(1-\gamma_{p,n})} \quad r_{p,n-1}^+ = \frac{r_{p,n-1}}{(1-\gamma_{p-1,n-1})} \tag{9.4.36}$$

and the updates are done using the previous PARCOR coefficients.

$$\varepsilon_{p+1,n}^+ = \varepsilon_{p,n}^+ - K_{p+1,n-1}^r r_{p,n-1}^+ \tag{9.4.37}$$

$$\varepsilon_{p+1,n}^{+g} = \varepsilon_{p,n}^{+g} - K_{p+1,n-1}^r r_{p+1,n}^+ \tag{9.4.38}$$

$$r_{p+1,n}^+ = r_{p,n-1}^+ - K_{p+1,n-1}^\varepsilon \varepsilon_{p,n}^+ \tag{9.4.39}$$

The a posteriori error signals are updated the conventional way.

$$\varepsilon_{p+1,n} = \varepsilon_{p,n} - K_{p+1,n}^r r_{p,n-1} \tag{9.4.40}$$

$$\varepsilon_{p+1,n}^g = \varepsilon_{p,n}^g - K_{p+1,n}^g r_{p+1,n} \tag{9.4.41}$$

$$r_{p+1,n} = r_{p,n-1} - K_{p+1,n}^\varepsilon \varepsilon_{p,n} \tag{9.4.42}$$

The a priori error signals are mixed into the conventional time, order and time and order updates to essentially eliminate the likelihood variable and its divisions from the algorithm. The likelihood variable can still be retrieved if desired by dividing the conventional a posteriori error signals by the a priori error signals as can be seen from Eq. (9.4.36). The direct update for the forward PARCOR coefficient is derived in Eq. (9.4.43). Elimination of the divides and the direct form of the PARCOR update are seen as the reasons for the improved numerical properties.

$$\begin{aligned}
K_{p+1,n}^\varepsilon = \Delta_{p+1,n} R_{p,n}^{-\varepsilon} &= \left[\alpha \Delta_{p+1,n-1} + \varepsilon_{p,n} r_{p,n-1}^{+H} \right] R_{p,n}^{-\varepsilon} \\
&= \left[\alpha K_{p+1,n-1}^\varepsilon R_{p,n-1}^\varepsilon + \varepsilon_{p,n} r_{p,n-1}^{+H} \right] R_{p,n}^{-\varepsilon} \\
&= \left[\alpha K_{p+1,n-1}^\varepsilon \left\{ R_{p,n}^\varepsilon - \varepsilon_{p,n} \varepsilon_{p,n}^{+H} \right\} \alpha^{-1} + \varepsilon_{p,n} r_{p,n-1}^{+H} \right] R_{p,n}^{-\varepsilon} \\
&= K_{p+1,n-1}^\varepsilon + \varepsilon_{p,n} \left(r_{p,n-1}^{+H} - \varepsilon_{p,n}^{+H} K_{p+1,n-1}^{\varepsilon H} \right) R_{p,n}^{-\varepsilon} \\
&= K_{p+1,n-1}^\varepsilon + \varepsilon_{p,n} r_{p+1,n}^{+H} R_{p,n}^{-\varepsilon}
\end{aligned} \tag{9.4.43}$$

The complete set of updates for the Double/Direct RLS Wiener lattice algorithm are given in Table 4. Considering the effect of the likelihood variable, which is not directly computed but rather *embedded* in the algorithm, it can be seen that there will be very little difference between the a priori and a posteriori error signals when the data is stationary and the lattice PARCOR coefficients have converged. When the data undergoes spectral changes the PARCORs adapt rapidly due to the difference in a priori and a posteriori error signals.

Table 4 The Least-Squares Lattice Algorithm

Description	Equation
Initialization at time $n = 0$ for $p = 0, 1, 2, \ldots, M$	$R_{p,0}^{\varepsilon} = R_{p,0}^{r} = E\{x_n x_n^H\}$
Input error signals at 1st stage input	$\varepsilon_{0,n} = r_{0,n} = \varepsilon_{0,n}^{+} = r_{0,n}^{+} = x_n$
A priori output error signal at 1st stage	$\varepsilon_{0,n}^{+g} = y_n - K_{0,n-1}^{g} r_{0,n}^{+}$
Output PARCOR update	$K_{0,n}^{g} = K_{0,n-1}^{g} + \varepsilon_{0,n}^{+g H} r_{0,n}^{+} R_{0,n}^{-r}$
A posteriori output error signal update at 1st stage	$\varepsilon_{0,n}^{g} = y_n - K_{0,n}^{g} r_{0,n}$
Lattice stage update sequence $\alpha = (N-1)/N$	$for \quad p = 0, 1, 2, \ldots, M$
A priori forward error	$\varepsilon_{p+1,n}^{+} = \varepsilon_{p,n}^{+} - K_{p+1,n-1}^{r} r_{p,n-1}^{+}$
A priori backward error	$r_{p+1,n}^{+} = r_{p,n-1}^{+} - K_{p+1,n-1}^{\varepsilon} \varepsilon_{p,n}^{+}$
Forward PARCOR update	$K_{p+1,n}^{\varepsilon} = K_{p+1,n-1}^{\varepsilon} + r_{p+1,n}^{+} \varepsilon_{p,n}^{+H} R_{p,n}^{-\varepsilon}$
Backward PARCOR update	$K_{p+1,n}^{r} = K_{p+1,n-1}^{r} + \varepsilon_{p+1,n}^{+} r_{p,n-1}^{+H} R_{p,n-1}^{-r}$
A posteriori forward error update	$\varepsilon_{p+1,n} = \varepsilon_{p,n} - K_{p+1,n}^{r} r_{p,n-1}$
A posteriori backward error update	$r_{p+1,n} = r_{p,n-1} - K_{p+1,n}^{\varepsilon} \varepsilon_{p,n}$
Forward error variance time update	$R_{p+1,n}^{\varepsilon} = \alpha R_{p+1,n-1}^{\varepsilon} + \varepsilon_{p+1,n} \varepsilon_{p+1,n}^{+H}$
Backward error variance time update	$R_{p+1,n}^{r} = \alpha R_{p+1,n-1}^{r} + r_{p+1,n} r_{p+1,n}^{+H}$
A priori output error update	$\varepsilon_{p+1,n}^{+g} = \varepsilon_{p,n}^{+g} - K_{p+1,n-1}^{g} r_{p+1,n}^{+}$
Output PARCOR coefficient update	$K_{p+1,n}^{g} = K_{p+1,n-1}^{g} + \varepsilon_{p+1,n}^{+g} r_{p+1,n}^{H} R_{p+1,n}^{-r}$
Output error a posteriori update	$\varepsilon_{p+1,n}^{g} = \varepsilon_{p,n}^{g} - K_{p+1,n}^{g} r_{p+1,n}$
Execute Levinson recursion in Table 9.4.2 then for $p = 0, 1, 2, \ldots, M$	$h_p = K_p^{g} + \sum_{i=p+1}^{M} a_{i,i-p}^{rH} K_i^{g}$

The Double/Direct algorithm may at first appear to be more complicated than the conventional least-squares lattice algorithm, but we are trading computationally expensive divides for far more efficient (and numerically pure) multiplies and adds. Most modern signal processing and computing chips can execute a combination multiply and accumulate (MAC operation) in a single clock cycle. Since a floating point divide is generally executed using a series approximation (integer divides often use a table look-up technique), a divide operation can take from 16 to 20 clock cycles to execute. The Double/Direct Wiener lattice in Table 4 requires 13 MACs and 3 divides per stage while the conventional Wiener lattice requires only 11 MACs, but 8 divides. The Double/Direct lattice algorithm is not only more numerically accurate, but is also more computationally efficient than the conventional lattice algorithm.

9.5 SUMMARY, PROBLEMS, AND BIBLIOGRAPHY

One of the truly fun aspects of digital signal processing is that the computer can be used not only to execute a digital filter for the signals, but also to implement a learning algorithm to adapt the filter coefficients towards the optimization of some criteria. Frequently, this criteria is expressed in terms of a cost function to be minimized. Since the digital filtering operation is linear, an error signal which is linear can be easily expressed in terms of the difference between the filter model's output and some desired output signal. The squared error can be seen as a quadratic surface which has a single minimum at the point which corresponds to the optimum set of filter coefficients. Solving for the least-squared error set of filter coefficients can be done using a "block" of N signal samples, or recursively using a sliding memory window where the optimum solution is valid only for the most recent signal samples. The latter technique is extremely useful for intelligent signal processing algorithms which adapt in real-time to changing environments or even commands (from humans or even other machines) changing the optimum criteria or cost function to be minimized.

The tradeoff between computational complexity and optimality is presented in detail. For the block solution with model order M and N data points, computing the optimal FIR filter coefficients requires $NM(1+NM)$ multiplies and adds just to compute the co-variance matrix and cross correlation vector. Inverting the co-variance matrix requires approximately M^3 plus an expensive divide operation. For a model order of 10 and a 100 sample rectangular data memory window, the block least-squares algorithm requires 1,002,000 multiplies and adds, plus one divide. Repeating this calculation for every time sample of the signals (to create a sliding rectangular memory window) would be extremely computationally expensive. For example, a 50 MFLOPS (50 million floating-point operations per sec) DSP chip could only run real-time for an unimpressive sample rate of about 49 Hz. Using the RLS algorithm with an exponentially-forgetting data memory window, only M^3+2M^2+2M multiples plus 2 divides are needed per time update. For the example model order of 10, this is about 1220 plus 2 divides per time update. The 50 MFLOPS DSP can now run real-time at a sample rate of about 40,000 samples per sec using the RLS algorithm getting almost exactly the same results.

Switching to the RLS Wiener lattice algorithm, we get exactly the same performance as the RLS algorithm at a cost of 13 multiplies plus 3 divides per stage, plus 4 multiplies and a divide at the input stage. For the model order of 10 example, and that a divide equals about 16 multiplies in terms of operations, the RLS Wiener lattice requires $61M+20$ operations, or 630 operations per time update. The 50 MFLOPS DSP can now run at about 79,000 samples per sec giving exactly the same results as the RLS algorithm. The improved performance of the lattice is not only due to its orthogonal order structure, it is due to the fact that the Levinson recursion (which requires M^2 operations) and output linear predictor generation (which requires $M^2 - M$ operations) algorithms need not be computed in real-time. If the linear predictors are needed in real time, the RLS Wiener lattice and RLS algorithms are nearly equal in computational complexity.

If convergence speed is less of a concern, we can trade adaptive performance for significant reductions in computational complexity. The shining example of this is the very popular LMS algorithm. Assuming a recursive update is used to estimate the signal power, the step size can be calculated with 3 multiples and a divide, or about 19 operations. The LMS update requires 3 multiplies and an add for each coefficient, or about $3M$ operations since multiplies and adds can happen in parallel. The normalized LMS filter can be seen to require only $3M+19$ operations per time update. For the example model order of 10, a 50 MFLOPS DSP can run real-time at a sample rate of over 1 million samples per sec. If the input signal is white noise, the LMS algorithm will perform exactly the same as the RLS algorithm, making it the clear choice for many system identification applications. For applications where the input signal always has a known power, the step size calculation can be eliminated making the LMS algorithm significantly more efficient.

PROBLEMS

1. Prove the matrix inversion lemma.
2. How many multiplies and adds/subtractions are needed to execute the Levinson recursion in Table 3 in terms of model order M?
3. Compare the required operations in the RLS, Lattice, and LMS algorithms, in terms of multiplies, additions/subtractions, and divides per time update for a $M = 64$ tap FIR whitening filter (including Levinson recursion for the lattice).
4. Compare the required operations in the RLS, Wiener Lattice, and LMS algorithms for Wiener filtering, in terms of multiplies, additions/subtractions, and divides per time update for a $M = 64$ tap FIR Wiener filter (including Levinson recursion and output predictor generation for the lattice).
5. An LMS Wiener filter is used for system identification of an FIR filter system. Should one use a large number of equally spaced in frequency sinusoids or random noise as a system input signal?
6. Given the autocorrelation data for a signal, calculate directly the whitening PARCOR coefficients.
7. Given the whitening filter input signal with samples $\{-2 \ +1 \ 0 \ -1 \ +2\}$, determine the PARCOR coefficients for the first 5 iterations of a single stage least squares lattice.
8. Compare the result in problem 7 to a single coefficient LMS whitening filter with $\mu = 0.1$.
    (a) Compare the PARCOR coefficient to the single LMS filter coefficient.
    (b) What is the maximum μ allowable (given the limited input data)?
    (c) Compare the PARCOR and LMS coefficients for $\mu = 5$.
9. Show that for a well-converged whitening filter lattice, the forward and backward PARCOR coefficients are approximately equal (for real data) and the likelihood variable approaches unity.
10. Can a Cholesky factorization of the autocorrelation matrix be done using the forward prediction error variances in place of the backward error variances?

BIBLIOGRAPHY

M. Bellanger Adaptive Digital Filters and Signal Analysis, Marcel-Dekker: New York, 1988.

W. S. Hodgkiss, J. A. Presley Jr. "Adaptive Tracking of Multiple Sinusoids whose Power Levels are Widely Separated," IEEE Trans. Acoust. Speech Sig. Proc., ASSP-29, 1891, pp. 710–721.

M. Morf, D. T. L. Lee "Recursive Least-Squares Ladder Forms for Fast Parameter Tracking," Proc. 17th IEEE Conf. Decision Control, 1979, p. 1326.

S. J. Orfandis Optimum Signal Processing, 2nd Ed., New York: McGraw-Hill, 1988.

B. Widrow, S. D. Sterns Adaptive Signal Processing, Englewood Cliffs: Prentice-Hall, 1985.

B. Widrow et al. "Adaptive Noise Cancelling: Principles and Applications," Proceedings of the IEEE, Vol. 63, No.12, 1975, pp. 1692–1716.

B. Widrow et al. "Stationary and Nonstationary Learning Characteristics of the LMS Adaptive Filter," Proceedings of the IEEE, Vol. 64, 1976, pp. 1151–1162.

REFERENCES

1. P. Strobach "Recursive Triangular Array Ladder Algorithms," IEEE Trans. Signal Processing, Vol. 39, No. 1, January 1991, pp. 122–136.

2. M. Morf "Fast Algorithms for Multivariate Systems," PhD dissertation, Stanford University, Stanford, CA, 1974.

3. B. Friedlander "Lattice Filters for Adaptive Processing," Proceedings of the IEEE, Vol. 70, No.8, August 1982, pp. 829–867.

4. S. J. Orfanidis "The Double/Direct RLS Lattice," Proc. 1988 Int. Conf. Acoust., Speech, and Sig. Proc., New York.

10

Recursive Adaptive Filtering

In this chapter we develop and demonstrate the use of some important applications of adaptive filters as well as extend the algorithms in Chapter 9 for multichannel processing and frequency domain processing. This book has many examples of applications of adaptive and non-adaptive signal processing, not just to present the reader with illustrative examples, but as a vehicle to demonstrate the theory of adaptive signal processing. Chapter 9 contains many examples of adaptive whitening filters and Wiener filtering for system identification. Part IV will focus entirely on adaptive beamforming and related processing. There are of course, many excellent texts which cover each of these areas in even greater detail. This chapter explores some adaptive filtering topics which are important to applied adaptive signal processing technology.

In Section 10.1 we present what is now well-known as adaptive Kalman filtering. Signal processing has its origins in telephony, radio and audio engineering. However, some of the most innovative advances in processing came with Bode's operational amplifier and early feedback control systems for Naval gun stabilization. Control systems based on electronics were being developed for everything from electric power generation to industrial processes. These systems were and still are quite complicated and are physically described by a set of signal states such as, temperature, temperature rate, temperature acceleration, etc. The various derivatives (or integrals) of the measured signal are processed using $+6$ dB/oct high pass filters (for derivatives) and or -6 dB/oct low pass filters (for integration). The ± 6 dB/oct slope of the filter represents a factor of $j\omega$ ($+6$ dB/oct for differentiation), or $1/j\omega$ (-6 dB/oct for integration), the plant to be controlled can be physically modeled with a set of partial differential equations. Then sensors are attached to the plant and the various sensor signal states are processed using high or low pass filters to create the necessary derivatives or integrals, respectively. Simple amplifiers and phase shifters supplied the required multiplies and divides. Current addition and subtraction completed the necessary signal processing operations. Signals based on control set points and sensor responses could then be processed in *an analog computer* literally built from vacuum tubes, inductors, capacitors, resistors, and transformers to execute closed-loop feedback control of very complex physical systems. This very elegant solution to complex system control is known as a "white-box" control problem, since the entire plant and input–output signals are completely known. (An elegant solution is defined as a solution one wishes one thought of first.)

By the end of the 1950s, detection and tracking of satellites was a national priority in the United States due to the near public panic over the Soviet Union's Sputnik satellite (the world's first man-made satellite). The implications for national defense and security were obvious and scientists and engineers throughout the world began to focus on new adaptive algorithms for reducing noise and estimating kinematic states such as position, velocity, acceleration, etc. This type of control problem is defined as a "black-box" control problem because nothing is known about the plant or input–output signals (there is also a "grey box" distinction, where one knows a little about the plant, a little about the signals, but enough to be quite dangerous). In 1958 R. E. Kalman (1) published a revolutionary paper in which sampled analog signals were actually processed in a vacuum tube analog computer. The work was funded by E.I du Pont de Nemours & Co., a world leader chemical manufacturer. Many of the complex processes in the manufacture of plastics, explosives, nylon, man-made textiles, etc., would obviously benefit from a "self-optimizing" controller. This is especially true if the modeled dynamics as described by the physical parameters in a benchtop process do not scale linearly to the full-scale production facility. Physical parameters such as viscosity, compressibility, temperature, and pressure do not scale at all while force, flow, and mass do. This kind of technical difficulty created a huge demand for what we now refer to as the Kalman filter where one simply commands a desired output (such as a temperature setpoint), and the controller does the rest, including figuring out what the plant is. Section 10.1 presents a derivation of the Kalman filter unified with the least-squared error approaches of Chapters 8 and 9. The recursive least-squares tracking solution is then applied to the derivation of the α-β-γ tracker (presented in Section 4.2) which is shown to be optimal if the system states have stationary covariances.

In Section 10.2 we extend the LMS and lattice filter structures to IIR forms, such as the all-pole and pole zero filters presented in Sections 3.1 and 3.2. Pole-zero filters, also known as autoregressive moving average ARMA filters, pose special convergence problems which must be carefully handled in the adaptive filters which whiten ARMA signals, or attempt to identify ARMA systems. These constraints are generally not a problem so long as stability constraints are strictly maintained in the adaptive algorithms during convergence and, the signals at hand are stable signals. Transients in the input–output signals can lead to incorrect signal modeling as well as inaccurate ARMA models. ARMA models are particularly sensitive to transients in that their poles can move slightly on or outside the unit circle on the complex z-plane, giving an unstable ARMA model. An embedding technique will be presented for both the LMS and lattice filters which also shows how many channels of signals may be processed in the algorithms to minimize a particular error signal.

Finally in this chapter we present frequency domain adaptive processing with the LMS algorithm. Frequency domain signals have the nice property of orthogonality, which means that an independent LMS filter can be assigned to each FFT frequency bin. Very fast convergence can be had for frequency domain processing except for the fact that the signals represent a time integral of the data. However, this integration is well-known to suppress random noise, making frequency domain adaptive processing very attractive to problems where the signals of interest are sinusoids in low signal-to-noise ratio (SNR). This scenario applies to a wide range of adaptive signal processing problems. However, the frequency domain adaptive filter transfer functions operations are eventually converted back to the time domain to give the filter coefficients. Spectral leakage from sinusoids which may not be perfectly aligned to a frequency bin must be eliminated using the technique presented in Section 5.4 to insure good results. Presentation of the Kalman filter, IIR and multichannel forms for adaptive LMS and lattice filters, and frequency domain processing complement the applications of signal processing presented throughout the rest of this book.

10.1 ADAPTIVE KALMAN FILTERING

Consider the problem of tracking the extremely high-velocity flight path of a low-flying satellite from its radio beacon. It is true that for a period of time around the launching of Sputnik, the world's first satellite, many ham radio enthusiasts provided very useful information on the time and position of the satellite from ground detections across the world. This information could be used to refine an orbit state model which in turn is used to predict the satellites position at future times. The reason for near public panic was that a satellite could potentially take pictures of sensitive defense installations or even deliver a nuclear bomb, which would certainly wreck one's day. So with public and government interest in high gear, the space race would soon join the arms race, and networks of tracking radio stations were constructed not only to detect and track enemy satellites, but also to communicate with one's own satellites. These now-familiar satellite parabolic dishes would swivel as needed to track and communicate with a satellite passing overhead at incredible speeds of over 17,500 miles per hour (for an altitude of about 50 miles). At about 25,000 miles altitude, the required orbital velocity drops to about 6500 mi/hr and the satellite is *geosynchronous*, meaning that it stays over the same position on the ground.

For clear surveillance pictures, the satellite must have a very low altitude. But, with a low altitude, high barometric pressure areas on the ground along the flight path will correspond to the atmosphere extending higher up into space. The variable atmosphere, as well as gravity variations due to the earth not being perfectly spherical, as well as the effects of the moon and tides, will cause the satellite's orbit to change significantly. A receiver dish can be pointed in the general direction of the satellite's path, but the precise path is not completely known until the tracking algorithm "locks in" on the trajectory. Therefore, today adaptive tracking systems around the world are used to keep up to date the trajectories of where satellites (as well as orbital debris) are at all times.

Tracking dynamic data is one of the most common uses of the Kalman filter. However, the distinction between Kalman filtering and the more general recursive least-squares becomes blurred when the state vector is replaced by other basis functions. In this text we specifically refer to the Kalman filter as a recursive least squares estimator of a state vector, most commonly used for a series of differentials depicting the state of a system. All other "non-state vector" basis functions in this text are referred to in the more general terms of a recursive least squares algorithm. The Kalman filter has enabled a vast array of 20th Century mainstay technologies such as: world-wide real-time communications for telephones, television, and the Internet; the global positioning system (GPS) which is revolutionizing navigation and surveying; satellite wireless telephones and computer networks; industrial and process controls; and environmental forecasting (weather and populations, diseases, etc); and even monetary values in stock and bond markets around the world. The Kalman filter is probably even more ubiquitous than the FFT in terms of its value and uses to society.

The Kalman filter starts with a simple state equation relating a column vector of n_z measurements, $z(t)$, at time t, to a column vector of n_x states, $x(t)$, as seen in Eq. (10.1.1). The measurement matrix, $H(t)$, has n_z rows and n_x columns and simply relates the system states linearly to the measurements. $H(t)$ usually will not change with time but it is left as a time variable for generality. The column vector $w(t)$ has n_z rows and represents the measurement noise, the expected value of which will play an important part in the sensitivity of the Kalman filter to the measurement data.

$$z(t) = H(t)x(t) + w(t) \tag{10.1.1}$$

It is not possible statistically to reduce the difference between $z(t)$ and $Hx(t)$ below the measurement noise described by the elements of $w(t)$. But, the states in $x(t)$ will generally have significantly less noise than the corresponding measurements due to the "smoothing" capability of the Kalman filter. We can therefore define a quadratic cost function in terms of the state to measurement error to be minimized by optimizing the state vector.

$$J(N) = \frac{1}{2} \left[\begin{bmatrix} z(1) \\ z(2) \\ \vdots \\ z(N) \end{bmatrix} - \begin{bmatrix} H(1) \\ H(2) \\ \vdots \\ H(N) \end{bmatrix} x \right]^H \begin{bmatrix} R(1) & 0 & \cdots & 0 & 0 \\ 0 & R(2) & 0 & \cdots & 0 \\ & & \ddots & & \\ 0 & 0 & \cdots & 0 & R(N) \end{bmatrix}^{-1}$$

$$\times \left[\begin{bmatrix} z(1) \\ z(2) \\ \vdots \\ z(N) \end{bmatrix} - \begin{bmatrix} H(1) \\ H(2) \\ \vdots \\ H(N) \end{bmatrix} x \right] \tag{10.1.2}$$

The cost function is defined over N iterations of the filter and can be more compactly written as

$$J(N) = \frac{1}{2} \left[z^N - H^N x \right]^H \left[R^N \right]^{-1} \left[z^N - H^N x \right] \tag{10.1.3}$$

where $J(N)$ is a scalar, z^N is Nn_z rows by 1 column ($Nn_z \times 1$), H^N is ($Nn_z \times n_x$), R^N is ($Nn_z \times Nn_z$), and x is ($n_x \times 1$). Clearly, R^N can be seen as $E\{w(t)w(t)^H\}$ giving a diagonal matrix for uncorrelated Gaussian white noise. Instead of solving for the FIR filter coefficients which best relates an input signal and output signal in the block least-squares algorithm presented in Section 8.1, we are solving for the optimum state vector which fits the measurements $z(t)$. Granted, this may not be very useful for sinusoidal states unless the frequency is very low such that the N measurements cover only a small part of the wavelength. It could also be said that for digital dynamic system modeling, one must significantly over sample the time updates for Kalman filters to get good results. This was presented for a mass-spring dynamic system in Section 4.1.

The cost function is a quadratic function of the state vector to be optimized as well as the error. Differentiating with respect to the state, we have

$$
\begin{aligned}
\frac{\partial J(N)}{\partial x} &= \frac{1}{2}[H^N]^H[R^N]^{-1}[z^N - H^N x] - \frac{1}{2}[z^N - H^N x]^H[R^N]^{-1}[H^N] \\
&= [H^N]^H[R^N]^{-1}[H^N]x - [H^N]^H[R^N]^{-1}z^N
\end{aligned}
\tag{10.1.4}
$$

where a second derivative with respect to x is seen to be positive definite, indicating a "concave up" error surface. The minimum error is found by solving for the least squared error estimate for the state vector based on N observations, $x'(N)$, which gives a zero first derivative in Eq. (10.1.4).

$$
x'(N) = \left\{[H^N]^H[R^N]^{-1}[H^N]\right\}^{-1}[H^N]^H[R^N]^{-1}z^N
\tag{10.1.5}
$$

The matrix term in the braces which is inverted in Eq. (10.1.5) can be seen as the covariance matrix of the state vector x prediction error. The state prediction error is simply

$$
\begin{aligned}
e(N) = x - x'(N) &= x - \left\{[H^N]^H[R^N]^{-1}[H^N]\right\}^{-1}[H^N]^H[R^N]^{-1}[H^N x + w^N] \\
&= x - [H^N]^{-1}[R^N][H^N]^{-H}[H^N]^H[R^N]^{-1}[H^N x + w^N] \\
&= \left\{[H^N]^H[R^N]^{-1}[H^N]\right\}^{-1}[H^N]^H[R^N]^{-1}w^N
\end{aligned}
\tag{10.1.6}
$$

making the covariance matrix of the state prediction error

$$
\begin{aligned}
P(N) = E\{e(N)e(N)^H\} &= \left\{[H^N]^H[R^N]^{-1}[H^N]\right\}^{-1}[H^N]^H[R^N]^{-1}R^N[R^N]^{-1}H^N \\
&\quad \times \left\{[H^N]^H[R^N]^{-1}[H^N]\right\}^{-1} \\
&= \left\{[H^N]^H[R^N]^{-1}[H^N]\right\}^{-1}
\end{aligned}
\tag{10.1.7}
$$

For a recursive update, we now augment the vectors and matrices in (10.1.2)–(10.1.3) for the $N+1$st iteration and write the state prediction error as an inverse covariance matrix.

$$
\begin{aligned}
P(N+1)^{-1} &= \left[H^{N+1}\right]^{H}\left[R^{N+1}\right]^{-1}\left[H^{N+1}\right] \\
&= \left[H^{NH}H(N+1)^{H}\right]\begin{bmatrix} R^{N} & 0 \\ 0 & R(N+1) \end{bmatrix}^{-1}\begin{bmatrix} H^{N} \\ H(N+1) \end{bmatrix} \\
&= \left[H^{N}\right]^{H}\left[R^{N}\right]^{-1}\left[H^{N}\right] + H(N+1)^{H}R(N+1)^{-1}H(N+1) \\
&= P(N)^{-1} + H(N+1)^{H}R(N+1)^{-1}H(N+1)
\end{aligned}
$$

$$(10.1.8)$$

Equation (10.1.8) is of the form where the matrix inversion lemma (Section 9.1) can be applied to produce a recursive update for $P(N+1)$, rather than its inverse.

$$
\begin{aligned}
P(N+1) = P(N) &- P(N)H^{H}(N+1)\left[R(N+1) + H(N+1)P(N)H^{H}(N+1)\right]^{-1} \\
&\times H(N+1)P(N)
\end{aligned}
$$

$$(10.1.9)$$

The quantity in the square brackets which is inverted is called the covariance of the measurement prediction error. The measurement prediction error is found by combining Eqs (10.1.1) and (10.1.5).

$$
\begin{aligned}
\zeta(N+1) &= z(N+1) - z'(N+1) \\
&= z(N+1) - H(N+1)x'(N) \\
&= H(N+1)x + w(N+1) - H(N+1)x'(N) \\
&= H(N+1)[x - x'(N)] + w(N+1) \\
&= H(N+1)e(N) + w(N+1)
\end{aligned}
$$

$$(10.1.10)$$

The covariance of the measurement prediction error, $S(N+1)$, (also known as the covariance of the innovation), is seen to be

$$
\begin{aligned}
S(N+1) &= E\left\{\zeta(N+1)\zeta^{H}(N+1)\right\} \\
&= E\left\{[H(N+1)e(N) + w(N+1)][e^{H}(N)H^{H}(N+1) + w^{H}(N+1)]\right\} \\
&= H(N+1)P(N)H^{H}(N+1) + R(N+1)
\end{aligned}
$$

$$(10.1.11)$$

We can also define the update gain, or Kalman gain, for the recursion as

$$
K(N+1) = P(N)H^{H}(N+1)[S(N+1)]^{-1}
$$

$$(10.1.12)$$

where Eqs (10.1.11) and (10.1.12) are used to simplify and add more intuitive meaning to the state error covariance inverse update in Eq. (10.1.9).

$$
P(N+1) = P(N) - K(N+1)S(N+1)K^{H}(N+1)
$$

$$(10.1.13)$$

There are several other forms the update for the inverse state error covariance. The state vector update is

$$
\begin{aligned}
x'(N+1) &= P(N+1)\big[H^{N+1}\big]^H\big[R^{N+1}\big]^{-1}z^{N+1} \\
&= P(N+1)\big[H^{NH}H(N+1)^H\big]
\begin{bmatrix} R^N & 0 \\ 0 & R(N+1) \end{bmatrix}^{-1}
\begin{bmatrix} z^N \\ z(N+1) \end{bmatrix} \\
&= P(N+1)\big[H^N\big]^H\big[R^N\big]^{-1}z^N \\
&\quad + P(N+1)H(N+1)^H R(N+1)^{-1}z(N+1) \\
&= [I - K(N+1)H(N+1)]P(N)\big[H^N\big]^H\big[R^N\big]^{-1}z^N \\
&\quad + P(N+1)H(N+1)^H R(N+1)^{-1}z(N+1) \\
&= [I - K(N+1)H(N+1)]x'(N) + K(N+1)z(N+1)
\end{aligned}
$$

$$(10.1.14)$$

where it is straightforward to show that $P(N+1)=[I - K(N+1)H(N+1)]\,P(N)$ and that the Kalman gain can be written as $K(N+1)=P(N+1)\,H^H(N+1)R^{-1}(N+1)$. The update recursion for the state vector can now be written in the familiar form of old state vector minus the Kalman gain times an error.

$$
x'(N+1) = x'(N) + K(N+1)\zeta(N+1) \tag{10.1.15}
$$

All we have done here is apply the recursive least squares algorithm to a state vector, rather than an FIR coefficient vector, as seen in Section 9.1. The basis function for the filter here is the measurement matrix defined in Eq. (10.1.1), and the filter outputs are the measurement predictions defined in Eq. (10.1.10). The RLS algorithm solves for the state vector which gives the least-squared error for measurement predictions. As seen in Eq. (10.1.5), the least-squares solution normalizes the known measurement error R^N, and as seen in Eq. (10.1.7), as the number of observations N becomes large, the variance of the state error becomes smaller than the variance of the measurement error. This smoothing action of the RLS algorithm on the state vector is a valuable technique to obtain good estimates of a state trajectory, such as velocity or acceleration, given a set of noisy measurements. A summary of the RLS state vector filter is given in Table 1.

The Kalman filter is really just a recursive kinematic time update form of an RLS filter for dynamic state vectors. Examples of dynamic state vectors include the position, velocity, acceleration, jerk, etc., for a moving object such as a satellite or aircraft. However, any dynamic data, such as process control information (boiler temperature, pressure, etc.), or even mechanical failure trajectories can be tracked using a dynamic state Kalman filter. The Kalman filter updates for the dynamic state vector are decidedly different from RLS, which is why we will refer to the dynamic state vector RLS filter specifically as Kalman filtering. It is assumed that an underlying kinematic model exists for the time update of the state vector. Using Newtonian physics, the kinematic model for a position-velocity-acceleration type state vector is well known. The modeled state vector update can be made for time $N+1$ using information available at time N.

$$
x(N+1) = F(N+1)x(N) + v(N) + G(N)u(N) \tag{10.1.16}
$$

Table 1   The RLS State Vector Filter

Description	Equation
Basis function (measurement matrix)	$H(N+1)$
Kalman gain update	$K(N+1) = \dfrac{P(N)H^H(N+1)}{R(N+1)+H(N+1)P(N)H^H(N+1)}$ $= P(N)H^H(N+1)[S(N+1)]^{-1}$
Inverse autocorrelation matrix update	$P(N+1) = (H^{N+1^H}[R^N]^{-1}H^{N+1})^{-1}$ $= P(N) - K(N+1)S(N+1)K^H(N+1)$
Measurement prediction error	$\zeta(N+1) = z(N+1) - z'(N+1)$ $= z(N+1) - H(N+1)x'(N)$
Optimal state vector update using prediction error	$x'(N+1) = x'(N) + K(N+1)\zeta(N+1)$

The term $v(N)$ in Eq. (10.1.16) represents the *process noise* of the underlying kinematic model for the state vector updates, while $G(N)\, u(N)$ represent a control signal input, which will be left to Part V of this text. The matrix $F(N+1)$ is called *the state transition matrix* and is derived from the kinematic model. As with the measurement matrix $H(N+1)$, usually the state transition matrix is constant with time, but we allow the notation to support a more general context. Consider the case of a state vector comprised of a position, velocity, and acceleration. Proceeding without the control input, we note that our best "a priori" (meaning before the state is updated with the least-squared error recursion) update for the state at time $N+1$, given information at time N is

$$x'(N+1|N) = F(N)x(N|N)$$

$$= \begin{bmatrix} 1 & T & \frac{1}{2}T^2 \\ 0 & 1 & T \\ 0 & 0 & 1 \end{bmatrix} \begin{bmatrix} x_{N|N} \\ \dot{x}_{N|N} \\ \ddot{x}_{N|N} \end{bmatrix} \qquad (10.1.17)$$

where T is the time interval of the update in seconds. The process noise is not included in the a priori state vector prediction since the process noise is assumed uncorrelated from update to update. However, the state prediction error variance must include the process noise covariance. A typical kinematic model for a position, velocity, and acceleration state vector would assume a random change in acceleration from state update to state update. This corresponds to a constant white jerk variance (not to be confused with a relentless obnoxious Caucasian (see Section 4.2)) which results in an acceleration which is an integral of the zero-mean Gaussian white jerk. Assuming a piecewise constant acceleration with a white jerk allows the state transition in Eq. (10.1.17) where the process noise variance for the acceleration is $E\{v(N)v^H(N)\} = \sigma_v^2$. The acceleration process variance scales to velocity by a factor of T, and to position by a factor of $\frac{1}{2}T^2$ leading to the process noise covariance

$Q(N)$ in Eq. (10.1.18).

$$Q(N) = \begin{bmatrix} \frac{1}{2}T^2 \\ T \\ 1 \end{bmatrix} \sigma_v^2 [\frac{1}{2}T^2 \quad T \quad 1] = \begin{bmatrix} \frac{1}{4}T^4 & \frac{1}{2}T^3 & \frac{1}{2}T^2 \\ \frac{1}{2}T^3 & T^2 & T \\ \frac{1}{2}T^2 & T & 1 \end{bmatrix} \sigma_v^2 \qquad (10.1.18)$$

The process noise Q, state transition F and measurement matrix H are usually constants, but our notation allows time variability. If one were to change the update rate T for the Kalman filter, Q must be re-scaled. There are quite a number of process noise assumptions which can be implemented to approximate the real situation for the state kinematics. Bar-Shalom and Li (2) provide a detailed analysis of various process noise assumptions. An a priori state error covariance can be written as

$$P'(N + 1|N) = F(N)P(N|N)F^H(N) + Q(N) \qquad (10.1.19)$$

The a priori state error estimate is then used to update the innovation covariance

$$S(N + 1) = H(N + 1)P'(N + 1|N)H^H(N + 1) + R(N + 1) \qquad (10.1.20)$$

and finally to update the Kalman gain.

$$K(N + 1) = P'(N + 1|N)H^H(N + 1)[S(N + 1)]^{-1} \qquad (10.1.21)$$

Given the latest measurement $z(N + 1)$ at time $N + 1$, an a posteriori update of the least-squared error state vector and state vector error covariance is completed as given in Eqs (10.1.22) and (10.1.23).

$$x(N + 1|N + 1) = x'(N + 1|N) + K(N + 1)\{z(N + 1) - H(N + 1)x'(N + 1|N)\} \qquad (10.1.22)$$

$$P(N + 1|N + 1) = P'(N + 1|N) - K(N + 1)S(N + 1)K^H(N + 1) \qquad (10.1.23)$$

Equations (10.1.17)–(10.1.23) constitute a typical Kalman filter for tracking a parameter along with its velocity and acceleration as a function of time. There are countless applications and formulations of this fundamental processing solution which can be found in many excellent texts in detail well beyond the scope of this book. But, it is extremely useful to see a derivation of the Kalman filter in the context of the more general RLS algorithm, as well as its similarities and differences with other commonly used adaptive filtering algorithms.

One of the more useful aspects of Kalman filtering is found not in its ability to smooth the state vector (by reducing the affect of measurement noise), but rather in its predictive capabilities. Suppose we are trying to intercept an incoming missile and are tracking the missile trajectory from a series of position measurements obtained from a radar system. Our anti-missile requires about 30 sec lead time to fly into the preferred intercept area. The preferred intercept area is determined by the highest probability of hit along a range of possible intercept positions defined by the trajectories of the two missiles. Equations (10.1.17)–(10.1.19) are used to determine the incoming missile trajectory and variance of the trajectory error, which in three dimensions is an ellipsoid centered around a given predicted future position for the incoming missile. The further out in time one predicts the incoming missile's position, the larger the size of prediction error ellipsoid becomes. The state error prediction can be seen in this case as a three-dimensional Gaussian probability density "cloud", where the ellipsoid is a one-standard deviation contour. Using Baye's rule and other hypothesis scoring methods, one can solve the kinematic equations in a computer for the launch time giving rise to the most likely intercept. This is not always easy nor straightforward, since the tracking filter is adaptive and the states and measurement noise are dynamically changing with time.

Some of the best evidence of the difficulty of missile defense was seen with the Patriot system during the Gulf War of 1991. To be fair, the Patriot was designed to intercept much slower aircraft, rather than ballistic missiles. But, in spite of the extreme difficulty of the task, the missile defense system brought to the attention of the general public the concept of an intelligent adaptive homing system using Kalman filters. Track-intercept solutions are also part of aircraft collision avoidance systems and may soon be part of intelligent highway systems for motor vehicles. Using various kinematic models, Kalman filters have been used routinely in financial markets, prediction of electrical power demand, and in biological/environmental models for predicting the impact of pollution and species populations. If one can describe a phenomena using differential equations and measure quantities systematically related to those equations, tracking and prediction filters can be of great utility in the management and control of the phenomena.

Table 2 summarizes the Kalman filter. The type of measurements used determine the measurement matrix $H(N+1)$ and measurement noise covariance $R(N+1)$. The state kinematic model determines the process noise covariance $Q(N+1)$ and state transition matrix $F(N+1)$. As noted earlier, H, R, Q, F, σ_v^2, σ_w^2, and T are typically constant in the tracking filter algorithm. As can be seen from Table 2, the Kalman filter algorithm is very straightforward. One makes a priori state and state covariance predictions, measures the resulting measurement prediction error, and adjusts the state updates according to the error and the define measurement and process noise variances. However, the inverse state error covariance update given in the table can be reformulated to promote lower numerical roundoff error. The expression in Eq. (10.1.24) is algebraically equivalent to that given in

$$P(N + 1|N) = [I - K(N + 1)H(N + 1)]P(N + 1|N)[I - K(N + 1)H(N + 1)]^H$$
$$+ K(N + 1)R(N + 1)K(N + 1)^H$$

$$(10.1.24)$$

Table 2, but its structure, known as Joseph form, promotes symmetry and reduces numerical noise.

An example of the Kalman filter is presented following the simple rocket height tracking example given in Section 4.2 for the α-β-γ tracker. As will be seen below, the α-β-γ tracking filter is a simple Kalman filter where the Kalman gain has been fixed to the optimal value assuming a constant state error covariance. In the rocket example, we have height measurements every 100 msec with a standard deviation of 3 m. The maximum acceleration (actually a deceleration) occurs at burnout due to both gravity and drag forces and is about 13 m/sec^2. Setting the process noise standard deviation to 13 is seen as near optimal for the α-β-γ tracking filter, allowing

Table 2 The Kalman Filter

Description	Equation
Basis function (measurement matrix)	$H(N+1)$
State transition matrix	$F(N+1)$
Process noise covariance matrix	$Q(N+1)$
Measurement noise covariance matrix	$R(N+1)$
A priori state prediction	$x'(N+1|N) = F(N)x(N|N)$
A priori state error prediction	$P'(N+1|N) = F(N)P(N)F^H(N) + Q(N)$
A priori measurement prediction	$z'(N+1) = H(N+1)x'(N+1|N)$
Innovation covariance update	$S(N+1) = H(N+1)P'(N+1|N)H^H(N+1)$ $+ R(N+1)$
Kalman gain update	$K(N+1) = P'(N+1|N)H^H(N+1)[S(N+1)]^{-1}$
Measurement prediction error	$\zeta(N+1) = z(N+1) - H(N+1)x'(N+1|N)$
Optimal state vector update using prediction error	$x(N+1|N+1) = x'(N+1|N) + K(N+1)\zeta(N+1)$
Inverse state covariance matrix update	$P(N+1|N+1) = P'(N+1|N)$ $- K(N+1)S(N+1)K^H(N+1)$

non-adaptive tracking of the changes in acceleration, velocity, and position of the rocket's height. The adaptive Kalman gain, on the other hand, depends on the ratio of the state prediction error covariance to the measurement noise covariance. The Kalman gain will be high when the state error is large and the measurements accurate (σ_w and R small), and the gain will be low when the state error is small and/or the measurements inaccurate. The adaptive gain allows the Kalman filter to operate with a lower process noise than the non-adaptive α-β-γ tracking filter, since the gain will automatically increase or decrease with the state prediction error. In other words, if one knows the maneuverability of the target (the maximum acceleration, for example), and the measurements are generally noisy, the α-β-γ tracking filter is probably the best choice, since the measurement noise will not affect the filter gain too much. But if one does not know the target maneuverability, the Kalman filter offers fast convergence to give the least-squared error state vector covariance. However, when the measurements are noisy, the Kalman filter also suffers from more noise in the state vector estimates when the same process noise is used as in the α-β-γ tracking filter. Figures 1 and 2 illustrate the noise sensitivity for the rocket example given in Section 4.2.

Comparing the filter gains used in Figures 1 and 2 we see that the α-β-γ tracking filter used [0.2548 0.3741 0.2746] while the Kalman filter calculated a gain vector of [0.5046 1.7545 3.0499]. The higher Kalman gain fit the state vector more accurately to the measurements, but resulted in more velocity and acceleration noise. In the simulation, if we let the measurement noise approach zero, the responses of

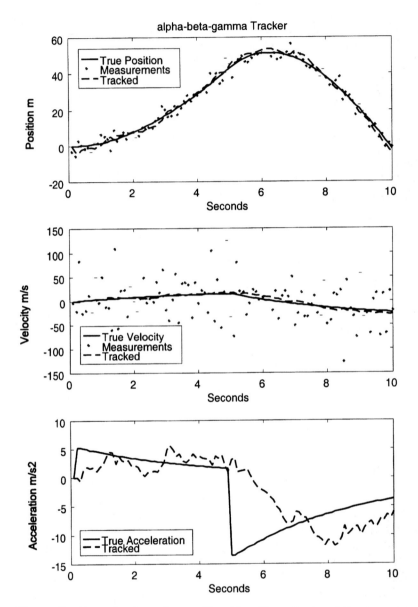

Figure 1 Non-adaptive α-β-γ tracking filter (from Section 4.2) where the measurement noise is 3 and process noise 13.

the two filters become identical and the gain vector converges to $[1 \ 1/T \ 1/(2T^2)]$, or $[1 \ 20 \ 200]$ when $T = 100$ msec. By comparing the gain vectors for a given measurement noise (and process noise for the α-β-γ tracking filter), we can adjust the Kalman filter process noise so that the filter gains are roughly the same when the measurement noise is still a relatively high 3 m. We find that the Kalman filter process noise can be reduced to about 1.0 to give a Kalman gain vector of $[0.2583 \ 0.3851 \ 0.2871]$, which is very close to the gain vector for the α-β-γ tracking filter. Figure 3 shows

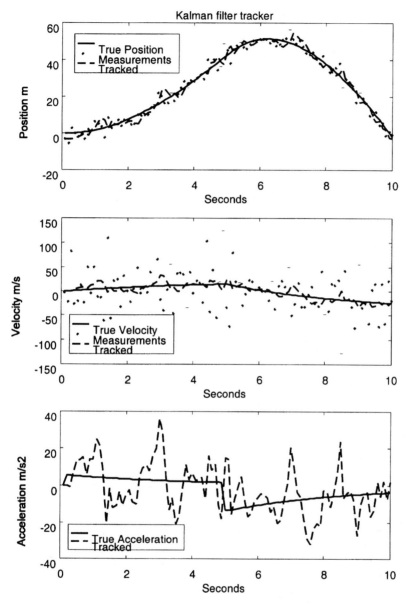

Figure 2 Kalman filter for the same measurement and process noise as seen in Figure 1.

the Kalman filter tracking results which are nearly identical to the α-β-γ tracking filter with a high process noise of 13.

The significance of determining the Kalman filter process noise which gives the same filter gain as the α-β-γ tracking filter when the measurement noise is high is that this gives an indication of the minimum process noise needed to track the target maneuvers without significant overshoot. Note that for a near zero measurement noise, the gain vectors are only identical for the same process noise. Given a reasonable minimum process noise for the Kalman filter, it gives superior tracking per-

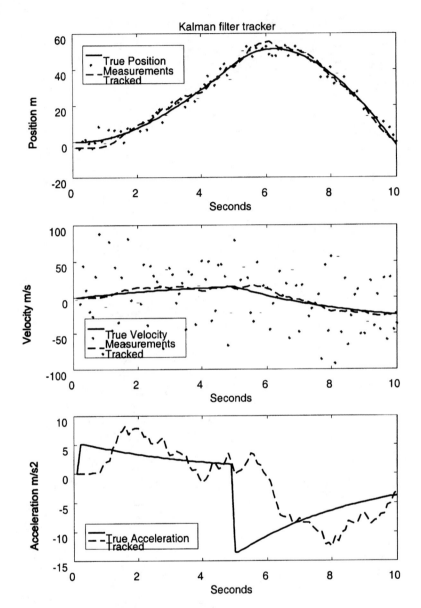

Figure 3   Kalman filter results using the same measurement data as seen in Figures 1–2, but with the process noise of 1.

formance to the α-β-γ tracking filter when the measurement noise is low as seen in Figures 4 and 5.

The examples in Figures 1–5 and in Section 4.2 indicate that the α-β-γ tracking filter is the algorithm of choice when the measurements are noisy. Unfortunately, there is not an algebraic expression which can tell us simply how to set the process noise in a Kalman filter for optimum performance. The reason is that the Kalman gain is not only a function of measurement noise, process noise, and time update

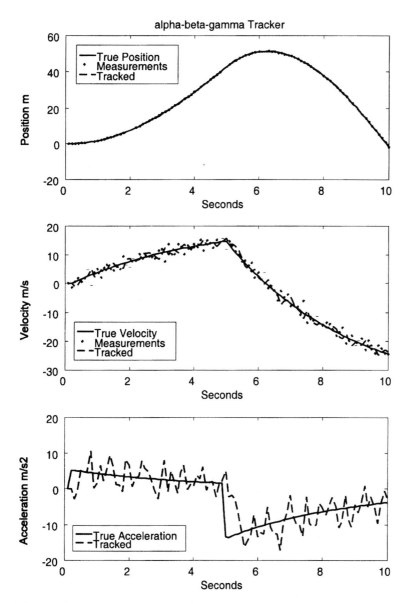

Figure 4 Low measurement noise (0.1) improves the α-β-γ tracking filter with the same process noise of 13.

(as is the case with the α-β-γ tracking filter), but it also depends on the state prediction error covariance, which is a function of target maneuvers. In most situations, one knows a reasonable expected target track maneuverability and the expected measurement noise a priori. This information allows one to execute simulations for a covariance analysis of the Kalman filter's performance on the expected data. With the measurement noise small (or near zero), one can set the Kalman filter process noise to almost any large value and the filter will perform nearly perfectly. With

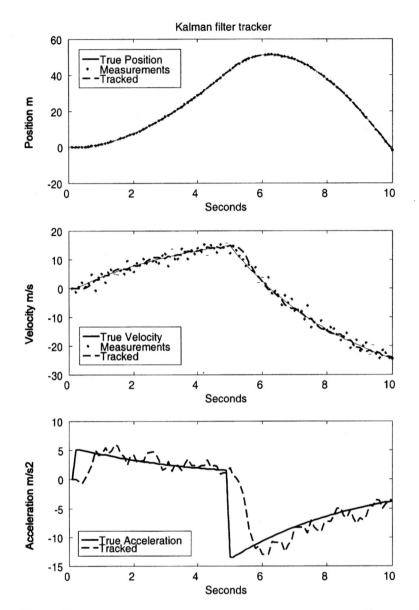

Figure 5 The Kalman filter with 0.1 measurement noise and a process noise of 1 gives improved performance over the α-β-γ tracking filter for the same data and a process noise of 13.

high measurement noise, it is easier to determine the minimum Kalman process noise empirically for a given target maneuver. Setting the process noise to the maximum expected acceleration is a good starting point (and ending point for the α-β-γ tracking filter), but the Kalman filter generally allows a lower process noise due to the enhanced ability of the adaptive updates to help the state error converge rapidly to target maneuvers.

The α-β-γ **tracking filter** is a Kalman filter with fixed state error covariance. In Section 4.2 we introduced the α-β-γ tracking filter as an important application of a state-variable filter and its derivation was referenced to this Section. Its derivation is rather straightforward, but algebraically tedious. We show the approach of Bar-Shalom (3) to demonstrate the technique. But more importantly, the solution for the α-β-γ tracking filter shows that α, β, and γ cannot be chosen independently. For a given amount of position noise reduction (as determined by $\alpha < 1$), β and γ are systematically determined from the measurement error σ_w and update time T. Setting α therefore also sets the process noise, and the state prediction error covariances to constant "steady-state" values. However, the filter will only converge to the prescribed steady-state error covariance if all the kinematic assumptions are true. If the tracking target makes an unexpected maneuver, the state error will increase for a while as the state vectors "overshoot" the measurements temporarily. This can be seen in Figure 4.5 where a "sluggish" track is produced from using too small a process noise in the α-β-γ tracking filter.

Combining Eqs (10.1.19), (10.1.21) and (10.1.23) we can write an expression for the updated state prediction error covariance P assuming steady-state conditions.

$$P = F^{-1}(P' - Q)F^{-H} = (I - KH)P' \tag{10.1.25}$$

If we simplify the problem by assuming only the α and β tracking gains and a piecewise constant acceleration model, we have the following matrix equation to solve.

$$\begin{bmatrix} 1 & -T \\ 0 & 1 \end{bmatrix} \begin{bmatrix} p'_{11} - \dfrac{T^4}{4}\sigma_v^2 & p'_{12} - \dfrac{T^3}{2}\sigma_v^2 \\ p'_{21} - \dfrac{T^3}{2}\sigma_v^2 & p'_{22} - T^2\sigma_v^2 \end{bmatrix} \begin{bmatrix} 1 & 0 \\ -T & 1 \end{bmatrix} = \begin{bmatrix} 1 - k_1 & 0 \\ -k_2 & 1 \end{bmatrix} \begin{bmatrix} p'_{11} & p_{12} \\ p'_{21} & p'_{22} \end{bmatrix} \tag{10.1.26}$$

Equation (10.1.26) is best solved by simply equating terms. After some simplification, we have

$$k_1 p'_{11} = 2T p'_{12} - T^2 p'_{22} + \frac{T^4}{4}\sigma_v^2 \tag{10.1.27}$$

$$k_1 p'_{12} = T^2 p'_{22} - \frac{T^3}{2}\sigma_v^2 \tag{10.1.28}$$

and

$$k_2 p_{12} = T^2 \sigma_v^2 \tag{10.1.29}$$

Solving for the predicted state covariance elements we have

$$p_{11} = \frac{k_1}{1 - k_1}\sigma_w^2 \tag{10.1.30}$$

and

$$p'_{12} = p'_{21} = \frac{k_2}{1-k_1}\sigma_w^2 \tag{10.1.31}$$

Finally, solving for the p'_{22} term we have

$$p'_{22} = \left(\frac{k_1}{T} + \frac{k_2}{2}\right)p'_{12} \tag{10.1.32}$$

Combining Eqs (10.1.27)–(10.1.32) and after some algebraic cancellations, we arrive at the following bi-quadratic equation for the filter gains.

$$k_1^2 - 2Tk_2 + Tk_1k_2 + \frac{T^2}{4}k_2^2 = 0 \tag{10.1.33}$$

The filter gains in terms of α and β are simply $k_1 = \alpha$ and $k_2 = \beta/T$. Equation (10.1.33) reduces to

$$\alpha^2 - 2\beta + \alpha\beta + \frac{\beta^2}{4} = 0 \tag{10.1.34}$$

Solving for α in terms of β using the quadratic formula we have

$$\alpha = \sqrt{2\beta} - \frac{\beta}{2} \tag{10.1.35}$$

and solving for β in terms of α we have

$$\beta = 4 - 2\alpha - 4\sqrt{1-\alpha} \tag{10.1.36}$$

For the α-β-γ filter, it can be shown (with some algebraic complexity) that $\gamma = \beta^2/\alpha$. Another important relation is seen when equating terms for p'_{12}.

$$p'_{12} = \frac{T^2\sigma_v^2}{k_2} = \frac{k_2}{1-k_1}\sigma_w^2$$

$$\frac{T^2\sigma_v^2}{\dfrac{\beta}{T}} = \frac{\dfrac{\beta}{T}}{1-\alpha}\sigma_w^2 \tag{10.1.37}$$

Equation (10.1.37) leads to an expression for the track maneuverability index λ_M.

$$\lambda_M = \frac{T^2\sigma_v}{\sigma_w} \tag{10.1.38}$$

The results of the derivation above are also presented in Eqs (4.2.7)–(4.2.11). The maneuverability index determines the "responsiveness" of the tracking filter to the measurements. For the fixed gain α-β-γ filter one estimates the maximum target acceleration to determine the process noise standard deviation σ_v. Given the measurement noise standard deviation σ_w and track update time T, the maneuverability index λ_M is determined as well as α, β, and γ (see Section 4.2).

The α-β-γ filter is an excellent algorithm choice for cases where the kinematics are known and the measurements are relatively noisy. Given unknown target maneuverability and/or low noise measurements, the adaptive Kalman filter is a better tracking algorithm choice, because the process noise can be significantly lowered without sacrificing responsiveness to give more noise reduction in the tracking state outputs. While derivation of the α-β-γ filter is quite tedious, knowledge of the Kalman filter equations allows a straightforward solution. Tracking filters are an extremely important and powerful signal processing tool. Whenever one measures a quantity with a known measurement error statistic, and observes the quantity changing deterministically over time, one is always interested in predicting when that quantity reaches a certain value, and with what statistical confidence.

10.2 IIR FORMS FOR LMS AND LATTICE FILTERS

The least-squared error system identification and signal modeling algorithms presented in Chapters 8 and 9 were limited to FIR filter structures. However, modeling systems and signals using infinite impulse response (IIR) digital filter is quite straightforward. Because an IIR filter with adaptive coefficients has the potential to become unstable, one must carefully constrain the adaptive IIR filter algorithm. If we are performing a system identification where both the input and output signals are completely known, stability is less of a concern provided that the unknown system to be identified is stable. When only the output signal is available and we are trying to model the signal as the output of an IIR filter driven by white noise, the stability issues can be considerably more difficult. Recall from Section 3.1 that FIR filters are often referred to as moving average, or MA, and are represented on the complex z-plane as a polynomial where the angles of the zeros determine the frequencies where the FIR filter response is attenuated, or has a spectral dip. In Section 3.2 the IIR filter is presented as a denominator polynomial in the z-domain. The angles of the zeros of the IIR polynomial determine the frequencies where the filter's response is amplified, or has a spectral peak. The IIR zeros are called the filter's poles because they represent the frequencies where the response has a peak. IIR filters with only a denominator polynomial feedback the past values of the output signal in the calculation of the current output, and as a consequence, are often called autoregressive, or AR filters.

The most general form of an IIR filter has both a numerator polynomial and a denominator polynomial, and is usually referred to as a pole-zero, or ARMA filter. The zeros of the numerator polynomial are the filter zeros and the zeros of the denominator polynomial are the filter poles. For the IIR filter to be stable, the magnitude of all the poles must be less than unity. This insures that the output feedback of some particular output signal sample eventually reverberates out to zero as it is fed back in the autoregressive generation of the current IIR output. Therefore, the denominator part of a stable IIR filter must be a minimum phase polynomial with all its zeros (the system poles) inside the unit circle.

Figure 6 depicts system identification of some unknown system represented by the ratio of z-domain polynomials $B[z]/A[z]$ where the zeros of $B[z]$ are the system zeros and the zeros of $A[z]$ are the system poles. The adaptive system identification is executed by a pair of adaptive filters which are synchronized by a common error signal $e[n]$, and the input $x[n]$ and output $y[n]$. Recall from Eq. (3.2.10) the difference

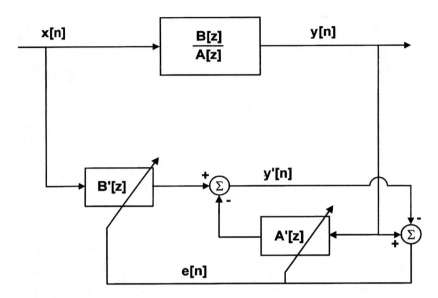

Figure 6 An ARMA LMS system identification operation given the unknown system input $x[n]$ and output $y[n]$.

equation for an ARMA filter assuming a numerator polynomial order of Q (Q zeros) and a denominator polynomial order P (P poles).

$$x[n]b_0 + x[n-1]b_1 + \ldots + x[n-Q]b_Q = y[n] + y[n-1]a_1 + \ldots + y[n-P]a_P$$

$$(10.2.1)$$

We can model the actual ARMA filter in Eq. (10.2.1) to make a linear prediction of the output $y'[n]$ using the available signals except the most recent output sample $y[n]$ (to avoid a trivial solution).

$$y'[n] = x[n]b'_0 + x[n-1]b'_1 + \ldots + x[n-Q]b'_Q - y[n-1]a'_1 - \ldots - y[n-P]a'_P$$

$$(10.2.2)$$

The prediction error is simply

$$
\begin{aligned}
e[n] &= y[n] - y'[n] \\
&= y[n] - x[n]b'_0 - x[n-1]b'_1 - \ldots - x[n-Q]b'_Q \\
&\quad + y[n-1]a'_1 + \ldots + y[n-P]a'_P
\end{aligned}
$$

$$(10.2.3)$$

The gradient of the error with respect to $B'[z] = b'_0 + b'_1 z^{-1} + \ldots + b'_Q z^{-Q}$ is negative and the gradient with respect to $A'[z] = a'_1 z^{-1} + a'_2 z^{-2} + \ldots + a'_P z^{-P}$ is positive. This sign distinction will be important to the LMS coefficient updates. However, we note that the model orders P and Q are assumed known and the AR model polynomial $A'[z]$ does not have the leading coefficient of unity seen in $A[z]$. The squared error for this case is a quadratic function of both sets of coefficients, and the second derivative is also positive indicating that the least-squared error solution corresponds to the set of ARMA coefficients which gives zero gradient. When the input and output signals are known, the least-squared error solution applies only to the chosen model orders P and Q. If P and Q are chosen not to both correspond to the actual unknown system's polynomial orders, one only has a least-squared error solution (best fit) for the chosen model orders, not overall. Recall that for the FIR system identification case, one could choose a model order much higher than the actual system's, and the unneeded coefficients would nicely converge to zero. This unfortunately is not the case for ARMA filter models. Each combination of P and Q will give a least-squared error solution, but only for that particular case. If all possible combinations of model orders are tried, the actual unknown system's polynomial orders will correspond to the overall least-squares solution. Obviously, it would be very valuable to have a straightforward methodology to determine the optimal ARMA model efficiently.

Since we have both input and output signals with known variances, we can directly apply an LMS algorithm to find the coefficients for the chosen model order. The FIR, or MA part of the model is updated at time n using

$$b'_{i,n} = b'_{i,n-1} + 2\mu_b x[n-i]e[n] \quad \mu_b \le \frac{1}{QE\{x[n]^2\}} \tag{10.2.4}$$

and the IIR, or AR part of the ARMA model is updated using

$$a'_{j,n} = a'_{j,n-1} - 2\mu_a y[n-j]e[n] \quad \mu_a \le \frac{1}{PE\{y[n]^2\}} \tag{10.2.5}$$

The sign of the updates in Eqs (10.2.4)–(10.2.5) are due to the need to adjust the coefficients in the opposite direction of the gradient as a means to move to the minimum squared error. Stability of the algorithm is simplified by knowing the variances of the input and output, and by the structure in Figure 6 where $y'[n]$ is not fed back into the adaptive filter. Doing so could cause an instability if one or more of the zeros of $A'[z]$ move outside the unit circle during adaptation. So long as the input $x[n]$ and output $y[n]$ are stationary allowing an accurate μ_b and μ_a, we are assured stable LMS updates and convergence to something close to the unknown system.

Error signal bootstrapping is used for ARMA signal modeling. When only the output signal $y[n]$ is available, we can model the signal as an ARMA process defined as an ARMA filter with unity variance white noise input, or innovation. This procedure is somewhat tricky, since both an input and output signal are needed for an ARMA model. The input is estimated using a heuristic procedure called error bootstrapping. Since the ARMA process innovation is a unity-variance white noise signal, we can estimate the input by calculating the prediction error in the absence of the actual input.

$$e^b[n] = y[n] - y^{b'}[n]$$
$$= y[n] - e^b[n-1]b'_1 - \ldots - e^b[n-Q]b'_Q + y[n-1]a'_1 + \ldots + y[n-P]a'_P \tag{10.2.6}$$

The bootstrapped error signal in Eq. (10.2.6) is very close, but not identical to, the linear prediction error in Eq. (10.2.3) except the ARMA input $x[n]$ is replaced by the bootstrapped error $e^b[n]$. Note that the bootstrapped output linear prediction $y^b[n]$ assumes a zero input for $e^b[n]$, allowing a prediction error prediction based on a least-squared error assumption. The bootstrapped ARMA input signal can be seen as an a priori estimate of the prediction error. The adaptive update for the MA coefficients must use an LMS step size based on the output power, since the bootstrapped error will initially be of magnitude on the order of the ARMA output.

$$b'_{i,n} = b'_{i,n-1} + 2\mu_a e^b[n-i]e[n] \quad \mu_a \le \frac{1}{QE\{y[n]^2\}} \tag{10.2.7}$$

The prediction error for the bootstrapped ARMA model $e[n]$, is calculated as

$$\begin{aligned} e[n] &= y[n] - y'[n] \\ &= y[n] - e^b[n]b'_0 - e^b[n-1]b'_1 - \ldots - e^b[n-Q]b'_Q \\ &\quad + y[n-1]a'_1 + \ldots + y[n-P]a'_P \end{aligned} \tag{10.2.8}$$

which is exactly as in the Wiener filtering case in Eq. (10.2.3) except the actual ARMA input $x[n]$ is replaced by the bootstrapped error signal $e^b[n]$. For an AR signal model, the MA order becomes $Q=0$, and the bootstrapping procedure reduces to a whitening filter algorithm.

To recover the MA coefficients for the ARMA bootstrapped model we must obtain the proper scaling by multiplying the MA coefficients by the standard deviation of the prediction error. This compensates for the premise that the ARMA signal is the output of a pole-zero filter driven by unity-variance zero-mean white Gaussian noise. For the LMS and RLS bootstrapped ARMA algorithms, a whitening filter structure is used where the bootstrapped error signal is used to simulate the unavailable input to the unknown ARMA system. Minimizing this error only guarantees a match to the real ARMA system within a linear scale factor. The scale factor is due to the variance of the bootstrapped error not being equal to unity. Multiplying the converged MA coefficients by the square-root of the bootstrapped error variance scales the whitening filter giving an ARMA signal model assuming a unity-variance white noise innovation. When the ARMA input signal is available (the Wiener filtering system identification problem), scaling is not necessary for the LMS and RLS algorithms. However, one will see a superior MA coefficient performance for the RLS bootstrapped ARMA algorithm when the bootstrapped error is normalized to unity variance. For the embedded ARMA lattice, a little more complexity is involved in calculating the MA scale.

Figures 7 and 8 show the results of an ARMA LMS system identification and error bootstrapping simulation. The actual ARMA filter has a pair of complex conjugate poles at ± 200 Hz with magnitude 0.98 (just inside the unit circle). A pair of conjugate zeros is at ± 400 Hz also of magnitude 0.95. In the simulation, the sample rate is 1024 Hz making the pole angles ± 1.227 radians and the zero angle ± 2.454 radians on the complex z-plane (see Sections 2.1–2.3 and Chapter 3 for more detail on digital filters). A net input–output gain of 5.0 is applied to the actual ARMA filter giving a numerator polynomial in z of $B[z] = 5.0 + 7.3436z^{-1} + 4.5125z^{-2}$. The actual ARMA denominator polynomial is $A[z] = 1.0 - 0.6603z^{-1} + 0.9604z^{-2}$ and 2000 output samples are generated using a unity variance white noise input

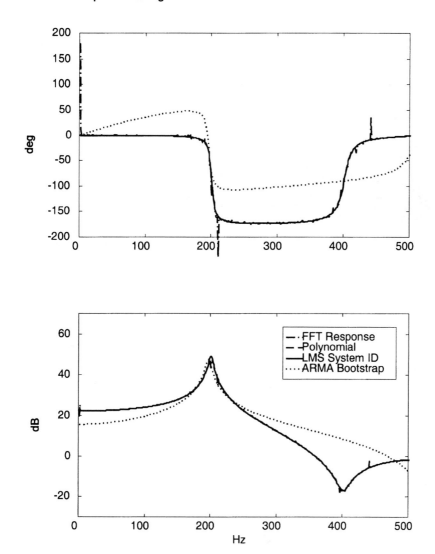

Figure 7 Frequency responses of actual ARMA filter and filter models for Wiener filtering and error Bootstrapping.

$x[n]$ and Eq. (10.2.9).

$$y[n] = 5.0x[n] + 7.3436x[n-1] + 4.5125x[n-2] + 0.6603y[n-1]$$
$$- 0.9604y[n-2] \tag{10.2.9}$$

The frequency responses shown in Figure 7 are generated two ways. First, the FFT of the output signal is divided by the FFT of the input signal and the quotient is time averaged (see Section 6.2 for details on transfer functions). Second, the response is calculated directly from the ARMA coefficients by computing the ratio of the zero-padded FFTs of the coefficients directly. This technique is highly efficient

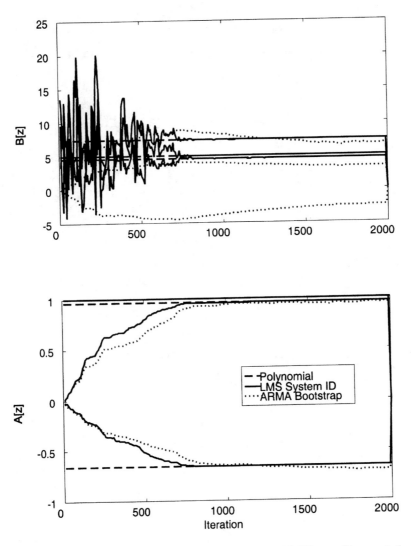

Figure 8 ARMA coefficient responses for the LMS Wiener filter and the LMS error bootstrapped ARMA model.

and is analogous to computing a z-transform where z is $e^{-j\Omega n}$ (see Section 3.1 for the frequency responses of digital filters). Figure 7 clearly shows the input–output signal FFT, the actual ARMA polynomial response, and the LMS Wiener filter response overlaying nearly perfectly. The LMS Wiener filter calculated polynomials are $B'[z] = 5.0000 + 7.3437z^{-1} + 4.5126z^{-2}$ and $A'[z] = 1.0000 - 0.6603z^{-1} + 0.9604z^{-2}$ which are very close to the actual ARMA coefficients.

The bootstrapped ARMA signal model is close near the ARMA peak created by the pole at 200 Hz, but completely misses the zero at 400 Hz. It will be seen below that proper normalization of the bootstrapped error signal to better model the ARMA innovation will result in improved MA coefficient performance for the RLS algorithm. In theory, the bootstrapped ARMA estimate may improve depending on the predictability of the innovation and the effective length of the data memory window. Performance limitations due to slow LMS algorithm convergence are the main reason for high interest in fast adaptive algorithms. The coefficient estimates as a function of algorithm iteration are seen in Figure 8.

A fast algorithm allows one to get the most information for a given input–output signal observation period. The LMS bootstrapped ARMA filter had to be slowed considerably by reducing the step size to about 1% of its theoretical maximum defined in Eqs (10.2.4)–(10.2.5) and (10.2.7) to get the performance seen in Figures 7 and 8. This corresponds to a data memory window for the LMS algorithm of about 100 samples. The bootstrapped ARMA performance seen in Figure 8 shows a reasonable AR coefficient response and a very poor MA coefficient response. This is also seen in Figure 7 in the reasonable peak alignment but completely missed zero. However, close examination of the ARMA coefficient response in Figure 8 shows a good convergence on the pole magnitude (0.98), but some drift in the pole angle (or frequency). This can be seen as caused by interaction of the pole with the zero. Indeed, the LMS algorithm for both the AR and MA parts are driven by a single error signal where the input data is not orthogonal in order. The interaction between the AR and MA part for the bootstrapped ARMA signal model is exacerbated by the coupling (due to the least-squared error approximations) in the LMS ARMA algorithm. The poorly-converged bootstrapped ARMA coefficients after 2000 iterations are $B'[z] = 6.7311 + 3.3271z^{-1} - 2.4981z^{-2}$ and $A'[z] = 1.0000 - 0.7046z^{-1} + 0.9537z^{-2}$. As noted earlier, the AR part of the match is reasonable.

The embedding technique allows multiple signal channels to be in the RLS or Lattice algorithms by vectorizing the equations. The RLS algorithm of Section 9.1 (see Table 9.1) is the most straightforward to embed an ARMA model. One simply extends the basis vector to include the ARMA filter input and output sequences and correspondingly, extends the coefficient vector to include the AR coefficients.

$$\varepsilon_n = y_n - \phi_n H$$

$$= y_n - \begin{bmatrix} x_n x_{n-1} \dots x_{n-Q} - y_{n-1} - y_{n-2} - \dots - y_{n-P} \end{bmatrix} \begin{bmatrix} b_0 \\ b_1 \\ \vdots \\ b_Q \\ a_1 \\ a_3 \\ \vdots \\ a_P \end{bmatrix} \qquad (10.2.10)$$

Using the basis function and coefficient vector in Eq. (10.2.10), one simply executes the RLS algorithm as seen in Table 9.1. For error bootstrapping for ARMA signal models, the bootstrapped error is simply

$$\varepsilon_n^b = y_n - \phi_n^b H$$

$$= y_n - \left[0\varepsilon_{n-1}^b \ldots \varepsilon_{n-Q}^b - y_{n-1} - y_{n-2} - \ldots - y_{n-P} \right] \begin{bmatrix} b_0 \\ b_1 \\ \vdots \\ b_Q \\ a_1 \\ a_3 \\ \vdots \\ a_P \end{bmatrix} \qquad (10.2.11)$$

where ε_n^b is zero on the right side of Eq. (10.2.11). Given the bootstrapped error to model the ARMA process innovation, the linear prediction error is simply

$$\varepsilon_n = y_n - \phi_n H$$

$$= y_n - \left[\varepsilon_n^b \varepsilon_{n-1}^b \ldots \varepsilon_{n-Q}^b - y_{n-1} - y_{n-2} - \ldots - y_{n-P} \right] \begin{bmatrix} b_0 \\ b_1 \\ \vdots \\ b_Q \\ a_1 \\ a_3 \\ \vdots \\ a_P \end{bmatrix} \qquad (10.2.12)$$

An example of the enhanced performance of the RLS ARMA algorithm is seen in Figures 9 and 10 where the RLS ARMA algorithm is applied to the same data as the LMS ARMA example seen in Figure 7 and 8. The RLS algorithm converges much faster than the LMS algorithm and appears to give generally better ARMA modeling results. This is true even for the bootstrapped error case, which still suffers from MA coefficient modeling difficulty. However, we see that there is much less coupling between the AR and MA parts of the model. Recall that the true ARMA polynomials are $B[z] = 5.0 + 7.3436z^{-1} + 4.5125z^{-2}$ and $A[z] = 1.0 - 0.6603z^{-1} + 0.9604z^{-2}$. The RLS Wiener filter ARMA model is $B[z] = 5.0 + 7.3436z^{-1} + 4.5125z^{-2}$ and $A[z] = 1.0 - 0.6603z^{-1} + 0.9604z^{-2}$. The RLS bootstrapped error ARMA signal model is $B[z] = 8.1381 + 1.6342z^{-1} + 0.4648z^{-2}$ and $A[z] = 1.0 - 0.7002z^{-1} + 0.9786z^{-2}$. The Wiener filter RLS results are practically zero error while the bootstrapped ARMA results still show a weak zero match in the MA part of the ARMA signal model.

Normalizing the bootstrapped error signal to unity variance in the RLS algorithm constrains the bootstrap process to better model the ARMA innovation. This also conditions the prediction error for the algorithm to be better balanced between the MA and AR coefficients, thus allowing vastly improved ARMA signal modeling.

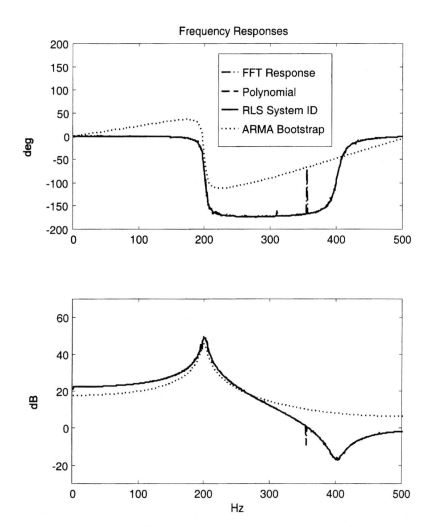

Figure 9 ARMA Frequency response results for the RLS algorithm using Wiener filtering and error bootstrapping.

The normalized bootstrap error performance can also be seen in the embedded ARMA lattice, presented below. The lattice PARCOR coefficients are naturally normalized by their respective error signal variances. The lattice ARMA requires simply multiplying the MA coefficients by the square-root of the linear prediction error to produce properly normalized MA coefficients (for $b_0 \neq 1$). Multiplying the MA coefficients by the standard deviation of the prediction error certainly scales the model so the spectral peaks match well. However, applying this technique directly to the LMS and RLS ARMA bootstrap algorithm does not result in a very good MA coefficient match. Normalizing the bootstrap error signal to unity variance achieves the same scaling effect in the ARMA model, but also greatly improves the MA coefficient match. It can be seen that normalizing the bootstrap error leads to improved conditioning in the adaptive algorithm to balance the adaptive effort

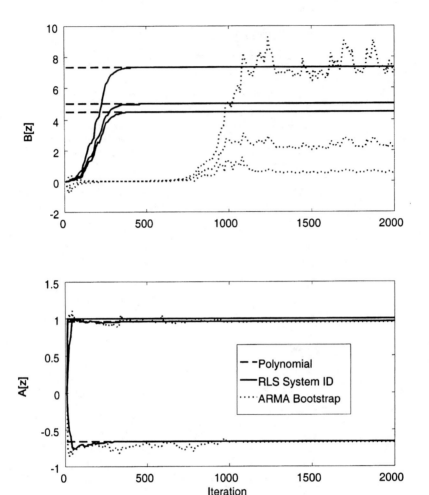

Figure 10 ARMA coefficient convergence results using the RLS algorithm for the Wiener filtering case and the error bootstrapping case for modeling ARMA processes.

between the MA and AR parts of the model. Equation (10.2.13) shows the basis function ϕ_n^b used in calculating the bootstrap error $\varepsilon_n^b = \phi_n^b H$, similar to the unnormalized bootstrap error in Eq. (10.2.11), except σ_b is the square root of the bootstrap error variance.

$$\phi_n^b = \left[0 \frac{\varepsilon_{n-1}^b}{\sigma_b} \dots \frac{\varepsilon_{n-Q}^b}{\sigma_b} - y_{n-1} - y_{n-2} - \dots - y_{n-P} \right] \tag{10.2.13}$$

The bootstrap error is then used in a recursive estimate for the bootstrap error variance, and subsequently, standard deviation can be calculated. The bootstrap error variance recursion is best initialized to unity as well. The RLS algorithm is very sensitive to the bootstrap error amplitude. The basis function used in the RLS algorithm is seen in Eq. (10.2.14) as is used to generate the linear prediction error in the RLS as seen in Eq. (10.2.12).

$$\phi'' = \left[\frac{\varepsilon_n^b}{\sigma_b} \frac{\varepsilon_{n-1}^b}{\sigma_b} \dots \frac{\varepsilon_{n-Q}^b}{\sigma_b} - y_{n-1} - y_{n-2} - \dots - y_{n-P} \right] \qquad (10.2.14)$$

It is both surprising and interesting that simply normalizing the bootstrap error improves the RLS ARMA model result as significantly as that demonstrated in Figures 11 and 12. Why should it matter whether $b_0 = 5$ and $\sigma_b = 1$ or $b_0 = 1$ and $\sigma_b = 5$? Consider that the RLS basis vector and linear prediction error must be separate for the net gain of the MA coefficients to be properly calculated. Since the RLS linear prediction error is not normalized, the MA coefficients are scaled properly in the RLS normalized ARMA bootstrap algorithm without the need to multiply the coefficients by some scale factor afterwards. The reason bootstrap error normalization does not help the LMS performance significantly is that much of

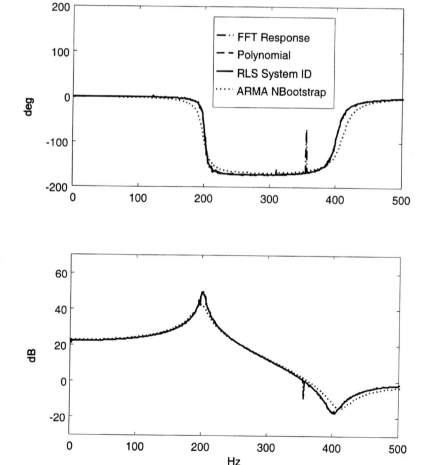

Figure 11   Frequency responses showing a close match between the actual and normalized bootstrap error Wiener filter.

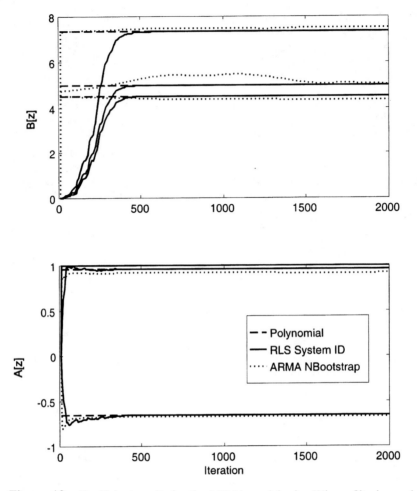

Figure 12   Coefficient results for the ARMA model using Wiener filtering and normalized ARMA bootstrap error.

the optimization in the RLS algorithm is lost in the approximations used to create the simple and robust LMS algorithm.

In this example of normalized bootstrap error RLS ARMA modeling, it is noted that the RLS bootstrap algorithm is very sensitive to initial conditions compared to the highly reproducible results for Wiener filtering, where both input and output are known. Initializing the bootstrap error variance to, say the output signal y_n variance, also gave improved MA coefficient matching results, but not as good as that with an initial unity variance. The ARMA normalized bootstrap results are $B[z] = 5.0114 + 7.5178z^{-1} + 4.3244z^{-2}$ and $A[z] = 1.0000 - 0.6738z^{-1} + 0.9149z^{-2}$ which are nearly as good as some of the Wiener filtering ARMA results.

Embedding an ARMA filter into a lattice structure is very straightforward once one has established a matrix difference equation. Recall that the projection operator framework in Chapter 8 was completely general and derived in complex matrix form (the superscript H denotes transpose plus complex conjugate). All of the lattice and Levinson recursions in Chapter 9 are also presented in complex matrix form. We start by writing the ARMA difference equations in matrix form.

$$\begin{bmatrix} \varepsilon_n^y \\ \varepsilon_n^x \end{bmatrix} = \begin{bmatrix} y_n \\ x_n \end{bmatrix} + \sum_{i=1}^{M} \begin{bmatrix} a_i & -b_i \\ -c_i & d_i \end{bmatrix} \begin{bmatrix} y_{n-i} \\ x_{n-i} \end{bmatrix} \tag{10.2.15}$$

It is straightforward to show that $d_i = a_i/b_0$ and $c_i = b_i/b_0$ where $i = 1, 2, ..., M$, $d_0 = 1/b_0$, and $a_0 = c_0 = 1$. The forward prediction error ARMA lattice recursion is seen to be

$$\begin{bmatrix} \varepsilon_{p+1,n}^y \\ \varepsilon_{p+1,n}^x \end{bmatrix} = \begin{bmatrix} \varepsilon_{p,n}^y \\ \varepsilon_{p,n}^x \end{bmatrix} - \begin{bmatrix} K_{p+1,n}^{ryy} & K_{p+1,n}^{rxy} \\ K_{p+1,n}^{ryx} & K_{p+1,n}^{rxx} \end{bmatrix} \begin{bmatrix} r_{p,n-1}^y \\ r_{p,n-1}^x \end{bmatrix} \tag{10.2.16}$$

and the backward error ARMA lattice recursion is

$$\begin{bmatrix} r_{p+1,n}^y \\ r_{p+1,n}^x \end{bmatrix} = \begin{bmatrix} r_{p,n-1}^y \\ r_{p,n-1}^x \end{bmatrix} - \begin{bmatrix} K_{p+1,n}^{\varepsilon yy} & K_{p+1,n}^{\varepsilon xy} \\ K_{p+1,n}^{\varepsilon yx} & K_{p+1,n}^{\varepsilon xx} \end{bmatrix} \begin{bmatrix} \varepsilon_{p,n}^y \\ \varepsilon_{p,n}^x \end{bmatrix} \tag{10.2.17}$$

Equations (10.2.16) and (10.2.17) lead to the ARMA lattice structure shown in detail for the $p + 1$st stage in Figure 13. The 3-dimensional sketch of the ARMA lattice stage shows how the 2-channel matrix equations can be layed out into an electrical circuit. The importance of the symmetry within each lattice stage, and

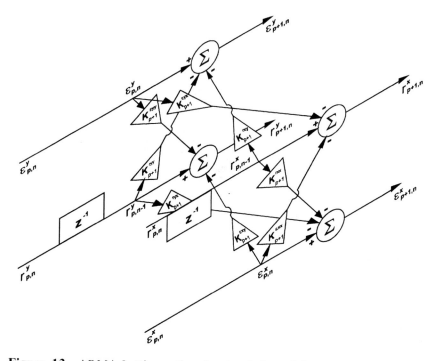

Figure 13 ARMA Lattice section showing 2-channel forward and backward prediction error for the embedded ARMA model.

among the various stages of the lattice filter is that a very complicated set of matrix updates can be executed systolically, dividing the total operational load between multiple processors in a logical manner.

Figure 14 shows the layout of three lattice stages to form a 3rd-order ARMA filter. The error bootstrap is shown for ARMA signal modeling where the innovation must be modeled. The unknown ARMA filter input and output are available, Wiener filtering is used to identify the system where the ARMA output y_n enters the lattice through $\varepsilon_{0,n}^y$ and the ARMA input x_n enters through $\varepsilon_{0,n}^x$ replacing the error bootstrap ε_n^b.

Figure 15 shows the frequency responses for the original ARMA system polynomials, the lattice results using Wiener filtering, and the ARMA lattice error bootstrap technique. Clearly, the ARMA modeling results using the lattice are quite good. Figure 16 shows the ARMA coefficient results.

The ARMA embedded lattice is essentially a 2-channel whitening filter. For the Wiener filtering case, the Levinson recursion supplies the coefficients of $A[z]$ and $B[z]$ assuming the lattice order M is greater than or equal to both P and Q, the ARMA polynomial orders, respectfully. The value of any coefficients from order P or Q up to M should converge to zero for the Wiener filtering case. However, we are missing b_0 (and $d_0 = 1/b_0$).

Given the forward prediction error variances leaving the Mth stage, the scale factor for b_0 is

$$b_0 = \sqrt{\frac{R_M^{\varepsilon yy}}{R_M^{\varepsilon xx}}} = \sqrt{\frac{R_M^\varepsilon(1,1)}{R_M^\varepsilon(2,2)}} \tag{10.2.18}$$

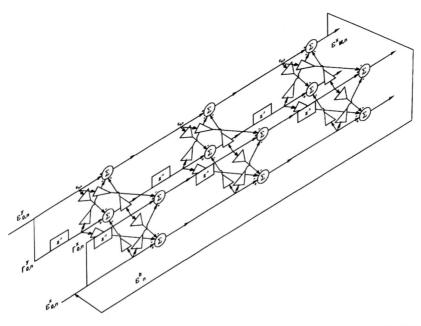

Figure 14   ARMA lattice for 3rd-order model showing bootstrap error path for innovation modeling.

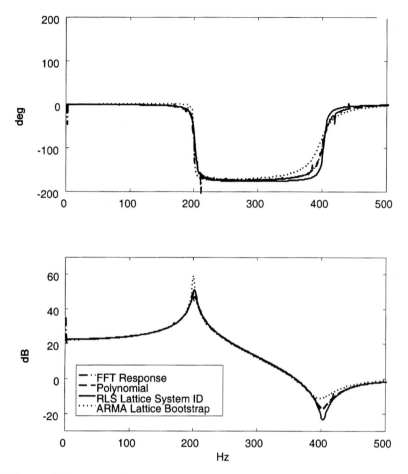

Figure 15 Frequency responses for ARMA modeling the embedded lattice structure.

which is only required for b_0 in the Wiener filtering case. The rest of the coefficients in $B[z]$ are actually properly scaled by the lattice and Levinson recursions. For the ARMA error bootstrap case, $B'[z]$ (b'_0 is assumed unity) is simply multiplied by the standard deviation of the bootstrap error.

$$b_0 = \sqrt{\frac{R_M^{\varepsilon yy}}{\bar{N}}}B[z] = \sqrt{\frac{E_M^{\varepsilon yy}}{\bar{N}}}B'[z] \tag{10.2.19}$$

The prediction error is divided by $\bar{N}$ before the square root because of the recursion

$$R_{M,n}^{\varepsilon} = \alpha R_{M,n-1}^{\varepsilon} + \varepsilon_{M,n}\varepsilon_{M,n}^{+H} \tag{10.2.20}$$

where $\alpha = 1 - 1/\bar{N}$. This removes a bias of $\bar{N}$ from the lattice forward prediction error variance. As seen in Figure 15, the scaling technique works quite well for the ARMA lattice.

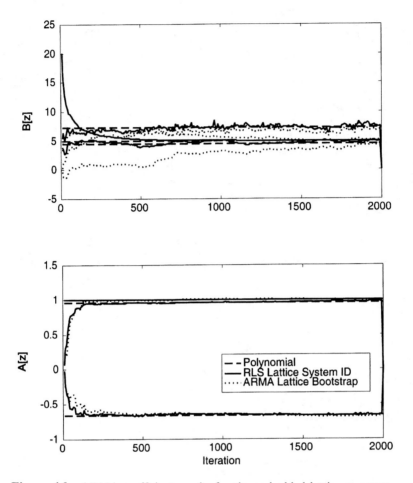

Figure 16 ARMA coefficient results for the embedded lattice structure.

Table 3 compares all of the ARMA filter and signal modeling performance results. By inspection of the converged ARMA coefficients, one can see that for Wiener filtering where both input and output signals are known, the ARMA coefficients can be estimated with great precision using either LMS, RLS, or the least-squares lattice. For ARMA bootstrapping, where one models the signal as an ARMA process with unity variance white noise innovation, the results using LMS and RLS with unnormalized bootstrap error are unacceptable (as indicated by the shaded table cells). By normalizing the bootstrap error to unity variance, good RLS results are obtained as indicated in the table by the "RLSN" algorithm. The ARMA bootstrapped lattice also gives good ARMA signal modeling perform- ance although Figure 16 clearly shows the sensitivity of the lattice to noise. The memory window $\bar{N}$ for the trials was generally kept at 200 samples for all cases except the LMS algorithm, which was too slow to converge in the space of 200 data samples. Reducing the LMS data memory window to 100 samples was necessary for comparable convergence.

Table 3 ARMA Filter Modeling Comparison

Algorithm	b_0	b_1	b_2	a_1	a_2	N
Actual	5.0000	7.3436	4.5125	−0.6603	0.9604	x
Wiener Filter						
LMS	5.0000	7.3437	4.5126	−0.6603	0.9604	100
RLS	5.0000	7.3436	4.5125	−0.6603	0.9604	200
Lattice	5.0047	7.5809	4.8011	−0.6565	0.9678	200
Bootstrapped ARMA Error						
LMS	6.7311	3.3271	−2.4981	−0.7046	0.9537	100
RLS	8.1381	1.6342	0.4648	−0.7002	0.9786	200
RLSN	5.0114	7.5178	4.3244	−0.6738	0.9149	200
Lattice	5.3203	7.2554	4.4180	−0.6791	0.9881	200

Generalized embedding of ℓ signal channels in the lattice algorithm can be done using the ARMA embedding methodology. This type of multichannel signal processor is useful when ARMA models are estimated simultaneously between many signals, multi-dimensional data (such as 2D and 3D imagery) is modeled, or when arrays of sensor data are processed. Examination of the signal flow paths naturally leads to a processor architecture which facilitates logical division of operations among multiple processors. To examine the generalized embedded multichannel lattice algorithm, consider ℓ input signals to be whitened $\bar{y}_n = [y_n^1 y_n^2 \ldots y_n^\ell]^T$. The forward error signal vector update is $\bar{\varepsilon}_{p+1,n} = \bar{\varepsilon}_{p,n} - \bar{K}_{p+1,n}^r \bar{r}_{p,n+1}$, or simply

$$\begin{bmatrix} \varepsilon_{p+1,n}^1 \\ \varepsilon_{p+1,n}^2 \\ \vdots \\ \varepsilon_{p+1,n}^\ell \end{bmatrix} \begin{bmatrix} \varepsilon_{p,n}^1 \\ \varepsilon_{p,n}^2 \\ \vdots \\ \varepsilon_{p,n}^\ell \end{bmatrix} - \begin{bmatrix} K_{p+1,n}^{r11} & K_{p+1,n}^{r12} & \cdots & K_{p+1,n}^{r1\ell} \\ K_{p+1,n}^{r21} & K_{p+1,n}^{r22} & \cdots & K_{p+1,n}^{r2\ell} \\ \vdots & \vdots & & \vdots \\ K_{p+1,n}^{r\ell1} & K_{p+1,n}^{r\ell1} & \cdots & K_{p+1,n}^{r\ell\ell} \end{bmatrix} \begin{bmatrix} r_{p,n-1}^1 \\ r_{p,n-1}^2 \\ \vdots \\ r_{p,n-1}^\ell \end{bmatrix} \tag{10.2.21}$$

and a similar expression for the backward prediction error vector.

$$\begin{bmatrix} r_{p+1,n}^1 \\ r_{p+1,n}^2 \\ \vdots \\ r_{p+1,n}^\ell \end{bmatrix} \begin{bmatrix} r_{p,n-1}^1 \\ r_{p,n-1}^2 \\ \vdots \\ r_{p,n-1}^\ell \end{bmatrix} - \begin{bmatrix} K_{p+1,n}^{\varepsilon11} & K_{p+1,n}^{\varepsilon12} & \cdots & K_{p+1,n}^{\varepsilon1\ell} \\ K_{p+1,n}^{\varepsilon21} & K_{p+1,n}^{\varepsilon22} & \cdots & K_{p+1,n}^{\varepsilon2\ell} \\ \vdots & \vdots & & \vdots \\ K_{p+1,n}^{\varepsilon\ell1} & K_{p+1,n}^{\varepsilon\ell2} & \cdots & K_{p+1,n}^{\varepsilon\ell\ell} \end{bmatrix} \begin{bmatrix} \varepsilon_{p,n}^1 \\ \varepsilon_{p,n}^2 \\ \vdots \\ \varepsilon_{p,n}^\ell \end{bmatrix} \tag{10.2.21}$$

While embedding the RLS basis vector with additional channel results in significant increase in the dimension (effective RLS model order) of the matrices and vectors, embedding additional signal channels into the lattice simply expands the size of each stage, leaving the effective lattice model order the same. Therefore, we expect significant computational efficiencies using the lattice structure over an RLS structure. The signal flow for a 8-channel lattice stage is seen in Figure 17.

The multichannel lattice stage in Figure 17 is a completely general adaptive processing structure allowing any number of signal channels to be embedded.

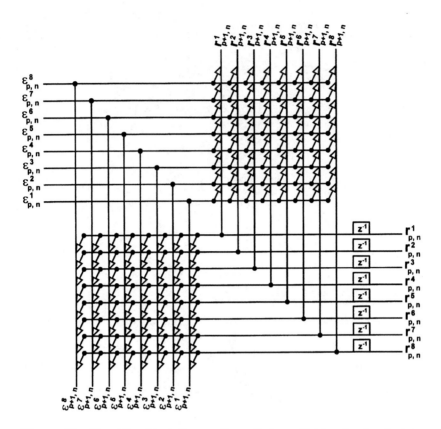

Figure 17 Signal flow block diagram for an 8-channel lattice showing the interconnections between the first four stages.

The solid "dots" represent connections between signal wires, or soldier joints, while the open triangles are the elements of the PARCOR coefficient matrix. The forward error signals enter at the top left side and are multiplied by the elements of $\bar{K}^{\varepsilon}_{p+1,n}$ in the upper right corner and summed with the backward error which exits at the top right. At the bottom left, the updated forward error signal vector is produced from the forward error input (top left side) summed with the product of $\bar{K}^{r}_{p+1,n}$ and the backward error which enters at the bottom right side. This curiously symmetric processing structure also has the nice property that successive stages can be easily "stacked" (allowing straight buss connections between them) by a simple 90° counter-clockwise rotation as seen in Figure 18.

Clearly, future developments in adaptive signal processing will be using structures like the multichannel lattice in Figure 18 for the most demanding processing requirements. The orthogonal decomposition of the signal subspace directly leads to the highly parallel processing architecture. It can be seen that correlations in both time and space (assuming the input signal vector is from an array of sensors) are processed allowing the PARCOR coefficient matrices to harvest the available signal information. Besides, the graphic makes a nice book cover.

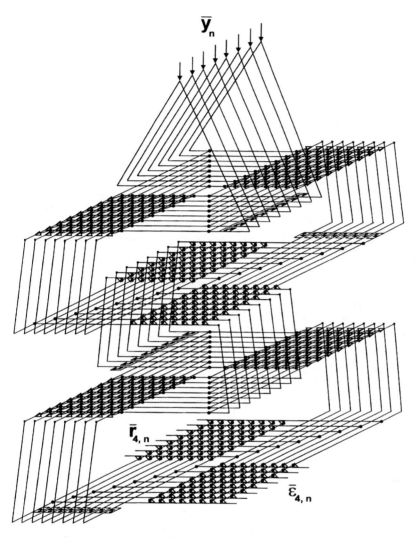

Figure 18 Graphical depiction of an 8-channel lattice showing the forward error entering from the left and leaving the bottom and backward error entering from the right, passing through the delay latch, and leaving at the top.

10.3 FREQUENCY DOMAIN ADAPTIVE FILTERS

Adaptive LMS processing in the frequency domain offers some advantages over the time domain LMS algorithm in performance, yet it is not as computationally complex as the RLS or lattice algorithms. The main advantage for the frequency domain LMS (FDLMS) comes from the orthogonality of spectral data, allowing the ability to implement a single complex coefficient LMS filter for each FFT bin independently. This results in the same fast convergence rate for all the FFT bins as compared to the time domain algorithm which only converges fast for the dominant eigenvalue of the input data. One can maintain a time domain filter for execution in parallel with the FDLMS, such that the input–output data and prediction error are transformed into the frequency domain, the adaptive FDLMS algorithm updates the optimal filter's frequency response, and the time domain filter coefficients are calculated using an inverse FFT and replace the existing filter coefficients. This sounds like, and is, a lot of computation. But, it does not need to be executed in real time to provide a filter coefficient update with every new input and output signal sample.

Because the FFT of a signal represents an integral of the signal over a fixed time interval, one should only consider "overlapping" the FFT buffers in time by no more than 50%. The reason for limited overlapping is that the error spectrum which causes the filter frequency response adaptation will not respond immediately to the new coefficients due to the time integral. If spectral updates to the filter coefficients continue before the new error response is seen in the error spectrum, the FDLMS algorithm will "over adapt" which generally will lead to oscillations (even instability) in the adaptive algorithm. For stationary input data, it can be seen that the point of diminishing return is about a 50% overlap (50% of the data buffer is old data). However, the error signal is not very stationary due to the changing filter coefficients. Therefore, a 50% data buffer overlap for the FFTs is seen as the maximum advisable overlap. For our work, we use an even more conservative 0% overlap so that the residual error from the previous adaptation is completely flushed from the error spectrum time buffer for a particular FDLMS filter frequency response update operation. While the FFT and update operations do not happen very often, the updates are quite spectacular in terms of the filter coefficient convergence. The FDLMS does not converge as fast as the RLS or lattice algorithms, but it is much less computationally complex and is seen as a reasonable choice for simplified LMS processing when the eigenvalue spread (frequencies and power levels) of the input data is wide.

The FDLMS algorithm is also of great interest for applications where the error and/or other signals are best expressed in the frequency domain. A good example of an appropriate FDLMS application for a spectral error is for an application of intensity error minimization. Sections 6.3 and 6.4 show representations of wave intensity as a spectral measurement. Other applications could be in the area of transfer function error spectra or even physical (electrical, mechanical, acoustic) impedance error spectra. These are largely adaptive control issues and will be discussed further in Section 15.4. However, some image processing applications could use frequency domain adaptive control on wavenumber data. Also, medical magnetic resonance imaging (MRI tomographic scanning) measures wavenumber data directly as a time signal. Finally, important modeling information such as spectral error, SNR, coherence, and confidence can be easily extracted from frequency domain data allowing the adaptive process to apply spectral weighting and other additional controls not readily available in the time domain.

Consider the standard filtered-x type LMS adaptive filter update on the finite impulse response (FIR) filter coefficients $h_{k,n}, k = 0, 1, 2, \ldots, M$. at times n

$$h_{k,n} = h_{k,n-1} - 2\mu x_{n-k}\varepsilon_n \qquad (10.3.1)$$

The parameter μ is $\mu_{\max}$ times μ_{rel}, where $\mu_{\max} = 1/\{\sigma_x^2\}$ and σ_x^2 is the variance of the input data x_n. As noted in Chapter 9 μ_{rel} is the step size component which creates an effective "data memory window" with an exponential forgetting property of approximate length $1/\mu_{rel}$. It can be seen that the shortest possible memory window length for a stable LMS algorithm is the filter length $M+1$ and occurs if $\mu_{rel} = 1/(M+1)$. The frequency domain version of Eq. (10.3.1) will provide the complex frequency response of the FIR filter, to which we subsequently apply inverse Fourier transform to get the time-domain FIR filter coefficients. To facilitate Fourier processing, we expand Eq. (10.3.1) to include a block of $N+1$ input and error data samples and the entire FIR filter vector of $M+1$ coefficients to be updated as a block every N_0 samples. N_0 is the sample offset between data blocks and was noted earlier to be no smaller than half the FFT buffer size (N_0 equal to the FFT buffer size is recommended). This prevents the FDLMS from over adapting to the error spectrum.

$$
\begin{bmatrix} h_{0,n} \\ h_{1,n} \\ h_{2,n} \\ \vdots \\ h_{M,n} \end{bmatrix} = \begin{bmatrix} h_{0,n-N_0} \\ h_{1,n-N_0} \\ h_{2,n-N_0} \\ \vdots \\ h_{M,n-N_0} \end{bmatrix} - 2\mu_{max}\mu_{rel} \begin{bmatrix} x_{n-N} & \cdots & x_{n-2}x_{n-1}x_n \\ x_{n-N-1} & \cdots & x_{n-3}x_{n-2}x_{n-1} \\ x_{n-N-2} & \cdots & x_{n-4}x_{n-3}x_{n-2} \\ \vdots & \vdots & \vdots & \vdots \\ x_{n-N-M} & \cdots & x_{n-2-M}x_{n-1-M}x_{n-M} \end{bmatrix} \begin{bmatrix} \varepsilon_{n-N} \\ \vdots \\ \varepsilon_{n-2} \\ \varepsilon_{n-1} \\ \varepsilon_n \end{bmatrix}
$$
$$(10.3.2)$$

It can be seen that the product of the $M+1$ by $N+1$ input data matrix and the error can be conveniently written as a cross-correlation.

$$
\begin{bmatrix} h_{0,n} \\ h_{1,n} \\ h_{2,n} \\ \vdots \\ h_{M,n} \end{bmatrix} = \begin{bmatrix} h_{0,n-1} \\ h_{1,n-1} \\ h_{2,n-1} \\ \vdots \\ h_{M,n-1} \end{bmatrix} 2\mu_{max}\mu_{rel} \begin{bmatrix} R_0^{x\varepsilon} \\ R_{-1}^{x\varepsilon} \\ R_{-2}^{x\varepsilon} \\ \vdots \\ R_{-M}^{x\varepsilon} \end{bmatrix}; \quad R_{-k}^{x\varepsilon} = \frac{1}{M}\sum_{j=0}^{M-1} x_{n-j-k}\varepsilon_{n-j} \quad (10.3.3)
$$

Equation (10.3.3) is expressed in the frequency domain by applying a Fourier transform. Equation (10.3.4) shows the Fourier transformed equivalent of Eq. (10.3.3).

$$
H_n(\omega) = H_{n-N_0}(\omega) - 2\mu_{max}\mu_{rel}X_n^*(\omega)E_n(\omega) \tag{10.3.4}
$$

The cross spectrum in Eq. (10.3.4) should raise some concern about the possibility of circular correlation errors when the input signal has sinusoids not bin-aligned with the Fourier transform. See Section 5.4 for details about circular correlation effects. The potential problem occurs because the finite length signal buffers in the FFT provide a spectrum which assumes periodicity outside the limits of the signal buffer. This is not a problem for random noise signals or sinusoidal signals where the frequencies lie exactly on one of the FFT frequency bins. Practical considerations make it prudent to develop an algorithm which is immune to circular correlation errors. As seen in Section 5.4, the precise correction for circular correlation error is to double the FFT input buffer sizes, while zero padding one buffer to shift the circular correlation errors to only half of the resulting inverse transformed FIR coefficient vector. Since we want the first $M+1$ coefficients of the FIR impulse response to be free of circular correlation errors, we double the

FFT buffer sizes to $2M+2$ samples, and replace the oldest $M+1$ error signal samples with zeros.

$$X_n^c(\omega) = \Im\{x_{n-2M-1} \ldots x_{n-M-1} x_{n-M} \ldots x_{n-1} x_n\} \tag{10.3.5}$$

$$E_n^c(\omega) = \Im\{0\ 0 \ldots 0 \varepsilon_{n-M} \ldots \varepsilon_{n-1} \varepsilon_n\} \tag{10.3.6}$$

Using the double-sized buffers in Eqs (10.3.5)–(10.3.6) in the FFTs, we can update a more robust filter frequency response (in terms of its inverse FFT) and also include a frequency dependent step size $\mu_{max}(\omega)$, which is the inverse of the signal power in the corresponding FFT frequency bin. It will be shown below that some spectral averaging of the power for the bin step size parameter will improve robustness. The parameter μ_{rel} can be set to unity since the memory window need not be longer than the integration already inherent in the FFTs.

$$\boldsymbol{H}_n^c(\omega) = \boldsymbol{H}_{n-N_0}^c(\omega) - 2\mu_{max}(\omega)\mu_{rel}\{X_n^c(\omega)\}^* E_n^c(\omega) \tag{10.3.7}$$

To recover the time domain FIR filter coefficients, an inverse FFT is executed which, due to the circular correlation correction described in Section 5.4 and Eqs (10.3.5)–(10.3.7), provides robust FIR filter coefficients in the leftmost $M+1$ elements of the FFT output buffer. The symbol in Eq. (10.3.8) depicts that one discards the corresponding coefficient which by design may have significant circular correlation error.

$$\boldsymbol{H}_n^c = \Im^{-1}\{H_n^c(\omega)\} = [h_{0,n} h_{1,n} \ldots h_{M,n} \cancel{N} \cancel{N} \ldots \cancel{N}] \tag{10.3.8}$$

An example is presented below comparing the RLS, LMS and FDLMS algorithms in a Weiner filtering application where the input, or reference data, has sinusoids plus white noise. With a white noise only input, the three algorithms perform the same (there is only one eigenvalue in that case). The Weiner filter to be identified is FIR and has 9 coefficients resulting from zeros at (magnitude and phase notation) $0.99\angle\pm0.2\pi$, $1.40\angle\pm0.35\pi$, $0.90\angle\pm0.6\pi$, and $0.95\angle\pm0.75\pi$ where the sample rate is 1024 Hz. The frequencies of the zeros are 109.4 Hz, 179.2 Hz, 307.2 Hz, and 384 Hz, respectively. The reference input signal has zero-mean Gaussian (ZMG) white noise with standard deviation 0.001, and three sinusoids of amplitude 10 at 80 Hz, amplitude 30 at 250 Hz, and amplitude 20 at 390 Hz. To facilitate the narrowband input, the FIR filter model has 18 coefficients, and the effective data memory window length is set at approximately 36 samples. This way, the RLS, LMS, and FDLMS algorithms all share the same effective memory window (the FDLMS doubles the FFT buffers to 36 samples). The FDLMS is updated every 36 samples so that no input or error data is shared between successive updates (0% buffer overlap). Because the input consists mainly of three sinusoids in a very small amount of broadband noise, we expect the model frequency response to be accurate only at the three sinusoid frequencies. As seen in Figure 19 for the RLS, LMS, and FDLMS algorithms, all three converge but the RLS shows the best results. Figures 20–22 show the short-time Fourier spectra of the error signals processed as 64-point blocks every iteration. In Figure 20 the RLS algorithm's superior performance is clear, but the FDLMS's performance in Figure 22 is also quite good.

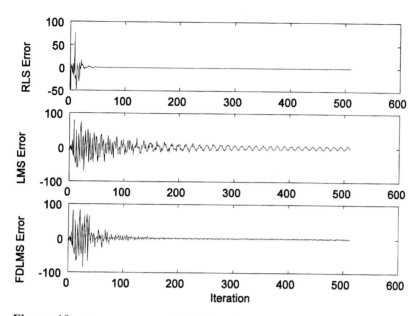

Figure 19 Error responses of RLS, LMS and FDLM algorithms for the three sinusoid input to the 9 coefficient FIR Weiner filter system identification problem.

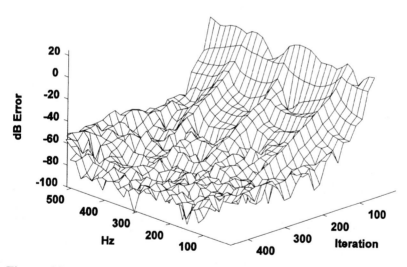

Figure 20 RLS system error spectra from 0–512 Hz for the first 450 iterations of the algorithm.

It was mentioned earlier that while the FDLMS allows each individual bin to operate as an independent adaptive filter with its own optimized step size, that this is generally not very robust for narrowband inputs. Recall that for a broadband ZMG input, the RLS, LMS, and FDLMS all perform the same. For the ZMG case, the FDLMS would have identical step sizes for each FFT bin equal to the inverse

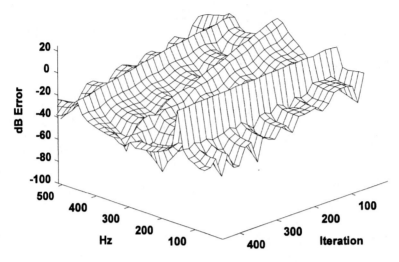

Figure 21   LMS system error spectra from 0–512 Hz for the first 450 iterations of the algorithm.

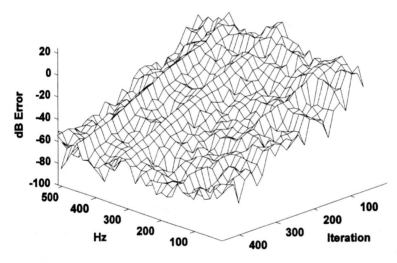

Figure 22   FDLMS system error spectra from 0–512 Hz for the first 450 iterations of the algorithm.

of the power of the input at that bin frequency. When the input consists of narrowband sinusoids, the power for some bins will be near zero. The small bin power gives a large gain for the adaptive update to that particular bin. Since the power estimate should be averaged over several adaptive updates (to minimize random fluctuations in the model estimate), occasional spectral leakage from a nearby bin with a high-amplitude sinusoid will drive the bin with the large step size to instability. This problem is solved by averaging the bin power with the power in adjacent bins. The number of adjacent bins to include in the average depends on the frequency spacing of the narrowband sinusoids in the input. If one cannot a priori determine the approximate spectral content of the input, a single step size can be used for the entire FDLMS based on the total power of the input signal. For a single step size for the entire spectrum, the FDLMS will perform with convergence rates similar to the LMS algorithm, where the strongest peaks converge the fastest while the weaker peaks converge much more slowly. Figure 23 shows the input spectral power, the bin-average power, and the step sizes used based on the bin-averaged power.

The frequency domain also offers us some very insightful ways to characterize the "goodness of fit" for the model given the available error and input signals. The FDLMS is not required to make this measurement, but the FDLMS allows a convenient way to implement a weighted least-squares algorithm using the frequency domain error. Consider the equation for the FDLMS error.

$$E_n(\omega) = X_n(\omega)\{H_{opt} - H'(\omega)\} \tag{10.3.9}$$

Multiplying both sides of Eq. (10.3.9) by $X^*(\omega)$ and rearranging, we can estimate the optimal filter from our current estimate and an error estimate based on the prediction error spectra.

$$|H_{opt}(\omega)| = |H'(\omega)| \pm \left|\frac{X^*(\omega)E(\omega)}{X^*(\omega)X(\omega)}\right| \tag{10.3.10}$$

For the modeling error depicted in Eq. (10.3.10) to be meaningful, $X(\omega)$ should be broadband, rather than narrowband. A broadband input insures that the complete frequency response of $H_{opt}(\omega)$ is excited and observed in the error spectrum $E(\omega)$. For a narrowband input, we only expect the response of the system to be excited and observed at the frequencies on the input. For the example seen in Figures

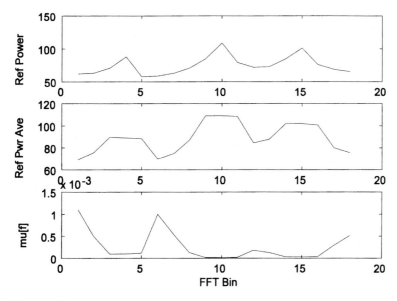

Figure 23 Input power, averaged input power, and step size for the FDLMS algorithm with 3-bin averages.

19–23 we expect precise modeling at 80 Hz, 250 Hz, and 390 Hz and a poor match between the frequency response of the actual system and the model elsewhere. However, increasing the broadband noise from an amplitude (standard deviation) of 0.001 to 5.0 yields the modeling result in Figure 24, which also includes and error measure as given in Eq. (10.3.10). The absolute values of the error are used as a "worst case" estimate error, but the error is not strictly a bound in the sense of a Cramer–Rao lower bound. The model is seen to be very precise at 80 Hz, 250 Hz, and 390 Hz and the error model correctly tracks the model errors in other frequency regions. However, if the input is not broadband, then the error measure will not be meaningful.

The spectral modeling error in Figure 24 is very useful for determining how close the model is to the optimum filter response. Both the input and error signals (the error signal in particular) can have interfering noise which will widen the error bounds as seen in Figure 24. Applying the probability density distributions presented in Sections 6.1 and 6.2, we can estimate a probability for the model to be some percentage above the actual, and so on. The statistic representation can be used to associate the probability of algorithm divergence, convergence, or confidence for the model estimate. While these measures can be applied to any adaptive algorithm, the FDLMS very conveniently allows adaptive weighing of the error spectrum to optimize performance in specific frequency ranges.

10.4 SUMMARY, PROBLEMS, AND BIBLIOGRAPHY

There are several major algorithm classes in adaptive filtering which can be seen to be recursive least squares (RLS), Kalman filtering, least-squares lattice algorithms, and

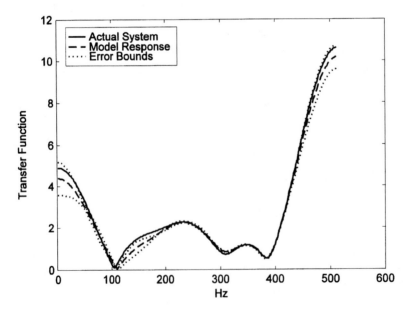

Figure 24 Actual system response, modeled response, and error measure based on the prediction error spectrum.

the simple and robust LMS algorithm. Each of the classes can be further broken down into algorithms optimized for specific applications including parameter tracking and prediction, system identification using Weiner filtering, and signal modeling and parameterization using ARMA filters. For each particular application, the literature has a substantial volume of articles with very detailed derivations and results limited to the specific application. The student can be overwhelmed by the number of different least-squared error algorithms published if attempting to study them individually. The approach of this book (in particular Chapters 9 and 10) is to assimilate least-squared error modeling of linear systems into three basic forms: recursive algorithms derived for a particular model order using the matrix inversion lemma, orthogonal projection based algorithms, and least-squared error modeling involving a state transition as part of the prediction which we refer to as Kalman filtering. The RLS and LMS algorithms are clearly based on application of the matrix inversion lemma to implement a recursive least-squared error solution for a specific model order. Using a projection operator framework, orthogonal updates in both time and/or model order produce the least-squares lattice algorithm. The projection operator update equation is surprisingly similar to the matrix inversion lemma Eq. (as seen in Chapter 8) but allows model order expansion in a least-squared sense as well as time updates. The Kalman filter is distinguished from recursive least-squares simply by the inclusion of the state transition before the prediction error is estimated and the state parameters adjusted to maintain a least-squared error state modeling performance.

Kalman filtering is one of the most important and prevalent adaptive filtering algorithms of the 20th century and there are literally hundreds of texts and thousands of articles available in the literature. The scope of this text can not possibly address the many subtle technical points of Kalman filtering and state variable theory. However, it is extremely useful to contrast a basic Kalman filter algorithm to RLS and other adaptive filtering derivations. The intended result is to present all least-squared algorithms in as unified and concise an approach as possible. In Section 10.1 we see that the algebraic Riccatti Eq. (combination of Eqs (10.1.19) and (10.1.23) for the update of the state error covariance), is really just an application of the matrix inversion lemma to the state error covariance including the state transition. The steady-state error covariance represents Kalman filter tracking performance when the underlying assumptions for the state transition (and number and type of states) and measurement noise are correct for the observed data.

In a steady-state situation, the required Kalman filter gain for the desired state process noise is a constant vector. We can calculate this gain vector in the form of, say, an α-β-γ filter (for 3 states). When the measurement noise is near zero, the α-β-γ and Kalman filters each give identical performance. However, the Kalman filter gain is computed not only from the measurement and process noises, but also by adaptive minimization of the state error covariance. Therefore for target maneuvers outside the range of the kinematic model used to determine the number and type of states, the Kalman filter gives superior performance compared to the α-β-γ tracker. This allows the Kalman filter to rapidly track target maneuvers with a smaller process noise than the α-β-γ tracker, giving more accurate state estimates overall. For high measurement noise levels, the Kalman filter will tend to over compensate to minimize the state error, making the state vector rather noisy. Because of the added power of the Kalman filter, a covariance performance analysis should be executed on simulated target maneuvers to determine the best designed process noise for the adaptive tracking system.

Section 10.2 introduced the embedding technique for the RLS and lattice algorithms allow the underlying FIR, or moving average filter (MA), to be enhanced to a pole-zero, or ARMA filter. In the RLS algorithm, the embedding of an Mth-order ARMA filter (M poles and M zeros) effectively doubles the length of the RLS algorithm coefficient vector, whose computations increase with the cube of the coefficient vector length. In the lattice algorithm, the PARCOR coefficients become 2 by 2 matrices resulting in an increase in complexity of approximately 4, where the lattice algorithm overall complexity increases linearly with model order. The lattice's huge advantage in complexity reduction is due in part to the fact that the Levinson recursions required to extract the linear prediction coefficients from the PARCOR coefficients are not required for each filter update. If the linear prediction coefficients are computed with each lattice filter update, the computations required is closer to the required RLS computations, but still less. Proper bootstrapping and prediction error normalization allows good performance for both Wiener filtering for system identification given input and output signals, and ARMA signal modeling. An interesting result of the examples given in Section 10.2 is that the LMS filter performs extremely well for the Weiner filtering problem and extremely poor for the bootstrapped ARMA signal model. It is also seen that the double-direct lattice form demonstrates superior numerical properties while also reducing computational load through the elimination of some divisions in the lattice updates.

The lattice structure allows simple embedding of multichannel data from sensor arrays or even image data. Figure 18 illustrates topology of time and order expansion in a lattice structure in three dimensions graphically. This elaborate adaptive filter structure can be clearly seen as a parallel process, where individual processors are assigned to each "layer" or lattice stage. However, for true parallel execution, a process time delay must be added between each stage making all the PARCOR coefficient matrices M time samples old for an Mth order parallel process lattice. A pure spatial adaptive filter could be computed using only a single multichannel lattice stage, as depicted in Figure 17. The multichannel lattice architecture represents the future for practical large-scale adaptive processing. While hardly the epitome of simplicity, the multichannel lattice structure completes our unified presentation of basic adaptive filtering structures.

Section 10.3 introduces the frequency-domain LMS (FDLMS) algorithm which offers several interesting features not available in the time-domain. First, we see that near RLS convergence performance is available for signals with wide eigenvalue spreads which would converge slowly in a time-domain LMS algorithm. The LMS step size can be set independently for each FDLMS filter frequency bin. However, the number of independent frequency bands where an independent step size can be used is best determined by the number of dominant eigenvalues in the input signal. This prevents signal bands with very low power from having overly sensitive adaptation from too large a step size. The bandwidth for each power estimate and corresponding step size can be simply determined from the spectral peak locations of the input signal, where one simply applies a step size for a particular peak over the entire band up to a band for an adjacent spectral peak. For a single sinusoid input, a fixed step size is used for the entire frequency range.

For multiple sinusoids, individual bands and step sizes are determined for each peak band. Because of the integration effects of the Fourier transform, the FDLMS algorithm need not be updated with each input time sample, but rather with a limited (say 50%) overlap of the input data buffers to prevent over-adaptation and amplitude modulation in the error-adaptation loop. The output of the FDLMS algorithm is a frequency response of the FIR optimal filter. To prevent circular correlation errors from spectral leakage, doubling the buffer sizes and zero-padding the later half of the error buffer shifts any circular correlation errors into the later half of the resulting FIR impulse response calculated from the inverse Fourier transform of the converged frequency response for the filter. Careful application of the FDLMS algorithm with circular correlation corrections, proper step size bandwidth, and update rates which do not allow over adaptation of the error spectrum, yields a remarkable and unique adaptive filter algorithm. Since the error is in the form of a spectrum, a wide range of unique applications can be done directly using physical error signals such as impedance, intensity, or even wavenumber spectra.

PROBLEMS

1. An airplane moves with constant velocity at a speed of 200 m/sec with a heading of 110 degrees (North is 0, East is 90, South 180, and West 270 degrees). The radar's measurement error standard deviation is 10 m in any direction. Calculate the α-β gains for a tracking filter designed with a 1 sec update rate and a process noise standard deviation of 1 m.

2. Determine the east–west and north–south (i.e. x and y components) of velocity state error for the airplane tracking data in problem 1.

3. Derive the Joseph form in Eq. (10.1.24).

4. Show that for $\mu_{rel} = 0.01$ the effective exponential data memory window is about 100 samples and for the lattice this corresponds to a forgetting factor α of 0.99.

5. For the joint process ARMA system model in Eq. (10.2.15) show that $d_i = a_i/b_0$ and $c_i = b_i/b_0$ where $i = 1, 2, ..., M$, $d_0 = 1/b_0$, and $a_0 = c_0 = 1$.

6. Prove Eq. (10.2.18) deriving the b_0 from the ratio of the forward prediction errors at the Mth ARMA Wiener lattice stage and from the normalized standard deviation of the forward prediction error for the bootstrapped error case.

7. Show that for the converged double/direct lattice, the forward a priori and a posteriori error signals are the same and the backward a priori and a posteriori error signals are the same.

8. Derive the likelihood parameter for a multichannel least squares lattice algorithm.

9. The time buffers in the FDLMS algorithm are $2M$ samples long where the oldest M error samples are replaced with zeros to correct for circular correlation errors due to possible spectral leakage. What other time buffer forms produce the same effect in the FDLMS algorithm?

10. Show that an FDLMS adaptive filter update every time sample could lead to over correction oscillations and that updating the FDLMS adaptive filter less often damps the oscillations.

BIBLIOGRAPHY

Y. Bar-Shalom, Xios-Rong Li Estimation and Tracking: Principles, Techniques, and Software, Norwood: Artech House, 1993.

B. Friedlander "System Identification Techniques for Adaptive Noise Cancelling," IEEE Trans. Acoustics, Speech, and Signal Processing, Vol. 30, No. 5, Part III, June, 1981, pp. 627–641.

B. Friedlander "Lattice Filters for Adaptive Processing," Proceedings IEEE, Vol 70, 1982, pp. 829–867.

Arthur Gelb, ed. Applied Optimal Estimation, Cambridge: MIT Press, 1974.

W. S. Hodgkiss Jr., J. A. Presley Jr. "Adaptive Tracking of Multiple Sinusoids Whose Power Levels are Widely Separated," IEEE Trans. Acoustics, Speech, and Signal Processing, Vol. 29, No. 3, Part III, June, 1981, pp. 710–721.

D. T. L. Lee, M. Morf, B. Friedlander "Recursive Ladder Algorithms for ARMA Modeling," IEEE Proceedings 19th Conference on Decision and Control, Albuquerque, MN), Dec 10–12, 1980, pp. 1225–1231.

D. T. L. Lee, M. Morf, B. Friedlander "Recursive Least-Squares Ladder Estimation Algorithms," IEEE Trans. Acoustics, Speech, and Signal Processing, Vol. 29, No. 3, Part III, June, 1981, pp. 627–641.

S. J. Orfanidis Optimum Signal Processing, New York: McGraw-Hill, 1988.

F. A. Reed, P. L. Feintuch "A Comparison of LMS Cancellers Implemented in the Frequency Domain and the Time Domain," IEEE Trans. Acoustics, Speech, and Signal Processing, Vol. 29, No. 3, Part III, June, 1981, pp. 770–775.

K. M. Reichard, D. C. Swanson "Frequency-domain implementation of the filtered-x algorithm with on-line system identification," Proceedings of The Second Conference on Recent Advances in Active Control of Sound and Vibration, 1993, pp. 562–573.

J. J. Shynk "Frequency-domain multirate adaptive filtering," IEEE Signal Processing Magazine, 1, 1992, pp. 14–37.

REFERENCES

1. R. E. Kalman "Design of a Self-Optimizing Control System," Transactions of the ASME, Vol. 80, Feb. 1958, pp. 468–478.

2. Y. Bar-Shalom, X. Li Estimation and Tracking: Principles, Techniques, and Software, Norwood: Artech House, Chapter 6, 1993.

3. Y. Bar-Shalom, X. Li Estimation and Tracking: Principles, Techniques, and Software, pp. 272–289.

Part IV

Wavenumber Sensor Systems

A wavenumber is a spatial representation of a propagating sinusoidal wave. It is generally referred to in the electromagnetics, vibration, and acoustics community with the symbol $k = \omega/c$, where ω is radian frequency and c is the wave propagation speed in m/sec, or $k = 2\pi/\lambda$, where λ is the wavelength in meters. Wavenumber sensor systems typically consist of arrays of sensors in a system designed to filter and detect waves from a particular direction (bearing estimation) or of a particular type (modal filtering). The most familiar device to the general public which employs wavenumber filtering is probably the medical ultrasound scanner. Medical ultrasound has become one of the most popular and inexpensive medical diagnosis tools due to the recent advancements in low-cost signal processing and the medical evidence suggesting that low-level ultrasonic acoustic waves are completely safe to the body (as compared to X-rays). The ability to see anatomical structures as well as measure blood velocity and cardiac output without insult to the body have saved millions of lives. Another example widely seen by the public at most airports is radar, which operates with nearly the same system principles as ultrasound, but typically scans the skies mechanically using a rotating parabolic reflector. Early medical ultrasound scanners also operated with a mechanical scan, but this proved too slow for real-time imaging in the body with its many movements. Electronic scanning of the beam requires no mechanical moving parts and thus is much faster and more reliable. The most advanced radar systems in use today also employ electronic scanning techniques.

The ultrasonic scanner is quite straightforward in operation. The transmitted frequency is quite high (typically in the 1–6 Mhz range) so that small structures on the order of a millimeter or larger will scatter the wave. A piezoelectric crystal on the order of a centimeter in size (15–100 wavelengths across the radiating surface) tends to transmit and receive ultrasound from a direction normal to its face. In other directions, the waves on the crystal surface tend to sum incoherently, greatly suppressing the amplitudes transmitted and received. An acoustic lens (typically a concave spherical surface) helps shape the sound beam into a hyperbola which narrows the beam further in the region from about 1–6 cm from the face. The lens has the effect of sharpening the resulting ultrasound image in the region below

323

the skin/fat layer. If one mechanically scanned a single ultrasonic transceiver the familiar gray-scale image could be constructed from the distance normalized back-scattered ultrasound received. However, using a line array of transceivers, the beam direction is controlled by delaying the transmitted signal appropriately for each array element, as well as delaying the received element signals before summing the beamformed output (see Section 7.1). The beamsteering can be electronically done quite fast enabling a clear image to be reconstructed in real-time for the medical practitioner. There are a wide range of industrial applications of ultrasound as well ranging from welding of plastics, cleaning, non-destructive testing and evaluation, and inspection.

Sonar is another ultrasonic technique, but one widely seen by the public mainly through movies and literature. The "ping" depicted in movies would never actually be heard by anyone except the sonar operator because in order to steer the sound beam to a specific direction, the frequency needs to be at a frequency higher than would be easily heard by a human (if at all). A frequency demodulator in the sonar operator's equipment shifts the frequency down into the audible range. Again, the ratio of transmitting array size (the apertures) to the sound wavelength deter-mines the beam width, and therefore, the angular resolution of the scanning operation. Small unmanned vehicles such as torpedoes/missiles use much higher sonar/radar transmitting frequencies to maintain a reasonable resolution with the smaller aperture transceiver array.

For very long range scanning, the time between transmitted pulses must be quite long to allow for wave propagation to the far off scatterer and back to the receiver. The long range to the scatterer also means that the signal received is generally going to have a limited or very low SNR. By increasing the length of the transmitted sonar pulse, more energy is transmitted increasing the received SNR, but the range resolution is decreased. The transmitted amplitude in an active sonar is limited by transducer distortion, hydrostatic pressure, and the formation of bubbles (cavitation) when large acoustic pressures are generated near the surface where the hydrostatic pressure is low. Radar transmitting power is only limited by the quality of the electrical insulators and available power. Low SNR radar and sonar processing generally involves some clever engineering of the transmitted signal and cross-correlating the transmitted and received signals to find the time delay, and thus range, of the scatterer in a particular direction. Cross correlating the transmitted and received signals is called a *matched filter* which is the optimum way to maximize the performance of a detection system when the background inter-ference is ZMG noise (spectrally white noise in the receiving band). For non-white noise in the receiver frequency band, an Eckart filter (Section 5.4) can be used to optimize detection. This becomes particularly challenging in propagation environments which have multipath. Our presentation here will focus on detection, bearing estimation, and field reconstruction and propagation techniques.

Wavenumber sensor systems detect and estimate a particular signals spatial wave shape. Foremost, this means detecting a particular signal in noise and depicting the associated probability confidences that you have, in fact, detected the signal of interest. Chapter 11 describes techniques for constant false alarm rate detection for both narrowband and broadband stationary signals using matched filter processing. For typical radar and sonar applications, wavenumber estimation pro-vides an estimate of the direction of arrival of a plane wavefront from a distance source or scatterer. However, as depicted in Chapter 12, wavenumber estimation can also be done close to the source for spherical waves, as a means to reconstruct the sound field in areas other than where the sensor array is located. Wave field reconstruction is correctly described as holography, and is extremely useful as a tool to measure how a particular source radiates waves. Some of these waves do not propagate to the far field as a complex exponential, but decay rapidly with distance as a real exponential. These "nearfield", or evanescent waves, are very important to understand because they can effect the efficiency of a source or array of sources, they contain a great deal of information about the surface response of the radiator or scatterer, and if ignored, the presence of evanescent waves can grossly distort any measurements close to the source. Using wavenumber domain Green's functions and a surface array of field measurements, the field can be propagated towards the source or away from the source, which provides a very useful analysis tool.

The presence of multiple wavenumbers for a particular temporal frequency is actually quite common and problematic in wavenumber detection systems. In the author's opinion, a view which may not be as popular among theorists as practitioners, there are two "white lies" in most adaptive beamforming mathematical presentations. The first white lie is that the background noise at each sensor position is incoherent with the background noise at the other sensor positions in the array. In turbulent fluids for acoustic waves, noise independence is not guaranteed and neither is signal coherence (1). The second white lie is that one can have multiple sources radiating the same frequency and yet be considered "incoherent" from each other. Incoherent background noise and sources makes the adaptive beamforming algorithm mathematical presentation clean and straightforward, but this is simply not real in a physical sense. One actually must go to great lengths to achieve the mathematical appearance of incoherence by spectral averaging of time or space array data "snapshots" to enforce noise independence and spatial coherence in the signal. The practical assumption is that over multiple data snapshots, the sources each drift in phase enough with respect to each other to allow eigenvector separation of the wavenumbers (bearing angles of arrival). Chapter 13 presents modern adaptive beamforming with emphasis on the physical application of the algorithms in real environments with real sources and coherent multipath. The problems due to coherent noise and source multipath are very significant issues in adaptive wavenumber processing and the techniques presented are very effective.

There are many other types of wavenumbers other than nearfield, spherical, and plane waves which require wavenumber filters to measure. These include resonant modes of cavities, waveguides, and enclosures, propagation modes in non-uniform flow or inhomogeneous media, and structural vibration responses. These modes all share the infamous distinction of pure temporal coherence among the modes. Although, for a single frequency excitation, the modes with resonant frequency closest to the excitation frequency will be excited the most, the other modes are also excited. Since the excitation is in many cases controllable, one can design a scanning system to detect and monitor the system modes using wavenumber sensing, or modal filtering. This new wavenumber filtering technology will likely be very important in industrial process controls and system condition based maintenance technology.

11

Narrowband Probability of Detection (PD) and False Alarm Rates (FAR)

Probability of detection P_d, and the probability of a false alarm P_{fa}, or false detection, define the essential receiver operating characteristics (ROC) for any wavenumber or other signal detection system. The ROC for a particular detection system is based on practical models for signals and noise using assumptions of underlying Gaussian probability density functions. When many random variables are included in a model, the central limit theorem leads us to assume an underlying Gaussian model because a large number of separate random events with arbitrary density functions will tend to be Gaussian when taken together (see Section 6.1). Therefore, it is reasonable to assume that the background noise is Gaussian at a particular sensor (but not necessarily independent from the background noise at the other sensors in the array). We will begin by assuming that our signal has a constant amplitude and it is combined with the background noise in the receiver data buffer. The job of the signal detector is to decide whether or not a signal is present in the receiver buffer. The decision algorithm will be based on the assumption that the receiver buffer data has been processed to help the signal stand out from the noise. A simple threshold above the estimated noise power defines a simple robust boundary between strong signals and noise. The underlying Gaussian noise process model will allow us to also label the signal-or-noise decision with an associated probability. The larger the signal in the receiver buffer is relative to the decision threshold, the greater the likelihood that it is in fact, signal and not a false detection. However, the closer the decision threshold is to the noise level, the greater the likelihood that we may call something that is actually noise a signal, giving a false alarm output from the detector.

The ROC are essentially defined by the decision threshold level relative to the noise power, which along with the noise mean and variance, define a probability of false alarm, P_{fa}. The probability of signal detection, P_d, depends on the receiver data buffer processing and the strength of the signal relative to the decision threshold. Processing of the receiver data such as averaging, cross correlations, Fourier transforms, etc., are used to improve the P_d without raising the P_{fa}. Section

11.2 describes techniques for constant false alarm rate (CFAR) detection which is highly desirable in environments with nonstationary background noise characteristics. CFAR signal detection is adaptive to the changing environment and permits a very robust ROC for system design.

However, in a multipath environment, the signal power may be either enhanced or reduced in level depending on the propagation path length differences. For active systems, the time delay and corresponding range estimate can be made ambiguous by a multipath environment. When the transceiver and/or scatterer are moving with respect to the multipath, or is the multipath is moving as is the case with turbulence, the multipath can be described statistically as outlined in Section 11.3. With a statistical description of multipath, we can enhance the value of a signal detection decision with statistical confidences or probabilities based on the ROC and on the underlying noise and multipath statistics. While multipath is a problem to be overcome in radar, ultrasonic imaging, and sonar applications, it can be a source of information for other applications such as meteorology, chemical process control, composite material evaluation, and combustion control.

11.1   THE RICIAN PROBABILITY DENSITY FUNCTION

The Rician probability density function describes the magnitude envelope (or rms) representation of a signal mixed with noise. Using statistical models to represent the signal and noise, one can design a signal detection algorithm with an associated P_d, for a given signal level, and a particular P_{fa}, for the particular background noise level and detection threshold. Before we present the Rician density function we will examine the details of straightforward pulse detection in a real waveform. Consider the decision algorithm for detecting a 1 μsec burst of a 3.5 Mhz sinusoid of amplitude 2 in unity variance zero mean Gaussian noise propagating in a lossless waveguide. This is a relatively low SNR situation but the received signal is aided by the fact that wave spreading and other wave losses do not occur in our theoretical waveguide. Figure 1 shows the transmitted waveform, the received waveform including the background noise, and the normalized cross correlation between the transmitted and received waveforms. Figure 2 shows the magnitude of the waveforms in Figure 1 which help distinguish the echo at 10 μsec delay. Note that for a water filled waveguide (assume the speed of sound is 1500 m/sec), the 10 μsec delay corresponds to the scatterer being about 7.5 mm away (5 μsec) from the source. Clearly from Figure 2, the cross-correlation of the transmitted and received waveforms significantly improves the likelihood of detecting the echo at 10 μsec. Why is this true? Let the transmitted source waveform be $s(t)$ and the received waveform be $r(t)$. The normalized cross-correlation of the transmitted and receiver waveforms is defined as

$$R(-\tau) = \frac{1}{\sigma_s^2} E\{s(t)r(t + \tau)\} \tag{11.1.1}$$

where σ_s^2 is the estimated variance of the transmitted waveform $s(t)$ over the recording time interval. If the background noise is ZMG, the correlation operation in Eq. (11.1.1) is often called a *matched filter*, since the frequency response of this "filter" (the cross-correlation corresponds to a cross-spectrum in the frequency

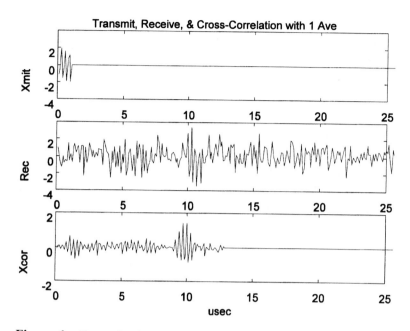

Figure 1 Transmitted, received, and cross-correlated signals for a simulated 3.6 MHz ultrasound pulse traveling about 7.5 mm in 1500 m/sec water-filled waveguide and reflecting back to the receiver.

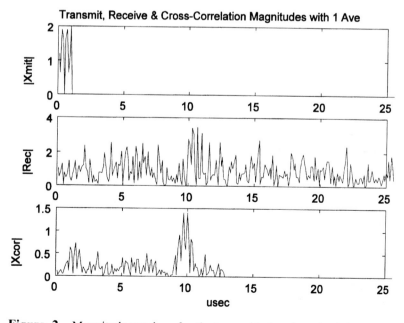

Figure 2 Magnitude envelope for the transmitted and received signals using only a single data buffer.

domain) matches the frequency response of the transmitted signal. The data in Figures 1 and 2 are simulated using 256 samples and a sampling rate of 10 Mhz. Each correlation lag τ is computed with $n/2$, or 128, sample averages. This averaging effect gives the result of reducing the randomness of the correlation of the background noise with the signal while maintaining the coherence of the correlation between the received and transmitted signals.

Time synchronous averaging is extremely effective at improving the SNR in periodic signals such as repetitive pulses in the sonar or radar. The background noise in our received waveform is zero-mean Gaussian (ZMG) with unity variance. The general expression for probability density function of a Gaussian random variable x with mean M_x and variance σ_s^2 is

$$p(x) = \frac{1}{\sigma_x \sqrt{2\pi}} e^{-\frac{(x-M_x)^2}{2\sigma_x^2}} \tag{11.1.2}$$

Suppose we scale our random variable as $y = ax$. We can easily determine the probability density for y using the following relationship.

$$p(y) = \sum_{i=1}^{N} p(x_i) \left| \frac{dx_i}{dy} \right| \tag{11.1.3}$$

where $x_1, x_2, ..., x_N$ are all solutions of $y = f(x)$, which in our case is only one solution

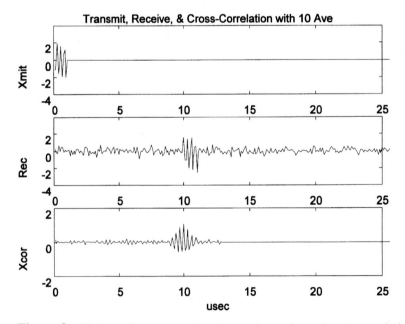

Figure 3 Time synchronous averaged transmit, receive and cross-correlation signals using 10 averages in the buffers before calculating the cross-correlation greatly enhances SNR.

$y = ax$. Since $y/a = x$, $dx/dy = 1/a$ and the probability density $p(y)$ is simply

$$p(y) = \frac{1}{a\sigma_x\sqrt{2\pi}} e^{\frac{\left(\frac{y}{a} - M_x\right)^2}{2\sigma_x^2}} = \frac{1}{a\sigma_x\sqrt{2\pi}} e^{\frac{(y - aM_x)^2}{2a^2\sigma_x^2}} \tag{11.1.4}$$

where clearly, $\sigma_y^2 = a^2\sigma_x^2$ and $M_y = aM_y$ in Eq. (11.1.4). Note that we are using a slightly different approach here to get the same statistical results given in Section 6.1 for averaged power spectra.

Consider the sum of a number of random variables $z = x_1 + x_2 + ... + x_N$. The probability density function for z is the convolution of the N density functions for $x_1, x_2, ..., x_N$. Noting that multiplications in the frequency domain are equivalent to convolutions in the time domain, the probability density function for z may be found easily using the density function's characteristic function. A Fourier integral kernel defines the relation between probability density $p(x)$ and its corresponding characteristic function $Q_x(\omega)$.

$$Q_x(\omega) = E\{e^{j\omega x}\} = \int_{-\infty}^{+\infty} p(x)e^{+j\omega x}dx$$

$$p(x) = \frac{1}{2\pi} \int_{-\infty}^{+\infty} Q_x(\omega)e^{-j\omega x}d\omega \tag{11.1.5}$$

Note that the sign of the exponential is positive for the forward transform of the density function. One of the many useful attributes of the characteristic function is that the moments of the random process (mean, variance, skewness, kurtosis, etc.) are easily obtainable from derivatives of the characteristic function. The nth moment for x is simply

$$\Box\{x^n\} = \frac{1}{j^n} \frac{d^n Q_x(\omega)}{d\omega^n}\bigg|_{\omega=0} \tag{11.1.6}$$

Returning to our new random variable $z = x_1 + x_2 + ... + x_N$, the density function $p(z)$ has the characteristic transform

$$Q_z(\omega) = \int_{-\infty}^{+\infty} p(z)e^{+j\omega z}dz$$

$$= \int_{-\infty}^{+\infty} [p(x_1) \circledast ... \circledast p(x_N)]e^{+j\omega z}dz \tag{11.1.7}$$

$$= Q_{x_1}(\omega)Q_{x_2}(\omega)...Q_{x_N}(\omega)$$

which is simply the product of each of the characteristic functions for each random variable x_i; $i = 1, 2, ..., N$. To derive the density function for z, we first derive the characteristic for the Gaussian density in Eq. (11.1.2).

$$Q_{x_i}(x_i) = \int\limits_{-\infty}^{+\infty} \frac{1}{\sigma_{x_i}\sqrt{2\pi}} e^{-\frac{(x_i - M_{x_i})^2}{2\sigma_{x_i}^2}} e^{+j\omega x_i} dx_i = e^{\left(jM_{x_i}\omega - \frac{\sigma_{x_i}^2 \omega^2}{2}\right)} \tag{11.1.8}$$

As can be seen when combining Eqs (11.1.7) and (11.1.8), the means and variances add for z, giving another Gaussian density function

$$p(z) = \frac{1}{\sigma_z \sqrt{2\pi}} e^{-\frac{(z - M_z)^2}{2\sigma_z^2}} \tag{11.1.9}$$

where $\sigma_z^2 = \sum_{i=1}^{N} \sigma_{x_i}^2$ and $M_z = \sum_{i=1}^{N} M_{x_i}$.

We can combine the summing and scaling to derive the density function for a random variable which is the average of a set of independent Gaussian variables with identical means and variances, such as is the case in the cross-correlation plot in Figure 1 in the regions away from the echo peak at 10 μsec. Let $z = (1/N) \sum_{i=1}^{N} x_i$. The mean for the resulting density stays the same as one of the original variables in the average, but the variance is reduced by a factor of N. Therefore, for the 128 samples included in the normalized cross-correlation in Figures 1–4, it can be seen that the variance of the background noise is reduced by about $1/128$, or the standard deviation by about $1/11.3$.

Time synchronous averaging of periodic data, such as that from a repetitive pulsing sonar or radar scanning system, is an important technique for reducing noise and improving detection statistics. Synchronous averaging can also be used to isolate

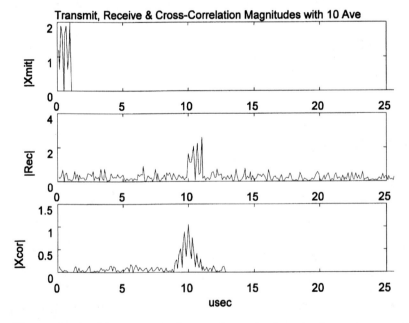

Figure 4 Applying a magnitude operation to the times synchronous averaged signals allow a straightforward probability model to be applied to the detection decision.

vibration frequencies associated with a particular shaft in a transmission or turbine. With the transmit and receiver waveforms recorded at the same time and synchronous to the transmit time, simply averaging the buffers coherently in the time domain before processing offers a significant improvement in SNR. Figure 3 shows the same time and correlation traces as in Figure 1, but with 10 synchronous averages. Figure 4 shows the result for the magnitudes of the transmit, received, and cross-correlated waveforms for 10 averages. With the signal coherently averaging and the noise averaging toward its mean, which is zero for ZMG noise, it is clear that this technique is extremely important to basic signal detection in the time domain. Figure 5 shows a block diagram of a pulse detection system using time synchronous averaging and the cross-correlation envelope.

Envelope Detection of a Signal in Gaussian Noise requires a very straightforward analysis of the underlying signal statistics. We develop the statistical models using a general complex waveform and its magnitude. Consider a zero mean signal (real part or imaginary part of a complex waveform) with ZMG noise, which after time synchronous and rms averaging, has the statistics of an rms signal level of S_0 and a noise standard deviation of σ_x. For the noise only case we apply the Gaussian density in Eqs (6.1.7) and (11.1.2) and Figure 6.1, with zero mean, to the magnitude-squared random variable $y = x^2$ using Eq. (11.1.3) to give the *Chi-Square 1 degree of freedom density function*

$$
\begin{aligned}
p(y) &= \frac{p(x)}{2x}\bigg|_{x=+\sqrt{y}} + \frac{p(x)}{2x}\bigg|_{x=-\sqrt{y}} \\
&= \frac{1}{\sigma_x\sqrt{2\pi y}}e^{-\frac{y}{2\sigma_x^2}}
\end{aligned}
\tag{11.1.10}
$$

where the mean is $E\{y\} = E\{x^2\} = \sigma_x^2$ assuming x is a ZMG random variable. This is the same density function as in Eq. (6.1.11) where the variance, $E\{y^2\}$ was found to be $2\sigma_x^4$ using the even central moments of a Gaussian density function in Eqs (6.1.12)–(6.1.13). This density function is known as a *Chi-Square probability density function* with one degree of freedom, as seen in Figure 6.2 with several other Chi-Square densities with more degrees of freedom.

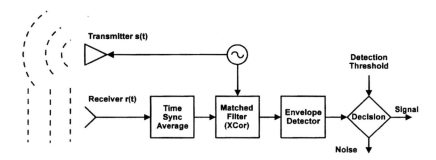

Figure 5 Block diagram of a typical matched filter enveloped detector showing a processing block for time synchronous averaging (transmitter sync), cross correlation of the transmitted and received signals in a matched filter, and envelope threshold detection based on an amplitude threshold criteria.

For summing the real part squared and imaginary part squared, $y = x_R^2 + x_I^2$, to get the magnitude squared of a complex random signal, we can either convolve two Chi-Square one degree of freedom densities or multiply the two corresponding characteristic functions. This results in a Chi-Square density with 2 degrees of freedom, better known as an *exponential probability density function*.

$$p(y) = \frac{1}{2\sigma_x^2} e^{-\frac{y}{2\sigma_x^2}} \tag{11.1.11}$$

To find the density function for the complex noise magnitude, we need to substitute a square-root for y in Eq. (11.1.11) as $z = y^{\frac{1}{2}} = (x_R^2 + x_I^2)^{\frac{1}{2}}$. Using Eq. (11.1.3) where $(dy/dz) = 2\sqrt{y}$ we have

$$p(z) = 2\sqrt{y} \frac{1}{2\sigma_x^2} e^{-\frac{y}{2\sigma_x^2}} \Big|_{y=z^2} = \frac{z}{\sigma_x^2} e^{-\frac{z^2}{2\sigma_x^2}} \tag{11.1.12}$$

which is known as a *Rayleigh probability density function*. This is the density function which describes the magnitude waveforms in Figures 2 and 4. Even though the Figures display real, rather than complex signals, the Rayleigh density function applies because the underlying mathematical time-harmonic signal structure is $e^{j\omega t}$. However, for other real signals, the magnitude of a ZMG random variable simply results in a Gaussian density which is zero for negative values and twice the normal amplitude for positive values with a mean of σ_x and a variance of σ_x^2.

The mean, mean-square, and variance for the Rayleigh density are found by simply evaluating the first and second moments of the probability density function. Unless you're really smart, a table of definite integrals will provide a relation such as the following.

$$\int_0^\infty x^n e^{-ax^p} dx = \frac{\Gamma\left(\frac{n+1}{p}\right)}{pa^{(n+1)/p}}; \; \Gamma(n+1) = n!; \tag{11.1.13}$$

$$\Gamma\left(m + \frac{1}{2}\right) = \frac{1 \cdot 3 \cdot 5 \ldots (2m-1)}{2m} \sqrt{\pi}$$

The mean of the Rayleigh density works out to be

$$\bar{z} = \int_0^\infty z \cdot \frac{z}{\sigma_x^2} e^{-\frac{z^2}{2\sigma_x^2}} dx = \sqrt{\frac{\pi}{2}} \sigma_x \approx 1.2533\sigma_x \tag{11.1.14}$$

and the mean square value is

$$\bar{z^2} = \int_0^\infty z^2 \cdot \frac{z^2}{\sigma_x^2} e^{-\frac{z^2}{2\sigma_x^2}} dx = 2\sigma_x^2 \tag{11.1.15}$$

so the variance of z is simply

$$\sigma_z^2 = \bar{z^2} - \bar{z}^2 = \left(2 - \frac{\pi}{2}\right)\sigma_x^2 \approx 0.4292\sigma_x^2 \qquad (11.1.16)$$

which is not quite half the variance of the ZMG random variable x. Figure 6 shows the unity-variance zero-mean Gaussian density, and the corresponding one degree of freedom Chi-Square, Exponential, and Rayleigh density functions.

The Rician probability density function is based on the envelope of a complex signal with rms amplitude S_0 and real noise x_R and imaginary noise x_I, where the envelope can be defined as

$$r = \sqrt{(S_0 + x_R)^2 + x_I^2}; \quad x_1 = r\cos\theta = S_0 + x_R; \quad x_2 = r\sin\theta = x_I;$$
$$\theta = \tan^{-1}(x_2/x_1)$$

$$(11.1.17)$$

where the angle θ is really of no consequence at this point. The Rice–Nagakami probability density function is often referred to as "Rician". The expected value of x_1 is S_0, while the expected value of x_2 is zero in Eq. (11.1.17). Assuming x_R and x_I are ZMG with identical variances σ_x^2, the variances of x_1 and x_2 are both σ_x^2. With x_1 and x_2 independent, the joint probability density for x_1 and x_2 can be written as

$$p(x_1 x_2) = \frac{1}{2\pi\sigma_x^2} e^{\frac{-\left[(x_1 - S_0)^2 + x_2^2\right]}{2\sigma_x^2}} \qquad (11.1.18)$$

To get the envelope density function, we convert the joint density in Eq. (11.1.18) to r and θ using the following Jacobian relationship (no sum is required because there is

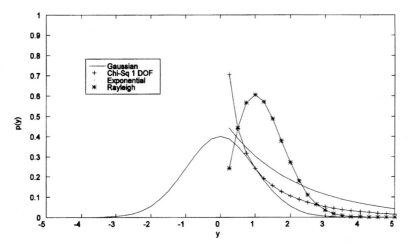

Figure 6   Gaussian, one degree of freedom Chi-Square, Exponential, and Rayleigh Density functions.

only one solution for r and θ in terms of x_1 and x_2).

$$p(r, \theta) = p(x_1, x_2) \begin{vmatrix} \dfrac{\partial x_1}{\partial r} & \dfrac{\partial x_1}{\partial \theta} \\[2mm] \dfrac{\partial x_2}{\partial r} & \dfrac{\partial x_2}{\partial \theta} \end{vmatrix}_{\substack{x_1=r\cos\theta \\ x_2=r\sin\theta}}$$

(11.1.19)

$$= \begin{vmatrix} \cos\theta & -r\sin\theta \\ \sin\theta & r\cos\theta \end{vmatrix} p(x_1, x_2) = rp(x_1, x_2) \Big|_{\substack{x_1=r\cos\theta \\ x_2=r\sin\theta}}$$

Evaluating the joint density in Eq. (11.1.19) gives

$$p(r, \theta) = \frac{r}{2\pi\sigma_x^2} e^{-\frac{r^2 + S_0^2 - 2rS_0\cos\theta}{2\sigma_x^2}}$$

(11.1.20)

where θ may be integrated out to give the envelope probability density function. This is why we were not concerned with the particular value of θ when the envelope detection model was set up.

$$p(r) = \frac{r}{\sigma_x^2} e^{-\frac{r^2 + S_0^2}{2\sigma_x^2}} \left[\frac{1}{2\pi} \int_0^{2\pi} e^{-\frac{S_0^2 r\cos\theta}{\sigma_x^2}} d\theta \right]$$

(11.1.21)

The square-bracketed term in Eq. (11.1.21) is recognized as a modified Bessel function of the first kind.

$$I_0(z) = \frac{1}{\pi} \int_0^{\pi} e^{\pm z\cos\theta} d\theta = \frac{1}{2\pi} \int_0^{2\pi} e^{\pm z\cos\theta} d\theta = J_0(jz); \quad j = \sqrt{-1}$$

(11.1.22)

The Rician envelope probability density function is therefore

$$p(r) = \frac{r}{\sigma_X^2} e^{-\frac{r^2 + S_0^2}{2\sigma_x^2}} I_0\left(\frac{S_0 r}{\sigma_x^2}\right)$$

(11.1.23)

It can be seen that for $S_0 \ll \sigma_x$, the Rician density becomes a Rayleigh density (the modified Bessel function approaches unity). For high SNR, $S_0 \gg \sigma_x$, and the modified Bessel function of the first kind may be approximated by

$$I_0\left(\frac{S_0 r}{\sigma_x^2}\right) \approx \frac{1}{\sqrt{2\pi \dfrac{S_0 r}{\sigma_x^2}}} e^{+\frac{S_0 r}{\sigma_x^2}} \quad S_0 \gg \sigma_x$$

(11.1.24)

Inserting the high SNR approximation for the modified Bessel function in the Rician density gives a very interesting result.

$$p(r) \approx \sqrt{\frac{r}{S_0}} \cdot \frac{1}{\sigma_x\sqrt{2\pi}} e^{-\frac{(r-S_0)^2}{2\sigma_x^2}} \quad S_0 \gg \sigma_x \tag{11.1.25}$$

The high SNR approximation to the Rician density shown in Eq. (11.1.25) indicates that the Rician has the shape of a Gaussian density function with mean at $r = S_0$ and variance σ_x^2, approximately. Figure 7 shows the Rician density for various SNRs. The Rician density function applies to complex signals where the envelope magnitude is found by summing the real and imaginary components of the signal. But what about purely real signals and noise?

For purely real signals in ZMG noise, lets assume the signal is a constant (delta function probability density) with rms level S_0 and the noise has variance σ_x^2. The received real waveform with both signal and noise present has the Gaussian density

$$p(x) = \frac{1}{\sigma_x\sqrt{2\pi}} e^{-\frac{(x-S_0)^2}{2\sigma_x^2}} \tag{11.1.26}$$

Squaring the non-zero mean Gaussian random variable x in Eq. (11.1.26) yields $y = x^2$ and the Chi-Square density

$$p(y) = \frac{1}{2\sigma_x\sqrt{2\pi y}} \left[e^{-\frac{(\sqrt{y}-S_0)^2}{2\sigma_x^2}} + e^{-\frac{(\sqrt{y'}-S_0)^2}{2\sigma_x^2}} \right] \tag{11.1.27}$$

which reduces to Eq. (11.1.10) when the signal level is zero. Note that y' is the sol-

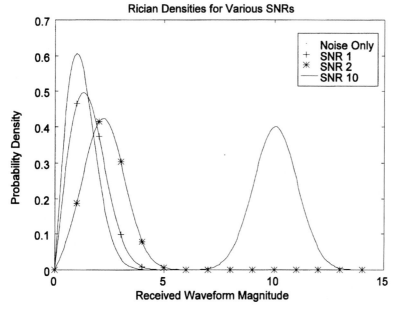

Figure 7 Rician probability density functions for linear rms SNRs of 0, 1, 2, and 10 which correspond to power SNRs of $-\infty$ dB, 0 dB, 6 dB, and 20 dB.

ution of $y = x^2$ which corresponds to negative values of $y^{1/2}$. The real waveform envelope is found from $z = y^{1/2}$ and the application of Eq. (11.1.3) where $dy = 2zdz$.

$$p(z) = \frac{1}{\sigma_x \sqrt{2\pi}} \left[e^{-\frac{(z-S_0)^2}{2\sigma_x^2}} + e^{-\frac{(-z'-S_0)^2}{2\sigma_x^2}} \right] = \frac{1}{\sigma_x \sqrt{2\pi}} \left[e^{-\frac{(z-S_0)^2}{2\sigma_x^2}} + e^{-\frac{(z+S_0)^2}{2\sigma_x^2}} \right] \quad (11.1.28)$$

The parameter z' in Eq. (11.1.28) corresponds to negative values of $y^{1/2}$ and the z' term represents a Gaussian density function over negative z with a mean at S_0. This is the same as a Gaussian density function over positive z with mean $-S_0$, so a simple change of variable of $z = -z' - S_0$ gives us a valid expression for the probability density function over positive z only. Figure 8 shows the Gaussian density functions for a real signal envelope and several SNRs. Note that for zero signal, we have twice a normal Gaussian density but for half the normal range (positive z only). The integral of these densities over positive z only are all unity. We will refer to this probability density function as the *Gaussian magnitude probability density function.*

$$p(z) = \frac{2}{\sigma_x \sqrt{2\pi}} e^{-\frac{(z-S_0)^2}{2\sigma_x^2}} \quad z \geq 0 \quad (11.1.29)$$

Suppose the "signal" is not a simple constant, but another random variable? The procedure for finding the density of this combination of random signal in random noise follows the procedure for summing two random variables. Specifically, the noise density function and the signal density function are convolved, or their

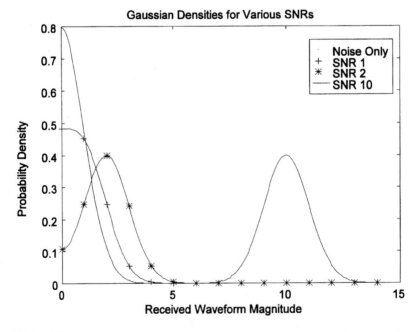

Figure 8   Gaussian density functions for various SNRs for received real signals where the magnitude is found without summing the squared real and imaginary signal components.

corresponding characteristic functions multiplied in the frequency domain to yield the resulting density via inverse Fourier transform. Indeed, the probability density function of a constant variable is a Dirac delta function. Convolving the delta function with a Gaussian function simply shifts the mean of the Gaussian density accordingly to correspond to the delta function position.

11.2 CONSTANT FALSE ALARM RATE DETECTION

In this section, we develop a straightforward decision algorithm to detect signals in noise. The decision is based on the obvious hypothesis that the signal is larger in amplitude than the noise making it detectable. Signal processing such as Fourier transforms, time synchronous averaging, and matched filtering are all designed to make signals as detectable as possible. But still, the decision is not entirely simple when the signal amplitude is close to the noise amplitude because the received signal waveform also has noise present. The "fuzziness" between signal plus noise and noise only is modeled physically and statistically using probability density functions. We want to associate the detection decision with a probability, and control the detection performance by maximizing the probability of detection, P_d, for a tolerable probability of false alarm, P_{fa}. These probabilities are found simply by integrating the probability density functions derived for signals and noise in Section 11.1.

Consider the simple Gaussian density for the magnitude of real data seen in Figure 9 (same as the noise only case in Figure 8). For the ZMG real noise with variance σ_x^2, the mean of the real magnitude random variable can be shown to be exactly σ_x. A detection threshold Λ is chosen at some level typically above the noise mean so that the P_{fa} is somewhat less than 0.5 (less than 50%). A more reasonable false

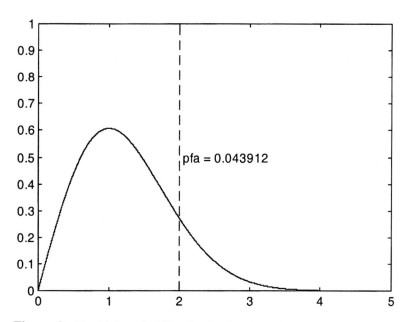

Figure 9 Rayleigh probability density function where the integral from $T = 2$ to infinity gives a probability of false alarm of about 4.4%.

alarm rate criterion might be something like 0.1%, or one wrong guess of calling a noise waveform a signal waveform in 1000 decisions. The probability of false alarm is evaluated by simply integrating the probability density function for the noise from the decision threshold to infinity. This represents the estimated probability that the noise waveform will be above the threshold and detected as signal.

$$P_{fa} = \int_{\Lambda}^{\infty} p(x)dx = 1 - \int_{0}^{\Lambda} p(x)dx \qquad (11.2.1)$$

Some probability density functions integrals can be evaluated in elegant algebraic equations. But, numerical integration can always be used to compute the probability.

$$P_{fa} = 1 - \sum_{n=1}^{N} p(x)\Big|_{x=n\Delta x} \Delta x \quad N\Delta x = \Lambda \qquad (11.2.2)$$

Constant false alarm rate (CFAR) detection is a classic example of sentient processing (the sensor system having the access to sensor data for environmental awareness) to adapt to things like changing background noise levels. Given an estimate of the background noise probability density function, the mean background noise ($\sqrt{(\pi/2)}\sigma_x$ for a Rayleigh and σ_x for a Gaussian-magnitude noise density), serves as a baseline for setting the signal detection threshold. Consider the Rayleigh density which describes the square-root of the sum of the real and imaginary parts of an FFT bin squared. We conveniently define a "floating" detection threshold, $\Lambda = T\sqrt{(\pi/2)}\sigma_x$ where T is set to a constant value, so if the background noise level increases or decreases, so does the absolute detection threshold Λ. As seen in Figure 9 for the Rayleigh noise magnitude distribution, the integral from the absolute threshold Λ to ∞ represents the probability that a background noise sample has magnitude above the threshold, and thus would be detected as signal rather than remain undetected as noise. The plot in Figure 9 is given in terms of the relative detection threshold T so that the Figure applies to any absolute noise level standard deviation σ_x. With the relative threshold set to twice the mean noise level of $\sqrt{(\pi/2)}\sigma_x$, the resulting probability of false alarm in Figure 9 is about 4.4%

When Λ is set equal to $T\sqrt{(\pi/2)}\sigma_x$, the integral from Λ to ∞ is always the same no matter what the mean level of the background noise. The corresponding absolute detection threshold Λ "floats" above the background noise mean $\sqrt{(\pi/2)}\sigma_x$ by the factor T. *The probability of a false alarm is therefore constant, even if the estimated mean background noise level changes.* Constant false alarm rate detection is very useful because the system performance can be easily controlled by design. For example, a CFAR detector with $P_{fa} = 0.02$, or 2%, would have on average 2 false alarms every 100 detection trials. If a sensor system executes a signal detection hypothesis once every second, a 2% false alarm rate would produce an erroneous signal detection a little more than once every 2 minutes. A false alarm rate of only 0.1% on a once-per-second detection trial would have a detection error of about once every 15 minutes.

Figure 10 shows the Gaussian magnitude density function where the relative threshold is set to twice the mean noise level of σ_x. In this scenario the probability

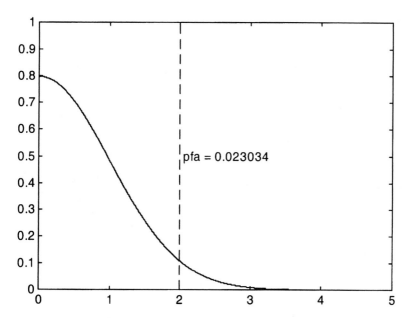

Figure 10 Gaussian magnitude probability density function where the integral from relative threshold $T = 2$ to infinity gives a false alarm rate of about 2.3%.

of false alarm is about 2.3%. The Gaussian magnitude probability density function is used when applying threshold detection on the magnitude of real signals in real ZMG noise while the Rayleigh probability density function applies to magnitude detection of complex signals. Figure 11 compares the P_{fa} for the Rayleigh and Gaussian magnitude density functions as a function of relative threshold T, which corresponds to an absolute threshold of $\Lambda = T\sqrt{(\pi/2)}\sigma_x$ for the Rayleigh density and $\Lambda = T\sigma_x$ for the Gaussian magnitude density. For high detection thresholds relative to the mean, the two densities are more comparable in terms of P_{fa}. However, one must keep in mind that the mean for the Rayleigh probability density is roughly half the mean of the Gaussian magnitude density.

Figure 12 shows some useful approximations for the false alarm rate as a function of relative threshold T for the Gaussian magnitude density and a direct realization of the Rayleigh false alarm rate. The integral of the Rayleigh density function actually gives another Gaussian-shaped function allowing a straightforward equation for the false alarm rate as a function of relative threshold.

$$P_{fa}(\Lambda) = \int_{\Lambda}^{\infty} \frac{x}{\sigma_x^2} e^{-\frac{x}{2\sigma_x^2}} \, dx = e^{-\frac{1}{2}\left(\frac{\Lambda}{\sigma_x}\right)^2} \tag{11.2.3}$$

For the Rayleigh distribution representing the magnitude of complex random numbers, one can also specify the desired false alarm rate and calculate the detection threshold relative to the noise mean.

$$T = \frac{\Lambda}{\sigma_x\sqrt{\frac{\pi}{2}}} = \sqrt{-\frac{4}{\pi}\ln(P_{fa})} \tag{11.2.4}$$

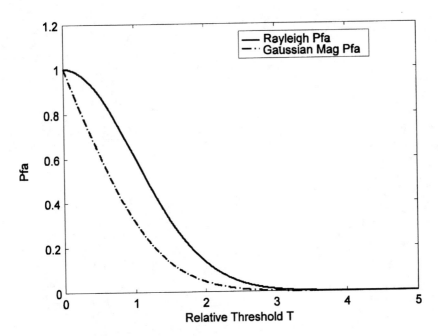

Figure 11   Probability of false alarm for both Rayleigh and Gaussian magnitude densities as a function of relative detection threshold T.

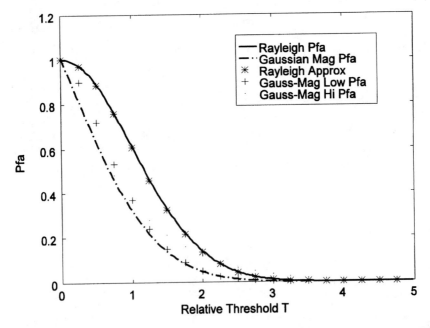

Figure 12   High and Low P_{fa} approximations to the Gaussian magnitude density P_{fa} as a function of relative threshold and a direct realization of the Rayleigh false alarm rate.

For the Gaussian magnitude false alarm rate, a bit more approximating must be done as seen in Figure 12.

$$P_{fa}(\Lambda) \approx e^{-T^{1.1}} \quad P_{fa} > 0.5$$
$$\approx e^{-T^{1.6}} \quad P_{fa} < 0.2 \qquad (11.2.5)$$

The false alarm rate approximations for the Gaussian magnitude density provide a very useful way to specify a P_{fa} and determine the detection threshold relative to the noise mean.

$$T \approx \left[-\ln(P_{fa})\right]^{1.1} \quad P_{fa} > 0.5$$
$$\approx \left[-\ln(P_{fa})\right]^{1.6} \quad P_{fa} < 0.2 \qquad (11.2.6)$$

Equations (11.2.5) and (11.2.6) are approximated by numerical inspection and are seen as very useful for setting a practical detection threshold for a desirable false alarm rate when the signals and noise are processed as the magnitude of a real quantity. Curve-fitting is very useful to relate the detection threshold to P_{fa}. However, one must be careful not to use Eqs (11.2.5) and (11.2.6) out of context. Equations (11.2.3) and (11.2.4) are exact direct expressions for the false alarm rate and relative threshold for the Rayleigh density function, which is used when the magnitude of a complex quantity is applied to the detection decision.

We now consider the common problem of detecting a sinusoid in ZMG noise using an FFT to enhance SNR, and a Hanning window to control spectral leakage. Let the real signal of interest be

$$y[n] = A \sin \Omega_0 n + w[n] \qquad (11.2.7)$$

where $\Omega_0 = 2\pi f_0 / f_s$ is the digital frequency and $w[n]$ is ZMG noise with variance σ_w^2. Executing a k-point FFT with no window (rectangular) and Ω_0 exactly aligned with an FFT bin, the expected value of the magnitude in the positive frequency bin is exactly $Ak/2 + \sigma_w$, where the real and imaginary components of the noise in the FFT bin are ZMG each with variance $\frac{1}{2}\sigma_w^2$. By carefully defining the noise mean and signal to noise ratio for the FFT bin, we can select a detection threshold for a desired P_{fa} and then determine the probability of detection for a given SNR.

Applying a Hanning window and normalizing the FFT by k, it can be seen from Section 5.3 that applying a narrowband correction factor of 2.0 will provide a magnitude of $A/2$ for the signal, but a broadband correction factor of $(8/3)^{1/2}$ (found from the square root of the window convolved with itself) would be needed to have the noise magnitude at σ_w/k. Since only one window power correction factor is applied, it can be seen that the use of the Hanning window lowers the SNR by a small factor of $(8/3)^{1/2}/2$, or about 0.8165 (SNR is about 82% of what it could be with a bin-aligned sinusoid using rectangular window) due to the scalloping loss of the Hanning window. The scalloping loss of the Hanning window should not be alarming, if the sinusoid happened to be in between two adjacent FFT bins the SNR loss could be nearly 50% and a large amount of spectral leakage would be present in the rest of the spectrum.

For a particular FFT spectrum we detect sinusoids by hypothesizing that when the spectral magnitude in a particular bin is larger than its two neighbors, it may be a peak. The hypothesis testing continues by estimating the background noise level in the vicinity of the peak. If the peak is relatively weak, the ratio of the peak magnitude to the local background noise will be small. Given the local background noise mean estimate, the detection threshold is set as a multiple T of the mean background noise level. If the peak candidate is larger than the threshold it is selected as a peak, otherwise it is considered noise. The relative threshold level allows for a known constant false alarm rate for the detection algorithm.

But, considering the Rician probability density function for a selected peak, the probability of detection for a particular peak can be determined by integrating the density from the absolute threshold $\Lambda = T\sqrt{(\pi/2)}\sigma_w$ to infinity.

$$P_d(\Lambda, A, \sigma_w) = \int_\Lambda^\infty \frac{r}{\sigma_w^2} e^{-\frac{(r^2+A^2)}{2\sigma_w^2}} I_0\left(\frac{Ar}{\sigma_w^2}\right) dr \tag{11.2.8}$$

Unfortunately, there is no simple analytical solution to Eq. (11.2.8) and the P_d is obviously a complicated function of signal level A, noise level σ_w, and detection threshold Λ. Figure 13 shows the calculated P_d as a function of relative detection threshold T for several SNRs. Clearly, it can be seen the probability curves all have nearly the same shape and pass through the 50% mark when the relative threshold is approximately at the signal level. This is to be expected especially for high SNR because the density function is nearly symmetric about the signal level. It would be very useful to have a straightforward algorithm to convert a measured signal peak level directly into a P_d, but looking at Eq. (11.2.8) this does not appear to be easy.

We can develop a P_d calculation algorithm in a heuristic manner by first shifting the P_d curves by their respective SNRs to give the plots in Figure 14, which

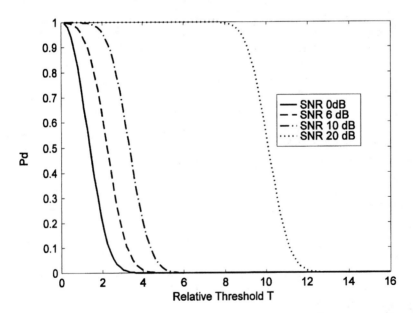

Figure 13 Pd Curves as a function of threshold T relative to the noise mean shown for several typical SNR levels.

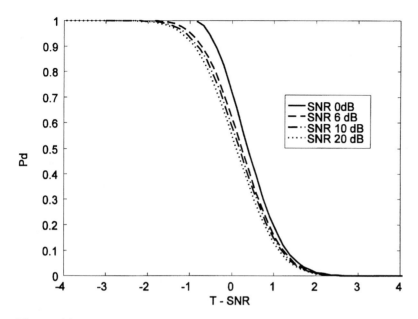

Figure 14 Shifting by the SNR allows the Pd curves to nearly overlay depending on the significance of the background noise level in the Rician probability density function.

except for the background noise contribution of the Rician density curves, nearly overlay one another in a range from 0 dB (SNR = 1) to 20 dB (SNR = 10). All of the curves are offset to the right slightly because of the Rayleigh noise magnitude combined with the signal in the Rician density function. Figure 15 shows how a moderately scaled hyperbolic tangent function nearly perfectly fits the P_d curves. We therefore apply this model to provide directly the P_d given the spectral peak SNR and the relative detection threshold T, which is actually the ratio of the absolute detection threshold to the mean Rayleigh noise. Equation (11.2.9) can be used directly to estimate the P_d given the relative threshold T and SNR. Again, we caution that Eq. (11.2.9) is just a heuristic curve fit which is neither physical nor optimal. However, it is much more convenient than generating a numerical table of the actual $P_d(T, SNR)$ for interpolation when such calculations are needed on a continuous basis in a real-time intelligent sensor system.

$$P_d(T, SNR) = \frac{1}{2} - \frac{1}{2}\tanh\left(\left[T - SNR - \frac{1}{2\,SNR}\right]\left[1 + \frac{3}{4\,SNR^2}\right]\right) \qquad (11.2.9)$$

Lastly, we have not discussed spectral averaging as it relates to signal density functions and detection theory. Note that time-synchronous averaging is an entirely different process which actually improves the physical SNR by letting incoherent noise and frequencies average to zero. Averaging Fourier spectra is quite different. The size of the Fourier integral (FFT size and physical integration time/space) determines the SNR improvement and spectral averaging does not change the mean values for the signal and noise in any given FFT bin. This is because the spectral density of the noise in the signal is constant in any given frequency range while

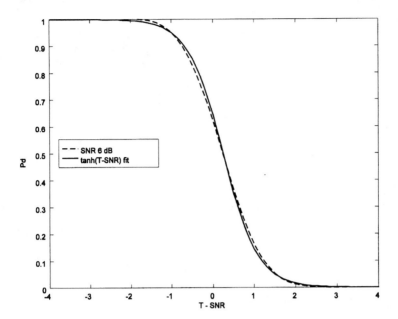

Figure 15 A hyperbolic tangent function provides a very useful algorithm for quickly recovering the Pd given the difference between T, the detection threshold relative to the noise mean, and SNR.

the theoretical spectral density of a sinusoid is infinite at its precise frequency and zero elsewhere. For the SNR in the FFT output to improve, one must integrate over a longer time interval (or spacial extent for beamforming). Increasing the digital sampling rate and FFT size together does not offer an improvement in SNR, just more high frequency bins. This is because the bin-width in Hz is constant when the FFT size and sample-rate are both increased or decreased proportionally.

The variance of the signal in the FFT bin would however decrease as the number of averages increases. Because the averaging process involves summing of random variables, the density function for an averaged FFT bin is derived from the convolution of the densities of the random variables used in the average (or correspondingly, the product of the characteristic functions inverse Fourier transformed). One can imagine that this complete derivation is quite difficult and is thus left as a subject beyond the scope of this book. However, by the central limit theorem, we can say that the density function of the averaged FFT bin will tend to be Gaussian and, that the variance will decrease in proportion to the number of averages while the estimated mean will converge to the true mean value. Therefore, the spectral averaging process tends to narrow the density functions of signal and noise, allowing a lower detection threshold for a given P_{fa} requirement and providing a higher P_d for a given signal level. If the signal and noise are stationary (and hence statistically ergodic), averaging is a really useful tool for improving performance.

The Rician density function is derived for a constant amplitude sinusoid in ZMG noise. As described in the Section 11.3, the signal can acquire statistical attributes in amplitude and phase due to the propagation channel. Convolving the density function of the signal or propagation channel with the Rician density will tend to broaden the probability density function for the signal plus noise. Averaging is generally the best remedy to narrow the densities back down to compensated for the signal fluctuations. With an infinite amount of averaging the noise and signal plus density functions become Dirac delta functions.

11.3   STATISTICAL MODELING OF MULTIPATH

For any wavenumber sensor system, whether it be a radar, sonar, ultrasonic images, or structural vibration modal filter, underlying assumptions of the wave shape, wavelength (or wave speed), and temporal frequency are required to reconstruct the wave field. A wave radiated from a distant point source (a source much smaller than the wavelength) in a homogeneous infinite-sized medium is essentially and plane-shaped wavefront beyond a hundred wavelengths as described in Section 6.3. Given a single plane wave propagating across an array of sensors, the angle of arrival can be easily determined. For a line array of sensors, the wave frequency and speed are needed to determine the angle of arrival relative to the line array axis. For a planar array, the angle of arrival can be determined in a half-space, and if the source is in the same plane as the sensors, the angle of arrival and wave speed can be determined explicitly. For a three-dimensional sensor array in an infinite homogeneous medium, any angle of arrival and wave speed can be measured explicitly for a simple plane wave. However, when more than one plane wave (from correspondingly different directions) but of the same frequency arrive at the sensor array, the direction finding problem becomes far more complicated. We term this situation multipath propagation, which occurs in real-world applications of sensor technology whenever the propagation medium is either inhomogeneous or has reflecting surfaces.

There are two general classes of multipath which can exist at any given instant of time. The first we call "coherent multipath" and results from a single source radiator and multiple ray paths to the sensors from either an inhomogeneous medium or reflectors or scatterers in the propagation medium. The second class of multipath is from multiple sources radiating the same frequency (or wavelength — the sources could be moving and radiating slightly different frequencies before Doppler) at the sensor array. This "multi-source" multipath carries an important physical distinction from coherent multipath in that over time, the phases of the sources will become incoherent, allowing the arrival angles to be measured. The same is true for coherent multipath when the source is moving relative to the sensor array, or if the multipath is changing due to a nonstationary inhomogeneous medium, reflecting surface, or scatterers.

Multi-source multipath results in statistically independent phases across the array. Chapter 13 presents adaptive beamforming algorithms to deal specifically with multi-source multipath. It is very important to fully understand the physics of the wave propagation before designing the sensor and signal processing machine to extract the propagation information. Both coherent and multi-source multipath have a significant impact on detection because the multiple waves at the sensor site interfere with each other to produce signal enhancement at some sensor site and signal cancellation at other sensor sites in the array. The wave interference throughout the sensor array allows one to detect the presence of multipath.

However, even in a homogeneous medium, one cannot resolve the angles of arrival unless one knows the amplitude, phase, and distance of each source (if multi-source), or the amplitude and phase differences between ray paths (if coherent multipath). Section 13.4 examines the case of known sources, where one is not so interested in localizing sources, but rather measuring the medium. From a signal detection point of view, the possibility of multipath interference impacts the detection algorithm with additional signal fluctuations at the receiver independent of the background noise.

Coherent multipath is characterized by the phases of the arrivals being dependent solely on the propagation channel, which in many cases is relatively stationary over the detection integration timescale. Consider the simple two-ray multipath situation depicted in Figure 16 where a point source and receiver are a distance h from a plane boundary and separated by a distance R in a homogeneous medium with constant wave speed c in all directions (if the field has a non-zero divergence, or net flow, the wave speed is directional). The field at the receiver can be exactly expressed using the sum of a direct and reflected ray path. For simplicity, we'll assume an acoustic wave in air where the boundary is perfectly reflecting (infinite acoustic impedance). This allows the incident and reflected wave to have the same amplitude on the boundary and our analysis to be considerably more straightforward. Figure 16 shows the reflected ray path using an equivalent "image source", which along with the real source, would produce the same field at the receiver with the boundary absent. If the impedance of the boundary were finite and complex, the image source would have an amplitude and phase shift applied to match the boundary conditions on the reflecting surface. The equation for the field at the receiver is

$$p(R) = Ae^{j\omega t}\left\{\frac{1}{R}e^{-jkR} + \frac{1}{R_f}e^{-jkR_f}\right\} \quad R_f = \sqrt{R^2 + 4h^2} \tag{11.3.1}$$

where A is the source amplitude, R is the direct path, R_f is the reflected path length, k is the wavenumber ($k = \omega/c = 2\pi/\lambda$), and ω is the radian frequency. Figure 17 shows

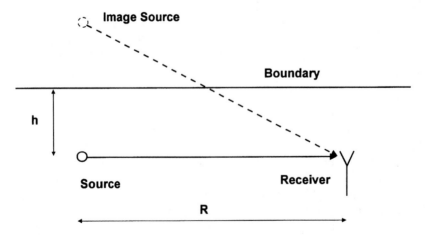

Figure 16   Direct and reflected ray paths depicted using an image source.

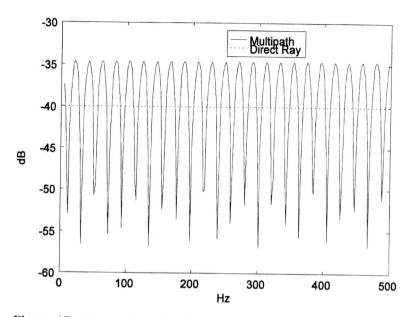

Figure 17 Direct and multipath responses for a source and receiver separated by 100 m and 30 m from an acoustically reflecting plane boundary in air.

the multipath frequency response at the receiver compared to the response without the reflected path assuming $R = 100$ m, $h = 30$, and an acoustic wave speed of 345 m/sec.

Equation (11.3.1) is exact for a perfectly reflecting planar surface. Before we consider a random surface (or random distribution of wave scatterers), we examine the case where we have a line array of sensors rather than a single receiver. Figure 18 shows how two waves with the same temporal frequency, arriving from two angles simultaneously, will cause an interference pattern spatially across a sensor array. In the time domain, each sensor detects the same frequency. But spatially, the amplitude and phase varies across the sensor locations due to the sum of the waves.

For example, if a line array observes a plane wave, the spatial amplitude will be constant and the wavelength "trace" will be representative of the frequency, wave speed, and angle of arrival. A plane wave passing the line array from a broadside direction will have a wavelength trace which looks like an infinite wavelength, while the same wave from the axial direction will have a wavelength trace equal to the free wave wavelength ($c/f = \lambda$). Now if two plane waves of the same temporal frequency arrive at the line array from different directions, the wavelength traces sum, giving an interference pattern where the "envelope wavelength" trace is half the difference of the respective traces for the two waves, and the "carrier wavelength" trace is half the sum (or the average) of the two respective wavelength traces. This is exactly the same mathematically as an amplitude modulated signal. If a direction-finding algorithm uses the spatial phase to detect the angle-of-arrival, it will calculate an angle exactly in between the two actual arrival angles (this would be weighted towards the stronger wave for two unequal amplitude plane waves). If a particular sensor happens to be near a cancellation node in the trace envelope, it will

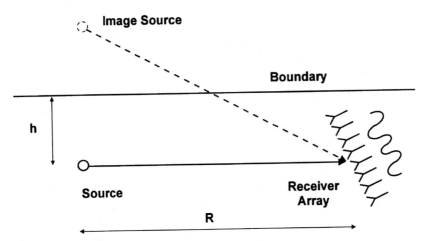

Figure 18 Coherent multipath showing wave interference at the array in the form of wavelength and enveloped distortion.

have very poor detection performance. This is often experienced with cordless or cellular telephones as well as automobile radios in a multipath environment. If the line array length is long enough to observe a significant portion of the envelope peaks and dips, the individual arrival angles can generally be determined using beamforming as described in Section 7.1.

Statistical representation of multipath is useful for situations where coherent reflections are coming from random surfaces or refraction in a random inhomogeneous wave propagation medium. Good examples of this are when the scanning system (radar or sonar) is traveling over or near a rough surface, or if the medium has turbulence or scatterers. Physically, we need to describe the variance of the multipath phase for a given frequency, and the timescale and or spatial scale for which ensemble observations will yield the modeled statistics for the multipath. When the medium is nonstationary, one also has to consider an outer timescale, beyond which one should not integrate to maintain a statistical representation of the multipath medium.

Consider multipath due to a large number of small scatterers with perfect reflecting properties. Using Babinet's principle in optics, our point scatterer will re-radiate the energy incident on its surface equally in all directions. If the scatterer is large or on the order of the wavelength, or if the scatterer is not perfectly reflecting, the subsequent re-radiation is quite complex and beyond the scope of this book. We can exploit Babinet's principle by approximating the reflection of a wave from a complicated boundary by replacing the boundary with a large number of point scatterers. The magnitude and phase of the waves re-radiated by the point scatterers is equal to the magnitude and phase of the incident field at the corresponding position along the boundary. This technique is well-known in acoustics as Huygen's principle which states that any wave can be approximated by an appropriate infinite distribution of point sources. Summing the responses of the Huygen's point sources over a surface, albeit a reflecting surface or a radiating surface, is called a Helmholtz integral. Therefore, we'll refer to approximating the reflection from a geometrically complicated boundary as a Helmholtz–Huygen technique.

Figure 19 compares the exact image source solution to the Helmholtz–Huygen's technique using 2000 point sources on the boundary $dl = 0.3$ m apart, where the wave speed is 345 m/sec, the boundary is 30 m from both source and receiver which are separated by 100 m as seen in Figure 16. Because 2000 point sources is hardly infinite and we are interested in only a frequency range from 0 Hz to 500 Hz, we also apply an empirical loudness adjustment over frequency so that the source energy per wavelength is constant over frequency at any one position on the boundary. This "normalizes" the channel frequency response in Figures 19 and 20. Using a Cartesian coordinate system with the direct path along the x-axis and the normal to the boundary along the y-axis, the magnitude and phase of each point source modeled on the boundary is

$$A_n = \frac{\lambda}{2dl R_{1,n}} e^{-jk R_{1,n}} \quad R_{1,n} = \sqrt{x^2 + h^2} \quad x = ndl \tag{11.3.2}$$

where $R_{1,n}$ is the ray from the main source to the nth point source model on the boundary. The Helmholtz–Huygen's pressure approximation at the receiver is

$$p(f) = \sum_{n=1}^{N} \frac{A_n}{R_{2,n}} e^{-jk R_{2,n}} \quad R_{2,n} = \sqrt{(R - x)^2 + h^2} \tag{11.3.3}$$

where $R_{2,n}$ is the ray from the nth boundary source to the receiver and R is the direct path length. The $N = 2000$ sources where spaced evenly $dl = h/R$ m apart symmetrically about the midpoint of the physical source and receiver to give a reasonable approximation for $h = 30$ m.

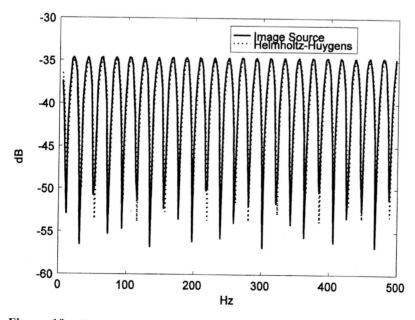

Figure 19 Comparison of exact image-source method to Helmholz–Huygens integral approximation.

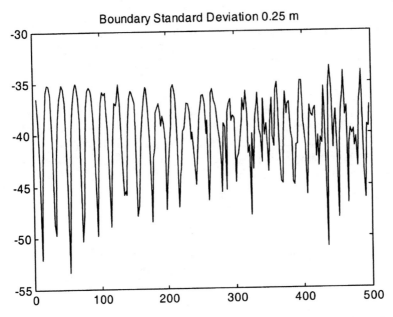

Figure 20   Direct and scattered acoustic field response when the boundary has a random surface (dB vs. Hz scale).

Figure 20 shows the utility of the Helmholtz–Huygen's technique for modeling rough boundaries. In the Figure, a ZMG distribution with a standard deviation of 0.25 m is used to randomly distribute the point sources on the y-axis centered around $y = h$, the boundary. The boundary sources therefore have a y coordinate of $y + \xi$. Figure 20 clearly shows the transition from deterministic to stochastic wave interference somewhere in the 200 Hz to 300 Hz range. This simple example of rough boundary scattering exposes a really interesting wave phenomena where the "roughness" of the reflecting boundary depends on wavelength along with the variance of the roughness. At low frequencies, the wavelength is quite large compared to ξ and the scattering is not detectable. But when the boundary roughness is approaching a quarter wavelength or more, the affect of the rough boundary is readily seen. When the boundary roughness is not ZMG, one would calculate a spatial Fourier transform of the rough boundary and use a power spectrum estimate to get the variance at a wavenumber appropriate for the boundary reflection to determine the scattering effect.

We can develop a model which relates the ZMG distribution in the y-direction along the reflecting surface to the reflected path length.

$$R_\xi = \sqrt{R^2 + 4(h + \xi)}$$

$$= R_f \sqrt{1 + \frac{8h\xi + 4\xi^2}{R^2 + 4h^2}}$$

(11.3.4)

Equation (11.3.4) is easily approximated noting that the second term in the square root is quite small.

$$\sqrt{1+\varepsilon} \approx 1 + \frac{\varepsilon}{2} \quad \varepsilon \ll 1 \tag{11.3.5}$$

The reflected path length including the random variation R_ξ is given in Eq. (11.3.6) in terms of the vertical distance to the boundary h, the reflected path length with no roughness R_f, and the roughness random variable ξ.

$$R_\xi \approx R_f + \frac{4h\xi}{R_f} \tag{113.6}$$

Clearly, the standard deviation of the reflected path length σ_ξ can be equated to the standard deviation of the boundary along the y-axis using Eq. (11.3.7).

$$\sigma_R \approx \frac{\sigma_\xi 4h}{R_f} \tag{11.3.7}$$

We can model the statistics of the acoustic pressure at the receiver by adding our random variable to Eq. (11.3.1).

$$p(R) + p' = \frac{A}{R} e^{j\omega t - kR} \left\{ 1 + \frac{R}{R_f} e^{-jk(\Delta R - \xi)} \right\} \quad \Delta R = R_f - R \tag{11.3.8}$$

The term p' in Eq. (11.3.8) represents the random part of the pressure at the receiver. We note that for the pressure to change from the midpoint loudness at -40 dB to the peak loudness at -34 dB in Figure 20, a phase change of $\pi/2$ is needed in the reflected path. If we consider a noticeable pressure fluctuation to be corresponding to a phase fluctuation of $\pi/4$, and note that $4h/R_f$ is about unity for $R = 100$ and $h = 30$, we need k on the order of π to have significant scattering effects. This corresponds to a frequency of about 170 Hz. Figure 20 clearly shows the stochastic effect of the random reflection boundary in this frequency range. If the source and receiver are moved closer to the boundary, the path randomness decreases. If h is increased, the path length variances increases to a limit of $2\sigma_\xi$ when $h \gg R$. However, for large h the reflected path is so much larger than the direct path that the pressure fluctuations again decrease. The maximum pressure fluctuations are for a ratio of $h/R = 0.5$. For low frequencies, k (the wavenumber) is small (wavelength large) and the fluctuations are also correspondingly small.

Random variations in refractive index (changes in wave speed relative to the mean) are also a very important concern in multipath propagation. We are not only interested in how refractive multipath affects detection, but also how received signal fluctuations can be used to measure the propagation medium. This physical effect occurs in electromagnetic propagation due to variations in propagation speed due to humidity, fluctuations in the earth's magnetic field, solar activity, and turbulence in the ionosphere. At night, when solar interference is low, long range AM and short-wave broadcasts have characteristic waxing and fading, and at times, unbelievably clear reception. The complexity of the propagation is due to multipath interference and the fluctuations are due to time varying things like winds, turbulence, and fluctuations in the earth's magnetic field (due to magma movement), rotation, and solar electromagnetic waves interacting with the ionosphere. One can think of the earth's atmosphere, with its charged particles in the ionosphere and fluctuating ground plane due to rain and weather at the surface, as a huge waveguide defined by two concentric spheres. Changes in the boundary conditions affect which modes propagate in the waveguide and at what effective speed (a ray which reflects off the ionosphere and ground travels slower than a ray traveling more parallel to the boundaries). Therefore, the receiver experiences multipath due to refraction of the waves as well as reflection.

In sonar, the refractive index of a sound wave in seawater is affected by the warm temperature near the surface, and changes in temperature, salinity, and pressure with depth and with ocean currents. Sound channels, or waveguides can form around layers of slow propagation media. The channel happens physically by considering that a plane wave leaving the layer will have the part of the wave in the faster media (outside the layer) outrun the part of the wave inside the layer, thus refracting the wave back into the layer. Whales are often observed using these undersea channels for long range communication. Clearly, it's a great place to be for a quiet surveillance submarine but not a good place to be if you're in a noisy submarine and want to avoid being detected acoustically. The seasonal changes in undersea propagation conditions are much more slow than in the atmosphere, where dramatic changes certainly occur diurnally, and can even occur in a matter of minutes due to a shift in wind direction. In general, sound propagation from sources on the ground to receivers on the ground is much better at night than during the day, and is always better in the downwind propagation direction. This is because at night the colder heavier air from the upper atmosphere which settles near the ground is unheated by the sun. The slower air near the ground traps sound waves since any wave propagating upward has the upper part of the wave outrunning the lower part, thus refracting the wave back down to the ground. This effect is easily heard on a clear night after a warm sunny day. Its a cool night in August when I'm writing this section. I can clearly hear trucks shifting gears on an interstate highway over 5 km away which would be impossible during a hot afternoon. However, the truck tire noise seems to fluctuate over periods of 10–15 sec. This is likely due to nocturnal turbulence from heavy parcels of air displacing warmer air near the ground. The same effect can be seen in the twinkle of lights from a distance at night. The same refractive effect happens when sound propagates downwind, since the wind speed increases with height and adds to the sound speed. However, like in the long-range short-wave radio broadcasts, the atmospheric turbulence due to wind and buoyancy, will cause acoustic propagation fluctuations due to stochastic variations in index of refraction.

Consider the case of acoustic propagation at night in the atmosphere. We will greatly simplify the propagation problem greatly by eliminating wind and splitting the atmosphere into two layers, the lower layer near the ground with a constant temperature, and an upper layer with a positive temperature gradient to represent the lighter warmer air which supports faster, downward refracting, sound propagation. Figure 21 shows graphically the direct and refracted sound rays where the direct ray is propagating in an air layer with constant sound speed, while the refracted ray propagates in an upper air layer where the air temperature (and sound speed) is increasing with increasing height. For our simplified model of a constant sound speed gradient, we can express the sound velocity profile as

$$c(z) = c_0 + z \frac{dc(z)}{dz} \tag{11.3.9}$$

where c_0 is the sound speed in the constant lower layer near the ground and z is the height in the upper layer. It is straightforward to show that for a linear gradient shape, the ray travels along the arc of a circle. The radius of this circle, R in Figure 21, is found by solving for $c(z) = 0$.

$$R = \frac{c_0}{\frac{dc(z)}{dz}} \tag{11.3.10}$$

For any length of direct ray x in Figure 21, there is a corresponding refracted ray s (it could be more than one ray for more complicated propagation) which intersects both the source and receiver at a "launch angle" of $\theta/2$.

$$\frac{\theta}{2} = \sin^{-1}\left(\frac{x}{2R}\right) \tag{11.3.11}$$

The refracted ray length is simply

$$s = \theta R \tag{11.3.12}$$

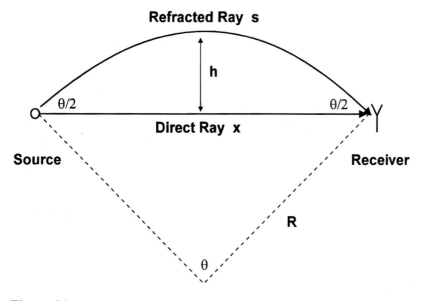

Figure 21 Depiction of downward refractive outdoor sound propagation showing a straight direct ray in a constant sound speed layer and a refracted ray in positive sound speed gradient.

and the ray maximum height is found by solving

$$h = R\left[1 \pm \sqrt{1 - \frac{1}{4}\left(\frac{x}{R}\right)} \right] \tag{11.3.13}$$

and can be useful in practical applications to see whether noise barriers of buildings effective block the sound path. This is one reason why noise barriers are really only helpful in blocking noise in the immediate vicinity of the barrier and not at long distances. As a practical example, if $c_0 = 345$ m/sec, $dc/dz = +0.1$ m/sec/m, and $x = 1$ km, we find $R = 3450$ m, $\theta/2 = 8.333°$, $h = 36.4$ m, and $s = 1003.5$ m, giving a path length difference for the two rays of only 3.5 m. These are very small differences and launch angles. However, when we consider a fluctuation in the sound speed gradient due to turbulence, a significant change in path length is possible.

Using a Taylor series approximation for the arc sine function, we can express a function for the refracted ray length as

$$s = \frac{2c_0}{\frac{dc}{dz}}\left\{ \frac{x}{2R}\left(1 + \frac{1}{6}\left(\frac{x}{2R}\right)^2 \right) \right\} \tag{11.3.14}$$

which can be reduced to show a path length difference of

$$s - x = \frac{x^3}{24c_0^2}\left(\frac{dc}{dz} + \zeta \right)^2 \tag{11.3.15}$$

where ζ is introduced as a random variable to account for fluctuations in the sound speed gradient. For the above practical example, Eq. (11.3.15) provides an estimated path difference of $s - x = 3.5 + 70\zeta + 350\zeta^2$ m. Clearly, a very small fluctuation in sound speed gradient leads to a very significant refracted ray path length.

As with the coherent reflection from the boundary, the sound pressure fluctuation scales with frequency since the phase difference between the direct and refracted path is the wavenumber times the path difference in meters. This means that low frequencies are much less impacted than high frequencies for a given overall distance. If we express the effect in terms of wavelengths, we can say that the pressure fluctuations will become severe when the path lengths exceed about a quarter wavelength. For small sound speed gradients this might require over a 1000 wavelengths propagation distance. But for larger sound speed gradients due to flow near a boundary, the effect may be seen over only a few dozen wavelengths. This technique of stochastically characterizing multipath could prove useful in medical imaging of arteries, chemical process sensors and control, and advanced sensor tasks such as pipe corrosion and lagging inspection by sensing the structure of the turbulent boundary layer. It may also be valuable as a signal processing tool for environmental sensing in the atmosphere and undersea, or for electromagnetic propagation channel studies.

11.4   SUMMARY, PROBLEMS, AND BIBLIOGRAPHY

Chapter 11 covers the important technology of a signal detection system, its design, and most importantly, the confidence of the detected signal based on signal and noise statistical models and direct measurements. The notion of *statistical confidence* is critical to an intelligent sensor system because it puts the detected information into proper context. When combined with other pieces of detected information, the confidence associated with each detection is extremely valuable to the process of *data fusion*, or the blending of pieces of information together to produce a *situational awareness*. Since we can robustly define the probability of signal detection P_d, and probability of false alarm (false detection) P_{fa}, for a particular signal in noise, it is prudent to compute these quantities even if they are not readily required for the detection process.

A *sentient* adaptive signal processing system is defined by having the power of perception by the senses, which includes a capability to combine and place weight on particular sensor information as well as assess the validity of a particular sensor's output. Our bodies provide a straightforward analogy. If one is sunburned on the face and arms, one resolves the local skin temperature differences from an awareness that: (1) there has been recent exposure to sunlight or heat increasing the likelihood of sensor damage; (2) the temperature differences are independent of body position or place (sensors are uncalibrated relative to one another); and (3) actual skin temperature in the exposed areas is elevated and appears to be more red (a symptom of damage). One realizes and deals with the situation of being sunburned by reducing confidence in temperatures from the burned areas, supporting healing of the burned skin, and avoiding any more exposure or damage. Our bodies are temperature regulated and thus, feeling too hot or too cold can sometimes be a very good indicator of pathology. Does one feel hot yet objects are cool to the touch? Does one feel cold yet objects do not feel cold? When feeling cold does the body perspire? One resolves these sensory contradictions by association with a known state of health. When the sensors are in agreement and match a pattern with other sensor types, one associates a known environmental state from experience and also that one's own sensors are operating correctly. In this self-perceived "healthy" state, one associates a very high confidence in the sensed information. This sentient process can be constructed, albeit crudely, in machines through the use of statistical measures of confidence and data and information fusion.

Section 11.1 presented the Rician probability distribution for a sinusoid in white noise along with the virtues of time synchronous averaging. Time is the one commodity which can be assumed to be known with great precision if needed. By averaging multiple buffers of the waveform which includes a known signal and random zero-mean noise over a period which is an integer multiple of the signal period, the noise in the averaged buffer tends toward its mean of zero while the signal is unaffected. *Time-synchronous averaging is one of the most simple and effective methods for signal-to-noise improvement in signal processing and should be exploited wherever possible.* When the signal is a sinusoid, the matched detection filter is also a sinusoid of the same frequency. When the signal has some bandwidth, the matched filter for detection also has the same spectrum as the signal when the background noise is white (see the example in Section 5.4 for non-white noise). Given an estimate of the background noise and the matched filter signal detection output, one can set a detection threshold above the mean background noise for a desirable low false alarm rate. The probability of detection, and thus signal detection confidence, can be estimated directly from the signal-to-noise ratio, as shown in Section 11.2. When the signal is really strong compared to the noise, the signal plus noise probability density function is essentially a Gaussian distribution with mean shifted up to the signal rms level. When the signal-to-noise ratio is closer to unity, the density function is Rician requiring greater care in estimating the probability of detection.

Section 11.3 examines the case where the signal propagation path can be modeled as a stochastic process. Such is the case with ground clutter in radar systems, scattering from tissue in medical ultrasound, and the twinkling of stars in astronomical observations due to turbulence multipath. In the situation of statistical multipath, the signal level is described by a non-zero mean probability density function. This impacts directly the calculation of the probability of detection because of the broadening of the density function (due to the convolution of the multipath density function with the signal-plus-noise density function). Given the physical distribution of scatterers we have shown how to calculate the signal probability density function. This is of course a very complicated task beyond the framework presented here for most real-world propagation situations.

Even more vexing, is the impact of time scale on the ergodicity of the multipath signal. Depending on the physics of the situation, the statistical distribution requires a certain amount of observation time (or distance for spatial scales) in order to manifest itself through calculation of a histogram to measure the probability density function. These time (or space) scales are extremely important in the proper assessment of detection confidence. Since our discussion continues towards the description of a sentient processor which depends on measurements of sensor signal confidence, it would appear prudent that direct statistical measures, such as histograms, should be used to validate the stochastic assumptions about the signal and noise.

PROBLEMS

1. A 1-sec real analog recording of a 1 vrms 50 Hz sinusoid signal in 1 vrms Gaussian white noise is available for digital processing using a spectrum analyzer.

   (a) If we low-pass filter at 400 Hz, sample the recording at 1024 samples/sec, and calculate a single 1024-point FFT, what is the spectral signal to noise ratio?

   (b) If we low pass filter at 3.2 kHz, sample the signal at 8192 samples/sec, and calculate a single 8192 point FFT, what is the spectral signal to noise ratio?

2. A seismic array geophones (measure surface velocity) for detecting rock slides needs to have no more than one false alarm per month on detection trial every 50 msec. Assume a signal magnitude detector in zero-mean Gaussian background noise with standard deviation of 1×10^{-7} m/s. What is the false alarm rate in % and how would you determine the detection threshold?

3. Derive the Rician probability density function. Show that for very high SNR, the Rician probability density function can be approximated by a Gaussian density function with mean equal to the signal level.

4. Show that the integral of the Rayleigh probability density function is proportional to a Gaussian function. Describe a detection system where this relationship would be very convenient.

5. Show that for M-averages of a Gaussian random variable, the mean stays the same while the standard deviation decreases by a factor of $M^{1/2}$.

6. Suppose one has a large number of complex FFT spectral buffers (of the same size and resolution) of a sinusoid in white noise.

   (a) If one simply added all the FFT buffers together into a single buffer, would one expect an increase in SNR?

   (b) If one first multiplied each FFT buffer by a complex number to make the phases at the frequency bin of the sinusoid identical and then added the buffers together, would the SNR increase?

   (c) If one first multiplied each FFT buffer by a complex number coresponding to the linear phase shift due to the time the FFT input buffer was recorded and then added the buffers together, would the SNR increase?

7. If one has a 1% false alarm rate on a magnitude detector of a signal peak in Gaussian noise with no averaging, how much averaging will reduce the false alarm rate to below 0.1% keeping the absolute detection threshold the same?

8. Describe qualitatively how one could model the detection statistics of a sinusoid in white Gaussian zero-mean noise where a reflected propagation path fluctuates with probability density $p(r)$ while the direct path fluctuates with density function $p(x)$. Neither density function is Gaussian, but histograms are available to numerically describe the path statistics.

BIBLIOGRAPHY

M. Abramowitz, I. A. Stegun Handbook of Mathematical Functions, Chapter 26, New York: Dover, 1972.

W. S. Burdic Underwater Acoustic System Analysis, 2nd Ed., Englewood Cliffs: Prentice-Hall, 1991.

G. R. Cooper and C. D. McGillem Probabilistic Methods of Signal and System Analysis, New York: HRW, 1971.

A. Papoulus Signal Analysis, Chapters 9–11, New York: McGraw-Hill, 1977.

W. H. Press, B. P. Flannery, S. A. Teukolsky, W. T. Vetterling Numerical Recipes: The Art of Scientific Computing, Chapter 7, Cambridge: Cambridge Univ. Press, 1986.

K. Sam Shanmugam Digital and Analog Communication Systems, Chapter 3, New York: Wiley, 1979.

H. L. Van Trees Detection, Estimation, and Modulation Theory, New York: Wiley, 1968.

D. Zwillinger, (ed.), Standard Mathematical Tables and Formulae, Chapter 7, New York: CRC Press, 1996.

REFERENCES

1. D. K. Wilson "Performance bounds for acoustic direction-of-arrival arrays operating in the atmosphere," Journal of The Acoustical Society of America, 103(3), March, 1998.

12

Wavenumber and Bearing Estimation

In this chapter we examine the fundamental techniques of measuring the spatial aspects of waves which can be propagating in a reflection free space from a distant source, reverberating in a confined space, or represent the complicated radiation in the nearfield of one or more sources. The types of wave of interest could be either mechanical (seismic, structural vibration, etc.), acoustic (waves in fluids), or electromagnetic. When the source of the waves is distant from the receiver array we can say that the wavefront is planar and the receiving array of sensors can estimate the direction of arrival, or bearing. This technique is fundamental to all passive and active sonar and radar systems for measuring the direction to a distant target either from its radiated waves or its reflections of the actively-transmitted sonar or radar wave. However, when more than one target is radiating the same frequency, the arriving waves at the receiver array can come from multiple directions at a given frequency.

To resolve multiple directions of arrival at the same frequency, the receiving array can process the data using a technique commonly known as beamforming. Beamforming is really an application of spatial wavenumber filtering (see Section 7.1). The waves from different directions represent different sampled wavelengths at the array sensor locations. An array beampattern steered in a particular "look" direction corresponds to a wavenumber filter which will pass the corresponding wavenumber to the look direction while attenuating all other wavenumbers. The "beam" notion follows from the analogy to a search light beam formed by a parabolic reflector or lens apparatus. The array beam can be "steered" electronically, and with parallel array processors, multiple beams can be formed and steered simultaneously, all without any mechanical systems to physically turn the array in the look direction. Electronic beam steering is obviously very useful, fast, and the lack of mechanical complexity is very robust. Also, electronic beamforming and steering allows multiple beams each in different look directions to exist simultaneously on the same sensor array.

The array can also be "focused" to a point in its immediate vicinity rather than a distant source. This application is fairly novel and useful, yet the technology simply involves the derivation of different wavenumber filters from the classic beamforming problem. We could simply refer to array nearfield focusing as "spherical beamforming" since we are filtering spherical, rather than planar wavenumbers. But, a more descriptive term would be holographic beamforming because the array sensor spatial sampling of the field for a source in the vicinity of the array can allow reconstruction of the wave field from measurements of both the propagating and non-propagating (evanescent) wavenumbers. Holographic beamforming implies measurement and reconstruction of the full three-dimensional field from scanning a surface around the source with the array. Analysis of the observed wavenumbers using wavenumber filtering is of great interest in the investigation of how a source of interest is radiating wave energy. For example, changes in the observed electromagnetic fields of a motor, generator, or electronic component could be used to pinpoint a pending problem from corrosion, circuit breakdown, or component wear out.

Wavenumber processing for fields in confined spaces is generally known in structural acoustics as modal filtering. The vibration field of a bounded space can be solved analytically in terms of a weighted sum of vibration modes, or structural resonances. Each structural resonance has a frequency and associated mode shape. When the structure is excited at a point with vibration (even a single frequency) all of the structural modes are excited to some extent. Therefore, in theory, a complete analysis of the structural response should allow one to both locate the source and filter out all the structural "reverberation", or standing wave fields. If there are changes in the structural integrity (say from corrosion or fatigue), changes in structural stiffness should be observable as changes in the mode shapes and frequencies. This should also be true for using microwaves to investigate corrosion or fatigue in metal structures. In theory, acoustical responses of rooms could be used by robotic vehicles to navigate interior spaces excited by known sources and waveforms. While these application ideas are futuristic, the reader should consider that they are all simply applications of wavenumber filtering for various geometries and wave types.

Section 12.1 presents the Cramer–Rao lower bound for parameter estimation. This general result applies not only to beamforming estimates, but actually any parameter estimate where one can describe the observable in terms of a probability density function. This important technique spans the probabilistic models of Chapter 11 and the adaptive filtering models in Chapter 8 and can be applied to any parameter estimate. For our immediate purposes, we present the Cramer–Rao lower bound for bearing estimates. This has been well-developed in the literature and is very useful as a performance measure for beamforming algorithms. In Section 12.2 we examine precision bearing estimation by array phase directly or as a "split-beam". In the split-beam algorithm, the array produces two beams steered close together, but not at exactly the same look direction. By applying a phase difference between the two beams, the pair can be "steered" precisely to put the target exactly in between the beams, thus allowing a precision bearing estimate. Section 12.3 presents the holographic beamforming technique and shows application in the analysis of acoustic fields, although this could be applied to any wave field of interest.

12.1 THE CRAMER–RAO LOWER BOUND

The Cramer–Rao lower bound (CRLB) (1,2) is a statistically-based parameter estimate measure which provides a basis for stating the best possible accuracy for a given parameter estimate based on the statistics of the observables and the number of observables used in the estimate. As will be seen below, the CRLB is very closely related to the least-squared error of a parameter estimate. The main difference between the CRLB and the least-squared error of a linear parameter estimate is that the CRLB represents the predictability of an estimate of a function's statistical value based on N observations. For example, one starts with a probability density model for the function of interest, say the bearing angle measured by a linear array of sensors. The array processing algorithm produces a time-difference of arrival, (or phase difference for a narrowband frequency), between various sensors which has a mean and a variance. If there are $N+1$ sensors, we has N observations of this time or phase difference. Because the signal-to-noise ratio is not infinite, the time delay or phase estimates come from a well-defined probability density function (see Section 11.1). The derivation of an angle-of-arrival, or bearing, requires translation of the probability density function from the raw sensor measurements, but this is also straightforward, albeit a bit tedious. With N statistical observations of the bearing for a given time interval from the array, we seek the mean bearing as the array output, and use the CRLB to estimate the minimum expected standard deviation of our mean bearing estimate. The CRLB provides an important measure of the expected accuracy of a parameter estimate. The derivation of the CRLB is quite interesting and also contains some rather innovative thinking on how to apply statistics to signal processing.

Consider a vector of N scalar observations, where each observation is from a normal probability distribution with mean m and variance σ^2.

$$Y = [y_1 y_2 \ldots y_N] \quad p(y_i) = \frac{1}{\sigma\sqrt{2\pi}} e^{-\frac{(y_i-m)^2}{2\sigma^2}} \quad i = 1, 2, \ldots, N \qquad (12.1.1)$$

We designate the parameter vector of interest to be $\lambda = [m\sigma^2]$ and the joint probability density function of the N observations to be

$$p(Y, \lambda) = \frac{1}{\sqrt{(2\pi\sigma^2)^N}} e^{-\frac{1}{2\sigma^2}\sum_{i=1}^{N}(y_i-m)^2} \qquad (12.1.2)$$

Suppose we have some arbitrary function $F(Y,\lambda)$, for example the bearing, for which we are interested in estimating the mean. Recall that the first moment is calculated as

$$m_F = E[F(Y, \lambda)] = \int_{-\infty}^{+\infty} p_F(Y, \lambda)F(Y, \lambda)dY \qquad (12.1.3)$$

For the statistical models of the observables and our arbitrary function described in Eqs (12.1.1)–(12.1.3), we will be interested in the gradient of m_F with respect to the parameter vector λ as well as the second derivative. This is because we are constructing a linear estimator which should have a linear parameter error. Following Chapter 8 we note that the error squared will be quadratic where the least-squared error will be the parameter values where the gradient is zero. The gradient of the expected value of F is

$$\frac{\partial}{\partial \lambda} E[F] = E\left[\frac{\partial F}{\partial \lambda}\right] + E[F\psi] \tag{12.1.4}$$

where

$$\psi(Y, \lambda) \doteq \frac{\partial \ln p_F}{\partial \lambda} = \frac{1}{p_F}\frac{\partial p_F}{\partial \lambda} \tag{12.1.5}$$

is the gradient of the log-likelihood function for the arbitrary function $F(Y,\lambda)$. The second derivative of Eq. (12.1.5) has an interesting relationship with the first derivative.

$$\begin{aligned} \frac{\partial \psi}{\partial \lambda} &= \frac{\partial^2 \ln p_F}{\partial \lambda^2} \\ &= \frac{\partial}{\partial \lambda}\left(\frac{1}{p_F}\frac{\partial p_F}{\partial \lambda}\right) \\ &= \frac{-1}{p_F^2}\frac{\partial^2 p_F}{\partial \lambda^2} \\ &= -\psi^2 = -\psi\psi^T \end{aligned} \tag{12.1.6}$$

Proof of Eq. (12.1.4) follows from a simple application of the chain rule.

$$\begin{aligned} \frac{\partial}{\partial \lambda} E[F] &= \frac{\partial}{\partial \lambda}\left[\int p_F F dY\right] = \int \left\{p_F \frac{\partial F}{\partial \lambda} + F\frac{\partial p_F}{\partial \lambda}\right\}dY \\ &= \int p_F \frac{\partial F}{\partial \lambda} dY + \int p_F\left(F\frac{\partial \ln p_F}{\partial \lambda}\right)dY \\ &= E\left[\frac{\partial F}{\partial \lambda}\right] + E[F\psi] \end{aligned} \tag{12.1.7}$$

Eq. (12.1.4) and its proof in (12.1.7) shows the intuitive nature of using the gradient of the log-likelihood function for our arbitrary function F. Since F is functionally a constant with respect to λ, it can be seen that for $F=1$, $E[\psi]=0$. For $F=\psi$ we obtain another important relation.

$$\begin{aligned} \frac{\partial}{\partial \lambda} E[\psi] &= \frac{\partial}{\partial \lambda}\left[\int p_F \psi dY\right] \\ 0 &= \int \left\{p_F \frac{\partial \psi}{\partial \lambda} + \psi\frac{\partial p_F}{\partial \lambda}\right\}dY \\ 0 &= \int p_F \frac{\partial \psi}{\partial \lambda} dY + \int p_F\left(\psi\frac{\partial \ln p_F}{\partial \lambda}\right)dY \\ 0 &= E\left[\frac{\partial \psi}{\partial \lambda}\right] + E[\psi\psi^T] \end{aligned} \tag{12.1.8}$$

Therefore,

$$-E\left[\frac{\partial \psi}{\partial \lambda}\right] = E[\psi \psi^T] = J \qquad (12.1.9)$$

where J in Eq. (12.1.9) is referred to as the *Fisher Information Matrix*.

We note that if the slope of the probability density function is very high in magnitude near the mean, there is not much "randomness" to the distribution. This corresponds to the elements of J being large in magnitude, and the norm of the Fisher information matrix to be large, hence, the observations contain significant information. Conversely, a broad probability density function corresponds to relatively low information in the observations. The elements of J are defined as

$$J_{ij} = E\left[\frac{-\partial^2 \ln p_F}{\partial \lambda_i \partial \lambda_j}\right] \qquad (12.1.10)$$

The reason we took the effort to derive Eq. (12.1.4) and (12.1.9) is we need to evaluate the statistics of a parameter estimate $\hat{\lambda}(Y)$ for the parameter $\lambda(Y)$. Since $\lambda(Y)$ and $\psi(Y)$ are correlated, we can write a linear parameter estimation error as

$$e(Y) = \lambda(Y) - \beta \psi(Y) \qquad (12.1.11)$$

Recall from Section 8.1 that minimization of the squared error is found by setting the gradient of the squared error with respect to β to zero and solving for β.

$$E[\beta] = E[\lambda \psi^T] E[\psi \psi^T]^{-1} \qquad (12.1.12)$$

making the parameter estimation error

$$e = \lambda - E[\lambda \psi] E[\psi \psi^T]^{-1} \psi \qquad (12.1.13)$$

We note that the error in Eq. (12.1.13) is uncorrelated with ψ, and since $E[\psi] = 0$, $E[e] = E[\lambda]$. Since we are interested in the variation between the actual parameter value and its expected value, we define the following two variational parameters.

$$\Delta \lambda = \lambda - E[\lambda] \qquad (12.1.14)$$

$$\Delta e = e - E[e] \qquad (12.1.15)$$

Note that subtracting a constant (the expected values λ) does not affect the cross-correlation

$$M = E[\lambda \psi^T] = E[\Delta \lambda \psi^T] \qquad (12.1.16)$$

where M in Eq. (12.1.16) is known as the *bias of the parameter estimation* and equals unity for an unbiased estimator. This will be described in more detail below. The variational error is therefore

$$\Delta e = \Delta \lambda - E[\lambda \psi^T] E[\psi \psi^T]^{-1} \psi$$
$$= \Delta \lambda - M J^{-1} \psi \qquad (12.1.17)$$

The expected value of the variance of the parameter estimation variational error is

$$
\begin{aligned}
E[\Delta e \Delta e^T] &= E\big[(\Delta\lambda - MJ^{-1}\psi)(\Delta\lambda^T - \psi^T J^{-1} M^T)\big] \\
&= E[\Delta\lambda\Delta\lambda^T] - MJ^{-1}E[\psi\Delta\lambda^T] - E[\Delta\lambda\psi^T]J^{-1}M^T + MJ^{-1}\psi\psi^T J^{-1} M^T \\
&= E[\Delta\lambda\Delta\lambda^T] - MJ^{-1}M^T - MJ^{-1}M^T + MJ^{-1}M^T \\
&= E[\Delta\lambda\Delta\lambda^T] - MJ^{-1}M^T
\end{aligned}
$$

$$(12.1.18)$$

Since $E[\Delta e \Delta e^T] \geq 0$ we can write a lower bound on the variance of our parameter estimate.

$$
\sigma^2(\hat{\lambda}) \doteq E[\Delta\lambda\Delta\lambda^T] \geq MJ^{-1}M^T \tag{12.1.19}
$$

Equation (12.1.19) is the result we have been looking for — a measure of the variance of our parameter estimate based on the statistics of the observables, with the exception of the bias term.

$$
\begin{aligned}
M = E[\lambda'\psi^T] &= \frac{\partial}{\partial\lambda}E[\lambda'] - E\left[\frac{\partial\lambda'}{\partial\lambda}\right] \\
&= I \quad \textit{for the unbiased case}
\end{aligned}
\tag{12.1.20}
$$

The bias in Eq. (12.1.20) is unity for the case where λ' has no explicit dependence on λ, making the partial derivative in the rightmost term zero. An unbiased parameter estimate will converge to the true parameter given an infinite number of observables. A biased estimate will not only converge to a value offset from the true parameter value, but the bias will also affect the variance of the parameter estimate depicted in Eq. (12.1.19). For the unbiased case

$$
\sigma^2(\hat{\lambda}) \doteq E[\Delta\lambda\Delta\lambda^T] \geq J^{-1} \tag{12.1.21}
$$

and in terms of N observations of the scalar probability density function in Eq. (12.1.1),

$$
\sigma^2(\hat{\lambda}) = \frac{-1}{NE\left[\dfrac{\partial^2}{\partial\lambda^2}\ln p(\lambda)\right]} = \frac{-1}{NE\left[\left(\dfrac{\partial^2}{\partial\lambda^2}\ln p(\lambda)\right)^2\right]} \tag{12.1.22}
$$

Equation (12.1.22) provides a simple way to estimate the CRLB for many parameter estimates. When the bias is unity, the estimator is called *efficient* because it meets the Cramer–Rao lower bound.

Consider an example of N observations of a Gaussian process described as a joint N-dimensional Gaussian probability density.

$$
p(Y, \lambda) = \frac{1}{\sqrt{(2\pi\sigma^2)^N}}\, e^{-\frac{1}{2\sigma^2}\sum\limits_{i=1}^{N}(y_i - m)^2} \tag{12.1.23}
$$

Our parameter vector is $\lambda = [m\ \sigma]^T$. To find the CRLB we first take the log of $p(Y, \lambda)$

$$\ln p(Y, \lambda) = -\frac{N}{2}\ln(2\pi) - N\ln\sigma - \frac{1}{2\sigma^2}\sum_{i=1}^{N}(y_i - M)^2 \quad (12.1.24)$$

and then differentiate

$$\psi(Y, \lambda) = \frac{\partial}{\partial\lambda}\ln p(Y, \lambda) = \begin{bmatrix} \dfrac{\partial \ln p}{\partial m} \\[2mm] \dfrac{\partial \ln p}{\partial(\sigma^2)} \end{bmatrix} = \begin{bmatrix} \dfrac{1}{\sigma^2}\sum_{i=1}^{N}(y_i - m) \\[4mm] -\dfrac{N}{2\sigma^2} + \dfrac{1}{2\sigma^4}\sum_{i=1}^{N}(y_i - m)^2 \end{bmatrix} \quad (12.1.25)$$

Differentiating again yields the elements of the matrix Ψ

$$\Psi(Y, \lambda) = \frac{\partial}{\partial\lambda}\ln p(Y, \lambda) = \begin{bmatrix} \dfrac{\partial \ln p}{\partial m \partial m} & \dfrac{\partial^2 \ln p}{\partial m \partial(\sigma^2)} \\[3mm] \dfrac{\partial^2 \ln p}{\partial m \partial(\sigma^2)} & \dfrac{\partial^2 \ln p}{\partial(\sigma^2)\partial(\sigma^2)} \end{bmatrix}$$

$$= \begin{bmatrix} \dfrac{N}{\sigma^2} & \dfrac{1}{\sigma^4}\sum_{i=1}^{N}(y_i - m) \\[4mm] \dfrac{1}{\sigma^4}\sum_{i=1}^{N}(y_i - m)^2 & -\dfrac{N}{2\sigma^4} + \dfrac{1}{\sigma^6}\sum_{i=1}^{N}(y_i - m)^2 \end{bmatrix}$$

$$(12.1.26)$$

and taking expected values gives the Fisher information matrix.

$$J = E\{\Psi(Y, \lambda)\} = \begin{bmatrix} \dfrac{N}{\sigma^2} & 0 \\[3mm] 0 & \dfrac{N}{2\sigma^4} \end{bmatrix} \quad (12.1.27)$$

The CRLB for an unbiased estimate of the mean and variance is therefore

$$\sigma^2(\hat{\lambda}) = \begin{bmatrix} E\{\Delta m \Delta m\} & E\{\Delta m \Delta \sigma^2\} \\ E\{\Delta m \Delta \sigma^2\} & E\{\Delta \sigma^2 \Delta \sigma^2\} \end{bmatrix} \geq \begin{bmatrix} \dfrac{N}{\sigma^2} & 0 \\[3mm] 0 & \dfrac{N}{2\sigma^4} \end{bmatrix} \quad (12.1.28)$$

An even more practical description can be seen if we consider a Gaussian distribution with say mean 25 and variance 9. How many observations are needed for an unbiased estimator to provide estimates within 1% of the actual values 63% of the time (i.e. 1 standard deviation of the estimate is 1% of its value)?

Solution: We note that the variance is 9 and that the variance of the mean estimate is $9/N$. The standard deviation of the mean estimate is $3/\sqrt{N}$. Therefore, to get a mean estimate where the standard deviation is 0.25 or less, N must be greater than 144 observations. To get a variance estimate with standard deviation 0.09, N must be over 1.6 million observations. To get a variance estimate where the variance of the variance estimate is 0.09, N must be greater than about 145,800 observations.

12.2   BEARING ESTIMATION AND BEAMSTEERING

In this section, we apply the technique of establishing the CRLB for a statistical representation of wavefront bearing, or the observed wavenumber by an array of sensors. This is presented both in terms of a direct bearing measurement for single arrival angles and later in this section by way of array beamforming and beam steering to determine source bearing. Direct bearing estimation using wavefront phase differences across an array is a mainstay process of passive narrowband sonar engineering, but also finds application in other areas of acoustics, phased array radar processing, as well as seismology and radio astronomy. By combining measured phase and/or time delay information from a sensor array with the array geometry and wave speed one can determine the direction of arrival of the wave from a single distant source. If more than one target is radiating the same frequency, or if propagation multipath exists, a beamforming and beam steering approach must be used to estimate the target bearings. This is the general passive sonar problem of determining the bearing of a plane wave passing the array to provide a direction to a distant target.

If the array size is quite large compared to the source distance, the array can actually be "focused" on the origin of a spherical wave radiating from the source allowing the source location to also be observed with some precision. When a complete surface enclosing the source(s) of interest is scanned by an array coherently, Gauss's theorem provides that the field can then be reconstructed on any other surface enclosing the same sources. The spherical and 3-dimensional field measurement representations will be left to the next section. This section will deal specifically with plane wave fields, which is always the case when the source is so far from the receiving array that the spherical wavefront observed by the array is essentially planar.

Lets begin by considering a simple 3-element array and a 2-dimensional bearing estimation problem for a single source and single sinusoid in white noise. To simplify our analysis even further, we place sensor 1 at the origin, sensor 2 is placed d units from the origin on the positive x-axis, and sensor 3 is placed d units from the origin on the positive y-axis. Figure 1 depicts the array configuration and the bearing of the plane wave of interest. For the plane wave arriving at the Cartesian-shaped array from the angle θ, one can write very simple expressions for the phase differences across the array of sensors

$$\Delta\phi_{21} = \phi_2 - \phi_1 = kd\cos\theta \tag{12.2.1}$$

$$\Delta\phi_{31} = \phi_3 - \phi_1 = kd\sin\theta \tag{12.2.2}$$

where ϕ_1, ϕ_2, and ϕ_3 are the phases of the particular FFT bin corresponding to the radian frequency ω and the wavenumber $k = \omega/c$, c being the wave propagation speed. The convenience of using a Cartesian-shaped array and the expressions for sine and cosine of the arrival angle are evident in the solution for θ in Eq. (12.2.3).

$$\theta = \tan^{-1}\left\{\frac{\Delta\phi_{31}}{\Delta\phi_{21}}\right\} \tag{12.2.3}$$

Note that for the Cartesian equal spaced array, the bearing angle is calculated independent of wave speed, frequency, and wavenumber. This can be particularly

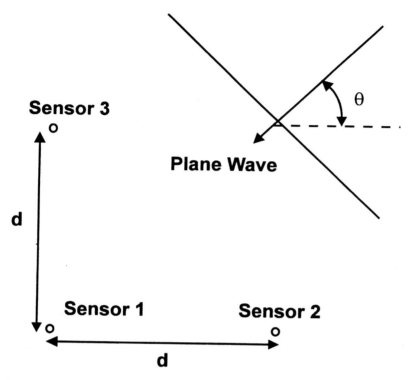

Figure 1 Array configuration for 2-dimensional phase difference estimation of bearing for a plane wave.

useful for dispersive waves, such as shear waves where high frequencies travel faster than low frequencies making bearing estimates from time delay estimation problematic. Clearly, one can also provide an estimate of bearing uncertainty given the probability density of the signal and noise in the FFT bin of interest. We will first generalize the bearing estimate to an arbitrary shaped array and then examine statistics of the bearing estimate.

Consider an arbitrary shaped planar array where each sensor position is defined by a distance and angle relative to the origin. Recall from problem 7.1 that

$$\Delta\phi_{jk} = \phi_j - \phi_k = \frac{\omega}{c}\{\cos\theta(r_j\cos\theta_j - r_k\cos\theta_k) + \sin\theta(r_j\sin\theta_j - r_k\sin\theta_k)\}$$

(12.2.4)

thus

$$\begin{bmatrix} \cos\theta \\ \sin\theta \end{bmatrix} = \begin{bmatrix} (r_3\cos\theta_3 - r_1\cos\theta_1) & (r_3\sin\theta_3 - r_1\sin\theta_1) \\ (r_2\cos\theta_2 - r_1\cos\theta_1) & (r_2\sin\theta_2 - r_1\sin\theta_1) \end{bmatrix}^{-1} \begin{bmatrix} \dfrac{\Delta\phi_{31}c}{\omega} \\ \dfrac{\Delta\phi_{21}c}{\omega} \end{bmatrix}$$

(12.2.5)

The inverted matrix in Eq. (12.2.5) contains terms associated with the position of the three sensors. It can be shown that the inverse exists if the three sensors define a plane. An arithmetic mean can be calculated for the sine and cosine of the arrival angle using a number of sensor pairings.

$$\begin{bmatrix} \cos\theta \\ \sin\theta \end{bmatrix} = \frac{1}{N^3 - 2N^2 + N} \sum_{i=1}^{N} \sum_{j=1}^{N} \sum_{k=1}^{N}$$

$$\times \begin{bmatrix} (r_i\cos\theta_i - r_j\cos\theta_j) & (r_i\sin\theta_i - r_j\sin\theta_j) \\ (r_k\cos\theta_k - r_j\cos\theta_j) & (r_k\sin\theta_k - r_j\sin\theta_j) \end{bmatrix}^{-1} \begin{bmatrix} \dfrac{\Delta\phi_{ij}c}{\omega} \\ \dfrac{\Delta\phi_{kj}c}{\omega} \end{bmatrix}$$

$$(12.2.6)$$

Equation (12.2.6) uses all possible pairings of sensors assuming all sensor pairs are separated by a distance less than one-half wavelength. For large arrays and relatively high wavenumbers (frequencies), this is not possible in general. However, averaging N-pairings which meet the requirement of less than a half-wavelength spacing will greatly reduce the variance of the bearing error.

To determine the CRLB for bearing error, we first make a systematic model for the statistics and then consider combining N observations for a bearing estimate. Consider the FFT bin complex number for each of the three sensors in Figure 1 for a particular frequency bin of interest. Figure 2 graphically shows how the stan-

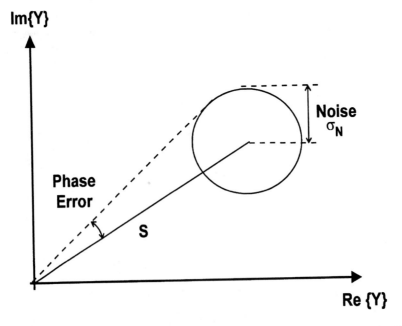

Figure 2 Graphical representation of the phase random error in a FFT bin due to background noise. The phase error can be seen as the arctangent of the inverse of the SNR.

dard deviation of the phase σ_ϕ, can be expressed as

$$\sigma_\phi = \tan^{-1}\left\{\frac{\sigma_N}{S}\right\} = \tan^{-1}\left\{\frac{1}{SNR}\right\} \tag{12.2.7}$$

where S is the signal amplitude and σ_N is the noise standard deviation in the FFT bin. While the envelope of the complex FFT bin probability density has been shown to be a Rician density function, the full complex density is actually a 2-dimensional Gaussian density where the mean is simply the complex number representing the true magnitude and phase of the signal.

The phase difference probability density results from the convolution of the two Gaussian densities for the real and imaginary part of the FFT bin. Therefore, the FFT phase variances for each sensor add to give the variance of the phase difference. Assuming the noise densities for the sensors are identical in a given FFT bin, the standard deviation for the phase difference is seen as

$$\sigma_{\Delta\phi} = \sqrt{2}\sigma_\phi \tag{12.2.8}$$

Normalizing the phase differences along the x and y axis by kd for our simple 3-sensor Cartesian shaped array we can consider the probability density functions for our estimates of the sine and cosine of the arrival angle θ for the plane wave. Figure 3 graphically shows the standard deviation for the bearing error.

Normalization by kd naturally scales the phase difference variance relative to the observable phase difference for the sensor separation. For example, suppose the standard deviation of the phase difference is 0.01 radians. If d/λ is smaller than

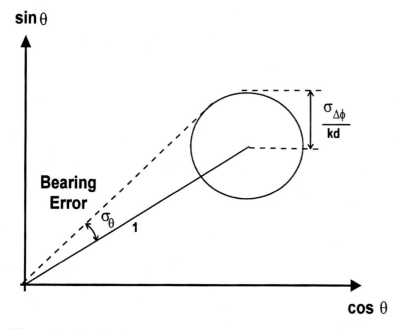

Figure 3 Graphical representation of the bearing random error given phase differences normalized by kd to provide representations for the sine and cosine of the bearing angle.

but close to 0.5, the random bearing error is quite small. But for a closer spacing, or lower frequency, d/λ might be 0.05, making the random bearing error considerably larger for the same phase difference variance. The standard deviation for the bearing error is therefore

$$\sigma_\theta = \tan^{-1}\left\{\frac{\sqrt{2}\sigma_\phi}{kd}\right\} = \tan^{-1}\left\{\frac{\sqrt{2}}{kd}\tan^{-1}\left\{\frac{1}{SNR}\right\}\right\} \tag{12.2.9}$$

If the SNR is large (say greater than 10), and $\sigma_\phi \ll kd$, the bearing error standard deviation is approximately

$$\sigma_\theta \approx \frac{\sqrt{2}}{SNR\,kd} = \frac{1}{\sqrt{2}\,\pi}\frac{1}{SNR}\frac{\lambda}{d} \tag{12.2.10}$$

which is in agreement with the literature. If we combined M sensor pairings with the same separation d as in Figure 1, the reduction in the CRLB is simply

$$\sigma_\theta \approx \frac{\lambda}{\pi\sqrt{2M}\,SNR\,d} \tag{12.2.11}$$

One can also average the expressions for sine and cosine in Eqs (12.2.6) using different frequencies and sensor spacings, but a corresponding SNR and wavelength/aperture weighting must be applied for each unique sensor pair and frequency bin. The expression for the CRLB tells us that to improve bearing accuracy one should increase SNR (integrate longer in time and/or space), choose sensor spacings near but less than $\lambda/2$ for the frequency, and combine as many sensor pairs and frequencies as is practical. When a wide range of frequencies are used in a non-white background noise, the Eckart filter (Section 5.4) can be applied to equalize SNR over a range of interest. This is equivalent to weighting various frequencies according to their SNR in the bearing estimate.

It can be shown that the CRLB for time delay estimation where the average of M observations is used is

$$\sigma_\tau \approx \frac{1}{\sqrt{M}\,SNR\,\beta^2} \tag{12.2.12}$$

where β is the bandwidth in Hz of the signal used in the time delay estimate. For estimation of Doppler frequency shift from a moving source based on the average of M observations, the CRLB can be shown to be

$$\sigma_{\Delta f} \approx \frac{1}{\sqrt{M}\,SNR\,T^2} \tag{12.2.13}$$

where T is the period of one observation in seconds. Time delay and frequency shift estimation are generally associated with target range and velocity along the bearing direction. However, it can be seen that for a given signal's time-bandwidth product and SNR, there are definable parameter estimation errors which cannot be less than the CRLB. Note that the SNR enhancement of an FFT defines a time-bandwidth product.

Suppose we have two or more sources at different bearing angles radiating the same frequency? What bearing would a direct phase-based bearing calculation predict? Clearly, the array would be exposed to a propagating wave field and an interference field from the multiple sources. If one were to decompose the waves from the sources into x and y axis components, summing the field of the sources results in the sum of various wavelengths on the x and y axis. Therefore, one can argue that along both the x and y axis, one has a linear combination of wave components from each source. For two sources of equal amplitude, phase, and distance from the array and bearings θ_a and θ_b the phase difference between two sensors separated by a distance d is

$$\frac{P_2(\omega)}{P_1(\omega)} = e^{j\Delta\phi_{21}} = e^{jkd\cos\theta_a} + e^{jkd\cos\theta_b} \tag{12.2.14}$$

Recalling that adding two sinusoids of different wavelengths gives an amplitude modulated signal where the "carrier wave" is the average of the two wave frequencies (or wavelengths) and the envelope is half the difference of the two wave frequencies.

$$\begin{aligned}\frac{P_2(\omega)}{P_1(\omega)} &= e^{j\frac{kd}{2}(\cos\theta_a+\cos\theta_b)}2\cos\left(\frac{kd}{2}[\cos\theta_a-\cos\theta_b]\right) \\ &= e^{j\frac{kd}{2}\cos\left(\frac{\theta_a+\theta_b}{2}\right)}\cos\left(\frac{\theta_a-\theta_b}{2}\right)2\cos\left(\frac{kd}{2}[\cos\theta_a-\cos\theta_b]\right)\end{aligned} \tag{12.2.15}$$

Equation (12.2.15) shows that for two arrival angles close together ($\theta_a \approx \theta_b$), the estimated bearing will be the average of the two arrival angles since the cosine of a small number is nearly unity. However, as the bearing difference increases, a very complicated angle of arrival results. When the two waves are of different amplitudes, the average and differences are weighted proportionately. Therefore, we can say with confidence that a direct bearing estimate from an array of sensors using only spatial phase will give a bearing estimate somewhere in between the sources and biased towards the source wave of higher amplitude. It would appear that measuring both the spatial phase and amplitude (interference envelope) should provide sufficient information to resolve the bearing angles. However, in practice the sources are both moving and not phase synchronized making the envelop field highly non-stationary and not linked physically to the direction of arrivals for the sources. The only practical way to resolve multiple source directions at the same frequency is to apply array beamforming and to steer the beam around in a search pattern for sources.

Using an array of sensors together to produce a beam-shaped directivity pattern, or beampattern, simply requires that all or a number of sensor output signals be linearly filtered with a magnitude and phase at the frequency of interest such that the sensor response to a plane wave from a particular bearing angle produces a filter output for each sensor channel that has the same phase. This phase coherent output from the sensor array has the nice property of very high signal output for the wave from the designed look-direction angle, and relatively low incoherent output from other directions. Thus, the sensor array behaves somewhat like a parabolic dish reflector. However, the array beamforming can "steer" the beam with no moving parts simply by changing the magnitude and phase for each frequency of interest to produce a summed output which is completely coherent in the new look direction. Even more useful is constructing beams which have zero output in the direction(s) of unwanted sources. Array "null-forming" is done adaptively, and several methods for adaptive beamforming will be discussed in Chapter 13.

Consider the beam pattern response for a simple theoretical line sensor as seen in Figure 4. The distant

$$D(\theta) = \int_{x=-L/2}^{x=L/2} e^{jkx\sin\theta}dx = \frac{e^{jkx\sin\theta}}{jk\sin\theta}\bigg|_{x=-L/2}^{L/2} = L\frac{\sin\left(\frac{kL}{2}\sin\theta\right)}{\frac{kL}{2}\sin\theta} \qquad (12.2.16)$$

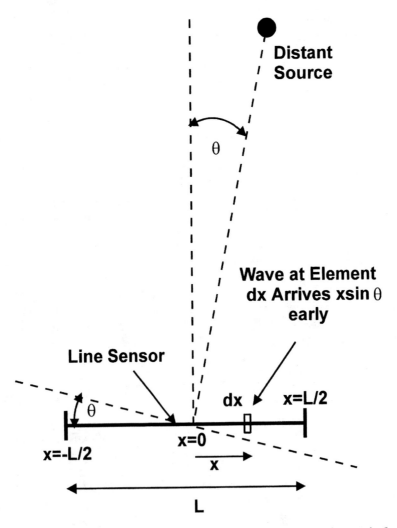

Figure 4 Line sensor configuration showing length L, bearing angle θ, and response at differential element dx where the wave arrives early compared to the line center at $x = 0$.

source essentially radiates a plane wave across the line sensor where, relative to the origin, we have early arrivals on the right and late arrivals on the left. If we sum all the differential elements along the line sensor, we can write the spatial response of the line sensor, relative to a single point source as seen in Eq. (12.2.16).

For an example using acoustic waves in air, $c = 350$ m/sec, $L = 2$ m, $f = 300$ Hz, and $k = 5.385$ m^{-1}. The beam response is seen in Figure 5 and in Figure 6 in polar form. Electronically steering the beam to a look direction θ' requires adjusting the phase at 300 Hz for each element so that the line sensor response at θ' is coherent.

$$D(\theta) = \int_{x=-L/2}^{x=L/2} e^{jkx\sin\theta} e^{-jkx\sin\theta'} dx = L\frac{\sin\left(\frac{kL}{2}[\sin\theta - \sin\theta']\right)}{\frac{kL}{2}[\sin\theta - \sin\theta']} \tag{12.2.17}$$

Figure 7 shows the 30 degree steered beam response in polar form. Note how the southern lobe also moves East by 30 degrees. This is because of the symmetry of the line sensor which is oriented along a East–West line. The reason the "South" beam also moves around to the East is that a line array cannot determine which side of the line the sources is on. In 3-dimensions, the beam response of the line array is a hollow cone shape which becomes a disk shape when no steering is applied.

We can explain beam forming in a much more interesting way using spatial Fourier transforms such as what was done in Section 7.2 for images to show high and low pass filtering. Consider a 256 point by 256 point spatial grid representing 32 m by 32 m of physical space. Placing our 2 m line sensor in the center of this space at row 128, we have 16 "pixels," each numerically unity, extending from $\langle x, y \rangle$ coordinate $\langle 56, 128 \rangle$ to $\langle 72, 128 \rangle$ representing the line sensor. Figure 8 shows

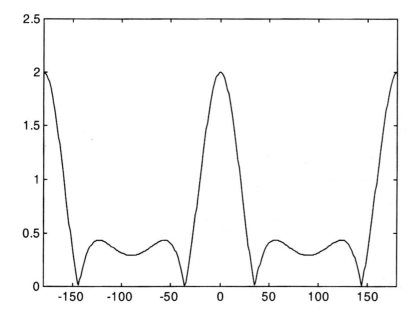

Figure 5 Beam response for a 2 m line sensor at 300 Hz in air (linear output vs. bearing degrees scale).

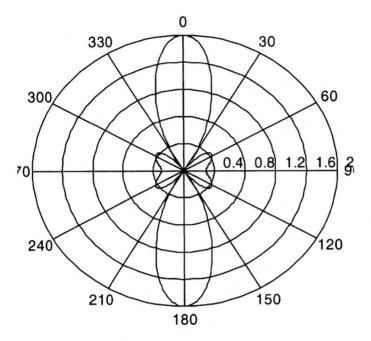

Figure 6   Polar plot of the 2 m line sensor showing the North–South beams expected to be symmetric about the horizontal axis of the line sensor (linear output vs. degrees).

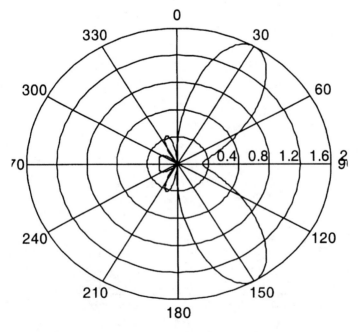

Figure 7   Beam response at 300 Hz for 2 m line sensor steered to 30 degrees East (linear output vs. degrees).

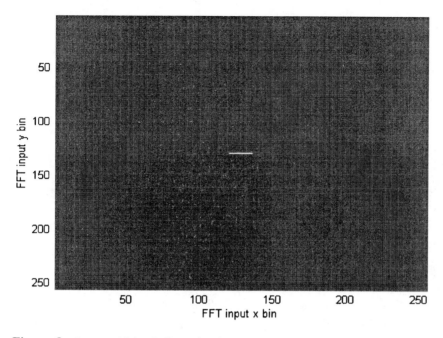

Figure 8 Input grid for 2-dimensional spatial FFT to represent line sensor response.

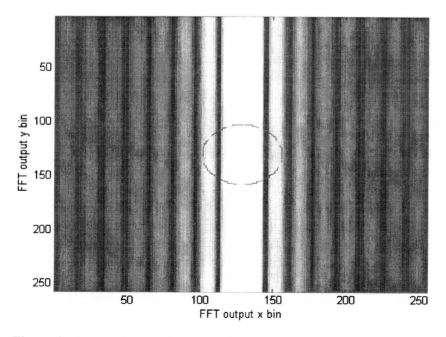

Figure 9 2-dimensional FFT of the 2 m line sensor showing wavenumbers from -8π to $+\pi$ where the spatial sample rate is 256/32 m or 8 samples/meter.

a place of our input Fourier space. Figure 9 shows the magnitude of a 2-dimensional FFT of the spatial array data with a dashed circle centered on the wavenumber space origin representing $k = 5.385$ m^{-1}. If we sample the wavenumber response along the $k = 5.385$ circle, we get the response shown in Figure 5. For frequencies below 300 Hz, the wavenumber response circle is smaller and the beam response is wider. At high frequencies well above 300 Hz, the the circle defined by k is much larger making the beamwidth much narrower.

Table 1 compares the discrete Fourier transform on temporal and spatial data. Determining the wavenumber range from the number of samples and the spatial range of the input can sometimes be confusing. Note that for say 1 sec of a real time signal sampled 1024 times, the sample rate is 1024 Hz and a 1024 point FFT will yield discrete frequencies 1 Hz apart from -512 Hz to $+512$ Hz. For our spatial FFT on the 2 m line sensor, we have a space 32 m by 32 m sampled 256 times in each direction. This gives a spatial sample rate k_s of 8 samples per meter. Since the wavenumber $k = 2\pi/\lambda$ and the FFT output has a digital frequency $\Omega = k/k_s$ span of $-\pi \leq \Omega \leq +\pi$, the physical wavenumber span is $-\pi k_s \leq k \leq +\pi k_s$, or -8π to $+8\pi$. This is perfectly analogous to the physical frequency span for temporal FFTs of real data being $-f_s/2 \leq f \leq +f_s/2$.

Our 2-dimensional wavenumber domain approach to beam forming is interesting when one considers the wavenumber response for a steered beam, such as the 30-degree beam steer in Figure 7. Figure 10 shows the effect of steering the line sensor's look direction beam 30 degrees to the East. Note how the wavenumber response circle appears shifted to the left relative to the main lobe of the array. It can be shown that for the "compass" (rather than trigonometric) bearing representation, the x and y wavenumber components are

$$k_x = k \sin \theta - k \sin \theta'$$
$$k_y = k \cos \theta \qquad\qquad (12.2.18)$$

where k is the wavenumber and θ' is the steered direction. This approach to beam forming is quite intuitive because we can define a wavenumber response for the array shape and then separately evaluate the beampattern for a specific wavenumber (temporal frequency and propagation speed) and steering direction.

We can also use the wavenumber approach to examine the effects of grating lobes which arise from the separation between array elements. With our line sensor

Table 1   Comparison of Spatial and Temporal FFT Parameters

	Spatial FFT	Temporal FFT
Input Buffer	Xmax meters	T sec
	N samples	N samples
Sample Rate	$k_s = N/X$max	$f_s = N/T$
	samples/meter	samples/sec
Digital Frequency	$-\pi \leq \Omega \leq +\pi, \ \Omega = k/k_s$	$-\pi \leq \Omega \leq +\pi, \ \Omega = f/f_s$
Range	$k = 2\pi/\lambda = \omega/c, \ \omega = 2\pi f$	$\Omega = 2\pi\omega/\omega_s, \ \omega = 2\pi f$
Physical Frequency	$-\pi k_s \leq k \leq +\pi k_s$	$-f_s/1 \leq f \leq +_s/2,$
Range		$-\omega_s/2 \leq \omega \leq \omega_s/2$

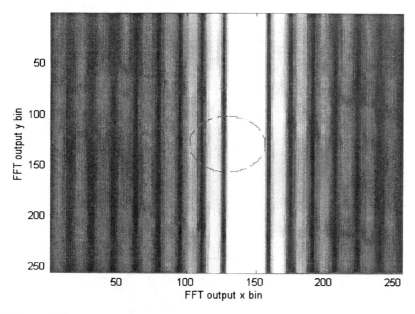

Figure 10   2-dimensional wavenumber response for the 2 m line sensor steered to 30 degrees East of North showing the $k = 5.3856$ circle corresponding to 300 Hz.

2 m long and FFT input space 32 m with 256 samples, our line array is actually 16 adjacent elements in the digital domain. This makes the response closely approximate a continuous line sensor. Figure 11 shows the wavenumber response for a 4-element line array still covering 2 m total aperture. The separation between sensors gives rise to "grating lobes" in the wavenumber transform. For low frequencies, the wavenumber is small and the circle representing the beam response is not significantly affected by the sensor element spacing. Figure 12 shows the polar response for 300 Hz, 30 degree steering, for the 4-element array. This response shows some significant leakage around 270 degrees (due West). Note that the sound speed is 350 m/sec, the wavelength at 300 Hz is 1.167 m while the element spacing is 2 m divided by 4 elements, or 0.5 m. In other words, the sensor spacing is slightly less than half a wavelength.

At 600 Hz, the array response is seen in Figure 13 where the larger circle represents the bigger wavenumber of $k = 10.771$ m^{-1}. At 600 Hz, the sensor spacing of 0.5 m is greater than a half wavelength (0.2917 m). The circle clearly traverses the grating lobes meaning that the array response now has multiple beams. Figure 14 shows the grating lobes at 600 Hz for the 2 m 4-element line array with the steering angle set to 30 degrees. Clearly, a beam pattern with grating lobes will not allow one to associate a target bearing with a large beam output when steered in a specific look direction.

Perhaps the most interesting application of our wavenumber approach to beamforming is seen when we consider 2-dimensional planar arrays. Consider an 8 element by 8 element square grid, 2 m on each side. Taking FFTs we have the wavenumber response seen in Figure 15 where the circle represents 300 Hz and a steering angle of 30 degrees. Figure 16 shows the corresponding polar response

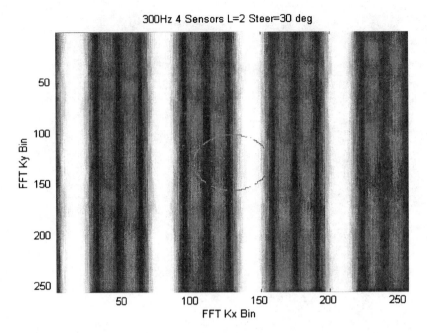

Figure 11 Wavenumber response for 4-element 2 m line array for steering angle of 30 degrees at 300 Hz showing multiple beam lobes (grating lobes) due to the element spacing of 0.5 m.

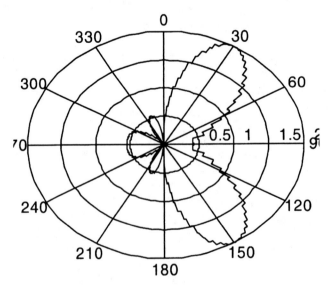

Figure 12 Polar response of 4-element 2 m line array at 300 Hz with a steering angle of 30 degrees East of North showing small grating lobe leakage (linear output vs. degrees).

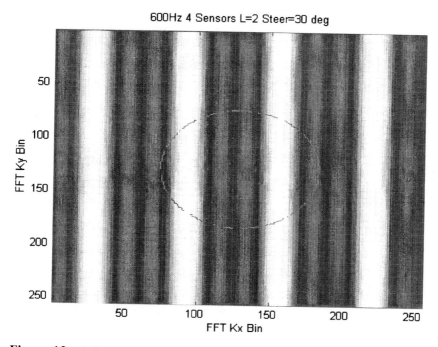

Figure 13 4-element 2 m line array response circle for 600 Hz showing full grating lobes.

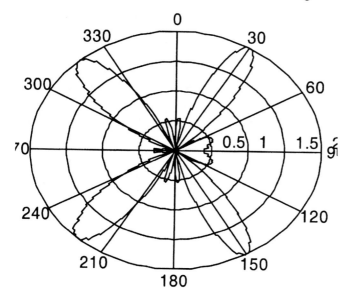

Figure 14 Full grating lobes are seen about 315 and 225 degrees for 600 Hz, 4 sensors, and a steering angle of 30 degrees (linear output vs. degrees).

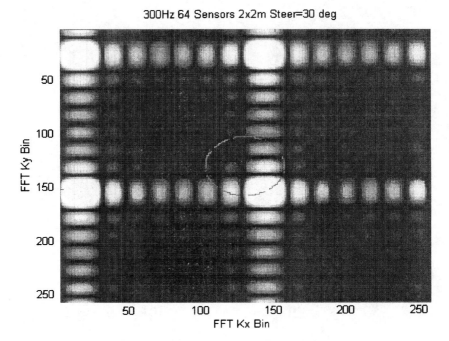

Figure 15   Wavenumber response for 8 by 8 element, 2 m by 2 m, grid array steered to 30 degrees at 300 Hz showing beam response circle shifted up and to the left by the beam steering.

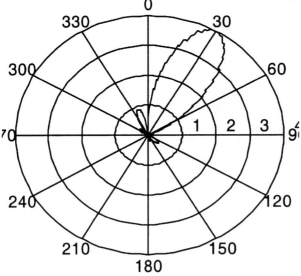

Figure 16   Polar response for the 8 by 8 element, 2 m by 2 m grid array at 300 Hz and a steering angle of 30 degrees (linear output vs. degrees).

in the circle in Figure 15. Note that the effect of steering is to shift the wavenumber circle up and to the left of the main array lobe in Figure 15. This is because the FFT data is displayed such that 0 degrees is down, 90 degrees is to the right, 180 is up, and 270 is to the left. In the polar plot, we display the beam using a "compass" bearing arrangement, which corresponds to how the bearing data is used in real-world systems. Therefore, we can calculate a generic 2-dimensional spatial FFT of the array, and place a circle on the array wavenumber response representing a wavenumber and steering direction of interest to observe the beam pattern. The wavenumber shifts follow the case for the line array

$$k_x = k \sin \theta - k \sin \theta'$$
$$k_y = k \cos \theta - k \cos \theta' \tag{12.2.19}$$

where k is the wavenumber, θ is the bearing, and θ' is the steered direction. Note that if one prefers a trigonometric circular coordinate system rather than compass bearings, all one needs to do is switch sines and cosines in Eqs (12.2.16) and (12.2.17).

Another physical effect which can cause the wavenumber circle to shift is a flow field which causes the wave propagation speed to be directional. This physical effect of flow is that the wave speed is now a function of direction. It also affects incoming waves differently than outgoing waves. For example, if winds are out of the East (90 degrees), the "listening" response of the array will be skewed slightly in the direction of the wind. Waves traveling towards the array will arrive faster from the East, just as if the beam were steered in that direction. For outgoing waves, an array of sources would beam slightly more towards the downwind direction, if effect, blowing the transmitted beam downstream somewhat. Outdoors this effect is very slight because the speed of sound is very high relative to a typical wind speed. But, in a jet airplane, it is the reason so much of the engine's noise radiates behind the plane.

Now that beam forming physics and processing have been established, we need to revisit the CRLB to include the SNR gains available from beam forming. For a beam forming based bearing estimate, the beam is swept around while the output is monitored for directions with high SNR. The CRLB for bearing is therefore tied to the array SNR gain in the look direction and the beamwidth. The SNR gain can be numerically computed by calculating a parameter called the directivity index.

$$DI = 10 \log_{10} \left\{ \frac{|D(\theta')|^2}{\frac{1}{4\pi} \int_{4\pi} |D(\theta)|^2 d\Omega} \right\} \tag{12.2.20}$$

Equation (12.2.18) must be integrated in 3 dimensions to properly model the isotropic noise rejection of the beam pattern. It represents the total response of the beam normalized by the gain in the look direction θ'. An omnidirectional point sensor has a DI of 0 dB. The higher the DI the more directional the beam and the greater the background noise rejection is. The DI can be numerically calculated, but in general, an analytic expression is foreboding without at least some approximations. We can use the following approximation based on a heuristic physical argument.

$$d_I = 10^{\frac{DI}{10}} \approx \frac{L}{\lambda/2} \tag{12.2.21}$$

A DI of 0 dB ($d_I = 1$) is nearly the case when the aperture of the line source $L = \lambda/2$ or less. One can argue that a sensor smaller in size than a half wavelength is essentially a point sensor with omnidirectional response. Since our line array has a maximum gain L in the look direction, we can assume $d_I = 2L/\lambda$, but it will be somewhat less than that for look directions steered off broadside to the array. However, the directivity index actually improves for a line array when the beam is steered along the axis (90 degrees from broadside). This is because the backward beam collapses leaving only one lobe pointing out from one end of the line array. The directivity has a similar effect on SNR as does an FFT for a sinusoid in white noise. This is because summing the array elements increases the amplitude of the spatially coherent wave from the look direction while not amplifying waves from other directions and the spatially incoherent background noise.

Finally, the CRLB is effected by beamwidth. The broader the beam width, the higher the CRLB because it will be more difficult to pin-point the precise bearing angle with a beam pattern which does not vary much with angle. Consider the angle off broadside where the directivity power gain is down by 1/2. We can estimate this by noting that $(\sin x)/x = 0.707$ for $x = 1.4$ approximately. Therefore,

$$\frac{\pi L}{\lambda} \sin \theta \approx 1.4$$

$$\theta \approx \sin^{-1} \left\{ \frac{1.4\lambda}{\pi L} \right\} \tag{12.2.22}$$

$$\theta \approx 0.456 \frac{\lambda}{L} \approx \frac{\lambda}{2L}$$

and we interestingly pick up another factor of $2L/\lambda$ from the beamwidth. Therefore, our approximate estimate for the CRLB for bearing error for a line array of length L and wavelength λ (near broadside arrival angle) is

$$\sigma_\theta \approx \frac{1}{SNR \left(\frac{2L}{\lambda} \right)^2} \tag{12.2.23}$$

Comparing the CRLB for phase difference in Eq. (12.2.11) to the CRLB for a line array, one might think that for $L > \lambda$, the bearing estimates for the line array beam pattern are better than a direct measurement of phase to get bearing. But, the CRLB for the beamforming estimator is actually not as good as a direct measurement. Recall that there are many elements available for phase difference pairings and the CRLB in Eq. (12.2.11) is for M observations for only one sensor element pair. When one has more than one arrival angle for a particular frequency, only beamforming techniques can provide correct bearing answers.

There are other physical problems which make large arrays with closely spaced sensors underperform the theoretical CRLB for bearing error. First, when the sensors are closely spaced, the noise is no longer incoherent from sensor to sensor. Thus, the beamforming algorithm does not reduce the background noise as much as planned. Second, for large arrays in inhomogeneous media (say acoustic arrays with flow and turbulence), the signal coherence from one end of the array to the other is not guaranteed to be unity. Therefore, it is unlikely that the array gain will be as high as theory predicts possible throughout the CRLB. The CRLB is *a lower bound* by definition, and is used to form a confidence estimate on the bearing error to go along with the bearing estimate itself. As many physicists and engineers know all too well, data with some confidence measure (such as error bars, variance, probability density, etc) is far more useful than raw data alone. For intelligent sensor and control systems, confidence measures are even more important to insure correctly weighted data fusion to produce artificially measured information and knowledge. Information is data with confidence metrics while knowledge is an identifiable pattern of information which can be associated with a particular state of interest for the environment. The CRLB is essential to produce bearing information, rather than bearing data.

12.3   FIELD RECONSTRUCTION TECHNIQUES

Sensor arrays can be used for much more than determining the directions of arrivals of plane wave radiated from distance sources. In this section we examine the use of array processing to measure very complicated fields in the vicinity of a source. Some useful applications are investigation of machine vibrations from radiated acoustic noise, condition monitoring electrical power generators or components, or even optical scattering from materials as a means of production quality control sensing. In all cases, a sensor array scans a surface to observe the field and relate the measurements on the array surface to what is happening where the sources are. For example, an acoustic intensity scan over a closed surface enclosing a sound source of interest can provide the net Watts of radiated power (Gauss's theorem). But the acoustic pressure and velocity on the scanning surface could also be used to reconstruct the acoustic field much closer to the source surface, allowing surface vibrations to be mapped without contact. It can be seen that this technique might be useful in the investigation of things like tire noise. For electromagnetic equipment, changes in the field could provide valuable precursors to component failure, allow one to locate areas of leakage/corrosion/damage, or measure the dynamic forces governing the operation of a motor or generator.

Field reconstruction is possible because of Green's integral formula. Green's integral formula can be seen as an extension to 3-dimensional fields of the well-known formula for integration by parts.

$$\int u dv = uv - \int v du \tag{12.3.1}$$

When we develop a measurement technique for the radiated waves from a source or group of sources, it is useful to write the field equations as a balance between the radiated power from the sources and the field flux through a surface enclosing the source(s) and the field space of interest. This follows from Gauss's law, which simply stated, says that the net electric flux through a surface enclosing a source(s) of charge is equal to the total charge enclosed by that surface. Gauss's law for electric fields is seen in Eq. (12.3.2)

$$\oint_S D_s \cdot dS = \int_{vol} \rho_e dv \tag{12.3.2}$$

where D_s is the electric field flux, dS is the surface area element, ρ_e is the charge density, and dv is the volume element. For the case of the wave field inside a closed surface due to a number of sources also inside the surface we have the following 3-dimensional equation for acoustic waves called the Helmholtz–Huygens integral.

$$\oint_S \left(p(X) \frac{\partial g(X|X')}{\partial n} - g(X|X') \frac{\partial p(X)}{\partial n} \right) dS = \int_v g(X|X') F(X') dv \qquad (12.3.3)$$

where $X = \langle x, y, z \rangle$ and $X' = \langle x', y', z' \rangle$ are the surface field and source points, $\partial/\partial n$ is the gradient normal to the surface, $p(X)$ is the acoustic pressure on the surface field point of interest X, and $g\,(X|X')$ is the free space Green's function for a source at X' and receiver at X in 3-dimensions given in Eq. (12.3.4).

$$g(X|X') - \frac{e^{jk|X-X'|}}{4\pi|X-X'|} \qquad (12.3.4)$$

The term "free-space" means that there are no reflections from distant boundaries, i.e. a reflection-free or anechoic space. However, if there were a reflection boundary of interest, the left-hand side of Eq. (12.3.3) would be used to define the pressure and velocity on the boundary surface allowing the field to be reconstructed on one side or the other. One could substitute any field quality such as velocity, electric potential, etc. for $p(X)$ with the appropriate change of units in $F(X')$.

Huygen principle states that a wave can be seen to be composed of an infinite number of point sources. The Helmholtz–Huygens integral establishes a field mathematical representation by an infinite number of sources (monopole velocity sources and dipole force sources) on a closed surface to allow the reconstruction of the field on one side of the boundary surface or the other. If we know the point source locations, strengths, and relative phases, one would simply use the right-hand side of Eq. (12.3.3) and sum all the source contributions from the locations X' for the field point of interest X. From an engineering perspective, we would like to measure the source locations, strengths, and relative phases from a sensor array which defines the field on a surface. However, this surface must separate the field point and the sources of interest to be of value mathematically, but this is easily achieved mathematically by separating a source or field point from the surface with a narrow tube and infinitely small sphere surrounding the field point. As will be seen later, the definition of the integration surface in Eq. (12.3.3) mainly has an effect on the sign of the derivatives with respect to the normal vector to the surface. In the midst of these powerful field equations, the reader should keep in mind foremost that the Green's function can be used with the field measured by an array of sensors to reconstruct the field on another surface of interest.

The physical significance of the left side of Eq. (12.2.3) is that the surface has both pressure $p(X)$ and velocity $(\partial p(X)/\partial n)$ (okay a quantity proportional to velocity) which can describe the field inside the surface (between the sources and bounding surface) due to the sources depicted by the right side of the equation. For a specific source distribution, there are an infinite number of combinations of pressure and velocity on the enclosing surface which give the same field inside. However, assuming one knows the approximate location of the sources (or a smaller volume within the integration volume where all sources are contained), and the one has measurements of the field on the outer surface, then an "image field" can be reconstructed on any surface of interest not containing a source. This 3-dimensional field reconstruction from a set of surface measurements is known in acoustics as acoustical holography (3). The physics are in fact quite similar when one considers the wave interference on the measurement surface and the spatially coherent processing to reconstruct the field. Field reconstruction using sensor arrays is an extremely powerful technique to analyze radiated waves.

Acoustic fields provide us with a nice example of how holographic reconstruction can be useful in measuring sound source distributions. For example, when an automobile engine has a loose or defective valve, a tapping sound is easily heard with every rotation of the cam shaft. It is nearly impossible to determine which valve using one's ears as detectors because the sound actually radiates to some extent from all surfaces of the engine. Mechanics sometimes use a modified stethoscope to probe around the engine to find the problem based on loudness. Given the valve tap radiates sound most efficiently in a given frequency range, one could use acoustic holography and a large planar array of microphones to find the bad valve. By measuring the field at a number of sensor positions in a plane, the field could be reconstructed in the plane just above the engine surface, revealing a "hot spot" of acoustic energy over the defective valve position. To do a similar mapping using intensity one would have to measure directly over the radiating surface in a very fine spacing. Using holographic field reconstruction, all field components can be calculated, which for acoustics means velocity, intensity, and impedance field can all be calculated from the pressure measurements. No professional mechanic in this millennium (or next) would consider this activity. The author is just trying to present a practical example for field reconstruction using holographic beamforming.

Unlike the time-average field intensity, one needs the precise magnitude and phase spatially for each frequency of interest in the measurement plane to reconstruct the field accurately in another plane. In acoustics, it is very useful in noise control engineering where one must locate noise "hot-spots" on equipment. The technique will likely also find uses in electromagnetics and structural vibrations, although full vector field reconstruction is significantly more complicated than the scalar-vector fields for acoustic waves in fluids. In machinery failure prognostics, field holography can be used as a measurement tool to detect precursors to failure and damage evolution from subtle changes in the spatial response. Changes in spatial response could provide precursors to equipment failure well before detectable changes in wave amplitude are seen at a single sensor.

We begin our development of the holographic field reconstruction technique by simply examining the free-space Green's function and its Fourier transform on a series of x–y planes at different distances z from the point source location. In cartesian coordinates, the 3-dimensional free-space Green's function for a receiver at $X = \langle x, y, z \rangle$ and point source at $X' = \langle x', y', z' \rangle$ is

$$g(x, y, z | x', y', z') = \frac{e^{j\sqrt{k_x^2 + k_y^2 + k_z^2}\sqrt{(x-x')^2 + (y-y')^2 + (z-z')^2}}}{4\pi\sqrt{(x - x')^2 + (y - y')^2 + (z - z')^2}} \qquad (12.3.5)$$

Figure 17 shows the spatial responses for the acoustic case of a 700 Hz ($c = 350$ m/sec, $k = 12.65$ rad m^{-1}) point source at $X' = \langle 0, 0, 0 \rangle$ where the field sufaces are planes located at $z = 0.001$ m, $z = 1.00$ m, and $z = 50.00$ m. The measurement planes are 10 m by 10 m and sampled on a 64 by 64 element grid. A practical implementation of this measurement would be to physically scan the plane with a smaller array of sensors, keeping on additional sensor fixed in position to provide a reference phase. One would calculate temporal FFTs and process the 700 Hz bin spatially as described here. On the right side of Figure 17 one sees the spatial Fourier transform of the Green's function, $G(k_x, k_y, z)$, on the corresponding z-plane for the left-hand column of surface plots. The spatial real pressure responses are plotted showing maximum positive pressure as white and maximum negative pressure as black to show the wave front structure. The wavenumber plots on the right column are shown

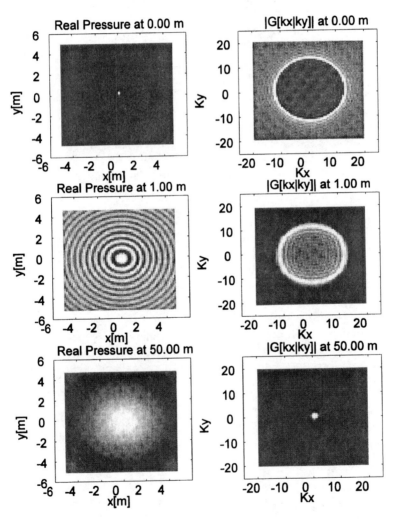

Figure 17 Real pressure responses of an acoustic free space Green's function in 3-dimensions showing corresponding wavenumber transforms for 3 z-axis planes at 700 Hz.

for the magnitude where white corresponse to maximum amplitude and black minimum amplitude. Each of the 6 plots are independently scaled.

The spatial and corresponding wavenumber plots in Figure 17 are extremely
interesting and intuitive. In the spatial plane just in front of the source seen in
the upper left plot, the singularity of the point source dominates the response at
$x = y = 0$. There is however, a wave structure in this plane where and
$k^2 \approx k_x^2 + k_y^2$ and $k_z \approx 0$. This is clearly seen in the wavenumber transform in the
upper right plot. The bright ring corresponds to the wavenumber $k = 12.65$. Since
we are practically in the plane with the point source, there is very little wave energy
at wavelengths longer than 0.5 m (the wavelength of $k = 12.65$). There is wave energy
at wavelengths shorter than 0.5 m, mainly due to the "point-like" spatial structure at
$z = 0.001$ m. This wave energy is called evanescent because it will not propagate very
far (the waves self-cancel). For propagation of these waves in the z-direction we have

$$\frac{e^{jk_z z}}{4\pi X} = \frac{e^{-\sqrt{k_x^2 + k_y^2 + k^2 z}}}{4\pi X} \qquad \sqrt{k_x^2 + k_y^2} > k \qquad (12.3.6)$$

causing the waves with wavelengths shorter than the free propagating wavelength to
exponentially decay in the positive z-direction. The middle row of plots clearly shows
the rapid decay of the evanescent field and the "leakage" of longer wavelengths
(smaller wavenumbers) into the center region of the wavenumber spectrum. One
can see in the spatial plot at $z = 1.00$ m that the wavelengths get longer slightly
as one approaches the center of the measurement plane. This is caused by the angle
between the normal to the surface of the spherical wave and the measurement plane.
The diameter of the bright ring in the wavenumber plot also is smaller. As z becomes
quite large ($z = 50$ m is 100 wavelengths), one sees a near uniform pressure spatially
and a wavenumber transform which reveals a near Dirac delta function. Recall
in Section 6.3 we showed a plane wave assumption could be used at ranges about
100 wavelengths from a point source using both geometry and intensity theory.

The ring-shaped peak energy "ridge" in the wavenumber plots in Figure 17
collapses into a delta function as z approaches infinity. The amount of the ring diameter collapse is a function of the measurement plane aperture and the distance from
the point source. If we call the equivalent wavenumber for this ridge diameter
k_d, it can be expressed as $k_d = k\sin\theta$, where $\theta = \tan^{-1}(\frac{1}{2}L/z)$ and L is the width
of the aperture. The aperture angle is an important physical quantity to the signal
processing. For example, if one wants to keep the ridge diameter wavenumber within
about 10% of the source plane value, the measurement aperture width needs to be
over 4 times the measurement plane distance z from the source plane. This can
be clearly seen in the wavenumber plots of Figure 17 where one can better
reconstruct the source-plane field at $z = 0.001$ m from a measurement at $z = 1$ m
than from measured data at $z = 50$ m. At 1 m, the evanescent field is significantly
attenuated, but still present along with all of the wavenumbers from the source.
The aperture angle for a 10 m by 10 m measurement plane can be seen as 90°
at $z = 0.001$ m, about 79° at 1.00 m, and about 5.7° at 50 m.

It should be possible to define a transfer function between the measurement
plane and the "image" plane so long as the SNR and dynamic range of the measurement plane wavenumber spectrum is adequate. Suppose our measurement plane
is parallel to the x–y plane at $z = z_m$. We wish to reconstruct the field in an image

plane also parallel to the measurement plane at $z = z_i$. Given the measured pressure wavenumber spectrum $P(k_x, k_y, z_m)$, the image plane wavenumber spectrum is found to be

$$P(k_x, k_y, z_i) = H(k_x, k_y, z_m)P(k_x, k_y, z_m) \qquad (12.3.7)$$

where

$$H(k_x, k_y, z_m) = \frac{G(k_x, k_y, z_i)}{G(k_x, k_y, z_m)} \qquad (12.3.8)$$

The Green's function wavenumber transform at z_m and z_i are defined as

$$G(k_x, k_y, z_k) = \int_{-L_x/2}^{+L_x/2} \int_{-L_y/2}^{+L_y/2} g(x, y, z_k)e^{-jk_x\hat{x}}e^{-jk_y\hat{y}}d\hat{x}d\hat{y} \quad z_k = z_i, z_m \qquad (12.3.9)$$

The Fourier transform in Eq. (12.3.9) is efficiently carried out using a 2-dimensional FFT where the measurement plane is sampled k_s samples per meter giving a wavenumber spectrum from $-k_s\pi$ to $+k_s\pi$. The larger L_x and L_y are the finer the wavenumber resolution will be. However, one should have at least the equivalent of two samples per wavelength in the source plane to avoid aliasing. Even though the 2-dimensional wavenumber FFT has complex data as input from the temporal FFT bin corresponding to the frequency of interest, the original time-domain signals from the array sensors are generally sampled as real digital numbers.

Figures 18 through 21 show results of holographic imaging of a quadrapole in a plane adjacent to the source plane from measurement planes a further distance away. As will be seen, the spatial resolution of the holographically-reconstructed image depends on wavelength, distance between the measurement and imaging planes, and the aperture of the measurement plane. We will use these 4 figures to help visualize the issue of holographic resolution. In Figure 18, the field from a symmetric 700 Hz quadrapole is measured at 1 m distance. Green's function wavenumber transforms for a point source at the origin are calculated numerically as seen in Eq. (12.3.9) for the measurement plane at $z = 1$ m and the image plane taken as $z = 0.001$ m. The transfer function $H(k_x, k_y, z_i, z_m)$ is computed as seen in Eq. (12.3.8) and the image wavenumber spectrum is computed using Eq. (12.3.7). The reconstructed field is seen in the upper left plot in Figure 18. The measurement field at 1 m distance is seen in the lower left plot. The corresponding wavenumber spectra are seen in the right-hand column. Note the substantial amount of evanescent field in the image plane. The equal spacing (the point sources are at $\langle 1, 1 \rangle$, $\langle -1, 1 \rangle$, $\langle -1, -1 \rangle$, and $\langle 1, -1 \rangle$ meters) and strengths of the quadrapole sources create symmetric interference patterns which are barely visible in the pressure field plots, but clearly seen in the wavenumber spectra. In Figure 19 we move the sources around just to show arrogance to the positions $\langle 1.2, 1.0 \rangle$, $\langle -1.0, 1.3 \rangle$, $\langle -1.8, -1.5 \rangle$, and $\langle 1.5, -0.5 \rangle$ meters. Actually, what is seen in the wavenumber spectra is an asymmetry to the wavenumber peaks and great complexity to the evanescent field. Also seen in Figures 18 and 19 are the fact that at 1 m it is impossible to determine that in fact 4 sources are present in the source plane. In Figure 20 the measurement field is move back to 5 m distance from the sources. In the reconstruction, the sources

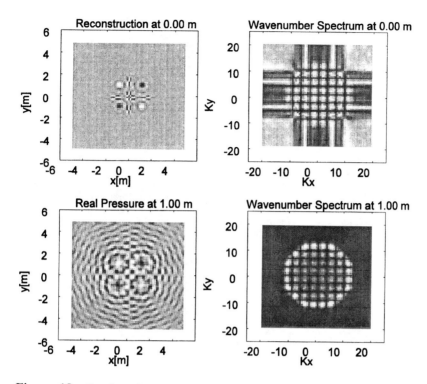

Figure 18 Quadrapole reconstruction from a 1 m measurement of 64 by 64 points with an aperture of 10 m by 10 m at 700 Hz.

are still detectable, but there is considerable loss of resolution. The "ridge diameter" for the wavenumber spectrum at 5 m is also smaller, indicating that the aperture angle may be too small or problematic. Figure 21 also has the measurement plane at 5 m distance, but this data is for a frequency of 1050 Hz, which translates into 30 wavelengths for the 10 m aperture of the measurement array. With the resolution nicely restored, one can see that there is an interesting relationship between aperture, measurement distance, frequency, and reconstruction resolution.

A model can be developed to estimate the available resolution for a particular array aperture, source wavelength, and measurement plane distance. We start by noting that the apparent wavelength in the measurement plane gets long as this plane is moved farther from the source plane. The wavelength in the measurement plane is $\lambda' = \lambda/\sin\theta$, where $\theta = \tan^{-1}(\frac{1}{2}L/z)$ and L is the width of the array aperture. As z gets large, θ tends to zero and λ' tends to infinity. From a beamforming point of view (see Section 12.2), the long wavelength trace in the measurement plane will translate into a limited ability to spectrally resolve the wavenumbers in the 2-dimensional wavenumber transform. We can estimate this resolution "beamwidth" as approximately $\beta = 2\sin^{-1}[\lambda/(L\sin\theta)]$. Given this beamwidth, the spatial resolution in the measurement plane is approximately

$$\Delta = z\sin 2\beta \approx 2z\frac{\lambda}{L\sin\theta} = 2\lambda\left(\frac{2z}{L}\right)\sqrt{\left(\frac{2z}{L}\right)^2 + 1} \tag{12.3.10}$$

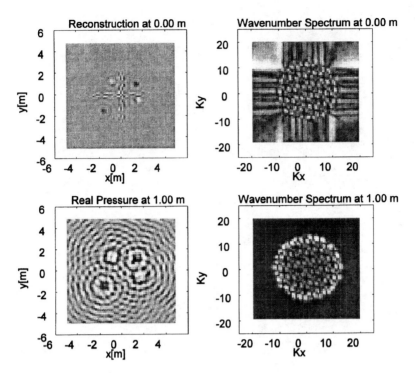

Figure 19 Quadrapole reconstruction at 700 Hz when sources are not spaced equally showing the changes in wavenumber patterns and some source visibility in the measurement plane at 1 m.

where Δ is the resolution in meters of the image plane available for the reconstruction geometry. Equation (12.3.10) clearly shows reconstruction resolution improving for higher frequencies, larger apertures, and measurement planes closer to the image plane. However, its not that simple because the measurement field is sampled spatially with a resolution of L/N, N being the number of samples in the N by N spatial FFTs, which in our case is 64 samples. One cannot get a higher resolution in the image reconstruction then is available in the measurements. However, for a given frequency and aperture one can find the distance z where the resolution begins to degrade seriously.

The resolution question then focuses on determining the maximum measurement plane distance where the resolution significantly starts to decrease. This is analogous to the depth of field (the depth where the view stays in focus) of a camera lens system, where the f-stop represents the ratio of the lens focal length to the aperture of the lens opening. The larger the lens aperture, the smaller the depth of field will be for a given focal length. To find the distance z_0 where resolution in the image plane is equal to the measurement plane resolution, we set Eq. (12.3.10) equal to L/N and solve for z_0.

$$z_0 = \frac{L}{2}\sqrt{\sqrt{\frac{1}{4} + \left(\frac{L}{\lambda N}\right)^2} - \frac{1}{2}} \qquad (12.3.11)$$

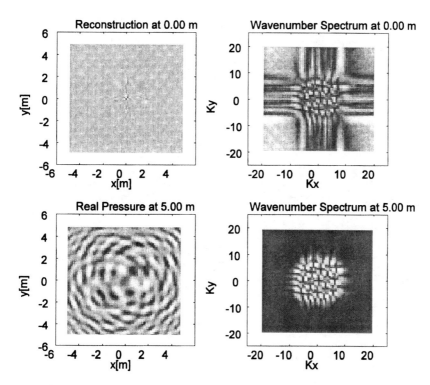

Figure 20 With a 5 m measurement plane distance the resolution limits for the 700 Hz quadrapole are close to the limit where the sources can be easily resolved.

Note that as N gets large z_0 approaches zero, which is counter intuitive. Usually more FFT resolution improves things. Using the camera analogy, increasing the lens aperture reduces the depth of field. This is good news for wave field holography, because fewer sensors in the array and smaller FFT sizes can actually improve resolution when the measurement plane is relatively far from the image plane. Figure 22 shows the 700 Hz quadrapole reconstructed from a 5 m measurement plane using a 32 by 32 measurement grid as compared to the 64 by 64 grid used in Figure 20. The reconstruction resolution actually improves with the smaller FFT as predicted by Eqs (12.3.10) and (12.3.11). For 700 Hz and 64 by 64 point FFTs, z_0 is about 1.5 m. Using 32 by 32 points, z_0 is about 2.75 m which means the resolution is degraded more for the reconstructed field using the larger spatial FFTs. However, using too small an FFT will again limit resolution at a particular frequency. Notice how with 32 by 32 points we have 3.2 samples per meter, reducing the wavenumber range to $\pm 3.2\pi$, or ± 10.05 m^{-1} from $\pm 6.4\pi$, or ± 21.11 m^{-1}. If our frequency were any higher there would be serious spatial aliasing. In fact, there is some aliasing going on since our spatial samples are 0.3125 m apart and the wavelength is 0.5 m. The aliasing is seen as the ripples near the origin in the reconstruction in the upper left of Figure 22 and also in the nearly out of band evanescent energy in the right-hand column of plots.

An analytical expression for the Green's function-based wavenumber transfer function can be approximated and is useful when one does not know where the

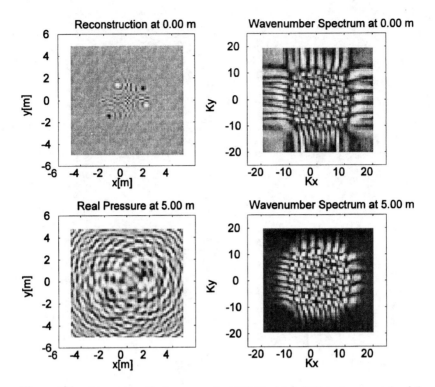

Figure 21 Increasing the frequency to 1050 Hz ($L = 30\lambda$) restores much of the resolution in the reconstruction plane when the measurement plane is at 5 m.

source plane is. In this regard, the field in the measurement plane can be translated a distance $d = |z_m - z_i|$ along the z-axis. This is not exactly the same as the transfer function method described above, but it is reasonable for many applications. The analytical solution is found by applying the Helmholtz–Huygens integral where the boundary between the source and field point is an infinite plane. We then assign a Green's function (a particular solution to the wave equation) of the form

$$g(x, y, z) = \frac{1}{2\pi} \frac{\partial}{\partial \alpha} \left\{ \frac{e^{jk\sqrt{x^2+y^2+\alpha^2}}}{\sqrt{x^2 + y^2 + \alpha^2}} \right\}_{\alpha=z} \tag{12.3.12}$$

The 2-dimensional Fourier transform of Eq. (12.3.12) can be found analytically (4) as

$$G(k_x, k_y, d) = e^{jd\sqrt{k^2-k_x^2-k_y^2}} \tag{12.3.13}$$

and we note that for the evanescent part of the wavenumber field Eq. (12.3.13) is equivalent to Eq. (12.3.6). The distance d in Eq. (12.3.13) represents the distance from the measurement plane in the direction away from the source one is calculating the wavenumber field in the image plane. This is considered wave propagation modeling and the pressure field in the image field plane which is farther away from

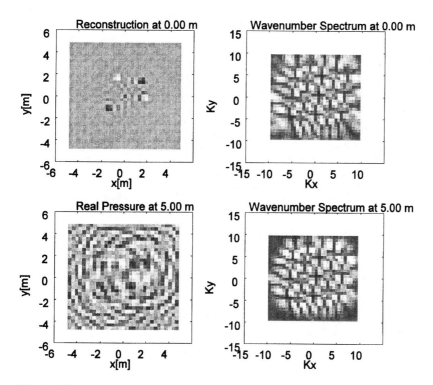

Figure 22   Reconstruction of the 700 Hz quadrapole measured at 5 m using a 32 by 32 sample grid actually improves reconstruction resolution.

the sources than the measurement plan is

$$p(x, y, z_i) = \mathscr{F}\left\{G(k_x, k_y d)P(k_x, k_y, z_m)\right\}^{-1} \quad d = z_i - z_m \tag{12.3.14}$$

where $\mathscr{F}\{\ \}^{-1}$ denotes an inverse 2-dimensional FFT to recover the spatial pressure in the image field from the wavenumber spectrum.

For holographic source imaging, we are generally interested in reconstructing the field very close to the source plane. Equation (12.3.15) shows the inverse Green's function wavenumber spectrum used to reconstruct the field.

$$p(x, y, z_i) = \mathscr{F}\left\{G(k_x, k_y, d)^{-1}P(k_x, k_y, z_m)\right\}^{-1} \quad d = z_i - z_m \tag{12.3.15}$$

where

$$G(k_x, k_y, d)^{-1} = e^{jd\sqrt{-k^2+k_x^2+k_y^2}} \tag{12.3.16}$$

Note the sign changes in the square-root exponent in Eq. (12.3.16). This very subtle change is the result of using the inverse wavenumber. Therefore, the region inside the free wavenumber circle is actually propagated back towards the source in a non-physical manner by having an evanescent-like exponential increase. This explains the apparent high frequency losses seen in the reconstructions in Figures 23 and 24

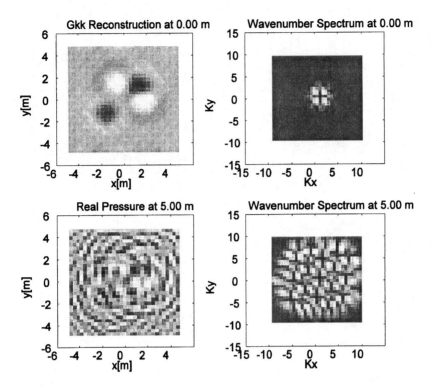

Figure 23 700 Hz quadrapole reconstruction for 32 by 32 samples at 5 m using the analytic Green's function wavenumber transfer function showing good results but a loss of high frequency.

The analytic Green's function technique works pretty well, even though some approximations are applied to its derivation. It can be seen as a "wavenumber filter" rather than a physical model for wave propagation, although its basis for development lies in the Helmholtz–Huygens integral equation. To calculate the wavenumber field a distance towards the source from the measurement plane, the sign of the exponent is changed simply by letting $d = z_i - z_m$, assuming the source plane is further in the negative direction on the z-axis than either the measurement or image planes.

12.4 WAVE PROPAGATION MODELING

Wave propagation modeling is vastly important for acoustics, electromagnetics, and material characterization. While this topic is generally considered a physics topic, the Green's function approach of the previous section brings us to the doorstep of presenting this important technique. Of particular interest is wave propagation in inhomogeneous media, that is media where the impedance or wave speed is not constant. From the sensor system perspective, wave propagation modeling along with spatial sensor measurement of the waves offer the interesting opportunity of *tomographic measurement of the media inhomogeneity*. For example, surface layer turbulence could be mapped in real-time for airport runway approaches using

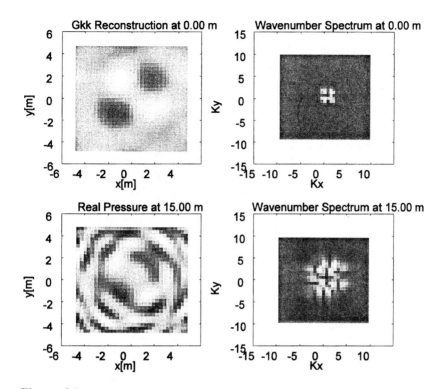

Figure 24 Even at 15 m (well beyond the resolution range of the transfer function method) the analytic Green's allows one to determine that a quadrapole is present in the source plane.

an array of acoustic sources and receivers. Another obvious example would be to use adaptive radio transmission frequencies to circumvent destructive interference predicted by a communications channel propagation model. Perhaps the most interesting application of wave propagation modeling may be in the area of materials characterization and even chemical process sensors and control. Chemical reactions and phase changes of materials results in wave speed and impedance changes whether the wave is vibration, acoustic, electromagnetic, or thermal/optical. Inexpensive and robust sensor systems which can tomographically map the state of the wave propagation media can be of enormous economical importance.

Our discussion begins by considering 2-dimensional propagation in cylindrical coordinates where our propagation direction is generally in the r-direction and our wavefronts are generally aligned in the z-direction. The homogeneous wave equation assuming field symmetry about the origin along the θ circle is simply

$$\frac{\partial \psi}{\partial z^2} + \frac{\partial \psi}{r^2} + k^2(r, z)\psi = 0 \tag{12.4.1}$$

where $k(r, z)$ is the wavenumber $\omega/c(r, z)$ which varies with both r and z for our inhomogeneous wave propagation media. The wave number can be decomposed a wave component in the r-direction which varies with z and a component in the z-direction which varies with r.

$$k^2(r, z) = k_r^2(z) + k_z^2(r) \tag{12.4.2}$$

If the wave speed is inhomogeneous, the wavefront from a point source will distort with refraction as some portions propagate faster than others. We note that if media impedance also changes, one would also have to include wave scattering in both forward and backward (towards the source) directions. Our discussion will be limited to the refraction case which is certainly important but will also allow a brief presentation. We also note that for 3-dimensional wave propagation in which the r–z plane represents symmetry for the source radiation, one can divide by the square-root of r as $\psi_{2D} = \sqrt{r}\,\psi_{3D}$.

The wave equation in Eq. (12.4.1) can be written in terms of an operator Q as

$$\left(\frac{\partial}{\partial r} + j\sqrt{Q}\right)\left(\frac{\partial}{\partial r} - j\sqrt{Q}\right)\psi = 0 \tag{12.4.3}$$

where $Q = (\partial^2/\partial z^2) + k^2(z)$ and we will assume $k(r,z)$ has only z-dependence for the moment (5). Given the homogeneous solution to the wave equation, one can find the complete particular solution by applying boundary, source, and initial conditions. However, our interests here assume one has the field measured over the z-direction at range r and is interested in the field over the z-direction at a farther distance $r + \Delta r$ away from the source. We will show that the solution to Eq. (12.4.1) will be in the same form as the Green's function used for holographic reconstruction in Section 12.3. Eq. (12.4.3) is generally known as a parabolic wave equation because of the operator product. Clearly, one can see that

$$\frac{\partial \psi}{\partial r} = \pm j\sqrt{Q}\psi \tag{12.4.4}$$

where waves represented temporally as $e^{j\omega t}$ traveling away from the source have the minus sign. Therefore, to propagate the wave a distance Δr, one must add the phase

$$\psi(r + \Delta r, z) = e^{j\Delta r\sqrt{Q}}\psi(r, z) \tag{12.4.5}$$

where the operator Q is constant over the range step Δr. But in the wavenumber domain, the spectral representation of the operator Q allows the wavenumber spectra in the z-planes at r and $r + \Delta r$ to be written as

$$\Psi(r + \Delta r, k_z) = e^{jk_r\Delta r}\Psi(r, k_z) \tag{12.4.6}$$

where

$$\Psi(r, k_z) = \int\limits_{-\infty}^{+\infty} \Psi(r, z)e^{-jk_z z}dz \tag{12.4.7}$$

Substituting Eq. (12.4.2) into (12.4.6) we have the same Green's function propagation expression as seen in Section 12.3 in Eqs (12.3.13)–(12.3.14).

$$\Psi(r + \Delta r, k_z) = e^{j\Delta r\sqrt{k^2 - k_z^2}}\Psi(r, k_z) \tag{12.4.8}$$

Equation (12.4.8) is not all that impressive in terms of inhomogeneous media propagation because we assumed that Q is constant over the range step Δr. The wavenumber k_z variation in z is due to an inhomogeneous medium is not explicitly seen in Eq. (12.4.8). One could inverse Fourier transform the wavenumber spectrum at $r + \Delta r$ and then multiply the field at each z by the appropriate phase to accommodate the wavefront refraction if a simple expression for this could be found. To accomplish this, we apply what is known as a "split-step approximation" based on the assumption that we have propagation mainly in the r-direction and that the variations in wavenumber along the z-direction are small. This is like assuming that the domain of interest is a narrow wedge with the waves propagating outward. We note that a wavenumber variation in the r-direction will speed up or slow down the wavefront, but will not refract the wave's direction.

Let the variation along the z-direction of the horizontal wavenumber $k_r(z)$ be described as

$$k_r^2(z) = k_r^2(0) + \delta k^2(z) \tag{12.4.9}$$

where $k_r^2(0)$ is simply a reference wavenumber taken at $z = 0$. Assuming we have mainly propagation in the r-direction, and we can make the following approximation.

$$
\begin{aligned}
k_r(z) &\approx \sqrt{k_r^2(0) + \delta k^2(z) - k_z(r)} \\
&\approx \sqrt{k_r^2(0) - k_z(r)} + \frac{\delta k^2(z)}{2k_r(0)}
\end{aligned}
\tag{12.4.10}
$$

Applying the split-step approximation to Eq. (12.4.5) we have

$$\psi(r + \Delta r, z) = e^{j\Delta r \sqrt{Q_r}} \psi(r, z) \tag{12.4.11}$$

where

$$\sqrt{Q_r} = \sqrt{\frac{\partial^2}{\partial z^2} + k_r^2(0) - k_z^2(r)} + \frac{\delta k^2(r)}{2k_r(0)}.$$

In the wavenumber domain the expression is a bit more straightforward, but requires an inverse Fourier transform.

$$\psi(r + \Delta r, z) = e^{j\Delta r \frac{\delta k^2(z)}{2k_r(0)}} \left\{ \frac{1}{2\pi} \int_{-\infty}^{+\infty} \left[\Psi(r, k_z) e^{j\Delta r \sqrt{k_r^2(0) - k_z^2(r)}} \right] e^{+jk_z z} dk_z \right\} \tag{12.4.12}$$

Equation (12.4.12) defines an algorithm for modeling the propagation of waves through an inhomogeneous media. The wavenumber spectrum is computed at range r and each wavenumber is multiplied by the appropriate phase in the square brackets to propagate the spectrum a distance Δr. Then the inverse Fourier transform is applied to give the field at range $r + \Delta r$ including any phase variations due to changes in horizontal wavenumber k_r. Finally, the phase is adjusted according to the wavenumber variations in z depicted in the term $(\delta k^2(z)/2k_r(0))$. To begin another cycle, the left-hand side of Eq. (12.4.12) would be Fourier transformed into the wavenumber spectrum, and so on.

This "spectral marching" solution to the problem of wave propagation has three significant numerical advantages over a more direct approach of applying a finite element method on the time-space wave equation. The first advantage is that the range steps Δr can be made larger than a wavelength and in line with the variations in the media, rather than taking many small steps per wavelength in a finite element computing method. The second advantage is that with so much less computation in a given propagation distance, the solution is much less susceptible to numerical errors growing within the calculated field solution. Finally, the wavenumber space approach also has the advantage of a convenient way to include an impedance boundary along the r-axis at $z = 0$. This is due to the ability to represent the field as the sum of a direct wave, plus a reflected wave, where the reflection factor can also be written as a LaPlace transform (6).

Considering the case of outdoor sound propagation, we have a stratified but turbulent atmosphere and a complex ground impedance $Z_g(r)$. The recursion for calculating the field is

$$\psi(r + \Delta r, z) = e^{j\Delta r \frac{\delta k^2(z)}{2k_r(0)}} \left\{ \frac{1}{2\pi} \int_{-\infty}^{+\infty} \left[\Psi(r, k_z) e^{j\Delta r \sqrt{k_r^2(0) - k_z^2(r)}} \right] e^{+jk_z z} dk_z \right.$$

$$+ \frac{1}{2\pi} \int_{-\infty}^{+\infty} \left[\Psi(r, -k_z) R(k_z) e^{j\Delta r \sqrt{k_r^2(0) - k_z^2(r)}} \right] e^{+jk_z z} dk_z \qquad (12.4.13)$$

$$\left. + 2j\beta(r) e^{-j\beta(r)z} e^{j\Delta r \sqrt{k_r^2(0) - \beta^2(r)}} \Psi(r, \beta) \right\}$$

where $\beta(r) = k_r(0)/Z_g(r)$ and $R(k_z) = [k_z(r)Z_g(r) - k_r(0)]/[k_z(r)Z_g(r) + k_r(0)]$. The ground impedance is normalized to $\rho c = 415$ Rayls, the specific acoustic impedance for air. As $Z_g \to 1$, $R(k_z) \to 0$ and no reflection occurs. This condition also drives $\beta(r) \to k_r(0)$, the reference horizontal wavenumber at ground level. This essentially places the horizontal wavenumber in the vertical direction and no phase change occurs as the result of the range step Δr. This high frequency spatial wave is then completely canceled in the next range step. Thus, as $Z_g(r) \to 1$ we have the no-boundary propagation case of Eq. (12.4.12).

The first term on the right-side of Eq. (12.4.13) represents the direct wave from the source, the second term is the reflected wave from the boundary (or the wave from the image source), and the third term is a surface wave which results from the ground impedance being complex and a spherical wave is interacting with a planar boundary surface representing the ground. The physics behind the third term are intriguing. One can see that the term $\Psi(r, \beta)$ is a single complex number found from the LaPlace transform of $\Psi(r, z)$ for the complex wavenumber β (assuming Z_g is also complex). This complex number is then phase shifted and attenuated for the range step (note that with β complex one has real and imaginary exponent elements). The complex result is then multiplied by $2j\beta$ and finally the function $e^{-j\beta z}$, which adds to the field at every elevation z. Generally, the amplitude of the surface wave decreases exponentially as one moves away from the surface. At the next range step iteration, the entire field is again used to derive the surface wave, which is then in turn added back into the field. It has been observed both numerically and experimentally that for some combinations of point source height, frequency, and ground impedance, the surface wave can propagate for substantial distances. We also note that because of the complex nature of β, the surface wave does not propagate as fast as the direct wave. All of this complexity makes measurement of inhomogeneous fields and impedance boundaries using arrays of sensors and propagation modeling an area for future scientific and economic exploitation.

Figure 25 shows the result of a Green's function-based parabolic wave equation in a "downward refracting" atmosphere. That is, one where sound travels faster in the propagation direction at higher altitudes than near the ground. This situation arises typically at night when cold air settles near the ground, (the wave speed is slower in colder air) or when sound is propagating in a downwind direction. The wind speed adds to the sound speed and winds are generally stronger at higher elevations. In the lower right corner, one can see the interaction of the downward refracting sound rays with the surface wave, causing a standing wave pattern. In the non-virtual world, turbulent fluctuations cause this standing wave field to move around randomly. To the listener on the ground, the source is heard to fade in and out of detection. The fluctuating multipath effect can also be experienced with short-wave or AM radio where the ionosphere and ground form a spherical annular duct which results in non-stationary multipath from transmitter to receiver.

Figure 26 shows a 100 Hz source in an upward refracting atmosphere. Such is the case during a sunny afternoon or when sound is propagating in the upwind direction. For upwind propagation, the wind speed is subtracted from the sound

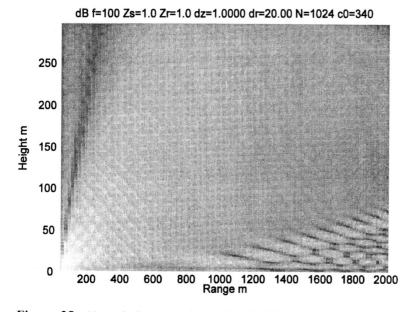

dB f=100 Zs=1.0 Zr=1.0 dz=1.0000 dr=20.00 N=1024 c0=340

Figure 25 Numerical propagation model of a 100 Hz point source at 1 m elevation over a $Z_g = 12.81 + j11.62$ ground impedance in an atmosphere where $c(z) = 340 + 0.1\,z\,\text{m/sec}$.

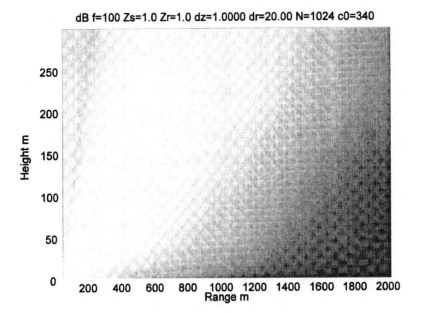

dB f=100 Zs=1.0 Zr=1.0 dz=1.0000 dr=20.00 N=1024 c0=340

Figure 26 Propagation model of the 100 Hz source with an upward refracting atmosphere where $c(z) = 340 - 0.1\, z$ m/sec.

speed and the wind speed generally increases with elevation. During a sunny afternoon, the air near the ground is heated significantly. Since the part of the sound wave close to the ground is opposing a slower wind and has a faster sound speed due to the high temperature, the near-ground part of the wave outruns the upper level part of the wave. Thus the wave refracts away from the ground, hence upward refracting propagation. Clearly, one can see a rather stark falling off of the wave loudness on the ground as compared to the downward refracting case. The dark, "quiet" area in the lower right of Figure 26 is known as a shadow zone where in theory, little or no sound from the source can penetrate. This is of extreme importance to submarine operations because shadow zones provide natural sonic hiding places. In the atmosphere, common sense tells us that is should be easier to hear a source when it is upwind of the listener, and that long range detection of sound in the atmosphere should be easier at night if for no other reason the background noise is low.

Figure 27 shows the effect of turbulence on propagation in an upward refracting environment. This is typical during a hot afternoon because the hot air near the ground becomes buoyantly unstable, and plumes upward drawing in cooler air from above to replace it. Thermal pluming generally leads to a build up of surface winds as various areas heat at differing rates. These surface winds are very turbulent due to the drag caused by the ground and its objects (trees, buildings, etc.). The parabolic wave equation propagation modeling technique allows the inclusion of turbulence as a simple "phase screen" to be added in at each range step in the algorithm. This alone illustrates the power of wavenumber filtering to achieve what otherwise would likely be an extraordinarily difficult and problematic modeling effort.

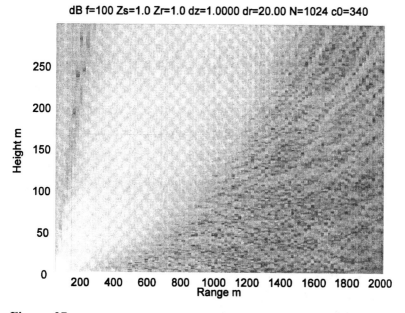

dB f=100 Zs=1.0 Zr=1.0 dz=1.0000 dr=20.00 N=1024 c0=340

Figure 27 100 Hz source in the upward refracting atmosphere including turbulence effects using a Gaussian phase fluctuation with a standard deviation of about 1% of the speed of sound.

Physical considerations for wave propagation modeling can be categorized into time scales, environmental models for wave speed profiles, knowledge of wave scattering attributes, and knowledge of impedances of the dominant boundaries. The model is only a good as the physical assumptions incorporated. For example, underwater sound propagation will have time scales that are mostly seasonal while atmospheric sound propagation will certainly have diurnal (daily) cycles, but also significant changes with local weather conditions. Radio communications channels will have a diurnal cycle, but also one synchronized with solar flares and the earth's magnetic field activity. For any given environmental state, one must have detailed models which allow an accurate wave speed profile. This requires expertise in surface-layer meteorology for atmospheric sound propagation, oceanography for underwater sound propagation, and oceanography, meteorology, and astrophysics and radio wave propagation modeling. Refraction is defined as an effect which changes the direction of a ray in a spatially coherent manner, such as seen in Figures 25 and 26 and in other devices such as a lens for an optical wave. Scattering is defined as an effect where the ray tends to be redirected in multiple directions in a spatially incoherent manner. In the non-virtual world, the physics which cause diffraction and scattering are often the same and it is very difficult to prove the precise nature of wave propagation in inhomogeneous media. However, from a wave processing perspective, we can say that modeling these effects with a wavenumber filter which "leaks" incoherent wave energy is a reasonable way to including the effects of random diffraction and scattering for inhomogeneous wave propagation modeling.

Numerical considerations of the Green's function parabolic equation are extremely important because the wave "step marching" approach carries a significant danger of numerical error accumulation. One of our initial assumptions was that the wave propagation is largely in the horizontal r-direction. With the point source near the left boundary in Figures 25–27, this means that the results in the upper left corner of the figures are likely in error due to the steep angle of the waves relative to the horizontal step direction. The assumption that the variations in wave speed are small compared to the mean wave speed are generally true, but one should certainly check to see if the square-root approximation in Eq. (12.4.10) is reasonable.

Another numerical consideration is the wave amplitude along the horizontal boundary, if present in the problem. For undersea acoustic propagation modeling the boundary condition at the sea surface is zero pressure (a pressure release boundary condition). However, for atmospheric sound propagation the acoustic pressure along the boundary is not zero and it is advisable to use a trapezoidal or higher order integration to prevent propagation of significant numerical error. At the upper elevation of the virtual wave propagation space, an artificial attenuation layer is needed to suppress wave "reflections" from the top of the space. Again, in underwater propagation one usually has a natural attenuation layer at the bottom sea floor which is usually a gradual mixture of water and sediment which does not reflect sound waves well. There are likely analogous electromagnetic boundary conditions which also require special handling of numerical error. These extremely important aspects of wave propagation are best handled by controlled experiments for validation.

12.5 SUMMARY, PROBLEMS, AND BIBLIOGRAPHY

The four sections of this Chapter bring together the theoretical and practical aspects of processing wavenumber signals as measured coherently by spatial arrays of sensors. Section 12.1 presented the derivation of the Cramer–Rao lower bound for error on statistical parameter estimates. We then used these results to analyze the precision of wave bearing angle estimates both as a direct phase measurement and as a beamformed result in Section 12.2. A new physics-based approach to describing array beamforming is seen using 2-dimensional wavenumber transforms of the array shape and then applying a wavenumber contour directly to the wavenumber spectrum to determine the array response. The interesting aspect of this presentation is that one can directly observe the beam response over a wide range of frequencies and steering angles in a single graph. Section 12.3 carries beamforming further to the problem of field reconstruction using wave field holography, a scanning array of sensors, and a wavenumber domain Green's function. By proper inversion of this Green's function, one can reconstruct the field in a plane in the region between the source(s) and the sensor array. This is useful for identifying wave sources and direct observations of the wave radiating mechanisms. Section 12.4 applies the wavenumber domain Green's function to model the propagation of waves in the direction away from the sources and sensor array surface. Wave propagation modeling has obvious value for many currently practical applications.

Fundamentally, measurements of spatial waves are all based on one's ability to observe spatially the time-of-arrival or phase at a given frequency. These types of measurements all have some SNR limitation which can be objectively determined.

As one then computes a physical quantity such as bearing, or beam output at a given look direction, the confidence of the time-of-arrival or phase is scaled according to the SNR enhancement of the array, array shape, and look direction. When the array beam is focused in a near range, rather than infinity (for plane waves from a distance source) one can apply the same principles to determine source localization error, or resolution of the holographic process. For the analytical Green's function, this resolution is mainly a function of array aperture and distance between the measurement plane and the reconstruction plane. However, simply increasing the number of scanning elements does not improve long distance resolution but rather improves resolution in the region close to the measurement array. Increasing sensor array aperture generally improves source localization accuracy at long distances for the same reason that it also narrows beam width for a plane wave beamforming array. Clearly, one can see from Chapter 12 a consistent framework for wavenumber-based measurements and processing in sensor systems. Since these measurements are ultimately used in an "intelligent" decision process, the associated statistical confidences are of as much importance as the measurement data itself to insure proper weighting of the various pieces of information used. What makes a sensor system output "intelligent" can be defined as the capability for high degrees of flexibility and adaptivity in the automated decision algorithms.

PROBLEMS

1. A noisy dc signal is measured to obtain the mean dc value and the rms value of the noise. The variance is assumed to be 100 mv^2 or 10 mv rms. How many samples of the voltage would be needed to be sure that the calculated dc voltage mean has a variance less than 1 mv^2?

2. A 5 m line array of 24 sensors is used to estimate the arrival angle of a 500 Hz sinusoid in air. Assuming the speed of sound is 345 m/sec and all sensors are used to estimate the bearing (assumed to be near broadside), what is the rms bearing error in degrees for one data snapshot?

3. A circular array 2 m in diameter has 16 equally-spaced sensors in water ($c = 1500$ m/s).

   (a) What is the highest frequency one can estimate bearings for?
   (b) Define an orientation and calculate the relative phases at 1 kHz for a bearing angle of $90°$.

4. Using the array in problem 3b, how many sources can be resolved at the same frequency?

5. What is the resolution beamwidth (assume -3 dB responses of the beam determine beamwidth) in degrees at 2 kHz, 1 kHz and 200 Hz for the arrays in problems 2 and 3?

6. An air conditioner is scanned by a line array of seven microphones where one microphone is held at a constant position as a reference field sensor to obtain an 8×8 grid of magnitude and phases at 120 Hz covering a 2 m by 2 m area. What is the spatial resolution of the holography system?

7. A survivor in a lifeboat has an air horn which is 100 dB at 1 m and 2 kHz. There are foggy (downward refracting conditions) such that there is

spherical wave spreading for the first 1000 m, then circular spreading of the wave in the duct just above the water. If the background noise ashore is 40 dB, how far out to sea can the horn be heard by a rescuer?

BIBLIOGRAPHY

W. S. Burdic Underwater Acoustic System Analysis, Englewood Cliffs: Prentice-Hall, 1991.

H. Cramer Mathematical Methods of Statistics, Princeton University Press, 1951, Section 32.3.

K. E. Gilbert, X. Di "A Fast Green's Function Method for One-Way Sound Propagation in the Atmosphere," Journal of The Acoustical Society of America, 94(4), Oct., 1993.

V. H. MacDonald, P. M. Schultheiss "Optimum Passive Bearing Estimation in a Spatially Incoherent Noise Environment," Journal of The Acoustical Society of America, 46(1), 1969.

S. J. Orfanidis Optimum signal Processing, 2nd ed., New York: McGraw-Hill, 1988.

C. R. Rao Linear Statistical Inference and Its Applications, 2nd ed., New York: Wiley, 1973.

E. Skudrzyk The Foundations of Acoustics, New York: Springer-Verlag, 1971.

S. Temkin Elements of Acoustics, New York: Wiley, 1981.

E. G. Williams, J. D. Maynard, E. Skudrzyk "Sound Source Reconstructions Using a Microphone Array," Journal of The Acoustical Society of America, 68(1), 1980.

D. K. Wilson "Performance bounds for direction-of-arrival arrays operating in the turbulent atmosphere," Journal of The Acoustical Society of America, 103(3), March, 1998.

REFERENCES

1. H. Cramer, Mathematical Methods of Statistics, Princeton University Press, 1951, section 32.3.

2. C. R. Rao, Linear Statistical Inference and Its Applications, 2nd ed, New York: Wiley, 1973.

3. E. G. Williams, J. D. Maynard, E. Skudrzyk, "Sound source reconstructions using a microphone array," Journal of The Acoustical society of America, 68(1), July 1980, pp. 340–344.

4. E. Skudrzyk The Foundation of Acoustics, New York: Springer-Verlag, 1971, Section 23.5.

5. K. E. Gilbert, X. Di "A Fast Green's Function Method for One-Way Sound Propagation in the Atmosphere," Journal of The Acoustical Society of America, 94(4), Oct., 1993, pp. 2343–2352.

6. X. Di, K. E. Gilbert "An Exact LaPlace Transform Formulation for a Point Source Above a Ground Surface," Journal of The Acoustical Society of America, 93(2), Feb, 1993, pp. 714–720.

13

Adaptive Beamforming

Adaptive beamforming is used to optimize the signal-to-noise ratio (SNR) in the look direction by carefully steering nulls in the beampattern towards the direction(s) of interference sources. From a physical point-of-view, one should be clear in understanding that at any given frequency the beamwidth and SNR for spatially incoherent noise (random noise waves from all directions) is dictated by physics, not signal processing algorithm. The larger the array the greater the SNR enhancement will be in the look direction provided all array elements are spaced less than a half-wavelength. The more array elements one has, the wider the temporal frequency bandwidth will be where wavenumbers can be uniquely specified without spatial aliasing. However, when multiple sources are radiating the same temporal frequency from multiple directions, it is highly desirable to resolve the sources and directions as well as provide a means to recover the individual source wavenumber amplitudes. Examples where multiple wavenumbers need to be resolved can be seen in sonar when large boundary is causing interfering reflections, in radio communications where an antenna array might be adapted to control multipath signal cancellation, and in structural acoustics where modes of vibration could be isolated by appropriate processing of accelerometer data.

Beamforming can be done on broadband as well as narrowband signals. The spatial cross correlation between array sensors is the source of directional (wavenumber) information, so it makes no physical difference whether the time-domain waveform is broadband (having many frequencies) or narrowband (such as a dominant sinusoid). However, from a signal processing point-of-view, narrowband signals are problematic because the spatial cross correlation functions are also sinusoidal. For broadband signals, the spatial cross correlation functions will yield a Dirac delta functions representing the time difference of arrivals across the sensor array. A narrowband spatial cross correlation is a phase-shifted wavenumber, where multiple arrival angles correspond to a sum of wavenumbers, or modes. This is why an eigenvalue solution is so straightforward. However, the sources radiating the narrowband frequency which arrive at the sensor array from multiple directions must be independent in phase for the spatial cross correlation to be unique. From a physical point-of-view, source independence is plausible for say, acoustics where the two sources are vehicles which coincidentally happen to have the same temporal frequency for a period of time. This might also be physically true for multipath propagation from a single source where the path lengths are randomly fluctuating. But, for vibration source multipath in a mechanical structure, or radio transmission multipath from reflections off of buildings, the multipath is *coherent*, meaning that the phases of the wave arrivals from different directions are not statistically independent. Thus, the coherent multipath situation gives sensor spatial cross correlations which depend on the phases of the wave arrivals, and thus, does not allow adaptive beamforming in the traditional sense.

We present adaptive beamforming using two distinct techniques which parallel the two adaptive filtering techniques (block least squares and projection-based least-squared error). The block technique is used in Section 13.1 to create a spatial whitening filter where an FIR filter is used to predict the signal output of one of the array sensors from the weighted sum of the other array sensors. The wavenumber response of the resulting spatial FIR whitening filter for the array will have nulls corresponding to the directions of arrival of any spatially coherent source. By relating the wavenumber of a null to a direction-of-arrival through the geometry of the array and the wave speed, a useful multipath measurement tool is made. If the sources and background noise are all broadband, such as Gaussian noise, the null-forming and direction-of-arrival problem is actually significantly easier to calculate numerically. This is because the covariance and cross-correlation matrices are better conditioned. When the sources are narrowband, calculation of the covariance and cross-correlation matrices can be problematic due to the unknown phases and amplitudes of the individual sources. This difficulty is overcome by compiling a random set of data "snapshots" in the estimation of the covariance and cross-correlation matrices. This reduces the likelihood that the phases of the individual sources will present a bias to the covariance and cross-correlation matrices, which should be measuring the amplitude phase due to the spatial interference, not the relative phases and amplitudes of the sources.

To be able to construct a beam for a desired look direction while also nulling any sources in other directions, we need to process the eigenvectors of the covariance matrix as seen in Section 13.2. This parallels the projection operator approach in Section 8.2 for calculating the least-squared error linear predictor. The covariance matrix has a Toeplitz structure, meaning that all the diagonals have the same number in each element. This numerical structure leads to a unique physical interpretation of the array signal covariance eigenvectors and eigenvalues. Assuming the signals from the sources in question have a reasonably high SNR (say 10 or better) the solved eigenvalues can be separated into a "signal subspace" and a "noise subspace", where the largest eigenvalues represent the signals. Each eigenvector represents an array FIR filter with a wavenumber response with nulls for all source locations except the one corresponding to itself. The "noise eigenvectors" will all have nulls to the signal source wavenumbers and have additional spurious nulls in their wavenumber responses if there are fewer than $M - 1$ sources for a M-element array. To obtain the whitening filter result one can simply add all the noise eigenvectors, each of which has the nulls to the sources. To obtain a beam in the desired look direction with nulls to all the other "interfering sources" one simply post-multiplies the inverse of the covariance matrix by the desired steering vector! This very elegant result is known as a minimum variance beamformer because it minimizes the background noise for the desired look direction.

We present another technique in Section 13.3 which is highly useful when one can control the transmitted signal such as in active radar/sonar and in communications. By encoding the phase of the signal with a random periodic sequence, one can uniquely define all the wavenumbers at the receiving array whether the effect is multisource multipath or propagation multipath. This is one reason why spread-spectrum signals are so useful in communication systems. By modulating the narrowband signal phase according to a prescribed unique repeating sequence, we are effectively spreading the signal energy over a wider bandwidth. A Fourier transform actually shows a reduced SNR, but this is recoverable at the receiver if one knows the phase modulation sequence. By cross-correlating the known signal's unique spread spectrum sequence with the received signal in a matched detection filter, the SNR is recovered and the resulting covariance is well-conditioned to extract the multipath information. This technique not only allows us to recover low SNR signals but also allows us to make multipath propagation measurements with an element of precision not available with passive narrowband techniques.

13.1 ARRAY "NULL-FORMING"

Fundamental to the traditional approach to adaptive beamforming is the idea of exploiting some of the received signal information to improve the SNR of the array output in the desired look direction. Generally, this is done by simply steering an array "null", or direction of zero-output response, to a useful direction other than the look direction where another "interfering" source can be suppressed from the array output to improve SNR (actually signal-to-interference ratio). The array SNR gain and look-direction beamwidth are still defined based on the wavelength to aperture ratio, but an additional interference rejection can be obtained by adaptively "nulling" the array output in the direction of an interfering source. When sweeping the look-direction beam around to detect possible target signals, the application of adaptive null-forming allows one to separate the target signals quite effectively. However, the spatial cross correlation of the array signals must have source independence and no source dependance on background noise for the process to work.

A straightforward example of the issue of source phase independence is given to make the physical details as clear as possible. This generally applies to narrowband temporal signals (sinusoids) although it is possible that one could have coherent broadband sources and apply the narrowband analysis frequency by frequency. Generally speaking, incoherent broadband sources such as multiple fluid-jet acoustic sources or electric arc radio-frequency electromagnetic sources will have both temporal and spatial correlation function approximating a Dirac delta function. For the narrowband case, consider two distant sources widely separated which radiate the same frequency to a sensor at the origin from two directions θ_1 and θ_2 (measured counter clockwise from the horizontal positive x-axis) with received amplitudes A_1 and A_2 and phases ϕ_1 and ϕ_2.

$$X_0 = A_1 e^{j\phi_1} + A_2 e^{j\phi_2} \tag{13.1.1}$$

If we have a line array of sensors along the positive x-axis each at location $x = d$, we can write the received signals from the two sources relative to the X_0 in terms of a simple phase-shift.

$$X_{d_\ell} = A_1 e^{j(\phi_1 + kd_\ell \cos\theta_1)} + A_2 e^{j(\phi_2 + kd_\ell \cos\theta_2)} \tag{13.1.2}$$

Clearly, an arrival angle of 90 degrees has the wavefront arriving simultaneously at all sensors from the broadside direction. The spatial cross correlation between X_0 and an arbitrary sensor out of the line array X_d is defined as

$$
\begin{aligned}
R_{-d_\ell} &= E\left\{X_0 X_{d_\ell}^*\right\} = E\left\{(A_1 e^{j\phi_1} + A_2 e^{j\phi_2})(A_1 e^{-j(\phi_1 + kd_\ell \cos\theta_1)} + A_2 e^{-j(\phi_2 + kd_\ell \cos\theta_2)}\right\} \\
&= E\left\{\left[A_1^2 + A_2 A_1 e^{-j(\phi_1 - \phi_2)}\right]e^{-jkd_\ell \cos\theta_1} + \left[A_2^2 + A_1 A_2 e^{-j(\phi_1 - \phi_2)}\right]e^{-jkd_\ell \cos\theta_2}\right\}
\end{aligned}
\tag{13.1.13}
$$

Note that the spatial cross correlation is a function of the amplitudes and phases of the two sources! This means that the spatial information for the two arrival angles and amplitudes is not recoverable unless one already knows the amplitude and phases of the two narrowband sources. However, if the sources are statistically independent, the phase difference between them will be a random angle between $\pm\pi$. Therefore, given a sampling of "snapshots" of spatial data from the array to estimate the expected value in Eq. (13.1.3) one obtains

$$R_{-d_\ell} = A_1^2 e^{-jkd_\ell \cos\theta_1} + A_2^2 e^{-jkd_\ell \cos\theta_2} \tag{13.1.4}$$

since the expected value of a uniformly distributed complex number of unity magnitude is zero. The wavenumber response of the spatial cross correlation given in Eq. (13.1.4) has the Dirac delta function structure (delta functions at $d_\ell \cos\theta_1$ and $d_\ell \cos\theta_2$) in the time-space domain as expected of two incoherent broadband sources. One can see that the linear combination of two or more sources for each spatial cross correlation can be resolved by either linear prediction or by an eigenvalue approach without much difficulty so long as there is at least one more sensor than sources.

In practice, one has to go to some effort to achieve this phase independence between sources. The term "spatially incoherent radiating the same frequency" is actually a contradiction. Since the time derivative of phase is frequency one would do better to describe the desired situation as having "sources of nearly identical frequency processed in a wider frequency band with multiple observations to achieve spatial phase independence in the average". For example, if one had two sources within about 1 Hz of each other to be resolved spatially using signals from a large sensor array with a total record length of a second, one would segregate the record into, say, 10 records of 100 msec length each, and then average the 10 available snapshots to obtain a low-bias spatial cross correlation for a 10 Hz band containing both sources. This is better than trying to resolve the sources in a 1 Hz band from a single snapshot because the phases of the sources will present a significant bias to the spatial cross correlation for the array. Another alternative (if the array is large enough) would be to divide the array snapshot into a number of "sub-arrays" and calculate an ensemble average spatially of each sub-array's cross correlation to get an overall cross correlation with any dependence of the source phases suppressed. This technique is known as the spatial smoothing method for dealing with coherent signals. It is very valuable for the design to understand the physical reason the correlation bias arises for coherent sources in the first place and the techniques required to suppress this coherence. For the rest of this section and Section 13.2, we will assume spatially incoherent sources. A better way to depict this assumption is to assume the sensor system has provided the additional algorithms necessary to insure incoherent sources in the covariance matrix data.

Consider the spatial "whitening" filter in Figure 1 which uses a weighted linear combination of the signals from sensors 1 through M to "predict" the signal from sensor 0. The filter coefficients (or weights in the linear combiner) are determined by minimizing the prediction error. Over N time snapshots, the prediction error is simply

$$
\begin{bmatrix} \varepsilon_t \\ \varepsilon_{t-1} \\ \varepsilon_{t-2} \\ \vdots \\ \varepsilon_{t-N+1} \end{bmatrix} = \begin{bmatrix} x_{0,t} \\ x_{0,t-1} \\ x_{0,t-2} \\ \vdots \\ \varepsilon_{t-N+1} \end{bmatrix}
$$
$$
- \begin{bmatrix} x_{1,t} & x_{2,t} & x_{3,t} & \cdots & x_{M,t} \\ x_{1,t-1} & x_{2,t-1} & x_{3,t-1} & \cdots & x_{M,t-1} \\ x_{1,t-2} & x_{2,t-2} & x_{3,t-2} & \cdots & x_{M,t-2} \\ \vdots & \vdots & \vdots & \cdots & \vdots \\ x_{1,t-N+1} & x_{2,t-N+1} & x_{3,t-N+1} & \cdots & x_{M,t-N+1} \end{bmatrix} \begin{bmatrix} h_1 \\ h_2 \\ h_3 \\ \vdots \\ h_M \end{bmatrix}
$$

$$(13.1.5)$$

where our array signals could be either broadband real time data or complex narrowband FFT bin values for a particular frequency of interest. For the

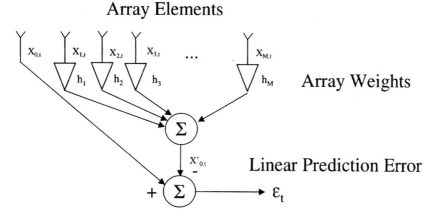

Figure 1 A spatial linear prediction filter can be used to determine the wavenumbers corresponding to the arrival angles of spatially incoherent source signals.

narrowband frequency-domain case, the coefficients are all complex. The relationship between these coefficients and the wavenumbers and physical arrivals angles of interest will be discussed momentarily. Equation (13.1.5) can be written compactly in matrix form as

$$\bar{\varepsilon} = \bar{x}_0 - \bar{X}H \tag{13.1.6}$$

Recalling from Section 8.1, the block least-squared error solution for the coefficients is simply

$$H = (\bar{X}^H \bar{X})^{-1} (\bar{X})^H \bar{x}_0^H \tag{13.1.7}$$

Applying the well-known least-squared error result in Eq. (13.1.7) to the spatial whitening problem can be seen by writing the array output as the spatial prediction error in one row of Eq. (13.1.6).

$$\varepsilon_t = x_{0,t} - \sum_{m=1}^{M} h_m x_{m,t} \tag{13.1.8}$$

Note that for the narrowband case, the t subscripts in Eq. (13.1.8) can refer to the complex value of an FFT bin at time t without any loss of generality. The array output can be written as the output of a spatial FIR filter.

$$\varepsilon_t = \sum_{m=0}^{M} A_m x_{m,t} \quad A_0 = 1, \quad A_m = -h_m \tag{13.1.9}$$

The spatial whitening FIR filter $A(z)$ can now be evaluated as a function of a complex variable $z = e^{-jkd_{\ell_m}\cos\theta}$ where d_{ℓ_m} is the x-coordinate of the mth sensor in the line array, k is the wavenumber, and θ is the direction of interest. This has an analogy in the frequency response being evaluated on the unit circle on the z-plane defined by $z = e^{j\omega T}$. However, while it was convenient to write $A(z)$ as a polynomial with integer powers of z when the time-domain signal samples are evenly spaced.

But, for our implementation of a spatial FIR filter, the sensors are not necessarily evenly spaced. The wavenumber response of our spatial whitening FIR filter is found by a straightforward Fourier sum using the appropriate wavenumber for the array geometry. For the line array, the wavenumber simply scales with the cosine of the arrival angle, which is limited to the unique range of 0–180 degrees by symmetry.

$$D(\theta) = \sum_{m=0}^{M} A_m e^{-jkd_{\ell_m}\cos\theta} \tag{13.1.10}$$

If the array is an evenly-spaced line array along the x-axis, the most efficient way to get the wavenumber response is to compute a zero-padded FFT where the array coefficients are treated as a finite impulse response (see Section 3.1). The digital domain wavenumber response from $-\pi$ to $+\pi$ is first divided by kd (wavenumber times element spacing) and then an inverse cosine is computed to relate the k to angle θ. If the array sensors are part of a two-dimensional array and each located at $x = d_{x_m}$ and $y = d_{y_m}$ the directional response of the spatial whitening filter is simply

$$D(\theta) = \sum_{m=0}^{M} A_m e^{-jkd_{x_m}\cos\theta} e^{-jkd_{y_m}\sin\theta} \qquad (13.1.11)$$

where k represents the wavenumber in the x–y plane. If the plane wave can arrive from any angle in three dimensions and we have a two-dimensional planar sensor array, the wavenumber in the x–y plane scales by $k' = k\sin\gamma$, where γ is the angle from the positive z-axis to the ray normal to the plane wavefront and k is the free propagation wavenumber for the wave ($k/c = 2\pi/\lambda$). If the wave source is on the x–y plane ($\gamma = 90$ degrees), $k' = k$. For a three-dimensional array and plane wave with arrival angles θ (in the x–y plane) and γ (angle from the positive z-axis), the directional response of the spatial whitening filter is given in Eq. (13.1.12) where d_{z_m} is the z-coordinate of the mth sensor in the three-dimensional array and γ is the arrival angle component from the positive z-axis.

$$D(\theta, \gamma) = \sum_{m=0}^{M} A_m e^{-jkd_{x_m}\cos\theta\sin\gamma} e^{-jkd_{y_m}\sin\theta\sin\gamma} e^{-jkd_{z_m}\cos\gamma} \qquad (13.1.12)$$

The coefficients of $A(z)$ are found by the least-squares whitening filter independent of the array geometry or wavenumber analysis. The spatial whitening filter is simply the optimum set of coefficients which predict the signal from one sensor given observations from the rest of the array. For the linear evenly-spaced array, one can compute the roots of the polynomial where the zeros closest to the unit circle produce the sharpest nulls, indicating that a distinct wavenumber (wave from a particular direction) has passed the array allowing cancellation in the whitening filter output. The angle of the polynomial zero on the z-plane is related to the physical arrival angle for the line array by the exponent in Eq. (13.1.10). In two and three dimensions where the array sensors are not evenly-spaced, root-finding is not so straightforward and a simple scanning approach is recommended to find the corresponding angles of the nulls. For example, the array response is spatially whitened at a particular free wavenumber k corresponding to radian frequency ω. The FFT bins corresponding to ω are collected for N snapshots and the spatial whitening filter coefficients are computed using Eqs (13.1.5)–(13.1.9). To determine the source arrival angles, one evaluates Eq. (13.1.12) for $0 < \theta < 360$ degrees and $0 < \gamma < 180$ degrees, keeping track of the dominant "nulls" in the array output. A straightforward way to find the nulls is to evaluate the inverse of the FIR filter for strong peaks. This has led to the very misleading term "super resolution array processing" since the nulls are very sharp and the inverse produces a graph with a sharp peak. But this is only graphics! The array beamwidth is still determined by wavelength and aperture and angle measurement is still a function of SNR and the Cramer–Rao lower bound.

To illustrate an example of adaptive "null-forming" Figure 2 shows the result of three 100 Hz sources at 20, 80, and 140 degrees bearing with a 10 m 8 element line array along the x-axis. The SNR in the 100 Hz FFT bin is 10 dB and the whitening coefficients are estimated using 10 snapshots. The bearings of the three 100 Hz sources are clearly seen and the relative phases for the three sources are assumed uniformly distributed between 0 and 2π for each of the 10 snapshots. With the source

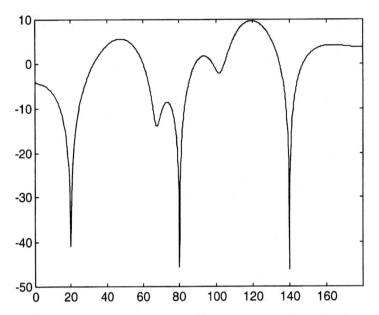

Figure 2 Spatial whitening filter directional response for three 10 dB SNR 100 Hz sources at bearing angles of 20, 80, and 140 degrees on a 8 element 10 m line array (dB vs. degrees).

phases distributed over a narrower range for each snapshot, the results are not nearly as consistent. One might also think that the higher the SNR, the better the result. But this too is not entirely true. Since for an $M + 1$ element line array one can resolve a maximum of M sources, some noise is required so that the $M \times M$ matrix inverse in Eq. (13.1.7) is well-conditioned. But in low SNR (0 dB or less), the magnitude and phase of the signal peak is not strong enough relative to the random background noise to cancel. Using more snapshots improves the whitening filter performance, since the background noise is incoherent among snapshots. Finally, the depth of the null (or height of the peak if an inverse FIR response is evaluated) is not reliable because once the source wavenumber is canceled, the residual signal at that wavenumber is random background noise causing the null depth to fluctuate.

The spatial resolution for the nulls is amazingly accurate. Figure 3 shows the detection of two sources only 2 degrees apart. However, when one considers the Cramer–Rao lower bound (Section 12.1) for a 10 m array with 8 elements, 10 dB SNR, and 10 snapshots, the bearing accuracy should be quite good. In practical implementations, the bearing accuracy is limited by more complicated physics, such as wave scattering from the array structure, coherent phase between sources, and even coherent background noise from wave turbulence or environmental noise.

13.2 EIGENVECTOR METHODS OF MUSIC AND MVDR

One can consider the wavenumber traces from multiple sources arriving at an array of sensors from different angles at the same temporal frequency as a sum of modes, or eigenvectors. In this section we explore the general eigenvector approach to resolving multiple (independent phase) sources. Two very popular algorithms for

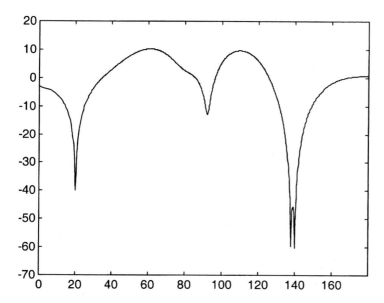

Figure 3 Whitening filter response with 10 dB SNR, 10 snapshots, and three 100 Hz sources at 20, 138, and 140 degrees bearing for the 10 m 8 element array (dB vs. degrees).

adaptive beamforming are MUSIC (MUltiple SIgnal Classification) and MVDR (Minimum Variance Distortionless Response). A simple example of the physical problem addressed by adaptive beamforming is a line array of evenly-spaced sensors where the arriving plane waves are measured by the magnitude and phase of each array sensor output for temporal FFT bin for the frequency of interest. If a plane wave from one distant source in a direction normal to the array (broadside) line is detected, the magnitude and phase observed from the array sensors should be identical. From a practical but very important standpoint, this assumes the sensors are calibrated in magnitude and phase and that no additional phase difference between sensor channels exists from a shared analog-to-digital convertor. The spatially constant magnitude and phase can be seen as a wavenumber at or near zero (wavelength infinite). But, if the array is rotated so that the propagation direction is along the line array axis (endfire), the wavenumber trace observed by the array matches the free wavenumber for the wave as it propagates from the source. Therefore, we can say in general for a line array of sensors that the observed wavenumber trace $k' = k \cos \theta$, where k is the free wavenumber and θ is the angle of arrival for the wave relative to the line array axis. Figure 4 shows the spatial phase response for a 150 Hz plane wave arriving from 90 degrees (broadside), 60 degrees, and 0 degrees (endfire).

Even though multiple sources may be radiating the same temporal frequency, the wavenumber traces observed by the sensor array will be different depending on the angle of arrival for each source. From a spatial signal point-of-view, the multiple sources are actually producing different wavenumber traces across the array, which are observed as different wavelength traces. Therefore, the spatial response of the array can be seen as a sum of eigenvectors representing the mode "shapes" and corresponding eigenvalues representing the spatial frequency of

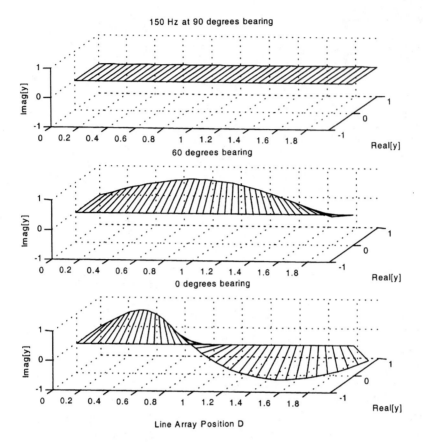

Figure 4 Complex spatial response of a 150 Hz plane wave in air ($c = 345$ m/sec) for broadside (top), 60 degrees bearing (middle), and endfire (bottom) directions.

the corresponding wavenumber trace observed by the array. Figure 5 shows the spatial complex response for the case where all three plane waves in Figure 4 are active simultaneously. Both the magnitude and phase are distorted by the combination of waves. This distortion changes over time as a result of the phase differences between the multiple sources as described in Eq. (13.1.3) for the spatial cross-correlation. Clearly, the complexity of the mixed waves is a challenge to decompose. However, if we assume either time or spatial averaging (moving the array in space while taking snapshots) where the multiple source bearings remain nearly constant, the covariance matrix of the received complex array sensor signals for a given FFT bin will represent the spatial signal correlations and not the individual source phase interference. Equation (13.2.1) depicts the covariance of the array signals.

$$
\bar{R} =
\begin{bmatrix}
R_0^x & R_1^x & R_2^x & \cdots & R_M^x \\
R_{-1}^x & R_0^x & R_1^x & \cdots & R_{M-1}^x \\
\cdot & & \cdot R_0^x & \cdot & \cdot \\
\cdot & & & \cdot & \cdot \\
R_{-M}^x & R_{-M+1}^x & \cdots & R_{-1}^x & R_0^x
\end{bmatrix}
\qquad
R_{i-j}^x = \sum_{k=0}^{N} x_{i,t-k}^* x_{j,t-k}
\qquad (13.2.1)
$$

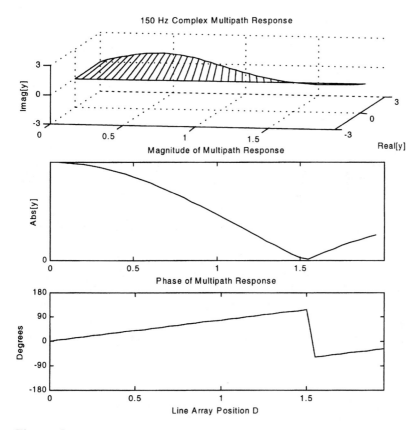

Figure 5 Complex spatial response of the multipath interference caused by three (equal amplitude and phase at sensor 1) 150 Hz plane waves arriving at broadside, 60 degrees and endfire directions.

Note that $x_{i,t}$ could be either a broadband time domain signal where the N time snapshots happen at the sampling frequency, or a narrowband FFT bin where the N snapshots correspond to N overlapped FFT blocks of time data. As suggested earlier, the N snapshots could also be accomplished by either moving the array around or by sampling "sub-arrays" from a much larger array. The snapshot averaging process is essential for the covariance matrix to represent the spatial array response properly. Note also that the covariance matrix in Eq. (13.2.1) is $M + 1$ rows by $M + 1$ columns for our $M + 1$ sensor array. For the nullforming case using a spatial whitening filter in Section 13.1, the inverted covariance matrix in Eq. (13.1.7) was M rows by M columns. In both cases, the covariance matrix has a Toeplitz structure, meaning that all elements on a given diagonal (from upper left to lower right) are the same. This structure is not necessary for matrix inversion, but does indicate that the matrix is well-conditioned for inversion so long as R_0^x on the main diagonal is non-zero. If there is any residual noise in the signals the main diagonal is always non-zero. Given that the inverse exists (the matrix is full rank), one can solve for the eigenvalues (which correspond approximately to the signal "power levels" for each mode), and eigenvectors (which approximately correspond to the mode shape for each mode). One can think of each eigenvector as an array beamsteering weight vector which, when multiplied by the array outputs and summed as a vector dot product, gives an array spatial response (beampattern) with the look direction aligned to the corresponding source direction. The corresponding eigenvalue represents the approximate signal power times the number of array elements and snapshots.

However, both of these physical claims are in practice affected negatively by a low SNR or phase coherence of the sources. "Incoherent sources radiating the same narrowband frequency" is physically impossible unless the "narrowband" frequency is really drifting around in the band randomly. Narrowband frequency modulation is also known as phase modulation (frequency is the time derivative of phase) because very small changes in frequency are seen as phase shifts in the time waveform. Therefore, if the sources or receiving array are moving, or if the sources are physically independent (such as separate vehicles), its not a bad assumption to assume source independence after N snapshots so long as the narrowband bandwidth is not too narrow. In fact, the broader the frequency bandwidth for the covariance matrix, the shorter the required time will be to observe a snapshot, and the faster one will be able to compile a number of snapshots to insure source independence.

Perhaps the single most useful feature of using an eigenvalue approach for adaptive beamforming is that the eigenvectors are *orthonormal*. The dot product of an eigenvector with itself gives unity and that the dot product of any two different eigenvectors is zero. Physically, this means that the spatial response of each eigenvector will have a main lobe in the corresponding source look direction and zeros in the directions of other sources (in theory), or in practice, a response that sums to near zero for other source directions. This is exactly the desired result for adaptive beamforming where one optimizes array performance by getting array SNR gain in the look direction while suppressing sources from other directions. The classic phased-array beam steering of Section 13.1 also provides SNR gain in the look direction of the main lobe, but beam response in the direction of other sources may not be zero leading to interference in the beamformer output.

Assuming the number of sensors is greater than the number of phase-independent sources (this intuitively implies a solution since there are more equations than unknowns), and each eigenvector and eigenvalue correspond to a source, what do the "extra" eigenvectors and eigenvalues correspond to? The extra eigenvector and eigenvalues will correspond to "spurious" sources which appear in the background noise. There will be some amount of spatially independent noise on each sensor output from electronic noise, turbulence (acoustic arrays), and even the least significant bit of the analog-to-digital conversion. This spatially independent noise can be seen as created by an infinite number of sources, so there is no way for the eigenvector approach to resolve them. The "noise eigenvectors" each have a look direction which is random, but will have a zero beam response in the directions of the real signal sources. This is a very nice property because if one sums all the noise eigenvectors, one is left with a beamresponse where the only nulls are in the real source directions, just like the nullforming spatial whitening filter of Section 13.1. This technique is called MUSIC (MUltiple SIgnal Classification). One can then, as in the spatial whitening filter nullforming case, evaluate the sharp nulls in the beam response for the sum of the noise eigenvectors to determine the directions of arrival of the sources. Subsequently, the optimal beam to look in a given direction while nulling all other sources is computed from the covariance matrix in the MVDR algorithm. To have a beam look in the direction of one physical signal source while nulling the other sources, one would use the corresponding eigenvector for the source of interest, or one could synthesize a steering vector based on knowledge of the source directions.

The general eigenvalue problem is defined by the solution of the following equation.

$$\bar{R}v_i = v_i\lambda_i \quad i = 0, 1, \ldots, M \tag{13.2.2}$$

where v_i is a column eigenvector and λ_i is the corresponding scalar eigenvalue. The problem is taught universally in all engineering and science undergraduate curricula because it provides an elegant numerical technique for inverting a matrix, an essential task for matrix linear algebra. From an algorithm point-of-view, Eq. (13.2.2) has three parameters and only one known. However, the structure of the eigenvectors is constrained to be orthonormal as depicted in Eq. (13.2.3) and each of the eigenvalues is constrained to be a scalar.

$$
\begin{aligned}
v_i^H v_j &= 1 \quad i = j \\
&= 0 \quad i \neq j
\end{aligned}
\tag{13.2.3}
$$

We can write Eq. (13.2.2) for all the eigenvectors and eigenvalues

$$\bar{R}A = A\Lambda \quad where \quad A = [v_0 \quad v_1 \quad v_2 \ldots v_M] \tag{13.2.4}$$

and the diagonal matrix of eigenvalues Λ is defined as

$$
\Lambda =
\begin{bmatrix}
\lambda_0 & 0 & 0 & 0 & \ldots & 0 \\
0 & \lambda_1 & 0 & 0 & \ldots & 0 \\
\vdots & & & \vdots & & \vdots \\
0 & \ldots & 0 & 0 & 0 & \lambda_M
\end{bmatrix}
\tag{13.2.5}
$$

Note that $A^H A = I$, which allows $A^{-1} = A^H$. The superscript "H" means Hermitian transpose where the elements are conjugated (imaginary part has sign changed) during the matrix transpose. We also refer to a Hermitian matrix as one which is conjugate symmetric.

We are left with the task of finding the eigenvectors in the columns of the matrix A defined in Eq. (13.2.4). This operation is done by "diagonalizing" the covariance matrix.

$$A^H \bar{R}A = A^H A\Lambda = \Lambda \quad since \quad A^H A = I \tag{13.2.6}$$

The matrix A which diagonalizes the covariance matrix can be found using a number of popular numerical methods. If the covariance is real and symmetric one simple sure-fire technique is the Jacobi Transformation method (1). This is also known in some texts as a Givens rotation. The rotation matrix A'_{kl} is an identity matrix where the zero in the kth row and lth column is replaced with $\sin\theta$, lth row and kth column with $-\sin\theta$, and the ones on the main diagonal in the kth and lth rows are replaced with $\cos\theta$. Pre and post multiplication of a (real symmetric) covariance matrix by the rotation matrix has the effect of zeroing the off-diagonal elements in the (k,l) and (l,k) positions when θ is chosen to satisfy

$$\tan 2\theta_{kl} = \frac{2R_{kl}}{R_{ll} - R_{kk}} \tag{13.2.7}$$

If the covariance matrix is 2-row, 2-column, the angle θ can be seen as a counter-clockwise rotation of the coordinate axis to be aligned with the major and minor axis of an ellipse (a circle if $R_{11} = R_{22}$ requires no rotation). The rotation also has the effect of increasing the corresponding main diagonal elements such that the norm (or trace) of the matrix stays the same. The residual covariance with the zeroed off-diagonal elements from the first Jacobi rotation is then pre and postmultiplied by another rotator matrix to zero another pair of off-diagonal elements, and so on, until all the off-diagonal elements are numerically near zero. The eigenvector matrix A is simply the product of all the Jacobi rotation matrices and the eigenvalue matrix in Eq. (13.2.5) is the diagonalized covariance matrix.

How can Jacobi transformations be applied to complex matrices? Suppose our covariance matrix is not composed of the broadband spatial cross correlation among element signals in the time domain, but rather the narrowband spatial cross correlation among the elements using a narrowband FFT bin complex signal (averaged over N snapshots of course). The covariance matrix is Hermitian (conjugate symmetric) and can be written in the form

$$\bar{R} = \bar{R}^R + j\bar{R}^I \tag{13.2.8}$$

where the real part is symmetric and the imaginary part is skew-symmetric (corresponding off-diagonals have same magnitude but opposite sign). The complex eigenvalue problem is constructed as a larger dimension real eigenvalue problem.

$$\begin{bmatrix} \bar{R}^R & -\bar{R}^I \\ \bar{R}^I & \bar{R}^R \end{bmatrix} \begin{bmatrix} v_i^R \\ v_i^I \end{bmatrix} = \lambda_i \begin{bmatrix} v_i^R \\ v_i^I \end{bmatrix} \tag{13.2.9}$$

The eigenvalues for the matrix in Eq. (13.2.9) are found in $M + 1$ pairs of real and imaginary components. For large matrices, there are a number of generalized eigenvalue algorithms with much greater efficiency and robustness. This is particularly of interest when the matrix is ill-conditioned. A good measure of effectiveness to see if the eigenvector/eigenvalue solution is robust is to simply compute the covariance matrix inverse and multiply it by the original covariance matrix to see how close the result comes to the identity matrix. The covariance matrix inverse is found trivially from the eigenvalue solution.

$$\bar{R}^{-1} = A\Lambda^{-1}A^H \quad where \quad \Lambda^{-1} = diag\{\lambda^{-1}\lambda_1^{-1} \dots \lambda_M^{-1}\} \tag{13.2.10}$$

There are also a number of highly efficient numerical algorithms for matrix inversion based on other approaches than the generalized eigenvalue problem. One of the more sophisticated techniques has already been presented in Section 9.4 of this book. Recall Eq. (9.4.31) relating the backward prediction coefficients and backward error covariance to the inverse of the signal covariance matrix R_M^x.

$$\left[R_M^x\right]^{-1} = L^H\left[R_M^r\right]^{-1}L \tag{13.2.11}$$

where L is a lower triangular matrix of backward predictors defined by Eq. (9.4.30) and Eq. (13.2.12) below and R_M^r is a diagonal matrix of the orthogonal backward prediction error variances.

$$\begin{bmatrix} r_{0,n} \\ r_{1,n} \\ \vdots \\ r_{M-1,n} \\ r_{M,n} \end{bmatrix} = \begin{bmatrix} 1 & 0 & \cdots & & 0 \\ a_{1,1}^r & 1 & 0 & \cdots & 0 \\ \vdots & & \ddots & & \vdots \\ a_{M-1,M-1}^r & \cdots & & 1 & 0 \\ a_{M,M}^r & a_{M,M-1}^r & \cdots & a_{M,1}^r & 1 \end{bmatrix} \begin{bmatrix} x_n \\ x_{n-1} \\ \vdots \\ x_{n-M+1} \\ x_{n-M} \end{bmatrix} = \bar{r}_{0,n:M,n} = L\phi_n^T$$

(13.2.12)

Note that L^H is upper triangular and L is lower triangular. If the elements of L and L^H are appropriately normalized by the square-root of the corresponding backward prediction error variance, the expression in Eq. (13.2.11) can be written as an *LU Cholesky factorization*, often used to define the square-root of a matrix. However, the L matrix in Eq. (13.2.12) differs significantly from the eigenvector matrix. While the backward predictors are orthogonal, they are not orthonormal. Also, the backward prediction error variance are not the eigenvectors of the signal. The eigenvector matrix and eigenvalue diagonal matrix can however be recovered from the Cholesky factorization, but the required substitutions are too tedious for presentation here.

Getting back to the physical problem at hand of optimizing the beam response for an array of sensors to separate the signals from several sources, we can again consider a line array of 8 equally-spaced sensors and plane waves from several distant sources. Our example has a single 100 Hz plane wave arriving at 90 degrees bearing. For simplicity, we'll consider a high SNR case, acoustic propagation in air ($c = 343$ m/sec), and an array spacing of 1.25 m (spatial aliasing occurs for frequencies above 137 Hz). A particular eigenvector is analyzed by a zero-padded FFT where the resulting digital spatial (wavenumber) frequency covers the range $-\pi \le \phi \le +\pi$. To relate this digital spatial angle to a physical bearing angle it can be seen that $\phi = kd\cos\theta$, where θ is the physical bearing angle. For unevenly spaced arrays, there is not such a direct mapping between the physical bearing angle and the digital spatial frequency. For random arrays, one uses the uneven Fourier transform (an FFT is not possible for uneven spaced samples) of Section 5.5 to evaluate the beam response wavenumber spectrum, but then uses an average spacing d' to relate the digital and physical angles. This is analogous to relating the digital frequency in the uneven Fourier transform to a physical frequency in Hz by using the average sample rate.

Figure 6 shows the eigenvector beam response calculated by simply zero-padding the eigenvector complex values to 1024 points and executing an FFT. The physical angle on the x-axis is recovered by dividing the digital frequency by kd and taking its inverse cosine. Obviously, $\theta = \cos^{-1}(\pi/kd)$ is not a linear function so relating the digital and physical bearing angles requires careful consideration even for evenly-spaced arrays. A total of 500 independent snapshots are used where the signal level is 3 and the noise standard deviation is 1×10^{-4}. The calculated eigenvalue for the signal is divided by the number of snapshots times the number of array elements (500 times 8 or 4000) to give a normalized value of 9.000006,

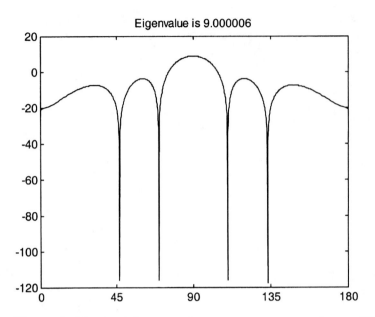

Figure 6   The signal eigenvector beam response corresponding to a 100 Hz sinusoidal plane wave of amplitude 3 at 90 degrees bearing in a low Gaussian background noise of 0.0001 standard deviation (dB vs. degrees).

or the signal amplitude-squared. For fewer snapshots and lower SNR there is less precision in this calculation. The normalized noise eigenvalues are all on the order of 1×10^{-8} or smaller. Clearly, one can simply examine the eigenvalues and separate them into *"signal eigenvalues"* and *"noise eigenvalues"*, each with respective eigenvectors. There are some algorithms such as the Akaike Information Criterion (AIC) and Minimum Description Length (MDL) which can help identify the division into the subspaces (2) when SNR is low or an eigenvalue threshold is difficult to automatically set or adapt in an algorithm.

$$[v_0 v_1 \ldots v_M] = A_S + A_N = \left[v_{S_1} v_{S_2} \ldots v_{S_{Ns}} | v_{N_1} v_{N_2} \ldots v_{N_{Nn}} \right] \tag{13.2.13}$$

The eigenvectors v_{Si}, $i = 1, 2, \ldots, N_s$ define what is called the *signal subspace A_S* while v_{Nj}, $j = 1, 2, \ldots, Nn$ define the *noise subspace A_N* spanned by the covariance matrix where $Nn + Ns = M + 1$, the number of elements in the array. Since the eigenvectors are orthonormal, the eigenvector beam response will have its main lobe aligned with the physical angle of arrival of the corresponding signal source (whether it be a real physical source or a spurious noise source) and nulls in the directions of all other corresponding eigenvector sources. If an eigenvector is part of the noise subspace, the associated source is not physical but rather "spurious" as the result of the algorithm trying to fit the eigenvector to independent noise across the array. Figure 7 shows the 7 noise eigenvector beam responses for our single physical source case. *The noise eigenvectors all have the same null(s) associated with the signal eigenvector(s).* Therefore, if one simply averages the noise eigenvectors the spurious nulls will be filled in while the signal null(s) will remain. This is the idea behind the MUSIC algorithm. Figure 8 shows the beam response of the music vector and

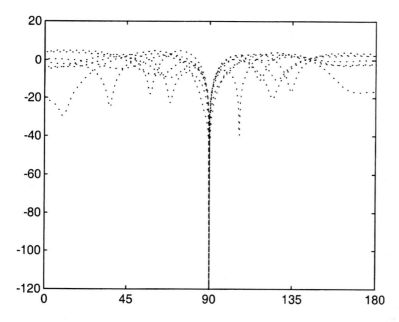

Figure 7 Beam responses of the remaining 7 "noise" eigenvectors which correspond to eigenvectors which correspond to eigenvalues on the order of 1×10^{-8} showing a common null at the source arrival angle of 90 degrees (dB vs. degrees).

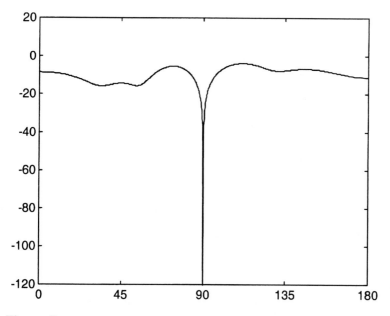

Figure 8 Beam response of the MUSIC vector found by averaging all the noise eigenvectors clearly showing the strong common null associated with a single signal angle of arrival of 90 degrees. (dB vs. degrees).

Eq. (13.2.14) shows the MUSIC vector calculation.

$$v_{MUSIC} = \frac{1}{Nn} \sum_{j=1}^{Nn} v_{N_j} \qquad (13.2.14)$$

Consider the case of three physical sources radiating plane waves of amplitudes 2, 3, and 1, at the array from the angles 45, 90, and 135 degrees, respectively, in the same low SNR of 1×10^{-4} standard deviation background noise. Figure 9 shows the eigenvector beam responses for the three signals. The normalized eigenvalues are 9.2026, 3.8384, and 0.9737 corresponding to the arrival angles 90, 45, and 135 degrees as seen in Figure 9 Most eigenvalue algorithms sort the eigenvalues and corresponding eigenvectors based on amplitude. The reason the eigenvalues don't work out to be exactly the square of the corresponding signal amplitudes is that there is still some residual spatial correlation, even though 500 independent snapshots are in the covariance simulation. It can be seen that the sum of the signal eigenvalues (normalized by the number of array elements times the number of snapshots, or $8 \times 500 = 4000$) is 14.0147 which is quite close to the expected 14. One should also note that the signal eigenvector responses do not show sharp nulls in the directions of the other sources due to the limited observed independence of the sources. It can be seen that for practical implementations, the signal eigenvectors are only approximately orthonormal. It is very important for the user of MUSIC and all eigenvector methods to understand the issue of source independence. The whole approach simply fails when the plane waves all have

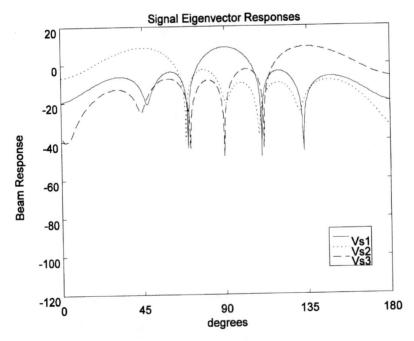

Figure 9 The three signal eigenvector beam responses for 100 Hz plane waves of amplitude 2, 3, and 1, and arrival angles of 45, 90, and 135 degrees, respectively (dB vs. degrees).

the same phase phase with respect to each other. Such is the case when multiple arrivals are from coherent reflections of wave from a single source mixing with the direct wave. Figure 10 shows the MUSIC vector beam response which clearly shows the presence of three sources at 45, 90, and 135 degrees.

If one wanted to observe source 1 while suppressing sources 2 and 3 with nulls, simply using the source 1 eigenvector, v_{S1}, would provide the optimal adaptive beam pattern if the source signals were completely independent. To steer an optimal adaptive beam to some other direction θ' we can use the inverse of the covariance matrix to place a normalized (to 0 dB gain) in the direction θ' while also keeping the sharp nulls of the MUSIC response beam. This is called a Minimum variance Distortionless Response (MVDR) adaptive beam pattern. The term "minimum variance" describes the maintaining of the sharp MUSIC nulls while "distortionless response" describes the normalization of the beam to unity gain in the look direction. The normalization does not affect SNR, it just lowers the beam gain in all directions to allow look-direction unity gain. If the delay-sum, or Bartlett beam, steering vector is defined as

$$S(\theta') = \left[1 e^{-jkd_1 \cos\theta'} e^{-jkd_2 \cos\theta'} \dots e^{-jkd_M \cos\theta'} \right]^H \tag{13.2.15}$$

where d_i $i = 1, 2, \dots, M$ is the coordinate along the line array and "H" denotes Hermitian transpose (the exponents are positive in the column vector), the MVDR steering vector is defined as

$$S_{MVDR}(\theta') = \frac{\bar{R}^{-1} S(\theta')}{S(\theta')^H \bar{R}^{-1} S(\theta')} \tag{13.2.16}$$

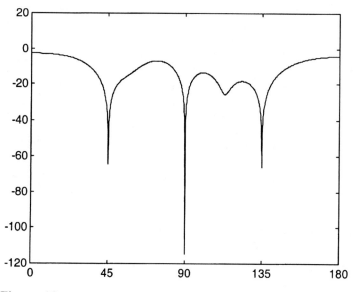

Figure 10 MUSIC beam response for the three 100 Hz plane waves at 45, 90, and 135 degrees (dB vs. degrees).

Note that the denominator of Eq. (13.2.16) is a scalar which normalizes the beam response to unity (0 dB gain) in the look direction. Even more fascinating is a comparison of the MVDR steering vector to the Kalman gain in the recursive least squares (RLS) algorithm summarized in Table 9.1. In this comparison, our "Bartlett" steering vector $S(\theta')$ corresponds to the RLS basis function and the MVDR steering vector represents the "optimal adaptive gain update" to the model, which is our array output. There is also a comparison of the projection operator update in Eq. (8.2.7) which is similar in mathematical structure. Figure 11 compares the Bartlett and MVDR responses for a steering angle of 112.5 degrees and our three sources at 45, 90, and 135 degrees. The MVDR beam effectively "blinds" the array to these signals while looking in the direction of 112.5 degrees. Note that the Bartlett beam response has a gain of 8 (+18 dBv) in the look direction from the 8 sensor elements of the linear array.

MVDR beamforming is great for suppressing interference from strong sources when beam steering to directions other than these sources. If one attempts an MVDR beam in the same direction as one of the sharp nulls in the MUSIC beam response, the algorithm breaks down due to an indeterminant condition in Eq. (13.2.16). Using the eigenvector corresponding to the signal look direction is theoretically the optimal choice, but in practice some intra-source coherence is observed (even if its only from a limited number of snapshots) making the nulls in the directions of the other sources weak. However, from the MUSIC response, we know the angles of the sources of interest. Why not manipulate the MVDR vector directly to synthesize the desired beam response?

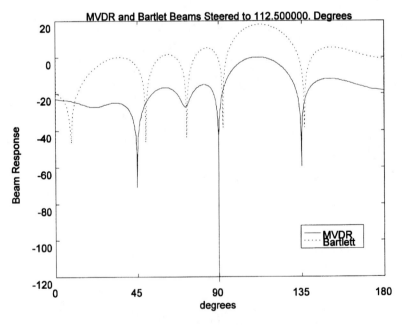

Figure 11 Comparison of the MVDR beam response to a delay-sum type, or Bartlett, beam response for a steering angle of 112.5 degrees showing the MVDR beam maintaining nulls in the source directions (dB vs. degrees).

Like any FIR filter, the coefficients can be handled as coefficients of a polynomial in z where $z = e^{jkd\cos\theta}$. For an evenly spaced linear array, the delay-sum or Bartlett steering vector seen in Eq. (13.2.5) can be written as $S(z) = [1 \; z^{-1} z^{-2} z^{-3} \ldots z^{-M}]^H$ where $z = e^{jkd\cos\theta}$. For a non-evenly-spaced array we can still treat the steering vector as an integer-order polynomial, but the interpretation of the physical angles from the digital angles is more difficult. This is not a problem because our plan is to identify the zeros of the MVDR steering vector associated with the signals, and then suppress the zero causing the null in the look direction. The resulting beam response will have a good (but not optimal) output of the source of interest and very high suppression of the other sources. The procedure will be referred to here as the "Pseudo Reduced-Order Technique" and is as follows:

1. Solve for the MVDR steering vector.
2. Solve for the zeros of the MVDR steering vector as if it were a polynomial in z.
3. Detect the zeros on or very close to the unit circle on the z-plane. These are the ones producing the sharp nulls for the signal directions. The number of these zeros equals the number of signal sources detected.
4. Noting that the polynomial zero angles will range from $-\pi$ to $+\pi$ while the physical wavenumber phase will range from $-kd$ to $+kd$, select the signal zero corresponding to the desired look direction and suppress it by moving it to the origin. Move the other spurious zeros further away from the unit circle (towards the origin or to some large magnitude).
5. Using the modified zeros calculate the corresponding steering vector by generating the polynomial from the new zeros. Repeat for all signal directions of interest.

The above technique can be thought of as similar to a reduced order method, but less mathematical in development. The threshold where one designates a zero as corresponding to a signal depends on how sharp a null is desired and the corresponding distance from the unit circle. Figure 12 shows the results of the pseudo reduced-order technique for the three beams steered to their respective sources. While the beams are rather broad, they maintain sharp nulls in the directions of the other sources which allows separation of the source signals. The corresponding MUSIC beam response using the steering vector modified by the pseudo reduced-order technique presented here is free of the spurious nulls because the those zeros have been suppressed.

We can go even farther towards improving the signal beams by making use of the suppressed zeros to sharpen the beam width in the look direction. We will refer to this technique as the "Null Synthesis Technique" since we are constructing the steering vector from the MVDR signal zeros and placing the remaining available zeros around the unit circle judiciously to improve our beam response. The usefulness of this is that the available (non signal related) zeros can be used to improve the beam response rather than simply be suppressed in the polynomial. The null synthesis technique is summarized as follows:

1. Solve for the MVDR steering vector.
2. Solve for the zeros of the MVDR steering vector as if it were a polynomial in z.

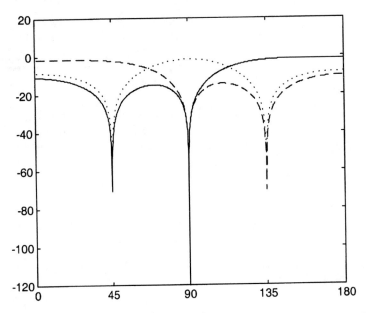

Figure 12   The pseudo reduced-order technique presented here cleanly separates the signals by suppressing spurious zeros and the MVDR zero associated with the particular signal direction (dashed 45 degrees, dotted 90 degrees, and solid 135 degrees—scale is dB vs. degrees).

3.   Detect the zeros on or very close to the unit circle on the z-plane. These are the ones producing the sharp nulls for the signal directions. The number of these zeros equals the number of signal sources detected.

4.   Distribute the number of remaining available (non signal related) zeros plus one around the unit circle from $-\pi$ to $+\pi$ avoiding a double zero at π. The extra zero nearest the main lobe will be eliminated to enhance the main lobe response.

5.   For the desired signal look direction and corresponding unit circle angle, find the closest zero you placed on the unit circle and eliminate it. Then add the zeros for the other source directions.

6.   Using the synthesized zeros calculate the corresponding steering vector by generating the polynomial from the new zeros. Repeat for all signal directions of interest.

Figure 13 shows the results using null synthesis. Excellent results are obtained even though the algorithm is highly heuristic and not optimal. The beam steer to 45 and 135 degrees could further benefit from the zeros at around 30 and 150 degrees being suppressed. The effect of the zeros placed on the unit circle is not optimal, but still quite useful. However, the nonlinearity of $\theta = \cos^{-1}(\phi/kd)$ means that the physical angles (bearings) of evenly spaced zeros on the unit circle (evenly spaced digital angles) do not correspond to evenly-spaced bearing angles. To make the placed zeros appear closer to evenly spaced in the physical beam response, we can "pre-map" them according to the known relationship $\theta = \cos^{-1}(\phi/kd)$ between the physical angle θ and the digital wavenumber angle π. If we let our

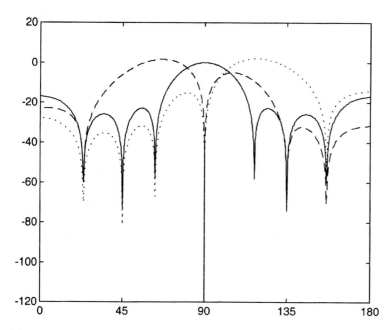

Figure 13 The null synthesis technique presented here sharpens the beam width in the signal direction (solid 90, dotted 135, and dashed 45 degrees source directions) by placing suppressed zeros at useful angles on the unit circle away from the look direction beam main lobe (scale is dB vs. degrees).

evenly-spaced zeros around the unit circle have angles ϕ', we can pre-map the z-plane zero angles by $\pi = \cos^{-1}(\phi'/kd)$. This is what was done for the null synthesis example in Figure 13 and it improved the beam symmetry and main lobe significantly. The nulls in Figure 13 also appear fairly evenly spaced except around the main lobes of the signal steering vectors. While the signal eigenvectors are the theoretical optimum beams for separating the source signals, the affect of source phase dependence limits the degree to which the signals can be separated. Heuristic techniques such as the pseudo reduced-order and null synthesis techniques illustrate what can be done to force the signal separation given the information provided by MUSIC. Both these techniques can also be applied to a spatial whitening filter for signal null forming as described in Section 13.1.

For N_S signal source eigenvectors in the matrix A_S, and $N_N = M + 1 - N_S$ noise eigenvectors in the matrix A_N, we note from Eq. (13.2.13) that since $A^H = A^{-1}$

$$
\begin{aligned}
I &= (A_S + A_N)^H (A_S + A_N) \\
&= (A_S + A_N)^H (A_S + A_N)^H \\
&= A_S A_S^H + A_S A_N^H + A_N A_S^H + A_N A_N^H \\
&= A_S A_S^H + A_N A_N^H
\end{aligned}
\tag{13.2.17}
$$

represents the sum of the projection matrices (Section 8.2) onto the noise and signal subspaces. This can be seen by considering a signal matrix Y which has $M + 1$ rows and N_S columns representing some linear combination of the signal eigenvectors $Y = A_S Q_S$, where Q_S is any invertible square matrix with N_S rows. Therefore, $Y^H Y = Q_S^H A_S^H A_S Q_S = Q_S^H Q_S$. The projector matrix onto the signal subspace can be seen to be

$$
\begin{aligned}
P_Y = A_S A_S^H &= (Y Q_S^{-1})(Q_S^{-H} Y^H) \\
&= Y(Q_S^H Q_S)^{-1} Y^H \\
&= Y(Y^H Y)^{-1} Y^H
\end{aligned}
\tag{13.2.18}
$$

which is a fancy way of saying that the signal subspace is expressible as the linear combination of the signal eigenvectors. So, if the signal levels of our plane wave sources change, the subspace spanned by the rows of Y does not change, but the eigenvalues do change according to the power relationship. If the bearing(s) of the sources change then the subspace also changes. The same is true for the noise subspace. If we define our noise as $W = A_N Q_N$,

$$
\begin{aligned}
P_W = A_N A_N^H &= (W Q_N^{-1})(Q_N^{-H} W^H) \\
&= W(Q_N^H Q_S)^{-1} W^H \\
&= W(W^H W)^{-1} W^H
\end{aligned}
\tag{13.2.19}
$$

and the combination of noise and signal projectors is

$$
P_Y + P_W = Y(Y^H Y)^{-1} Y^H + W(W^H W)^{-1} W^H = I
\tag{13.2.20}
$$

Equation (13.2.20) can be used to show that the noise is orthogonal to the signal subspace by $P_Y = I - P_W$. This orthogonal projection operator was used in Section 8.2, 9.3 and 9.4 to define a least-squared error update algorithm for the adaptive lattice filter, which results from the orthogonal decomposition of the subspace in time and model order. Because of the orthonormality of the eigenvectors, we could have divided the space spanned by the rows of the covariance matrix by any of the eigenvectors. It makes physical sense to break up the subspace into signal and noise to allow detection of the signal wavenumbers and the design of beam steering vectors to allow signal separation and SNR enhancement.

13.3 COHERENT MULTIPATH RESOLUTION TECHNIQUES

In the previous two sections we have seen that near optimal beam patterns can be designed using either adaptive spatial filter or by eigenvector processing. For narrowband phase-independent sources one can use the received signals from a sensor array of known geometry in a medium of known free wave propagation speed to design beams which identify the arrival angles of, and even separate, the source signals. However, the notion of "phase independent narrowband sources" is problematic in most practical applications. Many "snapshots" of the spectral data are averaged in the hope that the random phase differences of the "independent" sources will average to zero in the covariance matrix elements, as seen in Eq. (13.1.3). There are coherent multipath situations where the array records several wavenumber traces at a given temporal frequency where the snapshot-averaging technique simply will not work unless the individual phases and amplitudes of the sources are known. Furthermore, if it requires $1/\Delta f$ seconds of time data to produce the complex FFT bin Δf Hz wide, and then N FFT snapshots to be averaged to insure the spatial covariance matrix for that bin is independent of the source phases, a long net time interval is then required to do the problem. If the sources are moving significantly during that time the arrival angle information may not be meaningful. However, as one widens the FFT bin bandwidth by shorting the time integration, a large number of snapshots can be calculated in a very short time span. If the beam bandwidth approaches the Nyquist band (maximum available bandwidth), the snapshots become a single time sample and the "broadband" covariance matrix is calculated in the shortest possible time without the use of an FFT. Why not do broadband beamforming where the sources are more readily made independent? The answer is seen in the fact that a broadband beam may not be optimal for resolving specific narrowband frequencies, since it is optimized for the entire broad bandwidth. Therefore, a fundamental tradeoff exists between narrowband resolution in both frequency and wavenumber (arrival angle) and the amount of integration in time and space to achieve the desired resolution. As with all information processing, higher fidelity generally comes at an increased cost, unless of course human intelligence is getting in the way.

Suppose we know the amplitude and phase of each source and we have an extremely high fidelity propagation model which allows us to predict the precise amplitude and phase from each source across the array. The solution is now trivial. Consider an 8 element ($M = 7$) line array and three sources

$$\begin{bmatrix} \overline{s_1} \\ \overline{s_2} \\ \overline{s_3} \end{bmatrix} = \begin{bmatrix} s_1 & 0 & 0 \\ 0 & s_2 & 0 \\ 0 & 0 & s_3 \end{bmatrix} \begin{bmatrix} 1 & e^{jk_1d_1} e^{jk_1d_2} & \dots & e^{jk_1d_M} \\ 1 & e^{jk_2d_1} e^{jk_2d_2} & \dots & e^{jk_2d_M} \\ 1 & e^{jk_3d_1} e^{jk_3d_2} & \dots & e^{jk_3d_M} \end{bmatrix} \qquad (13.3.1)$$

where $d_i \, i = 1, 2, \dots, M$ is the position of the ith element along the line array relative to the first element. The wavenumber traces, $k_i \, i = 1, 2, 3$ are simply the plane wave projections on the line array. The right-most matrix in Eq. (13.3.1) is called a Vandermode matrix since the columns are all exponential multiples of each other (for an evenly-spaced line array). Clearly, the signal eigenvectors are found in the rows of the vandermode matrix and the eigenvalues are the magnitude-squared of the corresponding signal amplitudes and phases $s_i \, i = 1, 2, 3$. This is like saying if we know A (the signal amplitudes and phases) and B (the arrival angles or wavenumbers), we can get $C = AB$. But if we are only given C, the received waveform signals from the array, there are an infinite combination of signal amplitudes and phases (AB) which give the same signals for the array elements. Clearly, the problem of determining the angles of arrival from multiple narrowband sources and separating the signals using adaptive beamforming from the array signal data alone requires one to eliminate any coherence between the sources as well as the background noise, which should be independent at each array element.

Suppose one could control the source signal, which has analogy in active sonar and radar where one transmits a known signal and detects a reflection from an object, whose location, motion, and perhaps identity are of interest. Given broadband transducers and good propagation at all frequencies, a broadband (zero mean Gaussian, or ZMG) signal would be a good choice because the sources would be completely independent at all frequencies. An even better choice would be a broadband periodic waveform which repeats every N samples and has near zero cross correlation with the broadband periodic waveforms from the other sources. Time synchronous averaging (Chapter 11) of the received waveforms is done by simply summing the received signals in an N-point buffer. The coherent additions will cause the background noise signals to average to their mean of zero while coherently building up the desired signals with period N significantly improving SNR over the broad bandwidth. By computing transfer functions (Section 6.2) between the transmitted and received signals, the frequency response of the propagation channel is obtained along with the propagation delay. For multipath propagation, this transfer function will show frequency ranges of cancellation and frequency ranges of reinforcement due to the multipath and the corresponding impulse response will have several arrival times (seen as a series of impulses). Each element in the array will have a different frequency response and corresponding impulse response due to the array geometry and the different arrival angles and times. Therefore, the "coherent" multipath problem is solvable if a broadband signal is used. Solving the multipath problem no only allows us to measure the propagation media inhomogeneities, but also allows us to remove multipath interference in communication channels.

Maximal Length Sequences (MLS)

MLS are sequences of random bits generated by an algorithm which repeat every $2^N - 1$ bits, where N is the order of the MLS generator. The interesting signal properties of an MLS sequence are that its autocorrelation resembles a digital Dirac delta function (the 0th lag has amplitude $2^N - 1$ while the other $2^N - 2$ lags are equal to -1) and the cross correlation with other MLS sequences is nearly zero. The algorithm for generating the MLS sequence is based on primitive polynomials modulo 2 of order N. We owe the understanding of primitive polynomials to a young French mathematical genius named Èvariste Galios (1812–1832) who died as the unfortunate result of a duel. In the Romantic age, many lives were cut short by death from consumption or dueling. Fortunately for us, the night before Galios fate he sent a letter to a friend named Chevalier outlining his theories. Now known as Galios theory, it is considered one of the highly original contributions to algebra in the nineteenth century. Their application to MLS sequence generation has played an enabling role computer random number generation, digital spread-spectrum communications, the satellite global positioning system (GPS), and in data encryption standards. Clearly, this is an astonishing contribution to the world for a twenty-year-old.

   To illustrate the power of MLS, consider the following digital communications example. The MLS generating algorithm are known at a transmitting and receiver site, but the transmitter starts and stops transmission at various times. When the receiver detects the transmitter, it does so by detecting a Dirac-like peak in a correlation process. The time location of this peaks allows the receiver to "synchronize" its MLS correlator to the transmitter. The transmitter can send digital data by a simple modulo-2 addition of the data bit-stream to the transmitted MLS sequence. This appears as noise in the receiver's correlation detector. But once the transmitter and receiver are synchronized, a modulo-2 addition (an exclusive or operation or XOR) of the MLS sequence to the received bitstream gives the transmitted data! Since multiple MLS sequences of different generating algorithms are uncorrelated, many communication channels can co-exist in the same frequency band, each appearing to the other's as uncorrelated background noise. There are limits to how many independent MLS sequences can be generated for a particular length. However, a scientist named Gold (3) showed that one could add modulo-2 two generators of length $2^N - 1$ and different initial conditions to obtain $2^N - 1$ new sequences (which is not a MLS but close), plus the two original base sequences. The cross-correlation of these Gold sequences are shown to be bounded, but not zero, allowing many multiplexed MLS communication channels to co-exist with the same length in the same frequency channel. The "Gold codes" as they have become known, are the cornerstone of most wireless digital communications. For GPS systems, a single receiver can synchronize with multiple satellites, each of which also send their time and position (ephemeris), enabling the receiver to compute its position. GPS is another late 20th century technology which is finding its way into many commercial and military applications.

A typical linear MLS generator is seen in Figure 14 The notation describing the generator is $[N,i,j,k,...]$ where the sequence is $2^N - 1$ "chips" long, and i, j, k, etc., are taps from a 1-bit delay line where the bits ar XOR'd to produce the MLS output. For example, the generator depicted as $[7,3,2,1]$ has a 7 stage delay line, or register, where the bits in positions 7, 3, 2, and 1 are modulo-2 added (XOR'd) to produce the bit which goes into stage 1 at the next clock pulse. The next clock pulse shifts the bits in stages 1 through 6 to stages 2 through 7, where the bit in stage 7 is the MLS output, copies the previous modulo-2 addition result into stage 1, and calculates the next register input for stage 1 by the modulo-2 addition of the bits in elements 7, 3, 2, and 1. Thanks to Galios theory, there are many, many irreducible polynomials from which MLS sequences can be generated. The Gold codes allow a number of nearly maximal sequences to be generated from multiple base MLS sequences of the same length, but different initial conditions and generator stage combinations. The Gold codes simplify the electronics needed to synchronize transmitters and receivers and extract the communication data. A number of useful MLS generators are listed in Table 1 which many more generators can be found in an excellent book on the subject by Dixon (4).

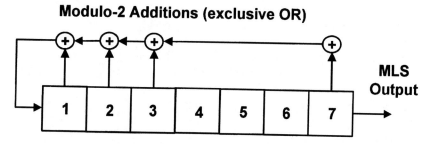

Modulo-2 Additions (exclusive OR)

Figure 14 Block diagram of a [7,3,2,1] MLS generator showing a 7 element tapped delay line for modulo-2 additions of the elements 7, 3, 2, and 1 to generate a $2^7 - 1$, or 127 bit random sequence.

Table 1 Simple MLS Generators

Number of Stages	Code Length	Maximal Tap Formula
4	15	[4,1]
5	31	[5,2]
6	63	[6,1]
7	127	[7,1]
8	255	[8,4,3,2]
9	511	[9,4]
10	1.023	[10,3]
11	2,047	[11,1]
12	4,095	[12,6,4,1]
13	8,191	[13,4,3,1]
14	16,383	[14,12,2,1]
15	32,767	[15,1]
16	65,535	[16,12,3,1]
18	262,143	[18,17]
24	16,777,215	[24,7,2,1]
32	4,294,967,295	[32,22,2,1]

Spread Spectrum Sinusoids

Our main interest in MLS originates in the desire to resolve coherent multipath for narrowband signals. This is also of interest in communication systems and radar/sonar where multipath interference can seriously degrade performance. By exploiting time synchronous averaging of received data buffers of length $2^N - 1$ the MLS sequence SNR is significantly improved and by exploiting the orthogonality of the MLS we can separate in time the multipath arrivals. For communications, one would simply synchronize with one of the arrivals while for a sonar or radar application, one would use the separated arrival times measured across the receiver array to determine the arrival angles and propagation times (if synchronized to a transmitter). Applying MLS encoding to narrowband signals is typically done by simply multiplying a sinusoid by a MLS sequence normalized so that a binary 1 is +1 and a binary 0 is −1. This type of MLS modulation is called *bi-phase modulation* for obvious reasons and is popular because it is very simple to implement electronically.

A bi-phase modulation example is seen in Figure 15 for a 64 Hz sinusoid sampled at 1024 Hz. The MLS sequence is generated with a [7,1] generator algorithm and the chip rate is set to 33 chips per sec. It can be seen that it is not a good idea for the chip rate and the carrier frequency to be harmonically related (such as chip rates of 16, 32, 64, etc., for a 64 Hz carrier) because the phase transitions will always occur at one of a few points on the carrier. Figure 16 shows the power spectrum of the transmitted spread spectrum waveform. The "spreading" of the narrowband 64 Hz signal is seen as about 33 Hz, the chip rate. At the carrier frequency plus the chip rate (97 Hz) and minus the chip rate (31 Hz) we see the nulls in the "$\sin x/x$" or sinc function. This sinc function is defined by the size of the chips. If the chip rate equals the time sample rate (1024 Hz in this example), the corresponding spectrum would be approximately white noise. Figure 16 shows a 200 Hz sinusoid with

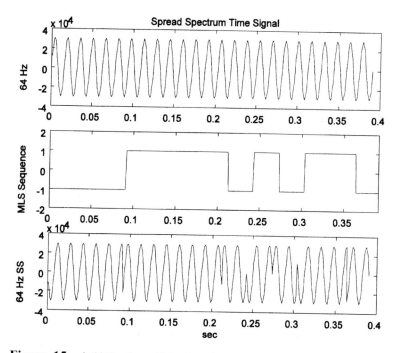

Figure 15 A 64 Hz sinusoid (top) multiplied by a [7,1] MLS sequence with 33 chips per sec (middle) yields the bi-phase modulated sinusoid in the bottom plot.

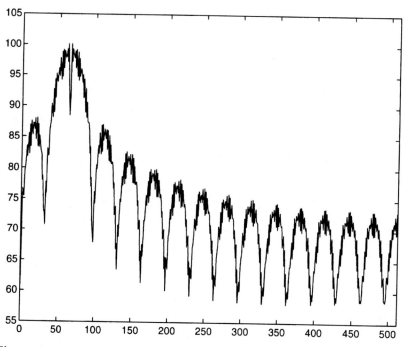

Figure 16 Transmitted spread spectrum signal obtained from a power spectrum average of the bi-phase modulated signal in the bottom of Figure 15 (scale is dB vs. Hz).

100 chips per sec. However, for any chip rate there is actually some energy of the sinusiodal carrier spread to every part of the spectrum from the sharp phase transitions in the time-domain spread spectrum signal. Note that because we are encrypting the sinusoid's phase with a known orthogonal pseudo-random MLS, we can recover the net propagation time delay as well as the magnitude and phase of the transfer function between a transmitter and receiver. This changes significantly what can be done in the adaptive beamforming problem of multipath coherence.

Resolving Coherent Multipath

Clearly, one can associate the propagating wave speed and chip time length (the inverse of the chip rate) to associate a propagation distance for one chip. For example, the acoustic wave speed in air is about 350 m/sec (depends on wind and temperature). If one transmitted a 500 Hz spread spectrum sinusoid with 35 chips per sec, the main sinc lobe would go from 465 Hz to 535 Hz in the frequency domain and the length of a chip in meters would be $c/f_c = 350/35$ or 10 m/chip where f_c is the chip rate and c is the wave speed. By cross correlating the transmitted and received waveforms one can resolve the propagation distance easily to within one chip, or 10 m if the propagation speed is known. If the distance is known then the cross correlation yields the propagation speed $c(1 \pm 1/[f_c T_p])$ where f_c is the chip rate and T_p is the net propagation time. Resolution in either propagation distance or in wave speed increases with increasing chip rate, or in other words, propagation time resolution increases with increasing spread spectrum bandwidth. This is in strong agreement with the Cramer–Rao lower bound estimate in Eq. (12.2.12) for time resolution in terms of signal bandwidth.

An efficient method for calculating the cross correlation between two signals is to compute the cross-spectrum (Section 6.2) and then computing an inverse FFT. Since the spectral leakage contains information, one is better off not using a data window (Section 5.3) or applying zero-padding to one buffer to correct for circular correlation (Section 5.4). Consider a coherent multipath example where the direct path is 33 m and the reflected path is 38 m. With a carrier sinusoid of 205 Hz and the 100 Hz chip rate (generated using a [11,1] MLS algorithm), the transmitted spectrum is seen similar to the one in Figure 17. The cross correlation of the received multipath is seen in Figure 18 where the time lag is presented in m based on the known sound speed of 350 m/sec. For this case, the length of a chip in m is $c/f_c = 350/100$ or 3.5 m. The 5 m separating the direct and reflected paths are resolved. Using a faster chip rate will allow even finer resolution in the cross correlation. However, using a slightly different processing approach will yield an improved resolution by using the entire bandwidth regardless of chip rate. This approach exploits the coherence of the multipath signals rather than try to integrate it out.

The Channel Impulse Response

The channel impulse response is found from the inverse FFT of the channel transfer function (Section 6.2) which is defined as the expected value of the cross spectrum divided by the expected value of the autospectrum of the signal transmitted into

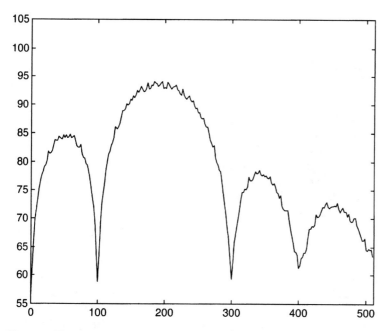

Figure 17 Spread spectrum signal generated from a 200 Hz sinusoid with 100 chips/sec from a [11,1] MLS sequence generator where $fs = 1024$ Hz and 16 sec of data are in the power spectrum (scale is dB vs. Hz).

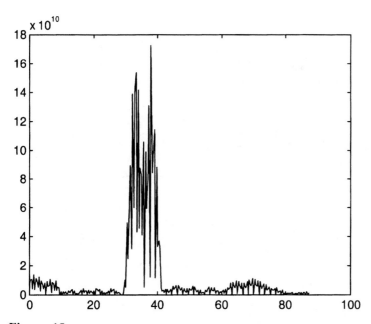

Figure 18 Cross correlation computed via cross spectrum of the 205 Hz carrier with 100 Hz chip rate generated by [11,1] showing the resolution of the 33 m and 38 m propagation paths (linear scale vs. m).

the channel. This simple modification gives the plot in Figure 19 which shows the broadband spectral effects of coherent multipath. Figure 20 shows the inverse FFT of the transfer function which easily resolves the 33 m and 38 m paths. The theoretical resolution for the channel impulse response is $c/f_s = 350/1024$ or about 0.342 m, where f_s is the sample rate rather than the chip rate. The transfer function measures the phase across the entire spectrum and its inverse FFT provides a much cleaner impulse response to characterize the multipath. To test this theory, let the reflected path be 34 m rather than 38 m giving a path difference of 1 m where our resolution is about 1/3 of a meter. The result of the resolution measurement for the 205 Hz sinusoid with the 100 Hz chip rate (generated by a [11,1] MLS) is seen in Figure 21 This performance is possible for very low chip rates as well. Figure 22 shows the 205 Hz sinusoid with a chip rate of 16 Hz while Figure 23 shows the corresponding 33 m and 34 m paths resolved.

In the previous examples, 512 point FFTs were used which have a corresponding buffer length of 175 m due to the 1024 Hz sample rate and the speed of sound being 350 m/sec the 100 Hz chip rate had a chip length of 3.5 m and the 16 Hz chip rate had a chip length of 21.875 m. What if the chip length were increased to about half the FFT buffer length? This way each FFT buffer would have no more than one phase transition from a chip change. As such, there would be no false correlations between phase changes in the transfer function. Consider a chip rate of 4 Hz where each chip is 87.5 m long or exactly half the FFT buffer length of 175 m. Some FFT buffers may not have a phase change if several chips in a row have the same sign. But, if there is a phase change, there is only one phase change in the FFT buffer. For the multipath in the received FFT buffer, each coherent path produces a phase change at slightly different times, each of which is highly

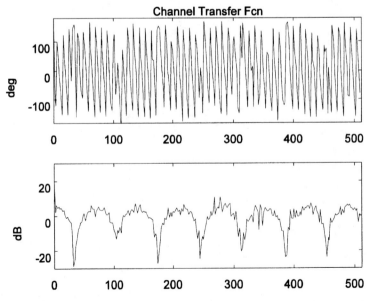

Figure 19 Channel transfer function clearly shows the broadband effects of coherent multipath (horizontal scale is Hz).

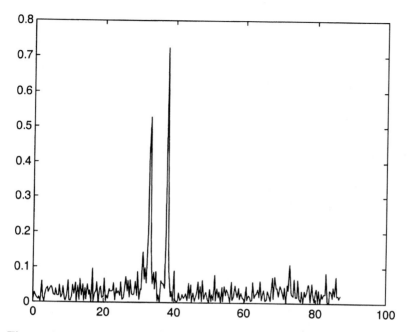

Figure 20 The channel impulse response found from the inverse FFT of the transfer function resolves the 33 m and 38 m paths much better than the cross correlation technique (linear scale vs. m).

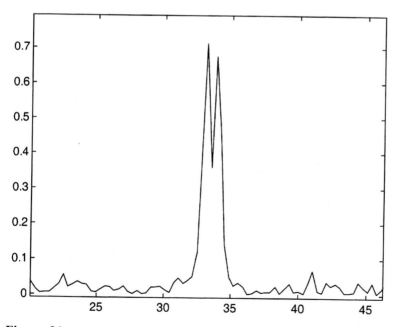

Figure 21 Resolution of a 33 m and 34 m propagation path is possible for a spread spectrum signal sampled at 1024 Hz (linear scale vs. m).

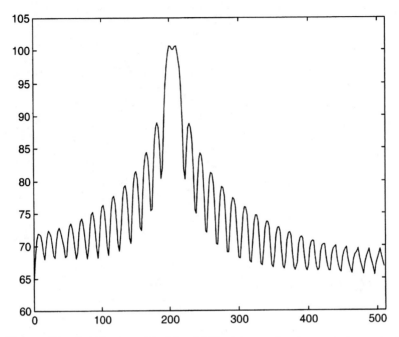

Figure 22   205 Hz sinusoid with a 16 Hz chip rate for the resolution test of the 33 m and 34 m paths (scale is dB vs. Hz).

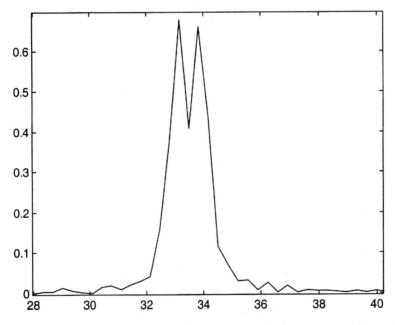

Figure 23   Resolution of the 33 m and 34 m paths using a 16 Hz chip rate (linear scale vs. m).

coherent with the transmitted signal. Figure 24 shows the multipath resolution using a chip rate of 4 Hz.

The multipath resolution demonstration can be carried to an extreme by lowering the carrier sinusoid to a mere 11 Hz (wavelength is 31.82 m) keeping the chip rate 4 Hz. Figure 25 shows the spread spectrum of the transmitted wave. Figure 26 shows the uncertainty of trying to resolve the multipath using only cross correlation. The chips have an equivalent length of 87.5 m with a 350 m/sec propagation speed. Its no wonder the multipath cannot be resolved using correlation alone. Figure 27 shows the transfer function for the channel with the transmitted 11 Hz carrier and 4 Hz chip rate. The inverse FFT of the transfer function yields the channel impulse response of Figure 28, which still nicely resolves the multipath, even though the wavelength is over 31 times this resolution and the chip length is over 87 times the demonstrated resolution! As amazing as this is, we should be quick to point out that low SNR will seriously degrade resolution performance. However, this can be at least partially remedied through the use of time-synchronous averaging in the FFT buffers before the FFTs and inverse FFTs are calculated. For synchronous averaging in the time domain to work, the transmitted signal must be exactly periodic in a precisely-known buffer length. Summing the time domain signals in this buffer repeatedly will cause the background noise to average to its mean of zero while the transmitted spread spectrum signal will coherently add.

One final note about MLS and spread spectrum signals. For very long MLS sequences, one can divide the sequences into multiple "sub-sequence" blocks of consecutive bits. While not maximal, these sub-MLS blocks are nearly uncorrelated with each other and also have an auto correlation approximating a digital Dirac delta function. The GPS system actually does this with all 24 satellite transmitters broad-

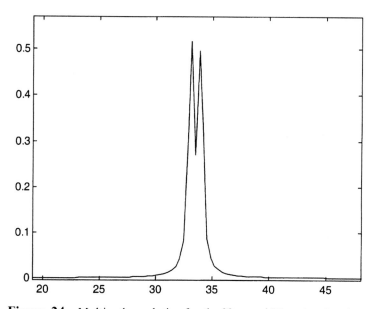

Figure 24 Multipath resolution for the 33 m and 34 m signals using a chip rate of 4 which places no more than one phase change per FFT buffer to enhance coherence (linear scale vs. m).

casting the same MLS code which has a period of 266 days. However, each satellite transmits the code offset by about one-week relative to the others. At the receiver, multiple correlators run in parallel on the composite MLS sequence received which allows rapid synchronization and simplified receiver circuitry. Once one of the

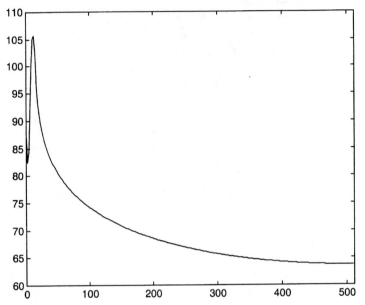

Figure 25   An 11 Hz carrier sinusoid with a 4 Hz chip rate generated by a [11,1] MLS (scale is dB vs. Hz).

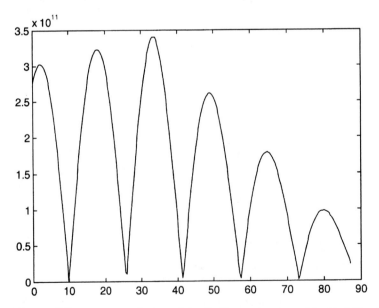

Figure 26   Cross correlation of the 11 Hz carrier sinusoid with 4 Hz chip rate showing an uncertain correlation peak around 33 m (linear scale vs. m).

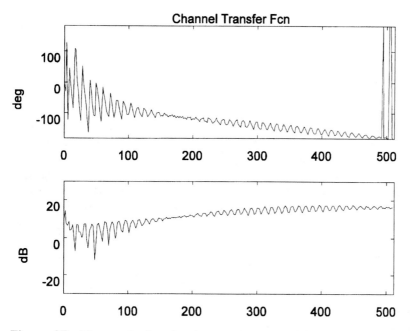

Figure 27 The transfer function for the 11 Hz carrier and 4 Hz chip rate show valuable information across the entire available bandwidth (horizontal scale is Hz).

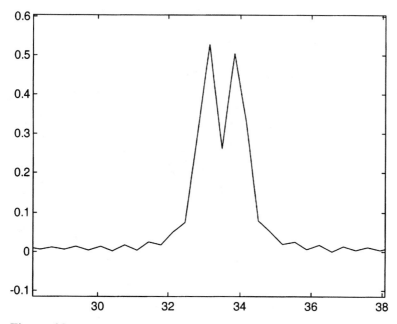

Figure 28 The 33 m and 34 m path are still resolvable using the channel impulse response technique even though the 11 Hz carrier has a wavelength of 31.8 m and the chip length is 87.5 m (linear scale vs. m).

receiver correlators is synchronized with a GPS satellite, it can start receiving the data sent via modulo-2 addition to the MLS sequence. This data identifies the satellite, its position (ephemeris), and its precise clock. When three or more satellites are in received by the GPS receiver a position on the ground and precise time is available. Combinations of MLS sequences (modulo-2 added together) are not maximal, but still can be used to rapidly synchronize communication while allowing long enough sequences for measuring large distances. Such is the case with the JPL (NASA's Jet Propulsion Laboratory) ranging codes. Other applications of spread spectrum technology include a slightly different technique called frequency hopping where the transmitter and receiver follow a sequence of frequencies like a musical arpeggio. The recently recognized inventors of the frequency hopping technique is Ms Hedy Lamarr, an Austrian-born singer/actress who with her second husband, musician George Antheil, developed a pseudo-random frequency hopping technique for secure radio using paper rolls like that of a player piano. See "http://www.microtimes.com/166/coverstory166.html" for a detailed historical account. Frequency hopping also uses pseudo-random numbers but is somewhat different from the direct sequence spread spectrum technique presented here as a means to resolve coherent multipath.

13.4 SUMMARY, PROBLEMS, AND BIBLIOGRAPHY

Chapter 13 presents a baseline set of approaches to adaptive beamforming. There are many additional algorithms in the literature, but they can be seen to fall within one of the three categories presented in Sections 13.1 through 13.3. Adaptive null-forming (Section 13.1) is simply a spatial correlation cancellor where the resulting spatial FIR response for a specific narrowband frequency represents a beam pattern which places nulls in the dominant source direction(s). The technique can also work for broadband sources, but the nulls will not be as sharp and precise in angle as in the narrowband case. It is also shown that for the spatial null forming to be precise, the phases of the sources must be independent so that the covariance matrix for the array has magnitudes and phases only associated with the spatial response and not the phases of the individual sources. The assumption of Phase independent sources is problematic for narrowband signals. To overcome this mathematical limitation, one averages many signal "snapshots" or time-domain signal buffers converted to the specific narrowband FFT bin for the covariance matrix signal input. Some presentations in the literature evaluate the inverse of the spatial FIR filter response, as if it were an IIR filter. The inverse response shows sharp peaks, rather than nulls in the beam response. This approach, sometimes called "super-resolution beamforming" or more simply "spectrum estimation" is specifically avoided here because it is physically misleading. An array's main beam lobe width and SNR gain is defined by the array aperture, number of elements, and the wavelength, which have nothing to do with the choice of mathematical display of the physical beam response with its nulls displayed as peaks.

Section 13.2 analyzes the array covariance data as an eigenvalue problem. This follows quite logically from the vandermode matrix seen in Eq. (13.3.1) where each source has an associated eigenvector representing the spatial phase due to the direction of arrival and array geometry, and a source eigenvalue associated with the magnitude-squared of the source signal at the array. The phase associated with a particular narrowband source is lost in this formulation, hence, the sources must be phase-independent and the background noise spatially independent for the eigenvalue approach to work. However, given phase independent sources and spatially incoherent background noise, solving for an eigenvalue representation of the spatial covariance matrix for the array provides a set of orthonormal beam steering vectors (the eigenvectors) and associated signal powers (the eigenvalues). If one assumes the source signals of interest are stronger than the spatially incoherent background noise, the eigenvalues can be separated into a signal subspace and a noise subspace. The eigenvectors associated with the signal subspace can be used as optimal beam steering vectors to detect the signal from the corresponding source while nulling the other sources since the eigenvectors are by definition orthonormal. However, in practice this theory is encumbered by the partial coherence between narrowband sources at the same frequency, even with snapshot averaging attempts to make the sources appear incoherent with each other.

One can safely assert that the noise eigenvectors each have nulls in their beam responses in the direction(s) of the signal source(s), plus "spurious" nulls in the directions of the estimated incoherent noise sources. Therefore, by summing the noise eigenvectors we are left with a beam steering vector which has nulls in the direction(s) of all the source(s) only. This beam response is called the minimum variance, or minimum norm beam response. Section 13.2 presents a technique for steering the array to a look direction other than a source direction and maintaining the sharp nulls to the dominant source directions and unity gain in the look direction. The beam is called the minimum variance distortionless response, or MVDR beam. It is very useful for suppressing interference noise in specific directions while also suppressing incoherent background noise in directions other than the look direction. But, the MVDR beam becomes indeterminant when the look direction is in the direction of one of the signal sources. As mentioned earlier, the associated eigenvector makes a pretty good beam steering vector for a source's look direction. We also show in Section 13.2 several techniques for synthesizing beams with forced sharp nulls in the other dominant source directions while maintaining unity gain in the desired look direction.

Section 13.3 presents a technique for resolving multipath which exploits multipath coherence to resolve signal arrival times at the array using broadband techniques where the transmitted signal is known. This actually covers a wide range of active sensor applications such as sonar, radar, and lidar. The transmitted spectrum is broadened by modulating a sinusoid with a maximum length sequence (MLS) which is generated via a specific psuedo-random bit generator algorithm based on primitive polynomials. The MLS bit sequence has some very interesting properties in that the autocorrelation of the sequence gives a perfect delta function and the cross correlation between different sequences is zero. These properties allow signals modulated with different MLS codes to be easily separated when they occupy the same frequency space in a signal propagation channel. Technically, this has allowed large number of communication channels to co-exist in the same bandwidth. Some of the best examples are seen in wireless Internet, digital cellular phones, and in the Global Positioning System (GPS). For our purposes in wavenumber processing, the spread spectrum technique allows coherent multipath to be resolved at each sensor in an array. Using straightforward cross correlations of transmitted and received signals, the propagation time for each path can be resolved to within on bit, or chip, of the MLS sequence. This is useful but requires a very high bandwidth defined by a high chip rate for precise resolution. Section 13.3 also shows a technique where one measures the transfer function from transmit to receive which produces a channel impulse response with resolution equal to the sample rate. This, like the high chip rate, uses the full available bandwidth of the signals. By slowing the chip rate to the point where no more than one phase transition occurs within a transfer function FFT buffer, we can maximize the coherence between the transmitted signal and the received multipath. This results in a very "clean" channel impulse response. The spatial covariance matrix of these measured impulse responses can be either analyzed in a broadband or narrowband sense to construct beams which allow separation of the signal paths.

PROBLEMS

1. For a linear array of 3 sensors spaced 1 m apart in a medium where the free plane wave speed is 10,000 m/sec, derive the spatial filter polynomial analytically for a plane wave of arbitrary frequency arriving from 60 degrees bearing where 90 degrees is broadside and 0 degrees is endfire.

2. Derive the steering vector to steer a circular (16 elements with diameter 2 m) array to 30 degrees if the wave speed is 1500 m/sec. Evaluate the beam response and determine the wavenumber range for propagating plane wave response (i.e. no evanescant wave responses) for a particular frequency.

3. Implement an adaptive null forming spatial FIR filter for the linear array in problem 1 but using an LMS adaptive filter. How is the step size set to guarantee convergence?

4. Using the Vandermode matrix to construct an "ideal" covariance matrix, show that the signal eigen vectors do not necessarily have nulls in the directions of the other sources.

5. Show that under ideal conditions (spatially independent background noise and phase independent sources) the MVDR optimal beam steering vector for looking at one of the signal sources while nulling the other sources is the eigenvector for the source in the look direction.

6. Given a linear but unequally-spaced array of sensor with position standard deviation σ_d determine the bearing accuracy at 500 Hz if the plane wave speed is 1500 m/sec, the wave arrives from a direction near broadside with SNR 10, and if there are 16 sensors covering an aperture of 2 m.

7. Show that the autocorrelation of the MLS sequence generated by [3, 1] is exactly 7 for the zeroth lag and -1 for all other lags.

8. Show analytically why a chip rate of say 20 chips per sec for bi-phase modulation of a sinusoid gives a sinc-function like spectral shape where the main lobe extends ± 20 Hz from the sinusoid center frequency.

9. Show that for bi-phase modulation of a sinusoidal carrier with a chip rate equal to the sample rate will produce a nearly white signal regardless of the carrier frequency.

10. What is the relationship between chip rate and resolution, and MLS sequence length and maximum range estimation?

BIBLIOGRAPHY

W. S. Burdic Underwater Acoustic System Analysis, Chapter 14, Englewood Cliffs: Prentice-Hall, 1991.

R. C. Dixon Spread Spectrum Systems with Commercial Applications, 3rd ed., Table 3.7, New York: Wiley, 1994.

S. J. Orfanidis Optimum Signal Processing, Chapter 6, New York: McGraw-Hill, 1988.

W. H. Press, B. P. Flannery, S. A. Teukolsky, W. T. Vetterling Numerical Recipes: The Art of Scientific Computing, Sections 7.4 and 11.1, New York: Cambridge University Press, 1986.

T. J. Shan, T. Kailath "Adaptive Beamforming for Coherent Signals and Interference," IEEE Transactions on Acoustics, Speech, and Signal Processing, Vol. ASSP-33(3), 1985, pp. 527–536.

R. J. Vaccaro (Ed.) "The Past, Present, and Future of Underwater Acoustic Signal Processing," UASP Technical Committee, IEEE Signal Processing Magazine, Vol. 15(4), July 1998.

B. D. Van Veen, K. M. Buckley "Beamforming: A versatile Approach to Spatial Filtering," IEEE ASSP Magazine, Vol. 5(2), April 1988.

REFERENCES

1. W. H. Press, B. P. Flannery, A. A. Teukolsky, W. T. Fetterling, Numerical Recipes, Section 11.1, New York: Cambridge University Press, 1986.

2. M. Wax, T. Kailath "Detection of Signals By Information Theoretic Criteria," IEEE Transactions on Acoustics, Speech, and Signal Processing, ASSP-33(2), 387–392, 1985.

3. R. Gold "Optimal Binary Sequences for Spread Spectrum Multiplexing," IEEE Transactions on Information Theory, October 1967.

4. R. C. Dixon Spread Spectrum Systems with Commercial Applications, 3rd ed., Section 3.6, New York: Wiley, 1994.

Part V

Signal Processing Applications

14

Intelligent Sensor Systems

There will always be a debate about what is "smart," "intelligent," or even "sentient" (which technically means having the five senses) in the context of artificial intelligence in computing systems. However, it is plausible to compare smart sensors to the pre-programmed and sensor-reactive behavior of insects. It's a safe argument that insects lack the mental processing to be compared to human intelligence but, insects have an amazing array of sensor capability, dwarfed only by their energy efficiency. A fly's life probably doesn't flash before its eyes when it detects a hand moving to swat it, it simply moves out of the way and continues its current "program." It probably doesn't even get depressed or worried when another fly gets spattered. It just follows its life function reacting to a changing environment. This "insect intelligence" example brings us to define an intelligent sensor system as having the following basic characteristics:

1. Intelligent sensor systems are adaptive to the environment, optimizing their sensor detection performance, power consumption, and communication activity
2. Intelligent sensor systems record raw data and extract information, which is defined as a measure of how well the data fit into information patterns, either pre-programmed or self-learned
3. Intelligent sensor systems have some degree of self-awareness through built-in calibration, internal process control checking and re-booting, and measures of "normal" or "abnormal" operation of its own processes
4. Intelligent sensor systems are re-programmable through their communications port and allow external access to raw data, program variables, and all levels of processed data
5. An intelligent sensor system can not only recognize patterns, but can also predict the future time evolution of patterns and provide meaningful confidence metrics of such predictions

The above five characteristics are a starting point for defining the smart sensor node on the wide-area network or intra-network (local network not accessible by the global Internet) for integrating large numbers of sensors into a control system for production, maintenance, monitoring, or planning systems. Smart sensors provide *information* rather than simply raw data. Consider Webster's dictionary definition of information:

> **information:** a quantitative measure of the content of information, specifically, a numerical quantity that measures the uncertainty in the outcome of an experiment to be performed

Clearly, when a sensor provides only raw waveforms, it is difficult to assess whether the sensor is operating properly unless the waveforms are as expected. This is a pretty weak position to be in if an important control decision is to be based on unusual sensor data. But, if the sensor provides a measure of how well the raw data fits a particular pattern, and if that pattern appears to be changing over time, one not only can extract confidence in the sensor information, but also the ability to predict how the sensor pattern will change in the future. Smart sensors provide information with enough detail to allow accurate *diagnosis of the current state* of the sensor's medium and signals, but also *prognosis of the expected future state*. In some cases, the transition of the pattern state over time will, in itself, become a pattern to be detected.

There are some excellent texts on pattern recognition (1,2) which clearly describe statistical, syntactic, and template type of pattern recognition algorithms. These are all part of an intelligent sensor system's ability to turn raw data into information and confidences. While beyond the already broad scope of this book, it is well worth a brief discussion of how pattern recognition, pattern tracking (prognosis), and adaptive algorithms are functional blocks of a smart sensor system. One of the most appealing aspects of Schalkoff's book is its balanced treatment of syntactical, statistical, and neural (adaptive template) pattern recognition algorithms. A syntactical pattern recognition scheme is based on human knowledge of the relevant syntax, or rules, of the information making up the pattern of interest. A statistical pattern recognition scheme is based on knowledge of the statistics (means, covariances, etc) of the patterns of interest. The neural network (or adaptive template) pattern recognition scheme requires no knowledge of pattern syntax or statistics.

Suppose one could construct a neural network to add any two integers and be right 95% of the time. That would be quite an achievement although it would require extensive training and memory by the network. The training of a neural network is analogous to a future non-science major memorizing addition tables in primary school. Indeed, the neural network began as a tool to model the brain and the reinforcement of electro-chemical connections between neurons as they are frequently used. Perhaps one of the more interesting aspects of our brains is how we forget data, yet somehow know we used to know it. Further, we can weight multiple pieces of information with confidence, such as the likelihood that our memory may not be accurate, in assembling the logic to reach a decision. Clearly, these would be extraordinary tasks for a software algorithm to achieve, and it is already being done in one form or another in many areas of computer science and signal processing.

A scientific approach to the addition problem is to memorize the algorithm, or syntax rules, for adding any two numbers, and then apply the rules to be correct all of the time. Using 2's compliment arithmetic as described in Section 1.1, electronic logic can add any two integer numbers within the range of the number of bits and never make a mistake. That's the power of syntax and the reason we have the scientific method to build upon and ultimately unify the laws of physics and mathematics. The only problem with syntactical pattern recognition is that one has to know the syntax. If an error is found in the syntax, ultimately one organizes experiments to quantify new, more detailed rules. But, to simply apply raw data to rules is risky, unless we change the data to information by also including measures of confidence.

Statistical pattern recognition is based on each data feature being represented by statistical moments such as means, variances, etc., and an underlying probability density function for the data. For Gaussian data, the mean and variance are adequate representations of the information if the underlying data is stationary. If it is moving, such as a rising temperature reading, a Kalman filter representation of the temperature state is the appropriate form to represent the information. In this case both the temperature reading and its velocity and acceleration have mean states and corresponding variances. The pattern now contains a current estimate and confidence, and a capability to predict the temperature in the future along with its confidence. Combining this representation of information with scientifically proven syntax, we have the basis for "fuzzy logic", except the fuzzyness is not defined arbitrarily, but rather by the observed physics and statistical models.

Neural networks are clearly powerful algorithms for letting the computer sort the data by brute force training and are a reasonable choice for many applications where human learning is not of interest. In other words, we cannot learn much about how or why the neural network separates the trained patterns or if the network will respond appropriately to patterns outside of its training set. However, the structure of the neural network is biologically inspired and the interconnections, weightings, and sigmoidal nonlinear functions support a very powerful capability to separate data using a training algorithm which optimizes the interconnection weights. Note that the interconnections can represent logical AND, OR, NOR, XOR, NAND, etc., and the weight amplitudes can represent data confidences. The sigmoidal function is not unlike the inverse hyperbolic tangent function used in Section 11.2 to model the probability of a random variable being above a given detection threshold. The biologically inspired neural network can carry not only brute-force machine learned pattern separation, but, can also carry embedded human intelligence in the form of constrained interconnections, and weights and sigmoidal functions based of adaptive measured statistics processed by the sensor system.

The anatomy of intelligent sensor systems can be seen in the sensor having the ability to produce information, not just raw data, and having the ability to detect and predict patterns in the data. Some of these data patterns include self-calibration and environmental information. These parameters are directly a part of the syntax for assessing the confidence in the data and subsequent derived information. As such, the smart sensor operates as a computing node in a network where the sensor extracts as much information as possible from its data and presents this information to the rest of the network. At higher levels in the network, information from multiple smart sensors can be combined in a hierarchical layer of "neural network with statistically fuzzy syntax" which combines information from multiple intelligent sensor nodes to extract yet more information and patterns. The structure of the hierarchical layers is defined by the need for information at various levels. All of what has just been stated is possible with 20th century technology and human knowledge of straightforward physics, signal processing, and statistical modeling. Given what we will learn about our world and our own thoughts from these intelligent sensor networks, it would appear that we are about to embark on an incredibly fascinating era for mankind.

Section 14.1 presents the three main techniques for associating information with known patterns. There are clearly many approaches besides statistical, neural, and syntactic pattern recognition and we hope that no reader gets offended at the lack of depth or conglomeration of these different techniques. Our view is to adopt the strong aspects of these well-accepted techniques and create a hybrid approach to the pattern recognition problem. Section 14.2 discusses features from a signal characteristic viewpoint. It will be left to the reader to assemble meaningful feature sets to address a given pattern recognition problem. Section 14.3 discusses the issue of pattern transition over time, and prognostic pattern recognition, that is, prediction what the pattern may be in the future. Prediction is the ultimate outcome of information processing. If the underlying physics are understood and algorithms correctly implemented, if the data is collected with high SNR and the feature information extracted with high confidence, our intelligent sensor system ought to be able to provide us with reasonable future situation predictions. This is the "holy grail" sought after in the development of intelligent sensor systems and networks.

14.1 AUTOMATIC TARGET RECOGNITION ALGORITHMS

Automatic target recognition (ATR) is a term most often applied to military weapon systems designed to "fly the ordnance automatically to the target". However, ATR should also be used to describe the process of detecting, locating, and identifying some information object automatically based on objective metrics executed by a computer algorithm. The ATR algorithm is actually many algorithms working together to support the decision process. But, before one can use a computer algorithm to recognize a target pattern, one must first define the pattern in terms of a collection of *features*. Features represent compressed information from the signal(s) of interest. In general, features should have some physical meaning or observable attribute, but this is not an explicit requirement. A collection of features which define a pattern can be defined statistically by measuring the mean and covariance, or higher order moments, of these features for a particular pattern observed many times. The idea is to characterize a "pattern template" in terms of the feature means and to experimentally define the randomness of the features in terms of their covariance. The feature data used to define the feature statistics is called the *training data set*. The training data can also be used to adapt a filter network to separate the pattern classes. The most popular type of ATR algorithm for adaptive pattern template separation is the *adaptive neural network* (ANN). The ANN is often a very effective algorithm at separating pattern classes. But, the ANN does not provide a physical meaning or statistical measure to its internal coefficients. This leaves ANN design very much an art. The last type of ATR we will consider is the *syntactic fuzzy logic* classifier. The difficulty with the syntactic classifier is that one must know the syntax (feature physics and decision rules) in advance. Fuzzy logic is used to tie the syntax together in the decision process such that lower confidence features can be "blended out" of the class decision in favor of higher confidence data. Our syntactic ATR exploits *data fusion* by enabling the fuzzy combination of many types of information including metrics of the information confidence.

Statistical Pattern Recognition is perhaps the most straightforward ATR concept. Patterns are recognized by comparing the test pattern with previously designated pattern classes defined by the statistics of the features, such as the mean and covariance. For example, one might characterize a "hot day" pattern class by temperatures averaging around 90°F with a standard deviation of say 10°. A "cold day" might be seen as a mean temperature of 40°F and a standard deviation of 15°. These classes could be defined using human survey data, where the individual circles either "hot day" or "cold day" at lunch time, records the actual temperature, and surveys are collected from, say one hundred people over a 12 month period. The 36,500 survey forms are collected and tabulated to give the mean and standard deviation for the temperature in the two classes. Given the temperature on any day thereafter, a statistical pattern recognition algorithm can make an objective guess of what kind of day people would say it is. The core to this algorithm is the characterization of the classes "hot day" and "cold day" as a simple Gaussian distribution with mean m_k and standard deviation σ_k for the kth class w_k.

$$p(x|w_k) = \frac{1}{\sqrt{2\pi}\sigma_k} e^{-\frac{1}{2}\left(\frac{x-m_k}{\sigma_k}\right)^2} \tag{14.1.1}$$

Equation (14.1.1) describes the probability density for our observed feature x (say the temperature for our example) given that class w_k is present. The term in parenthesis in Eq. (14.1.1) is called the *Mahalanobis distance* from x to m_k. Clearly, when this distance is small the feature data point matches the class pattern mean well. This density function is modeled after the observed temperatures associated by people to either "hot day" or "cold day" classes for our example. Why have we chosen Gaussian as the density function? Well for one reason, if we have a large number of random variables at work, the central limit theorem proves that the composite representative density function is Gaussian. However, we are assuming the our random data can be adequately described by a simple mean and variance rather than higher-order statistics as well. It is generally a very good practice to test the Gaussian assumption and use whatever density function is appropriate for the observed feature data. The answer we are seeking from the statistical classifier is given an observed feature x, what is the probability that we have class w_k. This is a subtle, but important difference from Eq. (14.1.1) which can be solved using Baye's theorem.

$$p(w_k|x) = \frac{p(x|w_k)p(w_k)}{p(x)} \tag{14.1.2}$$

Integrating the density functions over the available classes we can write Baye's theorem in terms of probabilities.

$$P(w_k|x) = \frac{p(x|w_k)P(w_k)}{p(x)} \tag{14.1.3}$$

The probability density for the feature x can be seen as the sum of all the class-conditional probability densities times their corresponding class probabilities.

$$p(x) = \sum_k p(x|w_k)P(w_k) \tag{14.1.4}$$

Equations (14.1.3) and (14.1.4) allow the user to *a priori* apply the likelihood of each individual class and incorporate this information in the *a posteriori* probability estimate of a particular class being present when the observed feature x is seen.

Figure 1 shows three Gaussian probability density functions with respective means of -2.0, 0.0, and $+3.0$, and respective standard deviations of 1.25, 1.00, and 1.50. The Figure also shows a density function for all three classes of data to be used to determine if the feature is an outlier of the observed training data. This context is important because Eq. (14.1.3) will give a very high likelihood to the nearest class, even if the feature data is well outside the range of the associated feature data used during training. To illustrate this and other points, three feature values of -2.00, 1.00, and 15.0 are considered for classification into one of the three given classes. Table 1 summarizes the classification results. The data point at -2.00 is clearly closest to class 1 both in terms of probability and Mahalanobis distance. Furthermore, the feature data point -2.00 (the * in Figure 1) fits quite well within the training data

It is tempting to simply use a small Mahalanobis distance as the metric for selecting the most likely class given the particular feature. However, the real question being asked of the ATR is which class is present given a priori feature likelihoods and the feature data. Case 2 where the data point in Figure 1 is 1.00, one can see in Table 1 that the highest probability class is class 2, while the smallest Mahalanobis distance is class 3. Each of the three classes is given a 33.33% likelihood for $P(w_k)$.

The third case shows the utility of maintaining a "training data" class to provide a measure of whether the feature data fits into the accepted data for class training. In other words, is this feature data something entirely new or does it

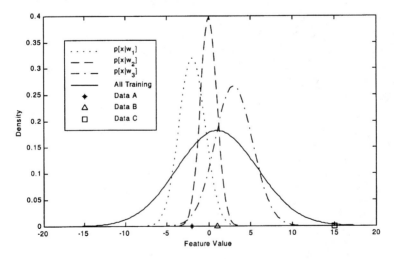

Figure 1   Three probability density functions describe three individual classes in which data points A, B, and C are to be classified. The solid curve shows the pdf for all the training data.

Table 1 Classification Results

Data	k	$P(w_k\|x)$	$(x - m_k)^2/\sigma_k^2$	$(x - m_k)/\sigma_k$	Result
A $= -2$	1	0.8066	0.0000	0.0000	Class 1
	2	0.1365	4.0000	2.0000	
	3	0.0569	4.9383	2.2222	
	Training Data	0.8234	0.3886	0.6234	
B $= 1$	1	0.1071	3.6864	1.9200	Class 2
	2	0.5130	1.0000	1.0000	
	3	0.3799	0.7901	0.8888	
	Training Data	1.0000	0.0000	0.0000	
C $= 15$	1	0.0000	118.37	10.879	Outlier
	2	0.0000	225.00	15.000	
	3	1.0000	28.444	5.3333	
	Training Data	0.0145	8.4628	2.9091	

fit within the range of the data we so cleverly used to train our classifier (determine the means and standard deviations for each of the classes). For the third feature data point of 15.0, Table 1 shows that class 3 is the best choice of the three classes, but the data is not part of the training data set, as indicated by the low 0.0145 probability for the "Training data" class. This low probability coupled with the large Mahalanobis distances to all classes clearly indicates that the feature data is an outlier corresponding to an "unknown" classification.

Multidimensional probability density functions are used to describe a number of feature elements together as they apply to a pattern class. Use of multiple signal features to associate a pattern to a particular class is very effective at providing robust pattern classification. To demonstrate multi-feature statistical pattern recognition we begin by define a M-element feature vector.

$$\bar{x} = [x_1 x_2 \ldots x_M] \tag{14.1.5}$$

The means of the features for the class w_k are found as before

$$\bar{m}_k = [m_{1,k} m_{2,k} \ldots m_{M,k}] = \frac{1}{N} \sum_{n=1}^{N} \bar{x}_n \tag{14.1.6}$$

but, the variance is now expressed as a covariance matrix for the class w_k

$$\Sigma_k = \frac{1}{N} \sum_{n=1}^{N} [\bar{x}_n - \bar{m}_k]^H [\bar{x}_n - \bar{m}_k] = \begin{bmatrix} \sigma_{1,1,k}^2 & \sigma_{1,2,k}^2 & \cdots & \sigma_{1,M,k}^2 \\ \sigma_{2,1,k}^2 & \sigma_{2,2,k}^2 & \cdots & \sigma_{2,M,k}^2 \\ \vdots & \vdots & & \vdots \\ \sigma_{M,1,k}^2 & \sigma_{M,2,k}^2 & \cdots & \sigma_{M,M,k}^2 \end{bmatrix} \tag{14.1.7}$$

where $\sigma_{i,j,k}^2$ is the covariance between feature i and j for class w_k.

$$\sigma_{i,j,k} = \frac{1}{N} \sum_{n=1}^{N} (x_{i,n} - m_{i,k})(x_{j,n} - m_{j,k}) \tag{14.1.8}$$

The multi-dimensional probability density function for the feature vector when class w_k is present is

$$p(\bar{x}|w_k) = \frac{1}{(2\pi)^{\frac{M}{2}}|\Sigma_k|^{\frac{1}{2}}} e^{-\frac{1}{2}[\bar{x}-\bar{m}_k]\Sigma_k^{-1}[\bar{x}-\bar{m}_k]^H} \tag{14.1.9}$$

where it can be seen that the determinant of the covariance must be non-zero for the inverse of the covariance to exist and for the density to be finite. This means that all the features must be linearly independent, and if not, the linearly dependent feature elements must be dropped from the feature vector. In other words, a given piece of feature information may only be included once.

Insuring that the covariance matrix is invertible (and has non-zero determinant) is a major concern for multi-dimensional statistical pattern recognition. There are several ways one can test for invertability. The most straightforward way is to diagonalize Σ_k (it already is symmetric) as done in an eigenvalue problem. The "principle eigenvalues" can be separated from the residual eigenvalues (which can be too difficult in matrix inversion) as part of a singular value decomposition (SVD) to reduce the matrix rank if necessary. However, an easier way to test for linear dependence is to simply normalize the rows and columns of Σ_k by their corresponding main diagonal square root value.

$$S_k = \begin{bmatrix} 1 & S_{1,2k} & S_{1,3,k} & \cdots & S_{1,M,k} \\ S_{2,1,k} & 1 & S_{2,3,k} & \cdots & S_{2,M,k} \\ S_{3,1,k} & S_{3,2,k} & 1 & \cdots & S_{3,M,k} \\ \vdots & \vdots & & \ddots & \vdots \\ S_{M,1,k} & S_{M,2,k} & S_{M,3,k} & \cdots & 1 \end{bmatrix} \qquad S_{i,j,k} = \frac{\sigma_{i,j,k}^2}{\sqrt{\sigma_{i,i,k}^2 \sigma_{j,j,k}^2}} \tag{14.1.10}$$

For completely statistically independent features both S_k and Σ_k are diagonal and full rank. The normalized matrix S_k makes it easy to spot which feature elements are statistically dependent because the off-diagonal element corresponding to feature i and feature j will tend towards unity when the features are linearly dependent. The matrix in Eq. (14.1.10) really has no other use than to identify linearly dependent features which should be dropped from the classification problem. However, a more robust approach is to solve for the eigenvalues of Σ_k and apply a SVD to identify and drop the linearly dependent features.

Consider a simple 2-class 2-feature statistical identification problem. A training set of hundreds of feature samples is used to determine the feature means and covariances for each class. The 2-dimensional problem is nice because its easy to graphically display the major concepts. However, a more typical statistical pattern problem will involve anywhere from one to dozens of features. The 1-σ bounds on a 2-feature probability density function can be seen as an ellipse centered over the feature means and rotated to some angle which accounts for the cross correlation between the features. For our example, class 1 has mean $m_1 = [7.5\ 8]$ and covariance matrix

$$\Sigma_1 = \begin{bmatrix} \cos\theta_1 & -\sin\theta_1 \\ \sin\theta_1 & \cos\theta_1 \end{bmatrix} \begin{bmatrix} 3.00 & 0.00 \\ 0.00 & 1.00 \end{bmatrix} \begin{bmatrix} \cos\theta_1 & \sin\theta_1 \\ -\sin\theta_1 & \cos\theta_1 \end{bmatrix}$$
$$= \begin{bmatrix} 1.2412 & -1.3681 \\ -1.3681 & 8.7588 \end{bmatrix}$$
(14.1.11)

where $\theta_1 = -80°$ and a positive angle is counter-clockwise from the x_1 axis. Class 2 has mean $m_2 = [7\ 7]$ and

$$\Sigma_2 = \begin{bmatrix} \cos\theta_2 & -\sin\theta_2 \\ \sin\theta_2 & \cos\theta_2 \end{bmatrix} \begin{bmatrix} 2.00 & 0.00 \\ 0.00 & 1.00 \end{bmatrix} \begin{bmatrix} \cos\theta_2 & \sin\theta_2 \\ -\sin\theta_2 & \cos\theta_2 \end{bmatrix}$$
$$= \begin{bmatrix} 3.6491 & -0.9642 \\ -0.9642 & 1.3509 \end{bmatrix}$$
(14.1.12)

where $\theta_2 = -20°$. The columns of the rotation matrices on the left in Eqs (14.1.11)–(14.1.12) can be easily seen as eigenvectors. The columns of the covariance matrix on the right are not linearly dependent, but are simply correlated through the coordinate rotation. Clearly, given the symmetric covariance matrix on the right, one can determine the principal variances and rotation orientation. In the 3-feature problem, the 1-σ bound on the Gaussian density function is an ellipsoid with three principal variances and two rotation angles. The reader is left to his or her own mental capacity to visualize the density graphics for 4 or more features, but one can clearly see the connection between eigenvalue analysis and feature independence. For the case of linearly dependent features, one would compute a near zero eigenvalue on the main diagonal of the principal variance matrix.

Figure 2 depicts 50 samples from each of our two classes and the 1-σ ellipses for the Gaussian density functions. The case shown has the two classes overlapping somewhat, which is not desirable but often the case for real statistical pattern recognition problems. The two classes are defined statistically based on the measured means and covariances associated with each class through analysis of a training data set. One can then do an eigenvalue analysis to determine if any of the features are linearly dependent and need to be eliminated. The ideal situation would be for the density functions for each class to completely separate in the feature space. In other words, we desire to have large distances between classes and small variances around each class. Given a situation where we do not have wide class separation, the issue of class discrimination because important. Following Eqs (14.1.3)–(14.1.4) we can estimate the probability of class w_k being present given feature set $\bar{x}$.

However, we also want to reject outliers which are well outside the range of the entire training samples for all classes. To do so, we derive a new function based on the probability density functions of the classes and total training data.

$$p^{\max}(w_k|\bar{x}) = \begin{cases} \max\{p(\bar{x}|w_1), p(\bar{x}|w_2)\} & \forall\, p(\bar{x}) > 0.01 \\ 0 & \forall\, p(\bar{x}) \leq 0.01 \end{cases}$$
(14..1.13)

The function in Eq. (14.1.13) is not a probability density, just a function to help us define the decision boundaries for our classes. Because we're dealing with only 2 dimensions, this approach is highly illustrative but not likely practical for larger

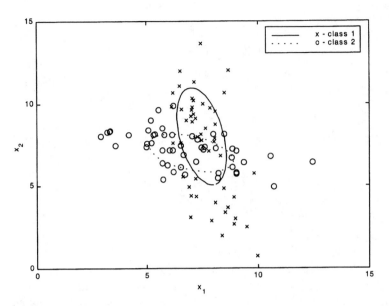

Figure 2 Samples of a 2-feature 2-class pattern set showing the 1-σ ellipses for the Gaussian density functions for each class and some of the samples in plotted in feature space.

numbers of features. At the boundaries between the two class density functions and at the ellipsoidal perimeter where the training density falls below 0.01 (about a 3-σ Mahalanobis distance) there will be abrupt changes in slope for $p^{\max}(\bar{x})$. Therefore, we can apply some straightforward image processing edge detection to highlight the decision boundaries. Now lets create an "edge" function as follows

$$p^{edge}(w_k|\bar{x}) = |\nabla^2 p^{\max}(w_k|\bar{x})|^{\frac{1}{8}} \tag{14.1.14}$$

where the 8th root is used to "flatten" the function towards unity — again to enhance our ability to see the decision lines, which are analytically very difficult to solve. Figure 3 shows the results of Eqs (14.1.13) and (14.1.14). The dark ellipsoidal rings are simply the areas where the class density functions have a near zero second derivative spatially. The bright lines clearly show the boundaries between classes, which in this case overlap, and the boundary for defining outliers. Figure 4 combines these edges with some logarithmically spaced contours for the function in Eq. (14.1.13). Depending on where the feature combination $[x_1 x_2]$ falls on this map, the statistical classifier will assign the sample to class 1, class 2, or outlier. In addition to making this decision, the algorithm can also provide probabilistic metrics for the confidence in the decision which is very valuable to the data fusion occurring in an intelligent sensor system.

Adaptive Neural Networks are a very popular approach to separating classes of data automatically. The concept is based on a biological model for neuron cells and the way electrochemical connections are made between these fascinating cells during learning. In the brain as connections are made between neurons at junctions called *synapses* to create neural pathways as part of learning, chemicals are deposited which either inhibit or enhance the connection. Each neuron can have

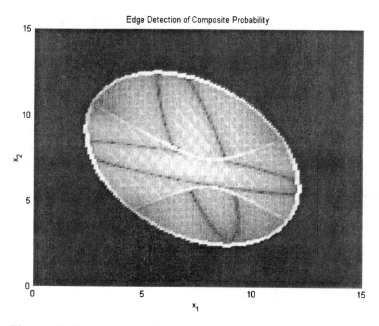

Figure 3 The thin bright lines of this 8th root LaPlacian of the composite maximum density shows the boundaries between the two classes as well as the boundary for a 3-σ total training data set.

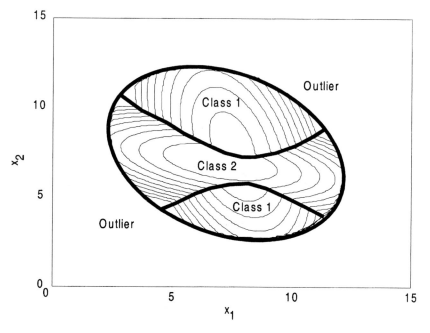

Figure 4 The outliers as well as class overlap is easily seen from simple edge-detection of the composite maximum density excluding any samples more than 3-σ from the total training set mean.

from 1,000 to 10,000 synapses. The human brain is composed of approximately 20 billion neurons. Beware, individual neural capacity will vary from one individual to the next. The effort-achievement response curve follows a natural logarithm, which means that all humans possess the ability to be infinitely stupid when zero effort is applied. The number of possible interconnections and feedback loops is obviously extraordinary. Even more fascinating, is the fact that many of these connections are preprogrammed genetically and can be "forgotten" and re-learned. While philosophically and biologically interesting, we will explore these issues no further here and concentrate specifically on the "artificial neuron" as part of the most basic adaptive neural network based on the *generalized delta rule* for adaptive learning. The reader should be advised that there is far more to adaptive neural networks for pattern recognition than presented here and many issues of network design and training which are beyond the scope of this book.

The basic artificial neuron is seen in Figure 5 where each of N inputs are individually weighted by a factor η_{jk} and summed with a bias term to produce the intermediate signal net_j. During the learning phase of the neural net, adaptive algorithms will be used to optimize the weights and bias at each node for every pattern of interest.

$$net_j = \sum_{k=1}^{N} \eta_{jk} i_{jk} + b_j \qquad (14.1.15)$$

The bias term in net_j is like a weight for an input which is always unity. This allows

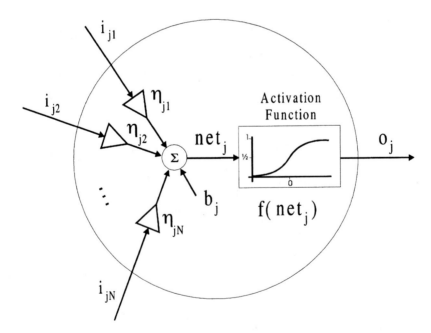

Figure 5 The "*j*th neuron" showing N inputs each with weight η_i a bias term b_j and the activation function $f(net_j)$ to nonlinearly limit the output o_j.

the neuron to have an output even if there are no inputs. The most typical of designs for an artificial neuron "squash" the signal net_j with a nonlinear *activation function*. This limits the final output of the neuron to a defined range (usually 0 to 1) and allows for a transition rather than a "hard clip" of the output between a zero or one state. By avoiding the "hard clip" or step response, and by limiting the range of the neural output, more information can be preserved and balanced across the outputs from other neurons such that no one pathway dominates. The most common activation function is the *sigmoid*.

$$o_j = f(net_j) = \frac{1}{1 + e^{-\varepsilon net_j}} \tag{14.1.16}$$

The output of the *j*th neuron o_j given in Eq. (14.1.16) and seen in Figure 5 is a nonlinear response of the inputs to the neuron i_{jk}, the weights η_{jk}, and the bias b_j. The parameter ε in Eq. (14.1.16) is a gain factor for the nonlinearity of the activation function. As ε becomes large even a small positive net_j will drive the output to unity or a small negative net_j will drive the output to zero. The choice of ε, the activation function, as well as the number and size of the hidden layers is up to the designer to choose. This coupled with the difficulty of a complete signal analysis in a neural network fuel the skeptics criticism of the approach. However, when applied appropriately, the adaptive neural network is a valuable and practical tool.

The optimization of the weights and bias terms for each neuron are calculated using error back propagation and the generalized delta rule (GDR). The GDR is very similar to the least-mean square (LMS) adaptive filter algorithm seen in Section 9.2. In fact, the man acknowledged with developing the LMS algorithm, Prof. Bernard Widrow of Stanford University, is also a pioneer in the development of adaptive neural networks. The back propagation algorithm is a little more subtle to understand. Consider the very simple network for classifying two features, x_1 and x_2, into two pattern classes, w_1 and w_2, seen in Figure 6 Each of the numbered circles represents a node seen in Figure 5. The two network outputs o_4 and o_5, are ideally [1, 0] when the features are from class 1, and [0, 1] when the features are from class 2. To make this happen, the weights and bias for each node must be adjusted to reduce, if not minimize the error at the network outputs. It is therefore straightforward to apply an LMS-like update to the weights and bias of each output node. But what about the hidden layer(s)? Back propagation of the error is used by taking the output layer node error, multiplying it by the weight between the output layer and the particular hidden layer node, and then summing all of the back propagated errors for that particular hidden layer node. The process continues using the output error for all the nodes in the hidden layer closest to the output layer. If additional hidden layers exist, the back propagation continues using the back propagated error rather than the actual output error. The weights and bias for each node are adjusted using the GDR until the output and back propagated errors becomes small, thus converging the neural network to a solution which separates the pattern classes of interest.

For the *p*th pattern class, we follow the nomenclature of the literature, we designate the node inputs i_{jk}^p and output o_j^p and the "training output" for class p as t^p. There are several philosophies about how the train the network, and contro-

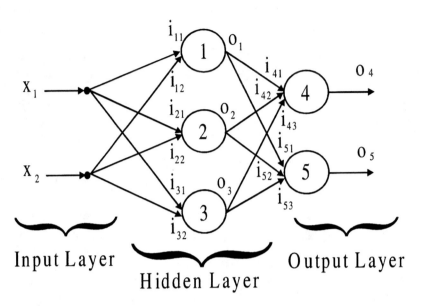

Figure 6 A simple 2-feature 2-output class neural network showing the 3-node hidden layer and the inputs and outputs of each node as part of a back propagation algorithm using the generalized delta rule.

versies with regard to over-training and extendability of the network result to data outside the training set. For example, one can train the net to identify only pattern p, and when testing for pattern p use those specific weights and biases. Therefore each class is tested using its optimum weights. Another more efficient approach is to train one set of network weights and biases to discriminate all the patterns of interest This is seen as more robust because the classes are directly compared. For an output node, the training error squared is seen to be

$$e_j^p = |t_j^p - o_j^p|^2 \qquad (14.1.17)$$

For a linear FIR filter, this error is a linear function of the filter coefficients. Therefore, the squared error can be seen as positive definite surface which has one minimum reachable by a series of weight adjustments in the direction opposite of the gradient of the squared error with respect to the weights. However, herein lies perhaps the biggest inconsistency of the neural network derivation: the error surface is rarely known and generally has multiple minima due to the nonlinear response of the sigmoids. Ignoring this point of lack of rigor, the neural network enthusiast is generally given to point out the success of the algorithm at separating classes. *We have no way of knowing whether the network is the optimum network, whether a different number of nodes and hidden layers will perform better, or whether the training of the network is complete.* One just has to accept this lack of mathematical rigor and optimization to move on and make use of this curious and useful algorithm. The gradient of the error is

$$\frac{\partial e^p}{\partial \eta_{jk}} = \frac{\partial e^p}{\partial net_j^p}\frac{\partial net_j^p}{\partial \eta_{jk}}$$

(14.1.18)

We can apply the chain rule again to break the error gradient down even further as

$$\frac{\partial e^p}{\partial \eta_{jk}} = \frac{\partial e^p}{\partial o_j^p}\frac{\partial o^p}{\partial net_j^p}\frac{\partial net_j^p}{\partial \eta_{jk}}$$

(14.1.19)

The gradient in Eq. (14.1.19) is broken down into its components. The gradient of the squared error with respect to output (Eq. (14.1.17)) is

$$\frac{\partial e^p}{\partial o_j^p} = -2|t_j^p - o_j^p|$$

(14.1.20)

The gradient of the output with respect to the weighted sum of inputs (Eq. (14.1.16)) is

$$\frac{\partial o^p}{\partial net_j^p} = f'(net_j^p)$$

$$= \frac{1}{1 + e^{-\varepsilon net_j^p}}\frac{\varepsilon\, e^{-\varepsilon net_j^p}}{1 + e^{-\varepsilon net_j^p}}$$

$$= o_j^p(1 - o_j^p)\varepsilon$$

(14.1.21)

and the gradient of the summed inputs and bias (Eq. (14.1.15)) is simply

$$\frac{\partial net_j^p}{\partial \eta_{jk}} = i_{jk}^p$$

(14.1.22)

The sensitivity of the pattern error on the net activation is defined as

$$\delta_j^p = -\frac{\partial e^p}{\partial net_j^p} = 2(t_j^p - o_j^p)[o_j^p(1 - o_j^p)\varepsilon]$$

(14.1.23)

and will be used for back propagation to the hidden layer(s). The weight adjustments using the GDR are simply

$$\eta_{jkt^+} = \eta_{jkt^-} + 2\mu\delta_j^p i_{jk}^p$$

(14.1.24)

where the parameter μ is analogous to the LMS step size, but here is referred to as the "learning rate." If the feature inputs to the neural network are scaled to be bounded with a ± 1 range, it's a fairly safe bet that choosing a learning rate less than, say 0.10, will yield a weight adjustment free of oscillation and allow the network to converge reasonably fast. The learning rate does effect the "memory" of the network during training such that a slow learner remembers nearly everything and a fast learner may forget the oldest training data. There's an interesting human analogy as well where many forms of mental retardation are characterized by slow highly repetitive learning and excellent long-term memory, while high mental capacity is often accompanied by very fast learning and surprising mid to long term memory "forgetfulness". Decreasing the learning rate (increasing the memory span) is advisable if the feature inputs are noisy.

The node bias term is updated as if its corresponding input is always unity.

$$b_{jt+} = b_{jt-} + 2\mu\delta_j^p \tag{14.1.25}$$

Each neuron receives a set of inputs, weights each one appropriately and sums the result in net_j, then passes the linear result through the activation function (the sigmoid in our example) to produce the neuron node output. At the output layer of the net, this is the neural network output used for class decision. For a node in a hidden layer, the neuron output is passed to the inputs of many other nodes in the next layer towards the output layer, if not the output layer. The pattern sensitivity for a hidden layer node is calculated by summing all the pattern sensitivities times their corresponding input weights for all the nodes in the next layer towards the output layer that its output is passed as input.

$$\delta_j^p = \sum_{n-N_1}^{N_2} \delta_n^p \eta_{nj} \tag{14.1.26}$$

In other words, if the output of node 15 is passed to nodes $N_1 = 37$ through $N_2 = 42$ in the next layer towards the output layer, the pattern sensitivity for node 15's weights and bias updates is found from the impact its sensitivity has on all the nodes it affects in the next layer. This is intuitively satisfying because the weight and bias adjustments near the feature input side of the network are specific to the feature inputs but affect most of the network outputs, while the weights and biases near the output layer are specific to the trained network output and encompass most of the feature inputs.

To summarize, given a trained neural network with k pattern classes and M features, one presents the M features for an unknown pattern observation to the input layer, computes the node outputs layer by layer until the network outputs are complete, and then chooses the largest output node as the class for the unknown pattern. Usually, the input features and trained outputs are bounded by unity to simplify the output comparison and net training. To train the neural network, the following procedure is acceptable:

1. The number of input layer nodes equals the number of features. There is at least one hidden layer with at least one more node than the input layer. The number of output nodes is equal to the number of patterns to be identified.
2. The weights and biases for each node are randomized to small values. Neglecting to randomize will lead to symmetry of the network weights and no real convergence. Small values are chosen to avoid "swamping" the sigmoidal functions or having one path dominate the network error. The inputs and outputs are bounded by unity to simply setting the learning rate and class selection from the maximum output node.
3. During training, the input features for the given class are presented to the input layer and the network outputs are computed layer by layer (Eqs (14.1.15)–(14.1.16)).

4. The "training error" is calculated for the output layer from the difference between the desired training output and the actual network outputs. The pattern sensitivity (Eq. (14.1.23)) is calculated for each output layer node.
5. The weights and biases for the inputs to the output layer nodes are adjusted (Eqs (14.1.24)–(14.1.25)).
6. The pattern sensitivity for the hidden layer nodes are calculated (Eq. (14.1.26)) and the corresponding weights and biases are adjusted (Eqs (14.1.24)–(14.1.25)).
7. Steps 3–6 may be repeated several times for the same input features until the error gets small, and/or steps 3–6 are repeated for a large number of known sample patterns constituting the pattern training set.

Figure 7 gives an example with the network output for class 1 shown as a surface plot where the input features form the x and y coordinates of the surface. There are two Gaussian data sets, class 1 with mean $\langle 4.5, 6.0 \rangle$ and class 2 with mean $\langle 8.0, 10.0 \rangle$. The neural network depicted in Figure 6 is trained with samples from class 1 normalized to the interval $\langle 0.0, 1.0 \rangle$ and the output for class 1 (node 4) set to unity and class 2 (node 5) set to zero. The training process is repeated with feature data from class 2 and the network output for class 1 zero and class 2 unity. The weights and biases for the network are then fixed and the output for class 1 is evaluated for every combination of features to produce the surface plot in Figure 7 For illustration, the Gaussian training data sets are superimposed over the surface. Clearly, the network does a good job separating the two classes. A more complex

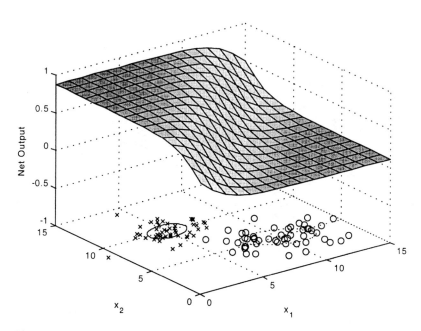

Figure 7 Neural network output "surface" for class 1 ($\times$) showing the scatter and Gaussian ellipse superimposed for both class where class 2 ($\bigcirc$) is deselected in the class 1 output.

network with more hidden layers and more nodes in the hidden layers could separate the classes with even greater deftness. This is more of an issue when the classes are closer together with larger standard deviations such that many of the training sample overlap. Surprisingly, the neural network still does a good job separating overlapping classes as can be seen in Figure 8 where the class 1 mean is $\langle 8.0, 9.0 \rangle$ and class 2 mean is $\langle 7.0, 6.0 \rangle$. The decision is clearly less confident due to the overlap, but it still appears reasonable by most metrics.

A much more sophisticated neural network (more layers, nodes, and more extensive training) might very well completely separate the training set classes with a decision boundary which winds its way between the samples in the overlap area. Therefore, if the training data is not random but rather simply complex, the sophisticated neural network with many layers and nodes should do an amazing job at memorizing which feature combinations go with which class. The converged network provides a fast nonlinear filter where the feature are inputs and the class decisions are outputs. This is of particular value when high quality training data is available, the class boundaries are far more complicated than Gaussian boundaries, and one is fairly confident that the actual feature data during usage is well represented by the available training data. Many scientists and engineers who have used neural networks can be described as enthusiasts because they are witnesses to the power of this adaptive nonlinear process to effectively separate complex classes. However, the extendability of the network performance to data covering a range outside of the training data is, like with statistical

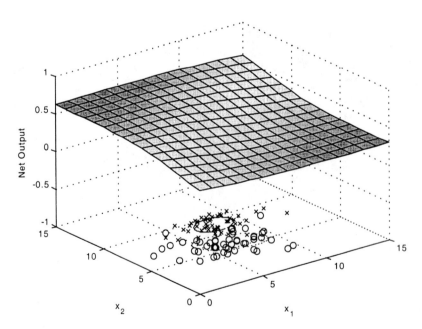

Figure 8 Neural network output "surface" for class 1 ($\times$) when the two classes overlap significantly in the training set showing a still operable, but less confident classification capability.

classification, questionable. To address the problem of extendability, one can use fuzzy logic and inferencing networks to build in human knowledge into parts of the network.

Syntactic Pattern Recognition is used when one can establish a *syntax*, or natural language to describe the way various pieces of information fit together in a pattern. Detection of the various information pieces can be done using statistical or neural network-based algorithms, but now we add a layer of logic where pattern data and confidences are fused together. Rather than insist that all class patterns be completely separate, we allow conflicting pattern estimates to co-exist for the benefit of subsequent data fusion to make a more balanced, informed decision. If all the information is binary (true or false), straightforward hard logic (and, or, if, then, else, etc., operations in software) can be used to construct a network for processing information. Given the human knowledge in the logic network, good performance can be expected outside the training data. If not, the flawed logic can be identified and corrected. You may remember this from grade school as *the scientific method* and it has worked so well for humans over the last few centuries that one should certainly consider employing it in computer artificial intelligence. *The main difficulty is that one must know the syntax.* Given the syntax rules, one can also employ the ability to discount information with low confidence in favor of higher confidence information and balance combinations of required and alternate information through weights and blending functions. Ultimately, the syntactic logic must have physical meaning to humans. This last point is quite important. For syntactic pattern recognition to reach its full potential in automatic target recognition, we humans need to be able to learn from flawed logic and correct the syntax to account for what has been learned. A flow diagram for the syntactic classifier can look very much like a neural network, but the weights, biases, activation functions, layers, and interconnections all have physical meaning and purpose. In a well understood syntactic classifier, no training is required to adjust weights or biases. However, it may be prudent to automatically adjust weights in accordance with the feature information confidence so that unreliable (low SNR for example) data can be automatically removed from the ATR decision. Syntactical classification algorithms can be found in commercial software for speech and handwriting recognition, decision aids for business logistics, autopilots, and even for product marketing.

The first algorithm needed to implement fuzzy logic is a *fuzzy set membership function* which allows one to control the transition between "true" or "false" for set membership. This transition can be anywhere from a step function for "hard logic" to the sigmoid given in Eq. (14.1.16) where epsilon is small. We call this a blend function $\beta(x)$. The blend function allows one to control fuzzy set membership. Why would one need to do this? Consider that we are combining, or fusing, multiple pieces of information together to achieve a more robust decision on the current pattern, or situation of patterns. This can be seen as a sort of *situational awareness* artificially estimated by the computer algorithm. Various pieces of information with individual confidences are combined where we want no one piece of information to dominate. We want a flexible outcome depending on the quality and availability of information. By blending or smoothing the decision thresholds, we can design the amount of "fuzziness" or flexibility desired in the data fusion and situational awareness algorithms. The following blend function (3) allows a great deal of design flexibility.

$$\beta(a, b, c, d, x) = \begin{cases} = b, x \le a \\ = \frac{1}{2}[d + b + (d - b)\sin\left(\frac{\pi(x - a)}{(c - a)} - \frac{\pi}{2}\right), a < x < c \\ = d, x \ge c \end{cases}$$

(14.1.27)

A plot of the blend function in Eq. (14.1.27) is seen in Figure 9 for values $a = 0.9$, $b = 1.5$, $c = 4.5$ and $d = 0.1$. This simple function allows one to easily set the maximum "confidence" for the blend function, the minimum "confidence", and the transition points for a sinusoid between the maximum and minimum. Furthermore, the blend function in Eq. (14.1.27) offers a great deal more flexibility than the sigmoid of Eq. (14.1.16) in controlling the transition between one state and the next. Figure 10 shows an example class pattern membership function derived from two blend functions to describe the average overweight person in terms of body fat. Note that the minimum and maximum confidence, as well as the transition zones are at the users discretion.

The next operator needed is a *fuzzy AND function* which will allow fusion of multiple pieces of information, each weighted by a corresponding "supportive" coefficient. We will use the same weight symbol as for the neural net η_{jk} (kth weight at the jth AND node) to keep terminology simple. The fuzzy AND function is used when one has the condition where any one information measure false (zero) will

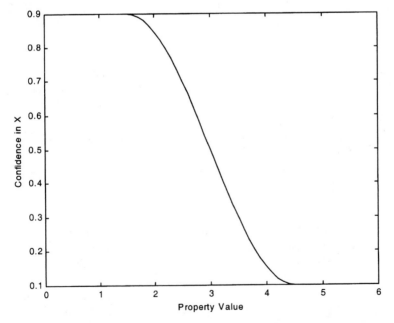

Figure 9 A blending function or "sigmoid" is used to map a property value to a confidence measure, which only in some well defined cases can be a statistical confidence.

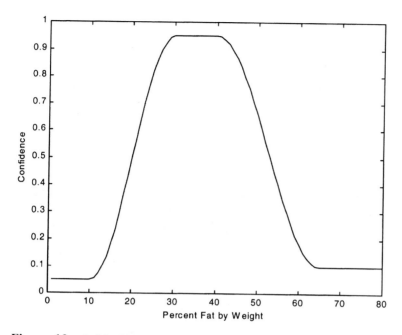

Figure 10 A "double blend" function is used to describe an example class membership confidence for typical overweight people in the developed world.

cause the resulting decision to be false (zero). If the supporting weight η_{jk} is close to unity, the corresponding information can be seen as required while if the supporting weight is small, the information is merely not important to the decision. To insure that lesser important information does not cause the AND to always produce a near false output, the fuzzy AND function is defined as

$$AND(\bar{X}_j, \bar{\eta}_j) = \left[\Pi_{i=1}^N \left(1 - \eta_{ji} + \eta_{ji}x_{ji} \right) \right]^u$$
$$u = 0.1 + 0.9e^{-0.3(\kappa-1)}$$
$$\kappa = \sum_{i=1}^{N} \eta_{ji} \tag{14.1.28}$$
$$\bar{X}_j = [x_{j1}x_{j2} \ldots s_{jN}]$$
$$\bar{\eta}_j = [\eta_{j1}\eta_{j2} \ldots \eta_{jN}]$$

Note how if the supporting weight is small, the corresponding term in the product does not become zero causing the result of the fuzzy AND to be false. Also note that if the sum of the supporting weights is large, the exponent u will tend towards 0.1. This has the effect of making the result of the fuzzy AND compressed towards unity if true, and zero if false.

Finally, we need a fuzzy OR function for cases where we are fusing information together where if any one of the pieces of information is true, the result is true. The intuitive mathematical operator to represent this is a Euclidian norm of each of the inputs to the fuzzy OR node times its corresponding supporting weight.

$$E_j = \sqrt{\sum_{i=1}^{N}(\eta_{ji}x_{ji})^2} \qquad (14.1.29)$$

The inputs and supporting weights are typically bounded between zero and unity. But, for any range of inputs and weights, the maximum value of any input-weight product is guaranteed to be less than or equal to the Euclidean norm. This is a reasonable place for the transition point of the blend function between true and false. If only one input is used, a step-like blend function is reasonable because no ORing is actually happening. As more inputs are used, we want the blend to span a wider range since more inputs increases the likelihood of a true result. Note that if all the inputs and supporting weights are unity, the Euclidean norm is simply $\sqrt{N}$. Therefore, our fuzzy OR function is described using a Euclidean norm in a blend function as given in Eq. (14.1.30).

$$OR(\bar{X}_j, \bar{\eta}_j) = \beta(a, b\sqrt{N}, 1, E)$$
$$u_j = \max\{(\eta_{ji}x_{ji})\; i = 1, 2, \ldots N\}$$
$$a = \begin{cases} 2u_j - \sqrt{N}; & N \le 4 \\ 0 & ; & else \end{cases} \qquad (14.1.30)$$
$$b = \min\{(\eta_{ji}x_{ji})\; i = 1, 2, \ldots N\}$$

The blend transition is symmetric about u_j only for $N \le 4$. If more than 4 inputs are used, we start the transition at zero and bring it to unity at $\sqrt{N}$, but it is less than 0.5 at u_j. This is largely the users choice. The blend implementation of the fuzzy OR keeps the output bounded between zero and unity.

When all the fuzzy logic inputs and outputs are bounded gracefully to a range between zero and unity, negating, or the NOT operator is a simple matter of subtracting the confidence (input or output) from unity. Now we have the capability to build all the usual logic operators AND, OR, NOT, NAND, NOR, XOR, etc, but in "fuzzified" format. Applying the blend function when the transition is rather sharp can be seen as adding back in the "crisp" to the logic. Humans build the logic to represent the known syntax and information. The fact that blend points, weights, and min/max values are set by the user should not imply a guess (at least not in all cases), but rather a placeholder to embed the physics the scientific method has taught us by derivation and experiment. Humans correct the logic based on observation and new physics built into the problem. In some cases it may be desirable as well to do some machine training and parameter optimization as seen for the neural network, which is always tempting. However, the real power of syntactic pattern recognition is possessing the correct knowledge that leads to the right syntax.

Finally, we can even employ a way to resolve a case where two or more inputs, representing the same general information, are conflicting thus giving rise to a state of confusion. Identifying a confusion pattern internal to the syntactic fuzzy network is extremely interesting and useful. For example, if A is true, AND B is false, AND C OR D is false, AND E is true, AND B AND C are not confused, THEN pattern Z is present. Measuring and using a confusion metric as part of the pattern recognition problem brings in a level of "self-objectivity" to the artificial intelligence. This can be an extremely important step in rejecting false alarms. Figure 11 shows a confusion blend function based on $\beta(-1,2\ 1, 0, dx_{ji})$, where dx_{ji} represents the difference in confidence (difference in input or output values) for two pieces of data representing the same information. When this difference is significant, the information is conflicting giving the syntactic algorithm confusion to deal with. The nonlinear blend function is useful in completely controlling the emphasis on confusion based on the information conflict.

To illustrate the power of syntactic pattern recognition, we present several spectrograms of bird calls seen in Figures 12–14 Several useful features can be extracted and sequenced from the spectrograms as seen in Figure 15 However, in order to correctly identify the bird calls, we must establish something called a *finite state grammar*. The strings of features must follow a specific order to match the bird call. This *natural language* is established from human application of the scientific method. Making the computer follow the same syntax is a matter of technique. Figure 16 shows one possible syntactic fuzzy logic to identify the Screaming Phiha bird song from the Blue Jay and Musician Wren. The weights are seen as the numbers next to the signal flow arrows and the boxed letter "B" depicts a blend function. The OR operator is assumed to also have its blend function built in. When the sequence is observed, the confidence output for the Screaming Phiha will be high. Dropping the "not quiet" confidences for simplicity, the

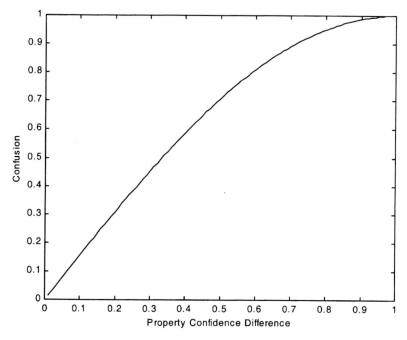

Figure 11 A confusion metric is derived by mapping the difference in confidences between two properties such one can control the amount of estimated confusion using a modified blend function.

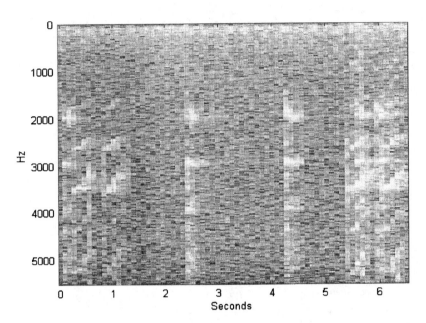

Figure 12 Spectrogram of Blue Jay showing quick pair of downward chirps (1 kHz fundamental weak, strong 2nd, 3rd, and 4th) followed by a wider spaced pair of constant harmonic bursts.

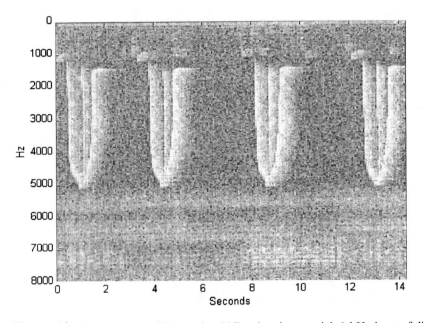

Figure 13 Spectrogram of Screaming Phiha showing a quick 1 kHz burst, followed by a chirp up to 5 kHz, then another quick chirp from 1.5 kHz to 5 kHz, followed by a down chirp to 1.5 kHz and a short hold.

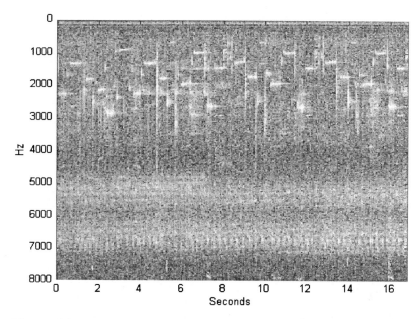

Figure 14   The remarkable signature of the Musician Wren showing a series of melodic steady "notes" from 1 kHz to 3 kHz which almost repeat as a pattern.

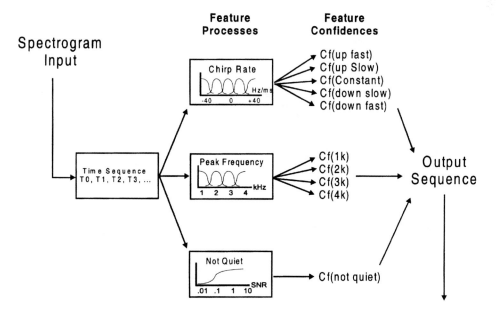

Figure 15   Block diagram showing the sequencing of features of the bird call spectrogram.

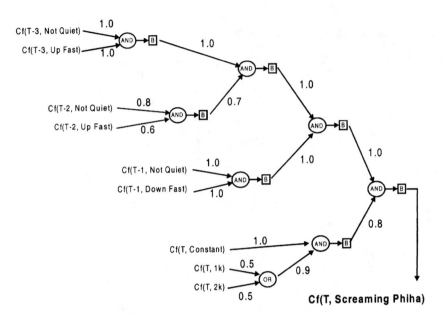

Figure 16 Fuzzy logic for computing the confidence that the spectrogram feature sequence matches the finite state grammar of the Screaming Phiha bird in Figure 13.

Screaming Phiha confidence at time T, $Cf(T, SP)$, fuzzy logic can be written as

$$Cf(T, SP) = 0.8 * \left\{ Cf(T, constant) \wedge 0.9* \right.$$
$$\times \left[0.5 * Cf(T, 1k) \vee 0.5 * Cf(T, 2k) \right] \right\} \wedge \left[Cf(T - 1, down\ fast) \right.$$
$$\wedge \left\{ 0.7 * Cf(T - 2, up\ fast) \wedge \ Cf(T - 3, upfast) \right\} \right]$$

$$(14.1.31)$$

Where the $\wedge$ symbol depicts a fuzzy AND and blend function and the symbol $\vee$ depicts a fuzzy OR function with its inherent blend operator. Similarly, the Blue Jay signature seen in Figure 12 confidence, $Cf(T, BJ)$, can be written as

$$Cf(T, BJ) = 0.75 * Cf(T, 3k)$$
$$\wedge \left\{ 0.9 * Cf(T, down\ slow) \vee 0.75 * Cf(T, constant) \right\} \qquad (14.1.32)$$

Note that one can invert the "fuzzy true-false" by simply subtracting the confidence from unity. The Musician Wren confidence (signature seen in Figure 14) is derived from a series of constant frequencies which are different from time to time. This confidence $Cf(T, MW)$ is a little more complicated to implement, but is still straightforward.

$$Cf(T, MW) = Cf(T, constant)$$

$$\wedge \left\{ Cf(T, 1k) \wedge [1 - Cf(T - 1, 1k)] \wedge [1 - Cf(T - 2, 1k)] \right.$$

$$\vee\ Cf(T, 2k) \wedge [1 - Cf(T - 1, 2k)] \wedge [1 - Cf(T - 2, 2k)]$$

$$\left. \vee\ Cf(T, 3k) \wedge [1 - Cf(T - 1, 3k)] \wedge [1 - Cf(T - 2, 3k)] \right\}$$

$$(14.1.33)$$

The output of the syntactic classifier is all of the pattern class confidences. One can choose the one with the highest confidence, say above 0.5, or choose none if the confidences for any one class are too low. The power of the syntactic technique is that the algorithm is a completely general way to implement the many forms of human logic on a computing machine. Inference engines which generate logic from rules are the basis for many powerful languages such as LISP and PROLOG. The main challenge for the programmer is that one has to master the syntax of the problem. In other words, for a solid solution, one needs to explicitly follow the physics and use credible signal processing technique to determine the data confidences going into the syntactic classifier. The information "supportive weights" and choices for blend, AND, and OR functions are more a function of the required logic than the dynamic data quality.

One other important technique in classification in the *hidden Markov model*. A Markov chain is a sequence of states of something where the state transition is defined using a probability. When a classifier is constructed based on a sequence of pattern class decisions, often achieving a finer or more specific resolution with each sequential decision, it is said that the classifier is based on a hidden Markov model. This technique is widely used in speech and handwriting recognition, and interpretation. The difficulty is in defining the transitional probabilities. The syntactic classifier technique described here can also be based on a hidden Markov model. Furthermore, its inputs could be generated from statistical classifier metrics or the outputs from a neural network.

The syntactic fuzzy logic classifier technique is completely general and can be the basis for data fusion of all types to achieve the response which best produces artificial intelligence. Defining machine intelligence is still a somewhat subjective thing to many people, even computer scientists. However, we can assert a "layered" vision of machine intelligence. At the base one has sensors capable of producing calibrated physical data with some confidence measure. At the next layer, various feature extractions occur with straightforward signal processing algorithms which carry forward the sensor confidence metrics. For example, an FFT and beamforming algorithm improves the SNR by a certain degree, thus allowing one to assert the probability of detection of a spectral peak on the output of the beamformer. At the third layer, basic pattern recognition occurs through the use of statistical or neural algorithms, which again carry along the confidence metrics as well as use them to discount unreliable data. The user employs whatever feature/pattern recognition algorithm which makes sense physically. Finally, a syntactic fuzzy logic classifier covers the top layer to fuse various pieces of information to produce an overall "situational awareness" which includes the state of the intelligent sensor system as well as the environmental and sensor data. In the future when humans interact with an intelligent sensor, they won't want to see waveforms, they'll want to know the situation the system is identifying and whether the system is in a reliable state. Perhaps the people who built the intelligent sensor system might be interested in the waveforms, but the general public and those who must act quickly in response to the intelligent sensor will likely never see or know the details of what data and processing are occurring in the system.

Like many of the topics covered in this book, we can only provide the reader with a very brief introduction to the important concepts. It is our hope that this encounter provides a fascinating view of the forest, with a more detailed view of a few typical and important trees. To get down to the tree bark and the rigor necessary to enhance or develop new algorithms in this area, the reader should consult one of the many excellent texts on fuzzy logic and natural artificial languages for more details. Syntactic pattern recognition is used for speech and handwriting recognition, as well as many other fascinating applications which are well beyond the scope of this book. However, it provides us an opportunity to examine our own logic, developed using the scientific method. By employing these techniques together, we stand to enjoy great triumphs in computing if we are willing to accept the inevitable humility.

14.2 SIGNAL AND IMAGE FEATURES

Features are distinguishing artifacts which can be assembled to define a useable pattern for classification of a signal. However, a better description of a feature is *an information concentrator*, which makes the job of detecting a pattern that much easier. For example, a sinusoid can have its information concentrated into one complex number (magnitude and phase) with a corresponding frequency by an FFT. If in the time domain, the signal is an impulse or burst waveform, there is no point in doing an FFT since it will only spread the signal over a wider space making feature detection more difficult. For an image with regular patterns across the focal plane, the spatial domain (the viewable image itself) is the obvious domain to compute features, since they are already concentrated by focusing of the lens-aperture system. The wavenumber (2-dimensional FFT) domain obviously offers advantages of spatial processing, filtering, etc., but not necessarily feature detection, unless the viewable image consists of regular periodic patterns or textures which are well-represented by a few FFT bins.

Our discussion will focus first on 1-dimensional signal features, such as those from acoustic, vibration, or electromagnetic sensors, or sensor arrays. Our fundamental signal is a "delta function-like" impulse, which can be integrated into step, ramp, quadratic, and higher-order functions. These signal classes are very important to identify for control systems as well as other applications. Next we consider periodic signals, starting of course with the sinusoid but extending into impulse trains. The impulse train can be seen as one case of a linear superposition of harmonically-related sinusoids. Another signal class is a signal distorted by a nonlinear operation, which also generates harmonics from a single sinusoid, and more complicated difference tones when several non-harmonic sinusoids are present. These nonlinearities can be detected using higher-order spectra as seen in Section 6.1. Finally, we will examine amplitude and frequency modulated sinusoid features.

Modulation of the amplitude (AM), phase (narrowband FM), or frequency (FM) of a high frequency radio wave "carrier" is a fundamental method to transmit information. Other modulation schemes can also be used such as generating digital maximal length sequences (a random sequence which repeats precisely) to modulate amplitude, phase, or cause "frequency hopping". These latter "spread-spectrum" techniques are the basis for secure digital communications and allow many information channels to share the same frequency bandwidth. Rather than pursue signal intelligence gathering techniques, we will focus more on the natural occurrence of AM and FM modulation in rotating machinery. Fatigue condition monitoring of machinery is becoming economically viable thanks to low cost microprocessors and networking equipment. As this technology becomes proficient at predicting remaining useful life for machinery as well as the current failure hazard, enormous sums of capital are to be saved by industry as well as the military.

It can be seen that a signal function can be uniquely defined if its value and all its derivatives are known for one input data point. It can also be said that if one has a large number of samples of the function at different input values, the function can be approximated accurately using least-squared error techniques or Fourier series. From physical modeling of the processes which give rise to the signal of interest, one can develop an understanding of the physical meaning of a parameter change in the signal function. So, given that one needs to detect some physical event, signal source type, or condition, the corresponding signal function parameter can be converted into a signal feature.

Basic statistical measures of signals are perhaps the common way to characterize a signal. The most basic time-domain 1-dimensional signal features are the mean, variance and *crest factor*, which is usually defined as the ratio of the peak to rms (standard deviation or square-root of the variance) signal parameters. However, the mean, standard deviation, and crest factor are generally the result of some degree of integration of the signal over a finite period of time. By integration we mean the sample interval over which the average is computed for the mean and standard deviation, and the interval over which the peak factor is obtained for the crest factor. Statistical measures can also include higher-order moments such as skewness and kurtosis (see Section 6.1). Yet another statistical signal feature is its histogram, which measures the number of occurrences of a particular value of the signal. This is used quite often in image signals as well as 1-dimensional signals to characterize the probability density function of the signal. There are a number of well documented distributions that can approximate a given signal by setting just a few parameters of the probability density function, or its frequency-domain equivalent, the characteristic function.

Signal characterization by derivative/integral relationships is a very generic way to extract primitive signal information. Consider the primitive signal functions given in Figure 17 One can clearly see going down either column the effect of integration. These integrals can have fairly simple features starting with whether they are "impulse-like", "step-like", "ramp-like", etc. Some useful impulsive signal features are: which integral produces the step function; the width of the impulse; the height of the step; the slope of the ramp, and so on. Together, these features can provide a set of primitive descriptors for an impulsive-like signal. For example, the popping sound made when speaking the letter "p" is recorded in the top waveform in Figure 18 The 1st integral shows a wide Dirac-like delta function rather

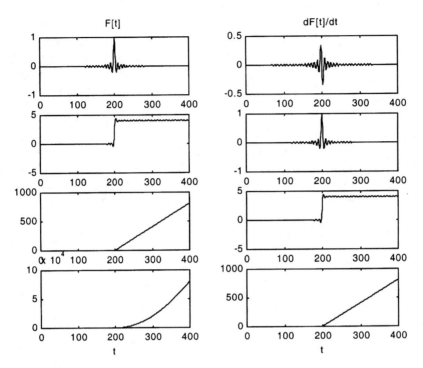

Figure 17 Primitive signal functions starting with the Dirac delta function (upper left) showing subsequent integration as one proceeds down, and corresponding differentiation in the right column (linear amplitude vs. time).

than the actual popping sound. This is somewhat typical for acoustic impulsive sounds which tend to be zero mean (unless there's some sort of explosion). The characteristics of the step and ramp functions (2nd and 3rd integrals) should distinguish this "pop" waveforms from other impulsive sounds. The integral is used to generate other features with the assumption that the high frequency components of the sound are not important for classification. This is certainly not universally true. However, signal integration is inherently less susceptible to noise than signal differentiation, and is generally preferred.

Periodic signals have unique harmonic patterns which can be identified. Applying derivatives and/or integration to periodic signals is not very interesting in general because the integral or derivative of a sinusoid is yet another sinusoid. Clearly, the periodic signal features are best concentrated in the periodogram of an FFT. This is straightforward enough for single sinusoids, where the amplitude and phase are sufficient descriptors. But for periodic waveforms in general, there is much going on which can be parameterized into features. Consider the following generic digital impulse train waveform.

$$y[n] = \frac{1}{\beta_p} \frac{\sin(\pi n / N_w)}{\sin(\pi n / N_r)} \qquad \beta_p = \frac{N_r}{N_w} \qquad (14.2.1)$$

To convert Eq. (14.2.1) to a physical time waveform, simply replace n by $t = nT_s$ where T_s is the sample interval in seconds.

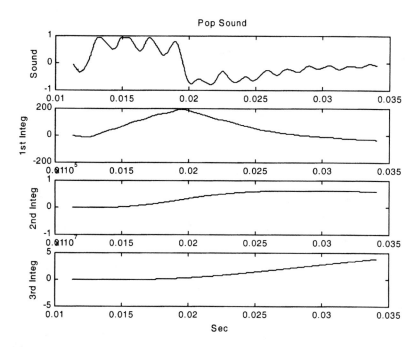

Figure 18 The popping sound in the letter "p" is used to generate some impulsive signal primitive features for analysis of the type of waveform showing the 1st integral giving a wide delta function.

The response of this impulse train, which repeats every N_r samples and has peak width N_w samples, is very interesting depending on the bandwidth factor β_p. If β_p is an even integer, one has an alternating impulse train as seen in the top row of plots in Figure 19. The bottom row of Figure 19 shows β_p chosen as an odd integer, giving rise to the even harmonics in the bottom right plot. In both cases, one sees either even or odd harmonic multiples up to the β_pth harmonic where every other harmonic is exactly zero. This has to do with the integer relationship between N_r and N_w. When β_p is irrational, the function in Eq. (14.2.1) is not realizable without spurious glitches. However, a nice result is given even if both N_r and N_w are irrational but β_p is an integer. Note that as N_w becomes small and β_p large, the number of harmonics becomes large. As β_p approaches 2 our pulse train becomes a sinusoid and at $\beta_p = 1$ the waveform is a constant dc-like signal. To avoid aliasing, N_w must be greater than or equal to 2. In Figure 19 the top pair of plots have $N_r = 100$ and $N_w = 10$, giving an even $\beta_p = 10$, thus causing alternating time domain peaks (Fs is 1000 Hz) and odd harmonics in the frequency domain. The bottom pair of plots have $N_r = 100$ and $N_w = 11.1111$, giving an odd $\beta_p = 9$, thus causing even harmonics in the frequency domain. A surprising number of periodic signals can be characterized in terms of even or odd harmonics.

The Fourier transform provides the spectral envelope of the signal defined by the magnitude frequency response or a discrete number of frequency-magnitude points defined by a series of narrowband peaks. This envelope can be fit to a polynomial where the zeros provide a basis for signal/system recognition, or one

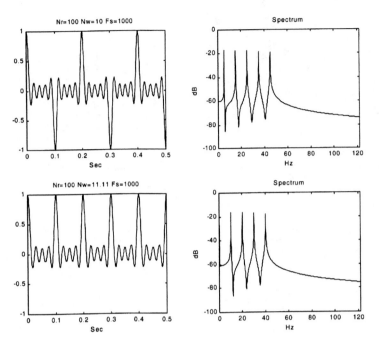

Figure 19 Even and odd harmonic series are generated by the appropriately repeated impulse trains.

can simply assemble the peak heights and frequencies as a pattern feature set for identification. Harmonic sets can be separated by a straightforward logic algorithm or cepstral techniques. The dominant Fourier coefficients naturally serve as a feature vector which can uniquely identify the signal.

Figure 20 compares the Fourier spectra of the spoken vowels "e" (top) and "o" (bottom). The narrowband peaks are detected as a means to minimize the number of data points to work with. One can clearly see from the graphs the higher harmonics in the 2 kHz to 3.5 kHz band which the mouth and nasal cavities radiate for the "e" sound. One can also see a difference in low frequency harmonic structure due to the acoustic radiation impedance at the mouth and its effect on the response of the throat and vocal chords. The details of this wonderfully complex sound generation model could allow one to build a sensor system which can identify a particular individual (speaker identification). Or, one could put together an "envelope" function for the overall spectral shape to simply identify the speech sound, or *phoneme*. There are over 40 such phonemes which make up the majority of spoken languages. Commercially available and surprisingly affordable speech recognition software detects the sequence of such phonemes and connects them together using combinations of fuzzy logic, statistical detection, and hidden Markov models. Text is then produced which best matches the speech recognition results. An even more interesting artificial intelligence problem is developing a way for the computer to understand and respond correctly to the spoken sentence. This is already a big enough problem for humans giving rise to the need for lawyers and politicians. Can one imagine a need for "legal" algorithms for arguing between intelligent computing systems? How about "political" algorithms to simply impose the syntax

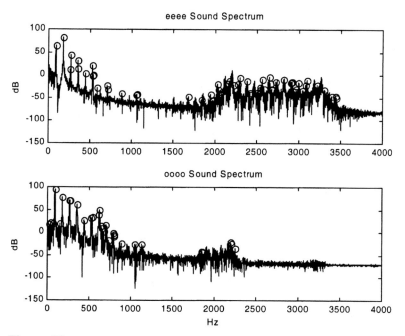

Figure 20 Peak detection of the spectra for the vowels "e" and "o" permits a very simple feature element construction representative of the important signal components for identification.

of one intelligent operating system on another? Such intelligent algorithms are being used to map legal and political strategies in simulations today.

The log-amplitude of the Fourier coefficients is a great choice for feature elements because a wide dynamic range of signal can be used to parse the feature space. The phase of the Fourier coefficients is also important, but only if this phase can be made time invariant from one FFT data buffer to the next. For example, the processed Fourier transform could represent a cross-spectrum with a particular signal, or a transfer function, or coherence. For the phase to be meaningful for an ordinary periodic signal, the FFT buffer size and sample rate should be synchronized with the harmonic of interest to produce a meaningful phase. Otherwise, the FFT phase will not likely be ergodic enough to converge with some averaging. However, one must have a linear, time invariant environment for the Fourier envelope to represent something meaningful physically. This approach to generating signal features is so intuitive and straightforward that we will not pursue its explanation further here. But, its usefulness and importance are only overshadowed by its simplicity of application. This is probably the easiest way to generate signal features in an objective way.

Distortions in signals can be characterized using higher-order spectra. This is because a transducer or system filter nonlinearity can cause frequencies to modulate each other, generating sum and difference frequencies as well as signal harmonics. Signal distortion is a very interesting and important identification problem. Signal distortion can be present for a wide range of signal levels, or only when the signal exceeds some loudness threshold, such as what occurs in amplifier "clipping" where the output voltage is limited to a fixed maximum range. Generally speaking, signal distortion effects always get stronger with increasing signal amplitude. This is because small signal level swings with a nonlinear input–output response can easily be linearized over small range.

A useful way to parameterize nonlinearity is the bi-spectrum. Recall from Section 6.1, Eq. (6.1.32), that an efficient way to compute the bispectrum is to first compute the FFT, then detect the dominant narrowband peaks, and finally directly compute the bispectrum on the relevant combinations of those peaks. Recall that

$$C_3^x(\omega_1, \omega_2) = E\{X(\omega_1)X(\omega_2)X^*(\omega_1 + \omega_2)\} \tag{14.2.2}$$

The interesting thing about signal nonlinearity is the "phase coupling" between frequencies generated by the nonlinearity and their principal linear components. This is because the nonlinear signal generation is occurring at precisely the same point in time for all frequencies in the time waveform. Therefore, for all applications of Eq. (14.2.2) we have coherence between the $X^*(\omega_1 + \omega_2)$ component and the $X(\omega_1)$ and $X(\omega_2)$ principal waveforms.

As an example, consider the case of two sinusoids.

$$x(t) = A\cos(\omega_1 + \theta_2) + B\cos(\omega_2 + \theta_2) \tag{14.2.3}$$

The signal in Eq. (14.2.3) is passed through a simple nonlinearity of the form

$$y(t) = x(t) + \varepsilon x^2(t) \tag{14.2.4}$$

By simply multiplying the signal input $x(t)$ by itself one finds the output of Eq. (14.3.4).

$$
\begin{aligned}
y(t) = {} & A\cos(\omega_1 + \theta_1) + B\cos(\omega_2 + \theta_2) + \varepsilon\left[\frac{A^2}{2} + \frac{B^2}{2}\frac{A^2}{2}\cos(2\omega_1 t + 2\theta_1)\right. \\
& + \frac{B^2}{2}\cos(2\omega_2 t + 2\theta_2) + AB\cos([\omega_1 - \omega_2]t + \theta_1 - \theta_2) \\
& \left. + AB\cos([\omega_1 + \omega_2]t + \theta_1 + \theta_2)\right]
\end{aligned}
\tag{14.2.5}
$$

It can be clearly seen in Eq. (14.2.5) that the phases are coupled between the ω_1 and $2\omega_1$ terms, the ω_2 and $2\omega_2$ terms, and the sum and difference frequency terms. The "phase coupling" is very interesting and clearly shows how the bispectrum in Eq. (14.2.2) can detect such coherence between different frequencies in the spectrum. Normally, (i.e. for linear time invariant systems), we expect the phase relationship between Fourier harmonics to be independent, and even orthogonal. Thus we would expect no "coherence" between different frequencies in a linear signal spectrum. But we do expect bispectral coherence between the phase coupled frequencies for a signal generated using a nonlinearity. Also note that Gaussian noise added to the signal but not part of the nonlinear filtering will average to zero in the bispectrum. The bispectrum is the feature detector of choice for nonlinear signals.

Amplitude and frequency modulation of signals is often detected as part of the spectrum of a signal of interest. But the physical features of amplitude modulation (AM) and frequency modulation (FM) can be broken down and extracted in a number of different ways. AM signals simply vary the amplitude of the "carrier" frequency with the message signal amplitude. It is useful to transmit messages in this way for radio because the use of a high frequency carrier allows one to efficiently use relatively small antennae (the wavelength at 300 Mhz is about 1 m). FM signals can be further categorized into narrowband FM, or phase modulation, and wideband FM. A formal definition of narrowband verses wideband FM is defined from the modulation index, which is the ratio of the maximum frequency deviation to the actual bandwidth of the message signal, or the signal doing the frequency modulation. The modulation index can be scaled any way desired because the frequency deviation of the FM signal from its carrier frequency is proportional to the amplitude of the message signal. But the message signal has its own bandwidth, which may be larger than or smaller than the bandwidth of the frequency modulation, which is proportional to the message signal amplitude. When the FM frequency deviation is smaller than the message signal bandwidth, we call the modulation narrowband FM. Likewise, when the FM frequency modulation is greater than the message signal bandwidth it is called wideband FM. Wideband FM uses much more bandwidth but offers a significantly greater signal to noise ratio than narrowband FM. Narrowband FM and AM use about the same bandwidth but the FM signal will have less susceptibility to background noise. For digital signal transmission, the transmitted bandwidth is roughly the maximum possible bit rate for the channel. It is generally on the order or 90% of the available bandwidth because of the need for error correction bits, synchronization, stop bits, etc.

While the general public is most familiar with AM and FM signals from their radio dial, AM and FM signals arise naturally in the vibration and electromagnetic signatures of rotating equipment. By applying a little physics and some straightforward communications signal processing, a great deal of useful information can be extracted. Information on machinery condition, fatigue, and failure hazard is worth a great deal of capital for the following (just to name a few) reasons: unexpected equipment failures cause substantial damage to the environment, personal injury, and even deaths; failures of critical machines can cause a sequence of much greater catastrophic failures; time based maintenance (fixing things before its really needed) is on average significantly more costly than condition-based maintenance (CBM).

A quick analysis of rotating equipment can start with three simple mechanical rules: (a) one cannot make a perfectly round object; (b) even if one could make a perfectly round object, one cannot drill a hole exactly in the center of it; (c) no two pieces of material are identical, respond to stress from heat or external force in the same way, nor will respond to damage the same way. Rolling elements will always have some degree of imbalance, misalignment, and non-constant contact forces with other rolling elements. These elements will produce vibration, acoustic, and in some cases electromagnetic signals which can be systematically related to the condition of the rotating elements. The contact forces between two rolling elements can be broken down into centripetal and tangential forces. For a pair of gears, the average vibration signal is a harmonic series of narrowband peaks with the dominant frequency corresponding to the gear tooth mesh frequency. The shaft frequency is found by dividing the fundamental gear mesh frequency by the number of teeth in the given gear. The centripetal forces end up on the shafts, propagate through the bearings, and are seen as amplitude modulated vibrations of the gear mesh frequency. The tangential dynamic forces become torsional shear on the shafts and are seen as a FM component of the gear mesh frequency. A typical transmission will have a large number of mesh harmonics and AM and FM sidebands, which can be explicitly attributed to specific rolling elements by a computer algorithm. As the surfaces and condition of each of the rolling elements change, the vibration signatures will also change. Therefore, one can monitor the condition of a piece of machinery through careful monitoring of the vibration and other signatures. Given this capability, one can then track changes in signature features to predict the number of hours until a failure condition is likely. Feature tracking will be discussed in more detail in the next section.

AM signals are common in vibration

Consider a gear mesh frequency of 5 kHz and a shaft frequency of 100 Hz (50 gear teeth). Our modeled AM modulated signal is

$$y(t) = [1 + \alpha \cos(\omega_{sh}t)]\cos(\omega_c t) \tag{14.2.6}$$

where α represents the amount of modulation for the 100 Hz shaft rate ω_{sh}, component and ω_c represents the 5 kHz carrier frequency component. Once-per-shaft-revolution modulations are seen to be physically due to out-of-round, imbalance, and/or misalignment problems. Figure 21 shows the time and frequency domain signals for Eq. (14.2.6) using a sampling frequency of 20 kHz and $\alpha = 1$.

Coefficients for twice-per-rev, three times per-rev, and so on, for the mesh frequency fundamental can be seen as Fourier coefficients for the overall surface shape of the rotating component modulating the sinusoidal fundamental component of the gear teeth Fourier transform. To observe pits, spalls, and cracks in the teeth, analysis is often done at 4 times the mesh frequency including the sidebands, if such high frequency vibration is detectable. For most turbine-driven transmissions, this type of vibration analysis extends well into the ultrasonic range of frequencies. Equation (14.2.7) depicts a general, multiple harmonic and sideband representation of an AM signal.

$$y(t) = \sum_{q=1}^{Q}\sum_{m=1}^{M} A_q[1 + \alpha_{q,m}\cos(m\omega_{sh}t)]\cos(q\omega_c t) \tag{14.2.7}$$

Figure 22 shows the time and frequency representation of Eq. (14.2.7) with $Q=4$ and $M=5$ and a shaft rate of 350 Hz which better shows the sidebands on the plot. The A_q coefficients are assumed unity in the response in Figure 22 and the $\alpha_{q,m}$ were taken as $1/\sqrt{qm}$ for demonstration purposes. One would expect that small surface wear would first be detectable up in the sidebands of the higher mesh frequency harmonics, which is the case. Many defects such as spalls, peels, pits, and cracks can be detected early in the high frequency range of the vibration spectrum. When a crack first occurs, a spontaneous ultrasonic acoustic emission is detectable which for some materials can also be detected in the electromagnetic spectrum. Complex machines such as helicopter transmissions have been monitored for fatigue at fre-

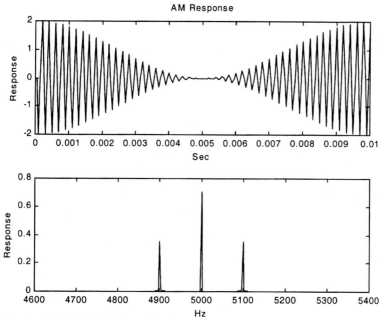

Figure 21 100% AM modulation of a 5 kHz carrier representing a gear mesh frequency and a 100 Hz modulation representing the shaft rate fundamental in the AM modulation model.

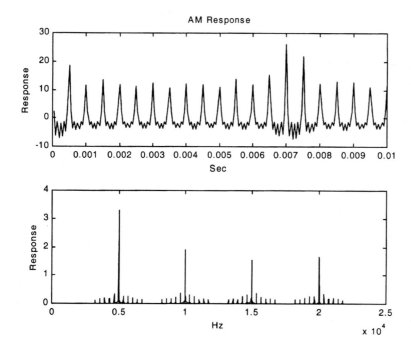

Figure 22 Time waveform (top) and linear spectrum (bottom) of a multiple harmonic 5 kHz AM signal with sideband at multiples of 350 Hz.

quencies up to four time the mesh frequency, or well into the ultrasonic frequency regime.

FM signals are more complicated but recently becoming of interest to CBM because the presence of internal cracks in gear teeth and shafts can decrease the mechanical stiffness when the gear is in a particular contact position. As such, as the gear rotates, a dynamic torsional rotation error signal is produced which is seen as an FM component on the gear mesh frequency. Detection of such defects before they reach the surface is of great value because they are precursors which can be used to predict the remaining useful life of the machine. As the crack grows, stiffness decreases more dramatically, the FM increases, and eventually the crack reaches the surface leading to high frequency features, but perhaps only shortly before component failure. In the case of helicopter gearboxes, a gear failure generally means a complete loss of the aircraft and crew and the aircraft needs to be flying for the gear to have sufficient applied stress to fail. To complicate matters more, the mere presence of FM does not imply imminent failure. It is when FM and other fatigue features are present and increasing that provides the basis for mechanical failure prognostics.

FM signals are a little more complicated than AM signals. For a FM signal, the frequency sweeps up and down proportional to the amplitude of the message signal. Consider a very simple case of a carrier frequency ω_c and FM message sinusoid frequency ω_m.

$$y(t) = \cos(\omega_c t + \beta \sin(\omega_m t)) \tag{14.2.8}$$

The term β in Eq. (14.2.8) is just a constant times the message signal maximum amplitude for what is called *phase modulation*, but is inversely proportional to the message signal frequency for FM. This is because the time derivative of the phase is frequency (or the time integral of the modulation frequency gives the phase). A more general representation of Eq. (14.2.8) is

$$y(t) = Re\{e^{j\omega_c t} e^{j\beta \sin(\omega_m t)}\} \tag{14.2.9}$$

where the right-most exponential is periodic with period $2\pi/\omega_m$ sec. Therefore, we can write this as a Fourier series.

$$e^{j\beta \sin(\omega_m t)} = \sum_{n=-\infty}^{+\infty} C_n e^{j\omega_m n t} \tag{14.2.10}$$

The coefficients of this Fourier series are

$$C_n = \frac{\omega_m}{2\pi} \int_{-\frac{\pi}{\omega_m}}^{+\frac{\pi}{\omega_m}} e^{j\beta \sin(\omega_m t)} e^{-jn\omega_m t} dt \tag{14.2.11}$$

$$= \int_{-\pi}^{+\pi} e^{j(\beta\theta - n\sin\theta)} d\theta = J_n(\beta)$$

making a frequency domain representation of our simple sinusoidal FM signal seen as an infinite sum of sidebands about our carrier frequency with carrier and sideband amplitudes dependent on the value of Bessel functions of the first kind, $J_n(\beta)$.

Figure 23 shows a fairly narrowband FM signal where the carrier is 5 kHz, the modulation frequency is 100 Hz, and β is 0.5. The sample rate of our digital signal model is 20 kHz, so the amount of modulation relative to the carrier phase is only around $\pi/2 : 0.5$, or about 30%. This can be seen as a very narrowband FM or phase modulation. Figure 24 shows the same carrier and FM modulation frequency, but with a modulation β of 40. In terms of phase shift, the FM can be seen as over 2400%. The maximum rate of this phase change gives the approximate maximum frequency shift of the carrier.

$$\frac{\Delta f}{f_s} = \beta \frac{fm}{f_s} \approx 20\% \tag{14.2.13}$$

The approximation in Eq. (14.2.13) tells us that the carrier frequency is warbling around $\pm 20\%$ of the sample frequency, or 4 kHz, spreading the signal spectrum from around 1 kHz to 9 kHz. The spectrum seen in the bottom plot of Figure 24 shows the spectrum used by the FM as approximately covering this range. Clearly, the greater the modulation, the more wideband FM looks "noise-like" in the frequency domain. The message signal can also be specified as a sum of Fourier components, which complicates the spectrum significantly more.

For FM radio, we all know the signal-to-noise ratio benefits from personal experiences listening to FM radio broadcasts of music and other information.

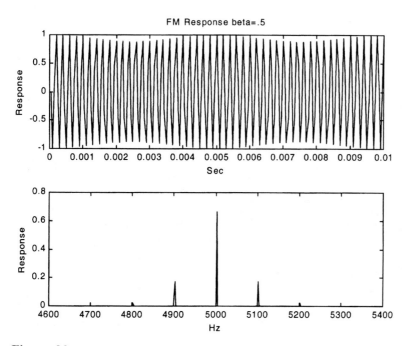

Figure 23   An FM signal with very small modulation has a barely noticeable frequency shift in the time domain and almost the same bandwidth requirements as AM in the frequency domain.

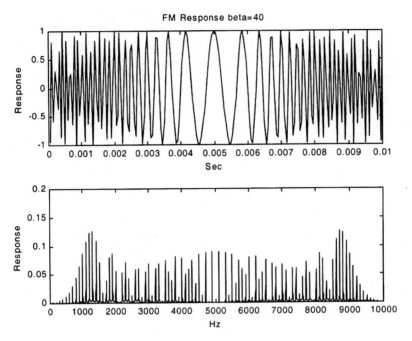

Figure 24 An FM signal with wideband modulation produces a very noticeable chirp in the time domain and gives a very complex noise-like spectrum.

The reason the received signal is nearly noise-free is because the FM carrier is converted in an nonlinear way into the signal amplitude. At the receiver, the carrier band (typically about 40 kHz wide) is bandpass filtered and connected to a frequency to voltage convertor, which is really just a high-pass filter with a constant slope in its frequency response, followed by an rms integrator, which provides the recovered amplitude signal information. Harry Armstrong, one of the 20th century's great engineers, developed this technology as far back as the 1930s but unfortunately did not live to enjoy the financial benefits of his invention. He committed suicide after long and costly legal battles over the patent rights to this brilliant invention. His legal battle eventually prevailed.

Rotational vibration signal often contain both AM and FM signals

This poses a significant challenge and opportunity to exploit the features of rotational vibration to measure the mechanical condition or rotating equipment. But how can one tell the two apart? The challenge is seen as how to get the AM and FM "message signals" into orthogonal spaces. This is done with a knowledge of the carrier frequency and the Hilbert transform, define as

$$Y_H(\omega) = Y(\omega) + Y^*(-\omega) \quad 0 < \omega \leq \frac{\omega_s}{2} \tag{14.2.14}$$

One can think of the Hilbert transform as a way of representing a real sinusoid as an equivalent complex sinusoid. For example, if one had a Fourier transform of a cosine function with frequency ω, ½ the spectral signal will be at $-\omega$ and ½ the spectral signal will be at $+\omega$. If the real input signal was a sine function, the $\frac{1}{2}$ of the spectral signal at $-\omega$ is multiplied by -1 and added into the complex sinusoidal representation. If the input to the Fourier transform is a complex sinusoid, no negative frequency component exists. The Hilbert transform provides a common basis for representing sinusoids for frequencies up to $\frac{1}{2}$ the sample rate for real signals, and up to the sample rate for complex time-domain signals. The inverse transform of the Hilbert transform gives us a complex sinusoid, which is what we need to recover the AM and FM signal components.

Consider our combination of AM and FM signal modulation.

$$y_{af}(t) = [1 + \alpha \cos(\omega_a t)] \cos(\omega_c t + \beta \sin(\omega_f t)) \qquad (14.2.15)$$

Figure 25 shows a time plot (top) and spectrum (bottom of an example combination AM and FM signal where the AM modulation frequency is 100 Hz with $\alpha = 0.9$ (90% AM modulation), and the FM modulation frequency is 40 Hz with $\beta = 50$ (± 2 kHz modulation) about a common carrier frequency of 5 kHz. By computing the Fourier transform of the real signal in Eq. (14.2.15), applying a Hilbert transform, and inverse Fourier transforming, one obtains a complex sinusoid $y_{af}^c(t)$, the real part of which matches Eq. (14.2.15). Therefore, we can demodulate the complex sinusoid by a complex carrier to obtain a "baseband" complex waveform.

$$y_{af}^b(t) = y_{af}^c(t)e^{-j\omega_c t} \qquad (14.2.16)$$

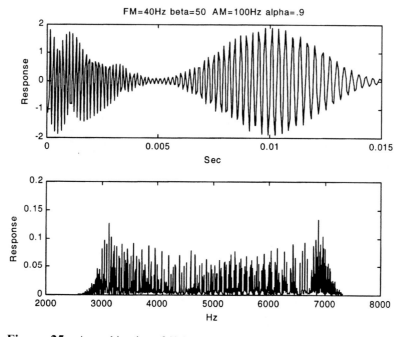

Figure 25 A combination of 40 Hz FM with $\beta = 50$ and 100 Hz AM with $\alpha = 0.9$ produces a very complicated waveform in the times domain as well as frequency domain spectrum.

Or, for digital waveforms sampled at $2\pi f_s = \omega_s$,

$$y_{af}^{b}[n] = y_{af}^{c}[n]e^{-j2\pi \frac{fc}{fs} n} \tag{14.2.17}$$

which has the effect of producing a waveform with frequency relative to the carrier frequency. The amplitude of this complex waveform is the AM part of the modulation signal, and the time derivative of the phase (unwrapped to remove crossover effects at $\pm \pi$) is the FM part of the modulation. Figure 26 depicts the recovered AM and FM waveforms clearly.

There several important physical insights to report on this interesting modulation feature extraction example. If the AM modulation is greater than 100%, a phase change will occur when the amplitude modulation crosses zero. This too can be recovered, but only if one is sure that no FM exists. For very complicated broadband signals as the modulation signal (also known as the message signal in communication theory), one is fundamentally limited by bandwidth and the carrier. If the message signal bandwidth is less than the carrier frequency, there is sufficient signal bandwidth for AM and narrowband FM modulation. The bandwidth of the FM signal is the bandwidth of the message signal *times β*, which has the potential to cover a very wide bandwidth. In the frequency domain, one can easily select either finite width bands or select spectral peaks within a finite band to focus the complex message signal on specific parts of the spectrum. If the carrier demodulating frequency in Eqs (14.2.16)–(14.2.17) are in error, a slope will be seen in the recovered FM waveform. One can adjust the demodulating carrier until this

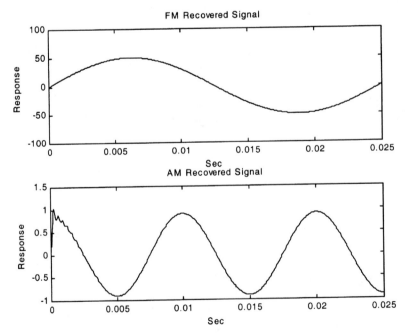

Figure 26 The AM and FM components are recovered using a Hilbert transform and demodulation in the time domain to given the envelope and phase signals.

FM slope is zero to be sure of the correct alignment for a given FFT buffer. Finally, it can be seen that extracting AM and FM signal parameters clearly has the effect of focusing important signal information spread over many time waveform samples and spectral bins into a few salient parameters for pattern recognition and tracking.

Lets test our modulation separation with a more complicated FM signal. Consider a pulse train where $N_w = 20$ and $N_r = 400$ as defined in Eq. (14.2.1), where the sample rate is 20,000 samples per sec. The fundamental of the pulse train is 25 Hz with odd harmonics extending up to 475 Hz. This gives our message signal a total bandwidth of about 1 kHz, which is readily seen in Figure 27 The recovered AM and FM signals are seen in Figure 28 where one can clearly seen the recovered FM-only pulse train signal. This is a narrowband FM case because the total bandwidth of the FM signal is approximately the same as the bandwidth of the message signal. In Figure 29, we increase β to 10 such that now the bandwidth of the FM signal requires most of the available bandwidth of our 20 kHz-sampled waveform with carrier frequency of 5 kHz. Figure 30 shows the recovered FM signal which appears correct with the exception of a bias in the modulation frequency. This bias is caused by the angle modulation in our digital signal.

$$y_{af}[n] = \left[1 + \alpha \cos\left(2\pi \frac{f_a}{f_s}n\right)\right] \cos\left(2\pi \frac{f_c}{f_s}n + \beta m[n]\right) \qquad (14.2.18)$$

If the FM message signal times β exceeds π in magnitude, the phase of the carrier wraps causing the bias error in the recovered FM message waveform. Curiously, this did not occur when the message signal was a pure sinusoid as seen

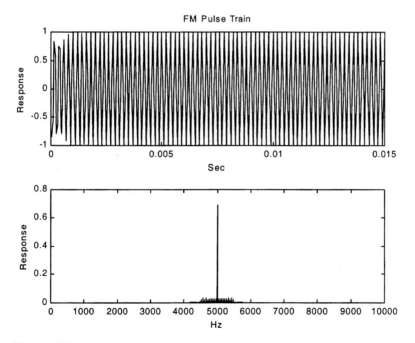

Figure 27 A narrowband FM signal where the message signal is a pulse train with about 1 kHz total bandwidth and β is 1 in the FM signal (no amplitude modulation is used).

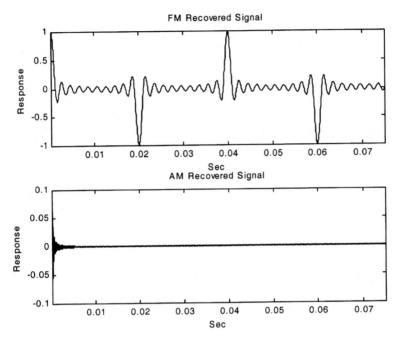

Figure 28 Recovered FM and AM signal components for the pulse train FM signal with $\beta = 1$.

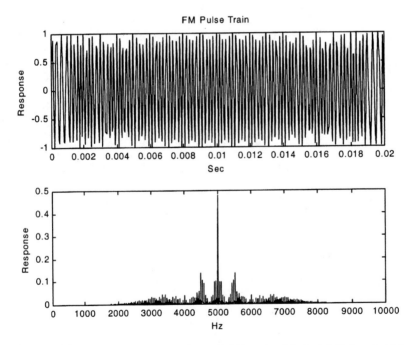

Figure 29 Wideband FM signal (no AM is used) with the 1 kHz bandwidth message signal and $\beta = 10$.

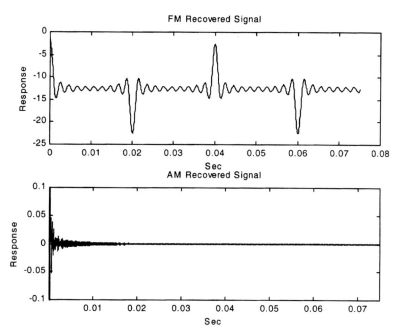

Figure 30 Recovered AM and AM components from the wideband $\beta = 10$ signal with the 1 kHz pulse train message signal showing a bias in the recovered FM signal.

in Figure 26. To understand why, consider that the frequency shift of the carrier is the derivative of the carrier phase in Eq. (14.2.18). The 40 Hz sinusoid FM message signal in Figure 26 with a $\beta = 50$ corresponds to a maximum frequency shift of 2 kHz, which in terms of digital frequency shift is range of 0.7π to 0.3π. For the case of a non-sinusoidal message signal, once β times the maximum of $m[n]$ in Eq. (14.2.18) exceeds or equals π, a bias error in the recovered FM message waveform is possible due to the higher harmonics of the message waveform. With $\beta = 10$ and a maximum harmonic of 475 Hz in the message signal, most of the available spectrum (5 kHz ± 4.75 kHz) is used for our pulse train message signal. If β were increased beyond around 15 for our example waveform, the AM signal recovery significantly breaks down, and if increased further beyond around 40, neither signal can be accurately recovered.

   Two-dimensional features can be extracted from image data as well as other two-dimensional signals. Our discussion will focus mainly on image data, but as the image is broken down into edges and other features, one can easily extend the analysis to any two dimensional data, or even higher dimensional data. Again, drawing on our philosophy that features should represent concentrated signal information, imagery data can be broken down into two main groups: *focal-plane features* and *texture features*. Focal plane features represent concentrated information on the focused image such as edges, shapes, local patterns, etc. Examples of focal plane features are generally the things we can easily recognize with our eyes and are improved by contrast, edge enhancement, etc. To help a computer algorithm recognize focused objects, one typically removes the textures by applying first and second derivatives (see Section 4.3) to enhance edges and lines. These lines may then be segmented into sequences of simple functions, like the pen strokes of an artist drawing the object. The functional segments can then be compared objectively to segments for known objects of interest. The functional representation of the object is useful because computing algorithms can be derived to take scale, rotation, and viewing aspect into account for object identification.

Texture features are patterns which are periodic and cover broad areas of the focal plane. Texture features are good candidates for the frequency domain where the broad areas of a periodic pattern on the focal plane can be concentrated into a few two-dimensional FFT bins. For example, inspection of a filled case of 24 beer bottles using an aligned digital picture will normally have a regular pattern of bottle caps. A broken or missing bottle is seen as a "glitch" in the pattern, or aberration of the texture. A simple 2-FFT of the image data and spectral subtraction, and an inverse FFT will quickly reveal the problem before the product is shipped from the factory. Texture patterns can also be applied to a wide range of material inspections including color, reflectance, laser scattering patterns, an so on.

One can examine what is known about the physiology of the eye for primates (monkeys, apes, and man) to gain some valuable insight into how biology has optimized our sight for detection and classification of objects (4). Insect, fish, and other animal eyes generally do not see the way we do, and might be useful for some specific machine vision tasks. Man's eyes and way of seeing (what is known of it) are truly amazing. By examining the psychophysics of the eye/brain operation we find some basic building blocks for the development of two-dimensional image features.

The optical properties of the eye are surprisingly poor. It has a single simple lens with a rather small focusing range controlled by a ring of muscles. The optical aberration can be rather large (as most of us with eyeglasses know all too well) and the aberration over color can be over 2 diopters from red to violet. With little physical correction possible, biology provides the necessary interconnections between receptors to turn this color "de-focusing" to an advantage which allows remarkable color recognition over a wide range of ambient light. There are four types of receptors, red, green, and blue *cones* with a significantly overlapping color (light frequency) response, and blue-green *rods* where the outputs are pooled to allow a lower resolution night vision. These retinal receptors are actually situated behind a layer of neural cells which contribute to a loss of acuity but allow more interconnections and additional protection from damage from intense light.

The eye uses tremor and hexagonal variable sampling to enhance acuity. The resolution of the eye is on the order of 30 seconds or arc (1 degree = 60 minutes or arc, 1 minute of arc = 60 seconds of arc), but the eye actually tremors at a rate from 10 Hz to 100 Hz with about the same 30 seconds of arc. These "matched res-olution tremors" have the effect of smoothing the image and removing optical effects of sampling. This is known as "anti-aliasing" filtering in signal processing and has the effect of spatially low pass filtering the optical signal. It is interesting to note that during motion sickness, extreme fatigue, mental impairment (say from drugs), or while telling a big fat lie, most humans eyes become noticeably "shifty" (perhaps the brain is too preoccupied). The arrangement of the receptors is hexagonal (honeycomb-like) which can be seen as naturally more optimal to the sampling of a point-spread function which is the fundamental resolution on the surface of the back of the eye to a pin-point of distant light. At the center of the field of view which corresponds to the center of the back of the eye called the *fovea*, the density of the receptors is highest and it decreases as one moves away from the fovea. This variable sampling density, along with the optics of the iris and lens, physically cause the acuity to diminish as one moves away from the fovea. In addition, the optics produce a smoothly diminishing acuity towards the other ranges which acts like a Gaussian window on the iris to suppress sidelobes in the point spread function (see Chapter 7). The result of this biology is that objects appear as "blobs" in our peripheral vision and the details are seen once we swivel our eyes around to bring the interesting part of the image over the fovea. This will translate into a useful strategy for image recognition where one scans the entire image with a small window to initially detect the "blobs" of interest, and then subsequently process those "blobs" with much higher resolution.

The structure of the receptors and neurons gives a parallel set of "pre-processed" image features. There are essentially four layers of neurons in the signal path between the rods and cones and the optic nerve. They are the *horizontal cells, the bipolar cells, the amacrine cells, and the ganglion cells*. In humans and other primates, there are additional cells called the *midget bipolar cells* which provide more direct connections to the ganglions and interactive feedback to other groups of receptors. Surprisingly, there are very few blue cones and fewer blue-green rods for low-light vision in the fovea area, perhaps to make space for a higher overall density of receptors. However, we "think" we see blue quite readily in our central field of view perhaps because of the poor optical color aberrations of the eye and the feedback interconnections provided by the midget ganglions. The amacrine cells provide further neural interconnections between spatially separated groups of cells which may explain the eye's superb *vernier acuity* (the ability to align parallel lines). Vernier acuity is about 6 times better (5 seconds of arc) than visual acuity and requires groups of spatially separated receptors to combine together their outputs. This is also likely to be a significant contributor to the brain's ability to judge distance with stereo vision, another superb human vision capability. From a Darwinistic point of view, without good stereo vision, one does not catch much food nor escape easily from being eaten — thus, these developments in primate eyes are seen as a result of either evolution or divine good fortune. Finally, the ganglion cells can be categorized into two (among many) groups: the X ganglions which have a sustained output which responds in about 0.06 seconds decaying in up to 2.5 seconds; and the Y ganglions which provide a transient output which responds in about 0.01 seconds and decays in about 0.1 seconds. Together, the X and Y ganglions provide the ability to detect motion quickly as well as "stabilize" an image for detailed sensing of textures, objects, and color.

So how can one exploit these building blocks to make computer vision possible? First, we recognize that many of the things we "see" are the result of changes, spatially and temporally. Temporal changes are either due to motion or changes in the scene — both of which are of great interest to situational awareness. Spatial changes provide clues for edge detection, lines, and boundaries. In the eye, the time constants of the X and Y ganglions, the spatial interconnections of the amacrine cells, and the receptor responses through the bipolar and horizontal cells provide a strong capability for providing *first and second difference* features. The spatial derivative images seen in Section 4.3 are an excellent example of techniques for detecting edges (such as a LaPlacian operator), and the direction or orientation of the edge (a derivative operator). Edge information and orientation can be seen as the primitives needed for object shape recognition. Much of this information is concentrated by the cellular structures and neural interconnections in the eye.

The eye amazingly exploits some of its optical imperfections and turns them to its advantage. From a best engineering practice perspective, one wants as stable and clear an image as possible with constant resolution spatially, temporally, and across the color spectrum. The eye actually exploits tremor and optical aberration to its advantage using the tremor for spatial antialiasing, using the color aberration to feed interconnections and spatially "fill in" for, say, the missing blue cones in the fovea area, and uses the blurry images in the off-fovea area to cue motion or image changes to subsequently swivel the fovea into view of the change. The eye distributes its receptors unevenly to meet a wide range of viewing goals. For example, night vision is done at much lower resolution using rods which are pooled together for signal gain in low light. There are practically no rods in the fovea area, which is optimized for intense detail spatially. This is probably where the edges and object shapes are detected through the spatial interconnections and subsequently processed in the visual cortex to associate the viewed shape with a known shape.

Finally, the receptors in the eye each have a nonlinear brightness response allowing detail in the shadows as well as bright areas of the image. Engineering practice to "linearize" the response of photodetectors on record and playback may not be such a bright idea. This is especially true now that analog circuits are largely a thing of the past. Linearization of the "gamma curve" in television cameras and receivers was very important to the standardization of television using analog circuits of the 1950s and 1960s. Digital systems can allow virtually any known nonlinearity in response to be completely corrected upon playback if desired. There is no technical reason why each pixel in a digital camera cannot have its own nonlinear response and automatic gain control such that the brightness dynamic range of the displayed image is compressed to meet the limits of the display device on playback only. The ability of the eye to see into shadows has long been a frustration of photographers using both chemical and digital film. However, such a change in digital image acquisition is so fundamentally different in terms of processing hardware that one might as well suggest the hexagonal pixel arrangement rather than the current square row-and-column pixel arrangement.

Figure 31 shows a logical way to layout the processing for computer vision. The reader should note that there are many ways of accomplishing this, but we will use the example in Figure 31 as a means to discuss two-dimensional features. First, we see that the camera output signal is received by a process which corrects any degrading optical aberrations, including color balance, brightness and contrast, but could also include focus, jitter, or blur corrections. The raw image can then be further processed in two main ways: matched field detection and segmentation.

One can do a "matched signal" search for any desired objects by correlating a desired scene in the box we call the "fovea window" which scans the entire image for the desired scene. This is definitely a brute force approach which can be made somewhat more efficient through the use of FFTs and inverse FFTs to speed the correlation process. A high correlation at a particular position in the raw image indicates the presence of the desired sub-scene there. This is useful for applications where scale and orientation are not a problem, such as identifying an individual's

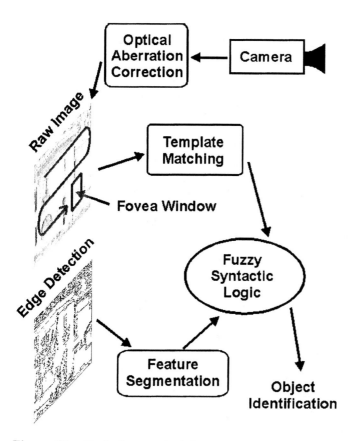

Figure 31 Block diagram depicting general computer vision using combinations of template matching and image segmentation.

retinal blood vessel pattern, identifying currency, or scanning a circuit board for a particular component.

When scale and orientation of the object in the image is part of the problem, a process called *image segmentation* is used to essentially convert the image to a maximum contrast black and white image where edges and lines are easy to detect. Use of first and second derivatives allow line segments to be detected and assigned a length and orientation angle. A cluster of line segments can then be sequenced a number of ways as if one were drawing the lines using a pencil to reproduce the object. This sequence of segments can be scales, rotated, and re-ordered to compare with a known object. A least-squared error technique can be used to select the most appropriate object class. Fuzzy logic can then be used to assemble the primitive objects into something of interest. For example, a cluster of rectangle objects with ellipses protruding from the bottom is a likely wheeled vehicle. The arrangement of the rectangles as well as their relative aspect ratios, number and arrangement of ellipses (wheels) are then used to identify the vehicle as a truck, automobile, trailer, etc. from a functional description, or *graph*.

Figure 32 depicts the process of line segmentation to create a two-dimensional object which can be described functionally to account for scale and orientation.

Window   Segmentation

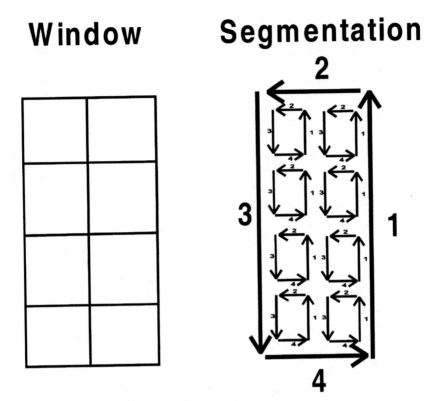

Figure 32 A sketch of a window and the corresponding line segmentation to describe the window as a series of lines sequenced to describe the window independent of scale and aspect.

For example, a rectangle can be described as a series of four line segments in a counter clockwise sequence where every other line is parallel. From one line segment to the next, a turn angle is noted and the even turn angles are equal and the odd turn angles are equal. The difference between the observed turn angles and a normal angle is due to the orientation and viewing aspect. Virtually any lined object can be segmented into a series of simple functions. These functions can be sequenced in different ways, scaled, skewed, and rotated using simple mathematical operations. Thus, one can systematically describe an object as a graph, or set of segmented functions in much the same way handwriting can be described as a series of pen strokes. As in handwriting recognition commonly seen in palm-sized computers, given a set of pen strokes, an algorithm can systematically test each known letter's segments, but that would be relatively slow. A more efficient process is to apply fuzzy logic to subdivide the alphabet into letters with curves only, straight lines only, combinations, etc. In this way, object recognition is similar to handwriting recognition, except the font set is much broader.

The image line segmentation can be seen as a hypothesis testing operation. Starting at a point along a difference-enhanced edge, one hypothesizes that the point is on a straight line whose direction is orthogonal to the spatial gradient at the point. The line is extended in either direction and if corresponding "edge" pixels are found, the line continues. At the end of the line, a search begins for a new "edge" to start a new line segment. Segments shorter than some prescribed minimum length are discarded (hypothesis proved false). Several possible hypothesis can be simultaneously alive until the final decision is made on the object. Such a final decision may include the context of other objects in the scene and allow obscured objects to still contribute to a correct image classification. This is an example of the power and flexibility of fuzzy logic.

Connecting edge pixels to a curve can be done by either direct least-squared error fit or by employing a cartesian tracking filter such as an α–β tracker or Kalman filter. Curves belonging to ellipses (a special case of which is a circle) can be described as particle motion with acceleration. By tracking the "particle" motion along the edge pixels, the position, velocity, and acceleration components of the curve can be estimated from the state vector. If the curve is part of an ellipse, this acceleration will change over time and point in a direction nearly orthogonal to the particle velocity direction. The curve can be segmented into sections which are described using specific functions, where the segmentation occurs as the particle "maneuvers" onto a new curve. The analytical representation of lines and curves to describe the salient features of an image allows one to detect objects when the scene is at an arbitrary scale and aspect. This is because our functions for the line and curve segments, and their relative position in the scene, can be easily scaled, skewed, and/or rotated for comparison to a functional description of a known object. The breaking down of an image into edges, lines, and ultimately functional line and curve segments is a computationally-intensive task. It is something that lends itself to parallel processing which is precisely how the brain processes signals from the optic nerve.

We have been careful in addressing signal and image features to stay as general as possible. Every application of this technology eventually ends up with very specific, proprietary, even classified (via government security requirements) features which work in unique special ways. The interested reader should consult the books listed in the bibliography of this chapter for more detailed information on image features and algorithms. From our perspective of intelligent sensor system design, it is most important to examine the *strategy* of signal and image features, using physics and biology as a guide.

14.3 DYNAMIC FEATURE TRACKING AND PREDICTION

In this section, we explore a new area for intelligent sensor systems: *prognostics*. Prognostics is the ability to predict a logical outcome at a given time in the future based on objective measures of the current and past situation. For example, a physician does an examination, checks the medical history of a patient, and then makes a careful prognosis of the outcome the patient can expect, and how long it will take to reach that outcome. Intelligent sensor systems can aspire to do the same sort of thing within limits. The limits are determined by our ability to create software algorithms which can accurately measure the current situation, track the changes in situation over time, and predict an outcome with reasonable certainty in both fidelity and time. What is most interesting about prognostics is that most pattern classification is based on a "snapshot" of data. That is, the pattern is assumed fixed and is to be identified as such. This "snapshot" pattern recognition is repeated again and again over time in case the pattern does change for some reason. However, in prognostics, the evolution of the pattern is generally very important. The "trajectory" of the pattern over time is, in itself a pattern to be identified and exploited to improve future predictions.

Take an example of the common cold verses a flu virus. Both can have similar symptoms at times, but the path of the sicknesses are quite telling. A cold is actually a combination of dozens or even hundreds of viruses combined with a bacterial injection. Generally, the body has had time to build up immunity to some of the infection and viruses, but has become overwhelmed. A virus on the other hand, attacks the body within hours often causing severe illness for a day or two before the antibodies gain the upper hand and lead to recovery. So, if one had a "yuk" feature which is zero when one feels great and 10 when one wants to welcome the grim reaper, both cold and virus would have the same feature at various times during the illness. However, the trajectory of the two illnesses are quite different. The cold comes on strong but then takes a while to eliminate while the virus attacks extremely fast and with great severity, but can often stabilize and fade away rather quickly. During the initial stages of a virus or bacterial infection when the severity increases, there is actually a chance that a life-threatening disease is at work. However, once the fever and other symptoms stabilize (stop rising or getting worse) the prognosis becomes much better. During recovery, the rate of improvement can be used to estimate when things will get back to normal.

This example illustrates an important point about prognosis. We know when things are deteriorating rapidly there is a significant health hazard. But as symptoms stabilize and begin to subside we can be assured that the prognosis is good if we are patient. To make a prognosis, a complete diagnosis of what is wrong is not necessary, only the likely causes of the current situation and their impact on the future progression of the situation. A partial diagnosis with several plausible causes should be used as working hypotheses with corresponding feature trajectories from the current situation to an ultimate outcome feature regime. If any of the possible outcomes presents itself as a significant hazard then corrective action is taken. With machinery health, we cannot expect anything like the "healing" of the body unless we shut down and repair the damage ourselves, although some symptoms may be reversible with rest (such as over temperature conditions), or resupply. The issue with machinery is the cost and health hazard of an unplanned failure.

Currently, most machinery maintenance is done based on a recommended schedule by the manufacturer. For example, automobile engine oil is changed every 3,000 miles or 3 months, whichever comes first. But what if you didn't drive the car for 3 months? Isn't the oil perfectly good? Isn't it unnecessary to replace the oil filter? Of course it is! Yet *time-based maintenance* is quite common because it is very easy to implement and plan around. However, some machinery is so complex and delicate that every time a maintenance action is done there is a significant chance of damaging some component. The old proverb "if it's not broken, don't try to fix it ..." carries a great deal of weight with mechanical systems because doing too much maintenance generally leads to more damage and repairs resulting from the original maintenance action. One way an equipment owner can save on maintenance costs is to simply not do it. Obviously, this looks pretty good for a while until the whole plant infrastructure falls apart. Some plant managers have managed to get promoted because of their "maintenance efficiency" before their lack of maintenance causes a complete shutdown while others aren't so lucky. A plant shutdown in big industry such as hydrocarbon production, refining, food production, textiles, etc., can amount to enormous profit losses per day and an unplanned, if not catastrophic, equipment failure can lead to many days of shutdown. This is because these industries operate on huge volumes of commodity product with small profit margins. Therefore, a $1 loss of profit may be equivalent to a $100 loss in sales. If the gross production volume falls below a break-even point, you are out of business very quickly.

Some plants are designed to be completely rebuilt after only a few years of operation. Examples are sulfuric acid production, semiconductor manufacture, or military weapons manufacture. An acid plant literally dissolves itself while semiconductor manufacture for particular microprocessor and memory products become obsolete. A military production plant is designed to produce a finite number of weapons, bought by mainly one customer (the government), and possibly some more for the government's allies. The industrial economics are such that these industries have a break-even date for each plant once it enters into production. When a plant continues in production past the break-even date, it starts to make a profit on the capital investment, and this profit increases the longer the plant can run. When the product is a commodity such as gasoline, the capital investment is huge with very significant risks, but once a well strikes oil and the gasoline flows to the customer, the profits are significant because of the enormous volumes of production. So how does one increase profits in gasoline sales? The price at the pump and at the crude well are determined by the market. Profits are increased by improving efficiency in production and distribution, both of which have margins affected by technology. Smart sensor networks developed to monitor and predict machinery health are one such technology which will have a profound impact on manufacturing profits in the 21st century. This type of maintenance is called *Condition-Based Maintenance*, or CBM and requires a broad spectrum of smart sensors, algorithms, pattern recognition, and prognostics algorithms for predicting machinery health.

The military has a slightly different need for CBM. Most weapon platforms such as ground vehicles, aircraft and ships are to be used for periods of time significantly beyond their designed lifetime. The B-52 bomber is a perfect example which is actually planned to stay in service well into the 21st century. Doing unneeded repairs to irreplaceable hardware like a B-52 or CH-47 Chinook helicopter places unnecessary risks on causing fatal damage from the repair, as well as requires unnecessary manpower to execute the repair. In addition, military commanders always want to know as objectively as possible the battle readiness of their forces. Clearly, CBM is an important technology for military operations as well as a means to cut costs and gain more effectiveness. Given the value and long life expectancy of military hardware, CBM using vast networks of smart sensors is in the future. For the Navy, this will have multiplicative effect on reducing the number of personnel on board a ship. If you can eliminate the sailors monitoring the equipment, you also eliminate the sailors needed to clean up, cook, do laundry, and attend the maintenance sailors. This has perhaps the biggest cost impact on the Navy because of the expense of supporting all those personnel on board a ship for extended periods of time.

In a CBM system there are essentially three "layers" to the network architecture. The bottom layer contains the smart sensor actually monitoring the machinery. This system is fully capable of pattern recognition and prognostics, but is also capable of self-diagnosis so that one also has a confidence measure that the sensors are calibrated and in proper working order. We'll call this bottom layer the Intelligent Node Health Monitor, or INHM, which is designed to be robust and low cost, and eventually become a part of the motor, pump, bearing, gearbox, etc., provided by the manufacturer. The INHM is either locally powered or self-powered using energy salvaging from local mechanical motion or thermal gradients. The INHM output to the network is wireless, most likely TCPIP (Transfer Communications Protocol — Internet Protocol) which is rapidly becoming the low-cost non-proprietary way to move packet data over large networks. The INHM has very limited storage to support its prognostics capability, so a hierarchal layer is established with Area Health Monitors, or AHMs, each directly connected to a group of INHMs and serving as their hub, router, and data archive server. The AHM watches over its group of INHMs to assure proper operations and manage their archival data. When an INHM signals an alert condition for machinery failure prognosis, or failure of one of its sensors, the AHM confirms the condition, assembles a report with the current and past history of INHM data, and sends an alert message to the next hierarchal layer, the Platform Level Monitor, or PLM. The PLM is where the human interfaces to the CBM system. To the PLM, the AHM is an unattended file server on the Internet (or Intranet — a private local network). The human responds to the alert from the AHM/INHM by reviewing and confirming the report, authorizing maintenance action, and assembling the information necessary, ordering parts and providing repair instructions. Note that the later data can be provided by the manufacturer through the INHM on the component expected to fail. At the PLM level, the repair decision can also be based on a wide range of factors beyond the information the AHM/INHM are providing. The human can "drill down" to the INHM level and re-program the alert thresholds to better meet the immediate information needs. For example, while in flight, a helicopter pilot does not need to be alerted about next week's maintenance. The pilot only needs to know if the current flight/mission needs to be aborted, and with enough time to abort the mission. The maintenance officer on the ground needs to know a great deal more.

Building a CBM system out of networks of smart sensors is straightforward, albeit challenging, but does require a capability in information prognostics. Recall that the difference between data and information is that information is data with a confidence metric and with context to an underlying physical model. Both of these important aspects of information are required for prognostics. The data confidence measures are included in a Kalman filter as part of the measurement error statistics and the contextual physical model is included in the state transition kinematic model. Given this Kalman filter, the "state" of either features, or a "situation" can be tracked over time to produce a smoothed estimate of the current situation and a way to predict the situation in the future. This prediction, based on the current state vector (often contains a "position, velocity, acceleration, etc" elements), into the future is called a linear prediction and it will be accurate if things continue as currently measured. But that's not good enough for some prognostics applications where a great deal is known about the failure mechanisms.

Given a well-understood set of failure trajectories of a state vector we can apply a pattern matching algorithm to the current observed trajectory and these possible outcome trajectories. In much the same way as multiple hypothesis testing is executed using syntactic fuzzy logic, we can continuously compare the current situational history defined by the observed track to each of the hypothesized trajectories, and use the nonlinear dynamics of the failure progression to determine which path we are on. Given the selected failure mode path, we can do a much better prediction of the remaining useful life and current failure hazard. This requires exquisite knowledge and scientific modeling of the underlying failure mechanisms which is where engineering should and will push the envelope of capability in the future. In the absence of such knowledge of failure mechanisms, we have linear prediction. In both cases we must define a "regime" for imminent failure so the remaining time and hazard rate can be computed. Defining a regime which signifies unacceptable risk for failure is much easier than defining the nonlinear fault trajectory from the current state to the failure point.

Figure 33 depicts graphically the difference between linear prediction prognostics and nonlinear fault trajectory prognostics. Time-temperature plots are given for three hypothetical cases: a healthy power supply; a supply with a defective resistor; and a supply with a failing voltage regulator. The three trajectories are quite different and nonlinear. They are identifiable based on trends and under-lying physical modeling of the failure mechanisms. However, at different times along any of the trajectories one or more of the situations can exist as components can and do begin to fail at any time. The prognostics task is to detect the precursors to event-ual component failure, and provide accurate prognostics for the current hazard and remaining useful life of the component. "Regimes" are identified on the right edge of Figure 33 to identify a set of meaningful situations based on temperature. For example, above 300°C a fire is assumed imminent. For temperatures above

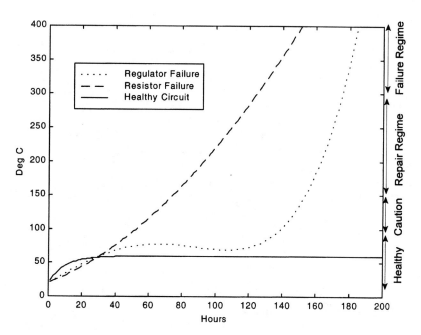

Figure 33 Temperature readings over time for three types of hypothetical failure of an electronic power supply showing the health, cautionary, repair, and failure regimes.

150°C the components are operating out of specification and an immediate repair is warranted. Between 85°C and 150°C caution is warranted because the situation could quickly deteriorate, but currently is tolerable. Below 85°C there is no immediate hazard of burning out the power supply. To complicate matters further, the power supply temperature will depend greatly on the load which can vary significantly as electronic devices are switched on and off, and catastrophically if an electronic system on the power bus fails, pulling down the power supply with it. With the exception of the latter case, which can often be handled using fuses or circuit breakers, the observed temperature can be expected to have many fluctuations. The job of the prognostic algorithm is to provide an accurate and meaningful prediction in the midst of all the fluctuation noise. This is why a combination of Kalman filtering based on physical modeling and fuzzy syntactic logic is needed to create machine prognostics.

Up until about 30 hours in Figure 33, any of the trajectories can exist and the supply is considered "healthy", but a linear prediction alone might prematurely predict a failure or cause an early repair. From 30 to about 130 hours the regulator fault trajectory and the healthy supply are not distinguishable, but when the regulator starts to fail, the temperature rises much quicker than the resistor failure. The differences in the nonlinear fault trajectory models is quite useful in improving the prognostics information to the user. We ought not expect the software of the INHM and AHM to make the full call, but rather to organize and prioritize the information for the human at the PLM level. For example, the alert message to the human might look like the following:

> Repair of component X requested in 10 hours
> Likelihood of failure 50% in 17 hr if current situation persists
> Likelihood of failure 90% in 17 hr if failure mode is Y
> Likelihood of failure 35% in 17 hr if failure mode is Z
> Failure mode probability is Y (80%), Z (50%), unknown (30%)
> INHM # AHM # reporting all sensors calibrated $\pm 5\%$

Notice how the sum of the failure mode probabilities is greater than 100%. This is an artifact of fuzzy logic which provides useful statistical measures to be associated with the multiple hypotheses. The human is now put into a desirable position of getting good information to contemplate for the repair decision from the CBM system. The diagnostic information while not complete, is very valuable to the human and may influence the repair decision. Finally, the self-diagnostics at the end of the message insures credibility of the message data and alert status. This information is straightforward for a smart sensor system to provide and creates a whole new economic basis for industrial production as well as military readiness and efficiency.

Figure 34 shows some temperature data for our power supply example. The data is observed every hour and has a standard deviation of 10°C mainly from the load variations. Understanding the variance of the measurements is very important to application of a tracking filter. Recall from Section 10.1 that the Kalman filter has both a *predicted* state vector and an *updated* state vector. The difference between them is that the one-step ahead prediction (T time units into the future) is a little less noisy than the updated state, which has been corrected using the Kalman gain and the prediction error. At any point in time, the predicted state vector can be estimated

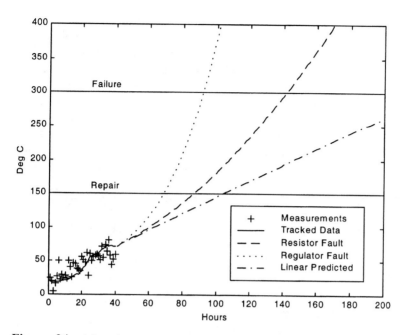

Figure 34 After about 40 hours one cannot diagnose the problem but can make prognosis estimates based on the current state (linearly produced), a resistor fault, or a regulator fault.

d steps into the future.

$$
\begin{aligned}
x'(N + d|N) &= F(N + d)x(N|N) \\
&= \begin{bmatrix} 1 & Td & \frac{1}{2}T^2d^2 \\ 0 & 1 & Td \\ 0 & 0 & 1 \end{bmatrix} \begin{bmatrix} x_{N|N} \\ \dot{x}_{N|N} \\ \ddot{x}_{N|N} \end{bmatrix}
\end{aligned}
\tag{14.3.1}
$$

This can also be done using a new state transition matrix where $T = Td$, or simply a one big step ahead prediction. This is a straightforward process using the kinematic state elements to determine the time and position of the state in the future. Therefore, it is really rather unremarkable that we can predict the future value of a state element once we have established a Kalman filter to track the trajectory of the feature of interest. This forward prediction is called a linear prediction. Its limitations are that it takes into account only the current state conditions and assumes that these conditions will hold over the future predicted step.

But, we know that as most things fail, the trajectory is likely to become very nonlinear, making linear predictions accurate only for very short time intervals into the future. To take into account several plausible failure trajectory models, we treat each trajectory as a hypothesis and assess the statistics accordingly. Figure 34 shows a linear prediction along side predictions assuming a resistor or regulator fault. These "tails" are pinned on the end of our Kalman filter predicted state by simply matching a position and slope (rate of change) from the terminal trajectory of our models to the most recent data smoothed by the Kalman filter. This is like postulating, "we know we're at this state, but if a resistor has failed, this is what the future looks like, or if the regulator has failed, then the future looks like this." At 40 hr into our example, one really cannot tell much of what is happening in the power supply. It might be normal or it might be developing a problem. The data points (+) and Kalman filter track (solid line plot) represent our best knowledge of the current condition. The linear prediction at 40 hr in Figure 34 shows a repair action needed after 100 hr, and failure expected after 240 hr. The hypothesis of a resistor failure says that a repair will be needed in 80 hr and failure is expected after 140 hr. The regulator failure hypothesis suggests a repair after 65 hr (25 hr from now) and that failure could occur after 90 hr. Since we have little information to judge what type of failure mode, if any is occurring, the linear predicted prognosis is accepted.

    At 140 hr into our example there is much more information to consider as seen in Figure 35. The temperature has stabilized for several days and for the most part, there is little cause for alarm. However, during the last 24 hr there has been a slight trend upward. The linear predicted prognosis suggests repair at 200 hr (60 hr from now) and failure may occur after 400 hr. The resistor failure hypothesis suggests repair after 180 hr and failure after 300 hr. The regulator failure hypothesis suggests repair after 175 hr and failure after 210 hr. It appears that either the resistor or regulator is failing, but its far enough into the future that we don't need to do anything at the moment. Figure 36 shows the results of our simulation at 150 hr. Failure now appears imminent, but more urgent is the need for a repair starting in a few hours. The linear predicted prognosis suggests repairing after 160 hr (just 10 hr from now) and failure after 220 hr. The resistor failure prognosis suggests repairing even sooner after 155 hr and failure at 180 hr. The regulator hypothesis suggests repair

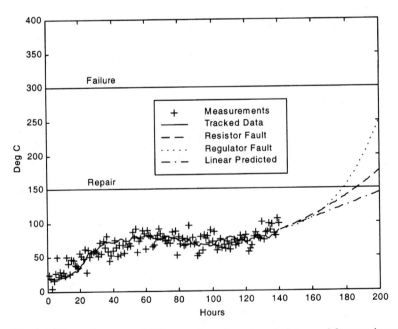

Figure 35   After about 140 hours, there is no immediate need for repair and the prognosis can be hypothesis based.

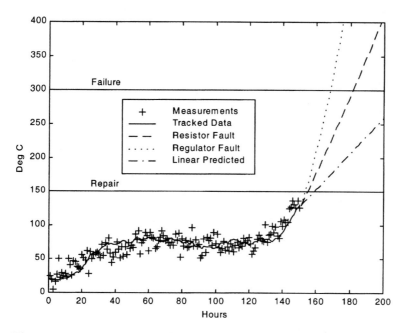

Figure 36 At 150 hours there appears to be an immediate need for repairs and the prognosis is that a regulator failure is likely in 10–12 hours, but if it's a resistor the failure may be 30 hours away.

now and that failure could occur after 170 hr, just 20 hr into the future. By failure, we mean the feature will cross the threshold we have defined as the failure regime, where failure can happen at any time. For our example, the regulator failed causing the failure threshold to actually be crossed at 180 hr. Even though our prognostic algorithm did not exactly pinpoint the precise cause nor the precise time of failure, it provided timely and accurate information necessary for the CBM actions to take place. As technology and science improve our ability to model fatigue and perform prognostics, the performance of such algorithms for predictive maintenance will certainly improve. But the basic idea and operations will likely follow our example here.

Figure 37 shows the results of the Kalman filtering. The measurements and predicted temperature state are seen in the top plot, the middle plot shows the estimated rate of temperature change, and the bottom plot shows the temperature "acceleration". These state components are used to do the forward prediction as shown in Eq. (14.3.1). However, as one predicts a future state, the state error covariance increases depending on the size of the step into the future. The process noise matrix is defined as

$$Q(N) = \begin{bmatrix} \frac{1}{2}T^2 \\ T \\ 1 \end{bmatrix} \sigma_v^2 \begin{bmatrix} \frac{1}{2}T^2 & T & 1 \end{bmatrix} = \begin{bmatrix} \frac{1}{4}T^4 & \frac{1}{2}T^3 & \frac{1}{2}T^2 \\ \frac{1}{2}T^3 & T^2 & T \\ \frac{1}{2}T^2 & T & 1 \end{bmatrix} \sigma_v^2 \qquad (14.3.2)$$

where σ_v is the minimum standard deviation of the "acceleration" element of the

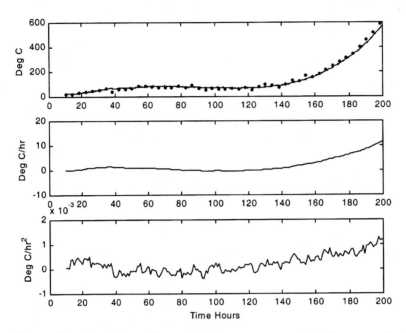

Figure 37   The Kalman filter states for our power supply failure example where the states will be used for failure prognosis and the associated time statistics and hazard prediction estimates.

state vector. Setting σ_v to a small value make a very smooth track, but also one which is very sluggish to react to dynamical changes in the data. When T represents the time interval of the future prediction step, the state error covariance is predicted according to

$$P'(N+1|N) = F(N)P(N|N)F^H(N) + Q(N) \tag{14.3.3}$$

Thus, the larger T is (the further one predicts into the future) the larger the state error covariance will be. One can extract the 2 by 2 submatrix containing the "position" and "velocity" state elements, diagonalize via rotation (see Eqs (14.1.11)–(14.1.12)), and identify the variance for position and the variance for velocity on the main diagonal.

   The variance for position is rather obvious but, to get an error estimate for our RUL time requires some analysis. Our prediction of the time until our state crosses some regime threshold is the result of a parametric equation in both time and space. Our state elements for position and velocity are random variables whose variances increase as one predicts further into the future. To parametrically translate the statistics to time, we note dimensionally that time is position divided by velocity. Therefore, the probability density function (pdf) for our RUL time prediction is going to likely be the result of the division of two Gaussian random variables (our state vector elements). This is a little tricky but worth showing the details here.

   A detailed derivation for the probability density of the ratio of two random variables is seen in Section 6.1. Let the random variable for our predicted RUL time, t_F, be z. Let the random variable representing the final "position" prediction be simply x with variance σ_F^2, and the random variable representing the "velocity" be simply y with variance σ_R^2. Our new random variable is therefore $z = x/y$. One derives the pdf by first comparing probabilities, and then, differentiating the probability function for z with respect to z to obtain the desired pdf. The probability that $z \le Z$, or $P(z \le Z)$, is equal to the probability that $x \le yZ$ for $y \ge 0$, plus the probability that $x \ge yZ$ for $y < 0$.

$$P(z \le Z) = \int_0^\infty \int_{-\infty}^{yZ} p(x, y)dxdy + \int_{-\infty}^0 \int_{yZ}^\infty p(x, y)dxdy \qquad (14.3.4)$$

Assuming the tracked position and velocity represented by x and y, respectively, are statistically independent, one differentiates Eq. (14.3.4) with respect to z using a simple change of variable $x = yz$ to obtain

$$p(z) = \int_0^\infty yp(x = yz)p(y)dy - \int_{-\infty}^0 yp(x = yz)p(y)dy$$

$$= \frac{2}{\sigma_F \sigma_R 2\pi} \int_0^\infty ye^{-\frac{y^2}{2}\left(\frac{z^2}{\sigma_F^2} + \frac{1}{\sigma_R^2}\right)} dy \qquad (14.3.5)$$

$$= \frac{1}{\pi}\left(\frac{\sigma_F}{\sigma_R}\right)\frac{1}{z^2 + \left(\frac{\sigma_F}{\sigma_R}\right)^2}$$

The pdf in Eq. (14.3.5) is zero mean. It is straightforward to obtain our physical pdf for the time prediction where the mean is t_F.

$$p(t) = \frac{1}{\pi}\left(\frac{\sigma_F}{\sigma_R}\right)\left[(t - t_F)^2 + \left(\frac{\sigma_F}{\sigma_R}\right)^2\right]^{-1} \qquad (14.3.6)$$

The mean of the pdf in Eq. (14.3.6) is t_F and Figure 38 compares the shape of the time prediction pdf to a Gaussian pdf where the standard deviation $\sigma = \sigma_F/\sigma_R = 10$. The density functions are clearly significantly different. While the second moment (the variance) of the time prediction pdf in Eq. (14.3.6) is infinite (therefore the use of the central limit theorem does not apply), one can simply integrate the pdf to obtain a confidence interval equivalent to the 68.4% probability range about the mean of a Gaussian pdf defined by $\pm \sigma$. This confidence range works out to be 1.836 times the Gaussian standard deviation. To present the confidence limits for our RUL time prediction, we can simply use 1.836 times the ratio of σ_F/σ_R. Figure 39 shows the RUL time based on linear prediction and also shows the 68.4% confidence limits. The straight dashed line represents real-time and shows the actual time the temperature crossed our failure threshold at 300°C was about 180 hr. Our RUL prediction converges nicely to the correct estimate around 140 hr, or nearly two days before the failure condition. However, it is only in the range beyond 170 hr that the confidence interval is very accurate. This is exactly what we want the prognostics algorithm to provide in terms of statistical data.

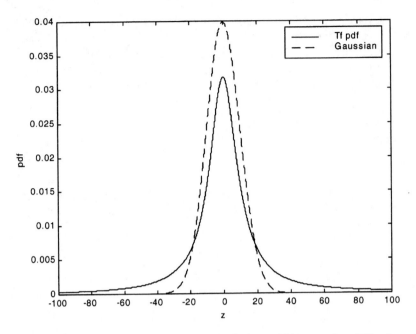

Figure 38 Comparison of the time prediction pdf in equation (10) where the standard deviation ration $\sigma_F/\sigma_R = 10$ to a Gaussian pfd with $\sigma = 10$.

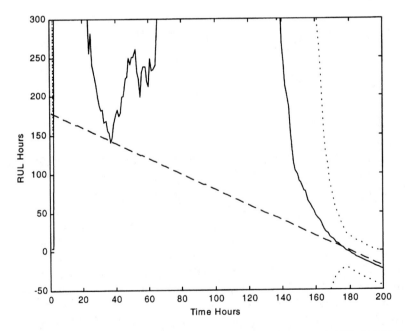

Figure 39 Using only linear prediction, the remaining useful life (RUL) is predicted and compared to actual remaining life (straight dashed line) showing the 63% confidence bounds on the RUL prediction (dotted line encompassing the RUL prediction plot).

Figure 40 is a three-dimensional rendering of Figure 39 showing the RUL pdf at several interesting times. The RUL pdf is centered over the RUL time prediction. But, for most of the time before 170 hr, the statistics on the RUL time are not very good. This is due mainly to the failure threshold being significantly in the future when the trajectory of the temperature plot is fairly flat from around 50–160 hr. During this period and situation, we should not expect any reliable prognostic RUL time predictions. The corresponding RUL time pdf's in the period leading up to 160 hr are almost completely flat and extremely broad. As the RUL time prediction converges, the corresponding RUL time pdf sharpens. As the temperature feature trajectory passes the threshold, the RUL time becomes negative (suggesting it should have failed by now), and the RUL time pdf is then fixed because we are interested in the time until the failure threshold is crossed and beyond as our failure regime. Again, this is seen as desirable algorithm behavior.

Consider Figure 41 which shows the probability of survival, which is the integral from infinity down to the current time. When the statistics of the RUL time prediction are accurate (giving a sharp pdf) and the failure time is well off into the future, the survivor probability would be near unity. When the time statistics are not well defined or the RUL time is in the immediate future, the survivor prob-

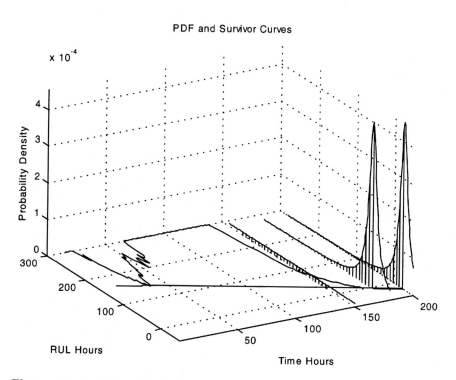

Figure 40 A 3-dimensional plot showing the RUL times probability density function centered over three points of interest near the end of the RUL showing the integral of the density functions (shaded with vertical lines) which provide the probability of surviving.

ability is near 50%. Equation (14.3.7) gives an analytical expression for the survivor rate.

$$R(t) = \int_t^\infty \frac{1}{\pi} \left(\frac{\sigma_F}{\sigma_R}\right) \left[(t - t_F)^2 + \left(\frac{\sigma_F}{\sigma_R}\right)^2\right]^{-1} dt$$

$$= \frac{1}{2} - \frac{1}{\pi} \tan^{-1}\left(\frac{t - t_F}{\sigma_F/\sigma_R}\right)$$

(14.3.7)

In Figure 41 the survivor probability is initially around 50% because it is so broad. As the statistics improve the survivor rate improves. This is desirable because we need assurances that failure is not immediate in the regime approaching the failure threshold. Then the RUL time prediction crosses the current time at 180 hr giving a 50% survivor rate. At this point we fix the pdf for the RUL time prediction since we have crossed the failure regime threshold. In other words, we do not want the RUL pdf to start flattening out again from a growing "backward time prediction". However, the survivor rate, which is the integral of the RUL pdf from infinity to the current time still changes as the fixed (in shape that is) RUL pdf slides by such that less and less of the pdf is integrated as time progresses. This causes the survivor rate to diminish towards zero as the hazard rate increases. The hazard rate $H(t)$ is defined as the RUL time pdf for the current time divided by the survivor

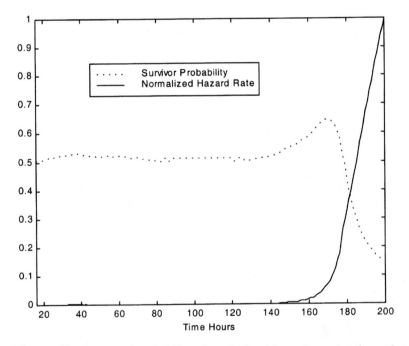

Figure 41   Estimated probability of survival and hazard rate showing a changing survivor probability over time due to the changing statistic of the Kalman state vector.

rate.

$$H(t) = 2\left(\frac{\sigma_F}{\sigma_R}\right)\left[\pi - 2\tan^{-1}\left(\frac{t - t_F}{\sigma_F/\sigma_R}\right)\right]^{-1}\left[(t - t_F)^2 + \left(\frac{\sigma_F}{\sigma_R}\right)^2\right]^{-1} \qquad (14.3.8)$$

The $(t - t_F)^2$ term in the right-most square-bracketed term in Eq. (14.3.8) is set to zero once the failure threshold is crossed (i.e. the term $t - t_F$ becomes positive). Again, this has the effect of stating that we've met the criteria for passing the failure regime threshold, now how far past that point are we? The hazard rate measures the likelihood of immediate failure by comparing the current probability density to the probability of surviving in the future. The hazard rate needs to be normalized to some point on the pdf (in our case the last point taken) because it will generally continue to increase as the survivor rate decreases. Clearly, we have an objective "red flag" which nicely appears when we cross into a prescribed regime we associate with failure.

We have one more metric seen in Figure 42 which helps us define component failure we call the stability metric. This term is design to measure how stable the feature dynamics are. When positive, it indicates the dynamics are converging. This means if the feature is moving in a positive direction, it is decelerating. If it is moving with negative velocity, it is accelerating. In both stable cases the velocity is heading towards zero. We can associate zero feature velocity as a "no change" condition, which if the machinery is running indicates that it might just keep on

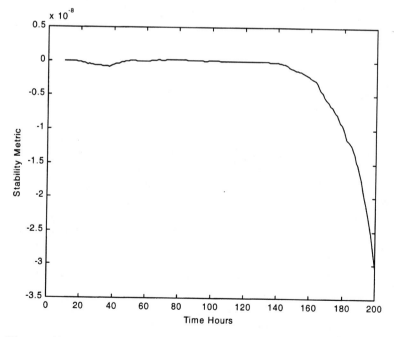

Figure 42 A simple stability metric is formed from minus one over the product of the feature velocity and feature acceleration where a negative indicates the system is not stable and will fail soon.

running under the same conditions. Obviously, the most stable situation is when the feature velocity is zero and there is no acceleration. Our stability metric is defined as

$$\Sigma(N|N) = -\langle \dot{x}_{N|N} \ddot{x}_{N|N} \rangle_t \qquad (14.3.9)$$

such that a zero or positive number indicates a stable situation. If the feature is moving in a positive direction and is accelerating, or if moving with negative velocity and decelerating, the stability metric in Eq. (14.3.9) will be negative indicating that the feature dynamics are not stable and are diverging from there current values. Change is not good for machinery once it has been operating smoothly for a time. Therefore, this simple stability metric, along with the hazard and survivor rates, are key objective measures of machinery health.

14.4 SUMMARY, PROBLEMS, AND BIBLIOGRAPHY

This Chapter was probably the most challenging to write of the entire book because of the breadth and depth of diverse information covered. As a result, a significant amount of important approaches to automatic target recognition had to be tersely presented, which provides a concise broad view of the area. In Section 14.1 we covered the three main types of decision algorithms used in automatic target recognition: statistical, neural network, and fuzzy logic-based classifier algorithms. Statistical recognition is based on the idea that the statistics of a pattern class of interest can be measured through training. Therefore, one can measure how close a particular pattern matches a known pattern class statistically. A pattern to be classified is then assigned to the known pattern class it most closely matches in a statistical sense. A neural network, on the other hand, does not require that one "knows" anything about the pattern classes of interest, nor the "unknown" pattern to be classified. It works by a brute force training of artificial neurons (weights and interconnections) to match a set of known patterns to a particular class output of the network. This amazing algorithm can truly conger the idea of a simple artificial brain which for some applications may be most appropriate. But its shortfall is that it teaches us little about why it works, and more importantly, what information in the patterns is causing the classification. Therefore, there is always a criticism of neural network about extendability to new patterns outside the training set. The syntactic fuzzy logic approach captures the best of both statistical and neural algorithms. But the difficulty is that one must know the syntax, design the fuzzy set membership criteria (blend functions, AND, OR, operations), and test (rather than train) the algorithm on real data to validate the logic. The beauty of syntactic fuzzy logic is that it allows us to use any combination of algorithms to produce inputs to the logic. As a result of testing the logic, flaws are trapped and re-wired (re-coded in software), and further testing is used to validate the response. Syntactic fuzzy logic embraces the scientific method for model development. It requires the most from us in terms of intelligence, and offers us the ability to refine our own intelligence. We apologize to the creators of those many ingenious algorithms which have been omitted due to space, since this is not a pattern recognition book. From the viewpoint of intelligent sensor systems, a basic view of automatic target recognition is needed to integrate this technology into the entire intelligent sensor system.

In Section 14.2, we presented some useful techniques for generating signal features in both one and two dimensions. A completely generic summary of feature strategies is given with emphasis on the concept that useful signal features *concentrate* information. This information should certainly be based on real physics rather than heuristics. But in the case of computer vision, a strategy of following the biology and psychophysics of the primate eye is suggested for developing image features. Vision is extraordinarily complex and an aspect of human performance in which machines may never really surpass. But we can train machines to see things, albeit crudely. And the suggested approach is to first determine whether one is interested in a texture or an object. Textures can be matched in the frequency domain while objects may best be identified via image segmentation. Segmentation of object shapes leads one to syntactic logic for combining segments into particular objects. This logic can include influence from the presence of other objects or data to suggest a particular scene. Segmentation also offers the ability to account for rotation, skewness, and scaling, which are major algorithm problems for computer vision.

Section 14.3 presents the concept of feature tracking, where the track over time can itself become a feature. This concept considers the evolution of information, rather than just situational awareness in a "snapshot" of data. The concept of prognostics is presented where the algorithm presents where the feature will be in the future, based of current information and reasonable hypotheses. This has significant economic value for machinery systems where maintenance is to be developed based on condition, rather than hours of operation. Condition-based maintenance (CBM) has the potential for significant economic impact on basic industry as well as increasing reliability of life-critical systems. The prognostics problem is of course helped by as much diagnostic information as possible as well as fault trajectory models for particular cases of damage. This follows closely the medical prognosis we all have experienced at some time. For example, the doctor says "You have a cold. You can expect to feel bad for a few days followed by slow improvement back to normal." Or, the doctor says, "You have liver cancer. I'm sorry but you only have a 50% chance of living more than 6 months." In both cases, the prognosis is based on current information and some projection (model) of how things will go in the future. Even when these prediction are not precisely correct, if they are interpreted correctly, the information is highly valuable. This is definitely the case with machinery CBM as the information regarding several prognoses can be very useful in determining the most economic repair actions.

The "Intelligent" adjective in the term "Intelligent Sensor Systems" is justified when the sensor has the capability of self-evaluation, self-organizing communications and operations, environmental adaption, situational awareness and prediction, and conversion of raw data into information. All five of these concepts require some degree of pattern recognition, feature extraction, and prediction. Earlier chapters in the book cover the basic signal processing details of sampling, linear systems, frequency domain processing, adaptive filtering, and array processing of wavenumber signals. Using these techniques, one can extract information from the raw data which can be associated with various features of a particular information pattern. This chapter's goal is to show the basic techniques of pattern recognition, feature extraction, and situational prognosis. We still need to know more about transducers, noise, and noise control techniques to complete our understanding of the raw signal physics, and this is the subject of the next chapter.

PROBLEMS

1. A pair of features are being considered for use in a statistical classifier. A training set for class A reveals a variance of feature 1 of 2.5 and mean 5, a variance for feature 2 of 5.5 and mean of 9, and a covariance of feature 1 and 2 of -3.2. What are the major and minor axis and rotation angle of the ellipse representing the 1-sigma bound for the training set?

2. A training set for class B reveals a variance of feature 1 of 1.75 and mean 2, a variance for feature 2 of 6.5 and mean of 5, and a covariance of feature 1 and 2 of $+0.25$. Sketch the ellipses defining class A (from question 1) and class B.

3. Suppose you are given a feature measurement of $\langle 4.35, 5.56 \rangle$ where 4.35 is feature 1 and 5.56 is feature 2. If the probability of class A existing is 75% and class B is 45%, which class should the feature pair be assigned to?

4. Consider a neural network with two input nodes, four nodes in one hidden layer, and three output nodes. The inputs range from 0 to 1 and the output classes represent the binary sum of the two inputs. In other words class A is for both inputs being "zero", class B is for one input "zero", the other "one", and class C is for both inputs "one". Train the network using a uniform distribution of inputs, testing each case for the correct class (i.e. if the input is > 0.5 it is "one", $\Leftarrow 0.5$ it is "zero").

5. Build the same classifier as in problem 4 but using syntactic fuzzy logic. Select the weights as unity and use identical sigmoidal functions as the neural network. Compare the two classifiers on a set of uniform inputs.

6. A condition monitoring vibration sensor is set to give an alarm if the crest factor is greater than 25. If the main vibration signal is a sinusoid of peak amplitude 100 mV, how big must the transients get to trip the alarm, assuming the sinusoid amplitude stays the same?

7. The harmonic distortion of a loudspeaker is used as a condition monitoring feature in a public address system. To test the distortion, two sinusoid signals drive the loudspeaker at 94 dB (1.0 Pascals rms) at 1 m distance at 100 Hz and 117 Hz. If distortion signals are seen at a level of 0.1 Pascals rms at 17 Hz and 217 Hz, and 0.05 Pascals rms at 200 Hz and 234 Hz, what is the % harmonic distortion?

8. Define a segmentation graph for the numbers 0 through 9 based on "unit-length" line equations. Show how the segmentation functions can be scaled and rotated using simple math operations.

9. The temperature of a process vessel is currently 800°C and has a standard deviation of 5°C. If the temperature is rising at a rate of 10°C per minute, where the velocity standard deviation is 1°C, how much time will it take for the vessel to get to 1000°C? What is the standard deviation of this time estimate? Hint: assume σ_v (the tracking filter process noise) and T are unity for simplicity.

10. Compute the hazard and survivor rates as a function of time for the temperature dynamics given in problem 9.

BIBLIOGRAPHY

Y. Bar-Shalom, X. R. Li Estimation and Tracking: Principles, Techniques, and Software, London: Artech House, 1993.

R. Billinton, R. N. Allan Reliability Evaluation of Engineering systems: Concepts and Techniques, New York: Plenum Press, 1983.

S. Bow Pattern Recognition and Image Processing, New York: Marcel Dekker, 1991.

Z. Chi, H. Yan, T. Pham Fuzzy Algorithms, River Edge, NJ: World Scientific, 1996.

H. Freeman, (Ed.), Machine Vision, New York: Academic Press, 1988.

L. A. Klein Millimeter-Wave and Infrared Multisensor Design and Signal Processing, Boston: Artech House, 1997.

I. Overington Computer Vision: A Unified, Biologically-Inspired Approach, New York: Elsevier, 1992.

M. Pavel Fundamentals of Pattern Recognition, 2nd ed., New York: Marcel Dekker, 1992.

D. E. Rumelhart, J. L. McClelland and the PDP Research Group, Parallel Distributed Processing: Explorations in the Microstructure of Cognition, Vol. I: Foundations, Cambridge: MIT Press, 1986.

R. Schalkoff Pattern Recognition: Statistical Structural, and Neural Approaches, New York: Wiley, 1992.

B. Souček and The IRIS Group, Fast Learning and Invariant Object Recognition: The 6th Generation Breakthrough, New York: Wiley, 1992.

J. T. Tou, R. C. Gonzalez Pattern Recognition Principles, Reading: Addison-Wesley, 1974.

REFERENCES

1. R. Schalkoff Pattern Recognition, New York: Wiley, 1992.
2. J. T. Tou, R. C. Gonzalez Pattern Recognition Principles, Reading: Addison-Wesley, 1974.
3. J. A. Stover, R. E. Gibson "Modeling Confusion for Autonomous Systems", SPIE Volume 1710, Science of Artificial Neural Networks, 1992, pp. 547–555.
4. I. Overington Computer Vision: A Unified, Biologically-Inspired Approach, New York: Elsevier, 1992.

15

Sensors, Electronics, and Noise Reduction Techniques

One of the characteristics of an *Intelligent Sensor System* is its ability to be self-aware of its performance based on physical models, analysis of raw sensor data, and data fusion with other sensor information. While this capability may sound like very futuristic, all of the required technology is currently available. However, very few sensor systems at the end of the 20th century exploit a significant level of *sentient processing*. By the end of the 21st century, it is very likely that all sensors will have the capability of *self-awareness*, and even *situational awareness*. The definition of "sentient" can be seen as "having the five senses of sight, hearing, smell, touch, and taste" according to most dictionaries. In general, "sentient" applies only to animal life and not machines as of the end of the 20th century. However, there is no reason why machines with commensurate sensors of video, acoustics, vibration, and chemical analysis can't aspire to the same definition of sentient, so long as we're not going as far as comparing the "intelligent sensor system" to a being with a con-science, feelings, and emotions. Some would argue that this too is within reach of humanity within the next few centuries, but we prefer to focus on a more straightforward physical scope of sensor intelligence. In other words, we're dis-cussing how to build the sensor intelligence of an "insect" into a real-time computing machine while extracting the sensor information into an objective form which includes statistical metrics, signal features and patterns, as well as pattern dynamics (how the patterns are changing). The information is not very useful unless we have some computer–human interface, so it can be seen that we are building the sensor part of an insect with an Ethernet port for information access!

In order to make an algorithm which can extract signal features from raw sensor data, match those features to known patterns, and then provide predictive capability of the pattern motion, we must have a solid foundation of sensor physics and electronic noise. Given this foundation, it is possible for the sensor algorithms to identify signal patterns which represent normal operation as well as patterns which indicate sensor failure. The ability to provide a signal quality, or sensor reliability, statistical metric along with the raw sensor data is extremely valuable to all subsequent information processing. The sensor system is broken down into the electronic interface, the physical sensor, and the environment. Each of these can be modeled physically and can provide a baseline for the expected signal background noise, signal-to-noise ratio (SNR), and ultimately the signal feature confidence. Our discussion of electronic noise will obviously apply to all types of sensor systems. However, our discussion of sensors will be limited to acoustic and vibration sensors, which are involved in most array signal processing and control systems. There are many important low bandwidth (measures signals which change slowly) sensors such as temperature, pressure, etc., which require a similar SNR analysis as well as a physical model for the sensor linearity over desired sensor range, which will not be explored in detail here. Our discussion is limited to sensors where we are interested in the SNR and response as a function of frequency. This is clearly the more complicated case as the basic presentation of electronic and transducer noise can be applied to any sensor system.

Finally, we present some very useful techniques for cancellation of noise coherent with an available reference signal. This can also be seen as adaptive signal separation, where an undesirable signal can be removed using adaptive filtering. Adaptive noise cancellation is presented in two forms: electronic noise cancellation and active noise cancellation. The distinction is very important. For active noise cancellation, one physically cancels the unwanted signal waves in the physical medium using an actuator, rather than electrically "on the wire" after the signal has been acquired. Active noise cancellation has wider uses beyond sensor systems and can provide a true physical improvement to sensor dynamic range and SNR if the unwanted noise waveform is much stronger than the desired signal waveform. The use of adaptive and active noise cancellation techniques give an intelligent sensor system the capability to countermeasure poor signal measuring environments. However, one must also model how the SNR and signal quality may be affected by noise cancellation techniques for these countermeasures to be truly useful to an intelligent sensor system.

15.1 ELECTRONIC NOISE

There are four main types of electronic noise to be considered along with signal shielding and grounding issues when dealing with the analog electronics of sensor systems. A well-designed sensor system will minimize all forms of noise including ground loops and cross-talk from poor signal shielding. But, an intelligent sensor system will also be able to identify these defects and provide useful information to the user to optimize installation and maintenance. As will be seen below, the noise patterns are straightforwardly identified from a basic understanding of the underlying physics. The four basic types of noise are *thermal* noise, *shot* noise, *contact* noise, and *popcorn* noise. They exist at some level in all electronic circuits and sensors. Another form of noise is interference from other signals due to inadequate shielding and/or ground loops. Ground loops can occur when a signal ground is present in more than one location. For example, at the data acquisition system and at the sensor location at the end of a long cable. The ground potential can be significantly different at these two locations causing small currents to flow between the ground points. These currents can cause interference with the signal from the sensor. Usually, ground loop interference is seen at the ac power line frequencies because these voltages are quite high and return to ground. Ground loops can be identified a number of ways and eliminated by proper shielding and floating of sensor signal grounds.

Thermal Noise results from the agitation of electrons due to temperature in conductors and resistors. The physical aspects of thermal noise were first reported by J. B. Johnson (1) and is often referred to as "Johnson noise". But it was Nyquist (2) who presented a mathematical model for the rms noise level. A little background in electronic material science is useful in describing how resistors and conductors can generate noise waveforms.

Recall that all solids can be described as a lattice of molecules or atoms, somewhat like a volume of various-sized spheres representing the molecules. The atoms are spatially charged where the nucleus is positive and the electron shells are negative. Atoms bonded into molecules also have spatial charges. The negative-charged electrons of one atom or molecule are attracted to the positive-charged nucleus of neighboring atoms or molecules, yet repelled by their electrons. One can visualize the "spheres" (atoms or molecules) connected together by virtual springs such that the motion of one atom or molecule effects the others. Because the spheres have thermal energy, they all vibrate randomly in time and with respect to each other. A solid crystalline material will generally have its molecules arranged in a nearly regular pattern, or lattice, the exceptions being either impurities or crystal defects called dislocations. These imperfections in the lattice cause "grain boundaries" in the material and also cause blockages to the natural lattice-formed pathways in the lattice for electrons to move. As two spheres move closer together, the repulsion increases the potential energy of the electrons. If the material is a conductor with unfilled electron shells (such as the metals of the periodic chart i.e. Al^{3+}), which tend to lose electrons readily in reactions. Metals tend to be good heat conductors, have a surface luster, and are ductile (bend easily without fracture). The unfilled electron shells allow a higher energy electron to enter and be easily transported across the lattice, but with an occasional collision with an impurity, dislocation, or vibrating atom. These collisions cause electrical resistance which varies with material type and geometry, temperature, and imperfections from dislocations and impurities. At absolute zero, a pure metal theoretically forms a perfect lattice with no dislocations or vibration, and therefore should have zero electrical resistance as well as zero thermal noise. Insulators, on the other hand, tend to have filled electron shells, are brittle and often amorphous molecular structures, and are often poor heat conductors. This binds the electrons to their respective atoms or molecules and eliminates the lattice pathways for electron travel. This is why oxides on conductor surfaces increase electrical resistance as well as cause other signal problems.

Nyquist's formula for thermal noise in rms volts is given in Eq. (15.1.1).

$$V_t = \sqrt{4kTBR} \tag{15.1.1}$$

The following parameters in Eq. (15.1.1) need to be defined:

k–Boltzmann's constant (1.38×10^{-23} Joules/°K)
T–Absolute temperature (°K)
B–Bandwidth of Interest (Hz)
R–Resistance in ohms (Ω)

Note that thermal noise can also apply to purely mechanical systems such that the ohms are mechanical and one ends up with an rms mechanical force from the dissipation effects of the damping material. The rms spectral density of the thermal noise in volts per $\sqrt{Hz}$ is

$$\sigma_t = \sqrt{4kTR} \tag{15.1.2}$$

which can be used to model the spectrum of circuit thermal noise.

As an example of the importance of thermal noise considerations, at room temperature (290°K) an accelerometer device with 1 MΩ resistance will produce about 10 μV rms thermal noise over a bandwidth of 10 kHz. If the accelerometer's sensitivity is a typical 10 mV/g (1 g is 9.81 m/sec^2), the broadband acceleration noise "floor" is only 0.001 g. This is not acceptable if one needs to measure broadband vibrations in the 100 μg range. One way to reduce thermal noise is to operate the circuitry at very low temperatures. Another more practical approach is to minimize the resistance in the sensor circuit. The structure of thermal noise is a Gaussian random process making the noise spectrally white. Thus, the noise is described by use of a spectral density (Section 6.1) which is independent of spectral resolution. 10 μV rms over 10 kHz bandwidth (0.01 pV^2/Hz) power density corresponds to 0.1 μV/$\sqrt{Hz}$ rms noise density, which is physically a reasonably small 1 ng/$\sqrt{Hz}$. For detection of a narrowband acceleration in a background noise of 1000 μg rms using a 1024-point FFT (spectral resolution of 10 Hz per bin), one can simply consider that the power spectrum SNR gain is about 30 dB, thus making a narrowband acceleration greater than about 1 μg rms detectable just above the noise floor. Note that thermal noise can also be generated mechanically from atomic and molecular motion, but instead of rms volts from electrical resistance the Boltzman energy one gets rms forces using mechanical resistance.

These background noise considerations are important because they define the best possible SNR for the sensor system. Of course, the environmental noise can be a greater source of noise. If this is the case, one can save money on electronics to achieve the same overall performance. To easily model a thermal noise spectrum, simply use the rms spectral density in Eq. (15.1.2) times the width of an FFT bin in Hz as the standard deviation of a zero-mean Gaussian (ZMG) random process for the bin (the real and imaginary parts of the bin each have ½ the variance of the magnitude-squared of the bin).

Shot Noise is also a spectrally-white Gaussian noise process, but it is the result of dc currents across potential barriers, such as in transistors and vacuum tubes. The broadband noise results from the random emission of electrons from the base of a transistor (or from the cathode of a vacuum tube). More specifically in a transistor, it comes from the random diffusion of carriers through the base and the random recombination of electron-hole pairs. A model for shot noise developed by Schottky is well-known to be

$$I_{sh} = \sqrt{2qI_{dc}B} \tag{15.1.3}$$

where I_{dc} is the average direct current across the potential barrier, B is the bandwidth in Hz, and q is the charge of an electron (1.6×10^{-19} Coulombs). The shot noise current can be minimized for a given bandwidth by either minimizing the dc current or the number of potential barriers (transistors or vacuum tubes) in a circuit.

To model the power spectrum of the shot noise, multiply Eq. (15.1.3) times the circuit resistance and divide by the square-root of total bandwidth in Hz to get the standard deviation of the noise voltage. For a power spectrum, each bin of an N-point FFT is modeled as an independent complex ZMG random process with $1/N$ of the noise variance. For a real noise time-domain waveform, the real and imaginary parts of the bin are also independent, each having $1/2N$ of the total noise variance. Figure 1 shows a ZMG noise waveform as a function of time and frequency.

Contact Noise is widely referred to as "one-over-f noise" because of the typical inverse frequency shape to the spectral density. Contact noise is directly proportional to the dc current across any contact in the circuit, such as solder joints, switches, connectors, and internal component materials. This is the result of fluctuating conductivity due to imperfect contact between any two materials. While connections like solder joints are considered quite robust, after years of vibration, heat cycle stress, and even corrosion, these robust connections can break down leading to contact noise. Like shot noise, it varies with the amount of dc current in the circuit. But unlike shot noise, its spectral density is not constant over frequency and the level varies directly with dc current rather than the square-root of dc current as seen in the model in Eq. (15.1.3).

$$I_c = KI_{dc}\sqrt{\frac{B}{f}} \qquad\qquad (15.1.4)$$

The constant K in Eq. (15.1.3) varies with the type of material and the geometry of the material. Clearly, this type of noise is very important to minimize in low fre-

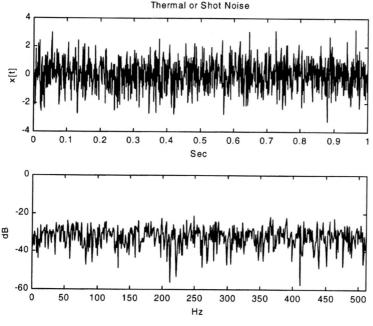

Figure 1 Both thermal and shot noise are zero-mean Gaussian and spectrally white.

quency sensor circuits. If we write the contact noise current as an equivalent voltage by multiplying by the circuit resistance R, and divide by the square-root of bandwidth, we get the rms spectral density in units of rms volts per $\sqrt{Hz}$.

$$\sigma_c(f) = RKI_{dc}\sqrt{\frac{1}{f}} \tag{15.1.5}$$

The rms spectral density in Eq. (15.1.5) is clearly a function of frequency and the variance represents the classic "$1/f$" noise often seen in real sensor systems. Contact noise is best modeled only in the frequency domain using a power spectrum. The time domain waveform can then be obtained via inverse Fourier transform. To get the power spectrum using an N-point FFT, first square Eq. (15.1.5) and then multiply by f_s/N, the bin width in Hz to get the total variance of the complex FFT bin magnitude-squared. The imaginary and real parts of this bin each are modeled with a ZMG process with ½ of the total bin variance for a real contact noise time-domain waveform. To insure that the time domain waveform is real, apply Hilbert transform symmetry to bins 0 through $N/2 - 1$ for the real and imaginary parts of each bin. In other words, the random number in the real part of bin k equals the real part of the random number in bin $N - k$, while the imaginary part of bin k equals the minus the imaginary part of bin $N - k$ for $0 \le k < N/2$. The inverse FFT of this spectrum will yield a real time noise waveform dominated by the low frequency $1/f$ contact noise as seen in Figure 2.

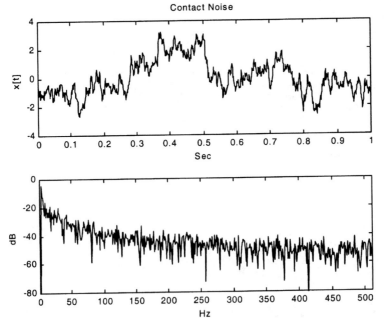

Figure 2 Contact noise is dominated by low frequencies and has a characteristic "$1/f$" shape in the frequency domain (note a log base 10 scale is displayed for dB).

Popcorn Noise is also known as burst noise, and gets its name from the sound heard when it is amplified and fed into a loudspeaker. Popcorn noise arises from manufacturing defects in semiconductors where impurities interfere with the normal operation. It is not clear whether popcorn noise is a constant characteristic of a semiconductor or if it can appear over time, perhaps from internal corrosion, damage, or wear-out of a semiconductor. Popcorn noise in the time domain looks like a temporary offset shift in the noise waveform. The amount of shift is fixed since it is due to the band gap potential energy of a semiconductor junction. The length of time the offset is shifted is random as is the frequency of the offset "bursts". The burst itself is a current offset, so the effect of popcorn noise can be minimized by keeping circuit impedances low. In the frequency domain, popcorn noise can have a spectral shape which averages to $1/f^n$, where n is usually 2. Figure 3 depicts popcorn noise and the corresponding power spectrum. Popcorn noise is most readily identified in the time domain.

Electronic Design Considerations to reduce intrinsic noise can be summarized into the following practical guidelines for any sensor system.

1. Physically locate the sensor and choose the sensor type to maximize SNR
2. Minimize circuit resistance to reduce thermal noise
3. Minimize dc currents and circuit resistance to reduce shot and contact noise
4. Choose high quality electronic components and test to verify low popcorn noise

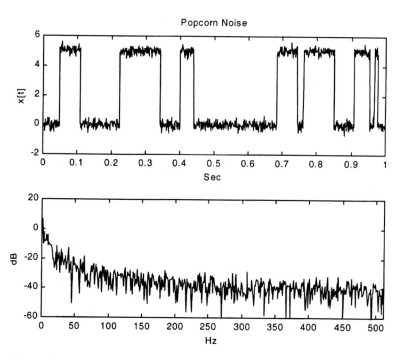

Figure 3 Popcorn noise (also known as burst noise) is due to a semiconductor defect where the plateau offsets are always the same but random in duration.

Following the above general guidelines, it can be seen that low noise sensor circuits will generally have low impedance, strong ac signal currents, and minimal dc currents. Low dc currents are obviously is not possible when the sensor is very low bandwidth, such as temperature, barometric pressure, etc. However, for dc-like sensor signals, the noise can be greatly reduced by simple averaging of the data. Depending on the desired sensor information, one would optimize the sensor circuit design as well as the data acquisition to enhance the SNR. For example, in meteorology, a precise pressure gauge (say $\pm 2''$ H_2O full scale) could measure barometric pressure relative to a sealed chamber as well as pressure fluctuations due to turbulence using two sensor circuits and data acquisition channels. If one wants the barometeric pressure, the turbulence needs to be averaged out of the measurement. The SNR of the pressure measurement is enhanced by low pass filtering a dc-coupled circuit while the turbulence SNR is enhanced by low pass filtering a separately sampled channel.

Controlling calibration gain in dc-circuits is particularly a challenge because small voltage or current offsets can drift over time, with supply voltage (say from batteries), or with environmental temperature. One power-hungry technique for controlling temperature drift is to locate the entire sensor circuit in a thermal mass which is heated to a constantly controlled temperature above the environmental temperature. Another approach is to purposely include a temperature or supply voltage sensitive part of the circuit and then compensate the sensor output bias due to temperature. However, such temperature compensation, which is often nonlinear over a wide range of environments, is very easily done in the digital domain. Finally, one should consider the slew rate and bandwidth of the electronic circuitry amplifiers. For example, if the gain-bandwidth product of a particular operational amplifier is 1 million and the sensor bandwidth is 20 kHz, the maximum theoretical gain is 50. If a gain of 160 ($+44$ dBV) is required, two amplifier stages each with a gain of 12.65 ($+22$ dBV) should be used. The electronics bandwidth now exceeds the required 20 kHz. In high gain ac-circuits, dc offsets may have to be compensated at each stage to prevent saturation.

Signal Shielding can improve SNR by reducing cross-talk between signal channels as well as suppressing environmental electric fields. A basic understanding of field theory suggests that wrapping the signal wire in a coaxial conductor will completely suppress all environmental electric fields. However, it is actually more complicated due to the wave impedance of the interfering signal and the shield impedance, which has a small amount of resistance. Recall that the characteristic impedance of a plane electromagnetic wave in air is

$$Z_W = Z_0 = \sqrt{\frac{\mu_0}{\varepsilon_0}} = 377 \, \Omega \tag{15.1.6}$$

where $\mu_0 = 4\pi \times 10^{-7}$ H/m and $\epsilon_0 = 8.85 \times 10^{-12}$ F/m are the characteristic permeability and dielectric constants for free space (a vacuum or dry air). This makes air a pretty good insulator given that copper is about 17 nΩ. The impedance of a conductive material ($\sigma \gg j\omega\epsilon$, where σ is the conductivity in mhos) is

$$Z_S = \sqrt{\frac{2\pi f \mu}{2\sigma}}(1 + j) \tag{15.1.7}$$

The loss due to plane wave reflection at the conducting shield surface is

$$R = 20 \log \frac{|Z_W|}{4|Z_S|}$$
$$= 168 - 10 \log \left(\frac{\mu_r f}{\sigma_r} \right) \, dB \qquad (15.1.8)$$

where $\mu_r = \mu / \mu_0$ and $\sigma_r = \sigma / \sigma_0$ are the relative permeability and conductivity, respectively and $\sigma_0 = 5.82 \times 10^7$ ohms is the conductivity for copper. Both μ_r and σ_r are unity for copper in Eq. (15.1.8). At 100 Hz, there is about 148 dB reflection loss to a plane wave impinging on a copper shield. But, at 1 MHz, this drops to 108 dB which is still pretty good shielding. But, a plane wave in free space occurs approximately only after several wavelengths from a point source, a hundred or more wavelengths to be precisely a plane wave (recall Section 6.3). Since the wave speed is 300×10^6 m/sec, a hundred wavelengths at 60 Hz is 5 million meters! Since the radius of the earth is 6.3 million meters, we can safely say that power line frequencies (50–60 Hz) from the other side of the planet are plane waves, but any power line electromagnetic fields within about 10 km of our circuit are definitely not. However, radio and television signals around 100 MHz can be considered plane waves if the source is 100 m away or more.

Non-plane wave shielding is much more difficult mainly due to currents induced in the signal wire from magnetic field and the asymmetry of the field around the signal wire. The electrical nearfield ($r < \lambda / 2\pi$) wave impedance can be approximated by

$$|Z_W|_E \approx \frac{1}{2\pi f \varepsilon_0 r} \qquad (15.1.9)$$

where r is the distance in meters from our signal wire to the field source. Applying Eq. (15.1.8) with our nearfield wave impedance one obtains the approximate electrical reflection loss for the shield.

$$R_E = 321 + 10 \log \left(\frac{\sigma_r}{\mu_r} \right) - 10 \log(f^3 r^2) \, dB \qquad (15.1.10)$$

Equation (15.1.10) tells us that electrical shielding is fairly strong even at close distances. But, one must also consider that the field strength is also increasing as the signal wire is moved closer to the source. Magnetic shielding is not nearly as strong in the nearfield. The magnetic nearfield wave impedance ($r < \lambda / 2\pi$) is approximately

$$|Z_W|_H \approx 2\pi f \mu_0 r \qquad (15.1.11)$$

in free space. Again, applying Eq. (15.1.8) gives the approximate nearfield magnetic reflection loss.

$$R_H = 14.6 + 10 \log \left(\frac{\sigma_r}{\mu_r} \right) + 10 \log(f r^2) \, dB \qquad (15.1.12)$$

However, for thin shields multiple reflections of the magnetic wave within the shield must be accounted by adding a correction factor which should only be used for thin shields in the nearfield.

$$R_H = 14.6 + 10\log\left(\frac{\sigma_r}{\mu_r}\right) + 10\log(fr^2) + 20\log(1 - e^{-2t/\delta}) \ dB \qquad (5.1.13)$$

Note that for really small thickness t, the correction may cause R_H to be negative in Eq. (15.1.13) which is erroneous. For the cases where R_H is either negative or near zero in both Eqs (15.1.12) and (15.1.13), it is reasonable to use $R_H = 0$ dB (no magnetic field shielding) as a further approximation. The skin depth, δ, in meters is defined as

$$\sigma = \sqrt{\frac{2}{2\pi f \mu \sigma}} \qquad (15.1.14)$$

and describes the exponential decay in amplitude as the field penetrates the conductor. The skin depth effect also results in an absorption loss to the wave given in Eq. (15.1.15).

$$A = 20\log(e^{-t/\delta}) = 8.7\left(\frac{t}{\delta}\right) \ dB \qquad (15.1.15)$$

Equation (15.1.15) says that as a rule of thumb one gets about 9 dB absorption loss per skin depth thickness shield. (The "rule of thumb" dates back to medieval times and specifies the maximum thickness stick one may use to beat one's spouse.) For copper at about 100 Hz, the skin depth is a little over 1 cm, but even so, in the nearfield this is the dominant effect of shielding at low frequencies when close to the source! Some additional shielding is possible by surrounding the signal wire with a shroud of low reluctance magnetic material. Magnetic materials, sometimes known as "mu-metals, ferrite beads, donuts, life-savers, etc," are sometimes seen as a loop around a cable leaving or entering a box. The mu-metal tends to attract magnetic fields conducted along the wire, and then dissipate the energy through the resistance of the currents induced into the loop. They are most effective however, for high frequencies (i.e. MHz) only. Magnetic shielding at low frequencies is extremely difficult. For all shielding of components in boxes where one must have access holes, it is better to have a lot of small holes than one big hole of the same total area. This is because the leakage of the shield is determined by the maximum dimension of the biggest opening rather than the total area of all the openings. In general, for low frequency signal wires one must always assume that some small currents will be induced by nearby sources even when shielding is used. To minimize this problem, one should make the sensor signal cable circuit as low impedance as practical and make the signal current as strong as possible.

   Ground Loops occur when more than one common signal ground is present in the sensor system. Grounding problems typically are seen as the power line frequency (60 Hz in North America, 50 Hz in Europe, etc) and its harmonics. Note that if your sensor is detecting acoustic or vibration waves and you see multiples of *twice* the power line frequency in the power spectrum (i.e. 120, 240, 360 Hz, etc, in North America), you may actually be sensing the thermal expansion noise from a transformer or motor which is a real wave and not a ground loop. Direct current powered circuits can also suffer from ground loops. This is because our power distribution system generates electricity relative to an earth ground which is local to each generating station. The distribution system is grounded at many places, including the main circuit breaker box in any given dwelling. If one simply attaches the positive and negative leads of a spectrum analyzer to two points on a non-conducting floor one would very likely see these stray power line voltages. These voltages, which may be less than 1 mV, could be due to a current from the local ground rod, or may be a magnetically induced current from a nearby power line. When the chassis of a sensor data acquisition system comes in contact with the ground, or a part of a structure with magnetically-induced currents, there will be some small voltages actually on its electrical ground reference. It is for this reason that most electrical measurement and high fidelity multimedia gear make a separate signal ground connection available. The problem is made even more vexing when multiple sensors cabled to a distance from the data acquisition system come in contact with yet a slightly different ground potential. In this situation ground loops can occur between sensors as well as each sensor and the data acquisition system. As noted in the shielding description above, the cable itself becomes a source for magnetically induced power line voltages especially if it is a high impedance circuit (signal currents are as small as the magnetically-induced ones).

The solution for almost all ground loop and power line isolation problems is seen graphically in Figure 4 and starts by selecting a common single point for all sensor signal grounds. This is most conveniently done at the data acquisition system. The grounds to all sensors must therefore be "floated". This means that for a 2-wire sensor, a 2-wire plus shield cable is needed to extend the ground shield all the way out to the sensor, *but this cable shield must not be connected to any conductor at the sensor.* Furthermore, the sensor "ground lead" must not be grounded, but rather brought back along with the signal lead inside the shielded cable to the data acquisition system. The sensor case if conductive must also be insulated from the ground and the cable shield. With both the sensor signal (+) and sensor

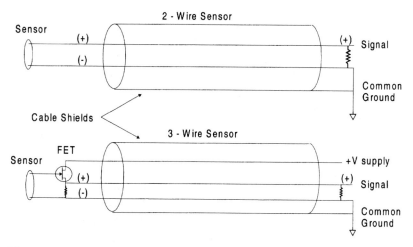

Figure 4 Proper ground isolation and shielding for 2-wire and 3-wire sensors to minimize interference and insure no ground loops are created.

ground (−) floated inside the shielded cable, any magnetically-induced currents are common to both wires, and will be small if the circuit impedance is low. At the front end of the data acquisition system, the common mode rejection of an instrumentation operational amplifier will significantly reduce or eliminate any induced currents, even if the sensor ground (−) is connected to the cable shield and data acquisition common ground at the acquisition input terminals. By floating all the sensor grounds inside a common shield and grounding all signals at the same point ground loops are eliminated and maximum shield effectiveness is achieved.

For a 3-wire sensor such as an electret microphone with FET preamplifier, one needs a 3-wired plus shield cable to float all 3 sensor wires (power, signal, and sensor "ground"). The FET should be wired as a current source to drive a low impedance cable circuit. At the data acquisition end, a small resistor is placed between the sensor signal (+) and sensor "ground" (−) wires to create a sensor current loop and to convert the current to a sensible voltage. Without this resistor, the impedance of the sensor signal-to-ground loop inside the cable shield would remain very high with a very weak sensor signal current. This weak sensor current could be corrupted by magnetic field-induced currents, thus, it is advisable to make the sensor signal current as strong as possible. Shot and contact noise-causing dc currents can be eliminated from the signal path by a simple blocking capacitor at one end of the sensor signal (+) wire. Using an FET at the sensor to transfer the high resistance of a transducer to the low resistance of a signal current loop is the preferred method for using high impedance sensors. That's why they're called transistors. The field-effect transistor (FET) is of particular value because its input impedance is very high. The resistors seen in Figure 4 should be chosen to match the impedance of the sensor (2-wire case) or the FET source resistor (3-wire case with a common drain n-channel FET shown). This maximizes the current in the twisted pair of conductors inside the cable shield, thus minimizing the relative strength of any interference. Note that if the shield ground isolation is lost at the sensor, interference voltages could appear from the potential difference between the ground at the data acquisition end and the sensor end of the cable.

15.2   ACOUSTICS AND VIBRATION SENSORS

Acoustics and vibration sensors are really only a small subset of sensors in use today, but they are by far the most interesting in so far as magnitude and phase calibration of wide band signals, transducer physics, and environmental effects. Our attention focuses on acoustics and vibration because antenna elements for electromagnetic wave sensing are basically a simple unshielded wire or coil, video sensors are just large arrays of photodiodes which integrate an rms measure of the photons, and low bandwidth sensors generally do not require a lot of sensor signal processing. The color correction/calibration problem for video signals is certainly interesting, but will be left for a future revision of this book. Correction of sensor nonlinearities (low bandwidth or video) is also an interesting and worthy problem, but it its really beyond the scope of this book, which is based almost totally on linear system signal processing. Acoustic and vibration sensors are used in a wide range of sensor signal processing applications and also require calibration periodically to insure sensor data integrity.

We must make a very useful distinction between sensors and transducers. A sensor detects a physical energy wave and produces a voltage proportional to the energy. But, sensors are not reversible, meaning that an electrical energy input also produces a proportional energy wave. A transducer passively converts physical wave energy from one form to another. Furthermore, a transducer is reversible, meaning for example, that if a mechanical wave input produces a voltage output, a voltage input produces a mechanical wave output. All linear passive transducers are reversible in this way. Transducers which are linear (meaning very low distortion but not necessarily constant frequency response) obey the laws of *reciprocity* for their respective transmitting and receiving sensitivities. In a conservative field if one transmits a wave at point A of amplitude X and receives a wave at point B of amplitude Y, the propagation loss being Y/X, one would have the same propagation loss Y/X by transmitting X from point B and receiving Y at point A. For reciprocal transducers, one could input a current I at point A and measure a voltage V at point B, and by subsequently inputting a current I in the same transducer at point B one would measure the same voltage V from the transducer still at point A.

Our discussion focuses initially on two main types of transducers. The first is the *electromagnetic mechanical* transducer and the second is the *electrostatic mechanical* transducer. These two types of transducer mechanisms cover the vast majority of microphones, loudspeakers, vibration sensors, and sonar systems in use today. Later, a brief discussion of *micro-electro-mechanical system* (*MEMS*) sensors will be given, which will also help clearly distinguish the physics between transducers and sensors.

The electromagnetic mechanical transducer is most commonly seen as a moving coil loudspeaker. There are literally billions of moving coil loudspeakers in the world today and has probably had one of the most significant impacts on society of any electronic device until the digital computer. The transduction mechanism in a loudspeaker is essentially the same as that in a "dynamic microphone", a geophone (used widely in seismic sensing and low frequency vibration), as well as electric motors and generators. Figure 5 shows the electromotive force (voltage) generated when a conductor has velocity u in a magnetic flux B in part (a), and the mechanical force f generated when a current i flows through the conductor in the magnetic field B in part (b). The orthogonal arrangement of the "voltage equals velocity times magnetic field" or "force equals current times magnetic field", follow the "right-hand-rule" orientation. This is where with the right-hand index finger, middle finger, and thumb pointing in orthogonal directions, the index finger represents velocity or current, the middle finger represents magnetic field, and the thumb points in the direction of the resulting voltage or force, respectively. The physical equations describing the relationship between electrical and mechanical parameters are

$$f = Bli \qquad e = Blu \qquad (15.2.1)$$

where i is the current in amperes, B is the magnetic field in Tesla, (1 Tesla equals 1 Weber per square meter, 1 Weber equals 1 Newton meter per ampere), l is the effective length of the conductor in the magnetic flux, e is the velocity-generated voltage, and u is the conductor velocity. One can see a very peculiar and important relation-

Electromagnetic - Mechanical Transducer

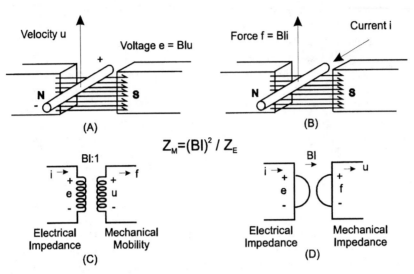

Figure 5 Voltage generated by moving conductor in magnetic flux (a), force generated by current in magnetic flux (b), transformer representation (c), and gyrator representation (d).

ship by examining the impedances defined by Eq. (15.2.1).

$$Z_M = \frac{f}{u} = (Bl)^2 \frac{1}{Z_E} = (Bl)^2 \frac{i}{e} \tag{15.2.2}$$

The mechanical impedance Z_M is seen to be proportional to the inverse of electrical impedance Z_E. This is not required for a transducer to be reversible, its just the result of the relationship between force and current, and voltage and velocity. It does create some challenging circuit analysis of the transducer as seen in Figure 6 for a complete loudspeaker system in a closed cabinet. Figure 7 shows a cutaway view of a typical loudspeaker.

Analysis of the loudspeaker is quite interesting. To maximize the efficiency of the force generation, the conductor is wound into a solenoid, or linear coil. This coil has electrical resistance Re and inductance Le as seen on the left in Figure 6 part (a). The movable coil surrounds either a permanent magnet or a ferrous pole piece, depending on the design. When a ceramic permanent magnet is used, it is usually a flat "washer-shaped" annular ring around the outside of the coil where a ferrous pole piece is inside the coil as seen in Figure 7. When a rare earth magnet such as Alnico (aluminum nickel and cobalt alloy) is used, its favored geometry usually allows it to be placed inside the coil and a ferrous return magnetic circuit is used on the outside. In some antique radios, an electromagnet is used in what was called an "electrodynamic loudspeaker". Clever circuit designs in the mid 1930s would use the high inductance of the electromagnet coil as part of the high voltage dc power supply, usually canceling ac power line hum in the process. A gyrator (there is no physical equivalent device of a "gyrator") is used in part (a) and (b) of Figure 6 to

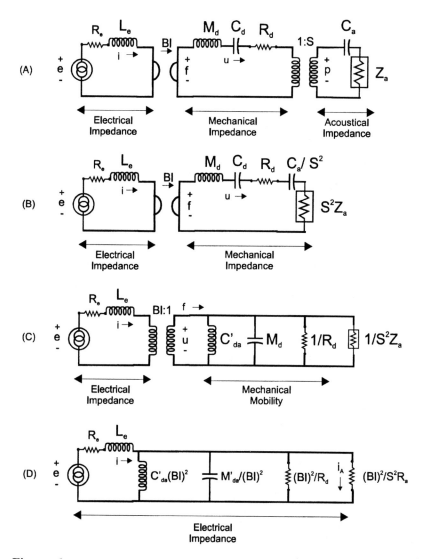

Figure 6 Equivalent circuit representations of a moving coil loudspeaker in a closed cabinet.

leave the mechanical and acoustical parts of the circuit in an impedance analogy. The coil is mechanically attached to a diaphragm, usually cone-shaped for high stiffness in the direction of motion. The diaphragm and coil assembly have mass Md. The diaphragm structure is attached to the frame of the loudspeaker, called the basket, at the coil through an accordion-like structure called the spider, and at the outer rim of the diaphragm through a structure called the surround, usually made of foam rubber. The spider and surround together has a compliance (inverse of mechanical stiffness) Cd, and mechanical damping Rd. The velocity of the diaphragm u produces an acoustic volume velocity $U = Su$ cubic meters per second, where S is the surface area of the diaphragm. The acoustic pressure p is the mech-

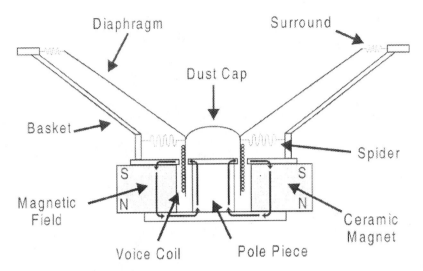

Figure 7 Typical loudspeaker cutaway showing the magnetic field, voice coil and diaphragm suspension where the driving force is created in the gap at the top of the pole piece.

anical force f divided by S. The air in the closed cabinet of the loudspeaker has acoustic compliance $C_a = V_b / \rho c^2$, where V_b is the volume in cubic meters, the density $\rho = 1.21 \, \mathrm{Kg/m^3}$ for air, and c is the speed of sound in air (typically 345 m/sec). Note that for an infinite-sized cabinet the capacitor equivalent is short circuited while if the cabinet were filled with cement (or other stiff material) the capacitor equivalent becomes an open circuit signifying no diaphragm motion. Finally, we have the acoustic radiation impedance Z_a defined as

$$ Z_a = \frac{\rho c}{S} \left(\frac{k^2 a^2}{4} + j0.6\,ka \right) = R_a + j\omega M_a \tag{15.2.3} $$

where $M_a = 0.6\rho a / S$ and $S = \pi a^2$ (diaphragm is assumed circular). Multiplying acoustical impedance by S^2 gives mechanical impedance, which can be seen on the right in part (b) of Figure 6. The mechanical compliance (the inverse of stiffness) of the cabinet air volume C_a / S^2 is combined with the mechanical compliance of the diaphragm C_d to give the total mechanical compliance C'_{da}.

$$ C'_{da} = \frac{C_d C_a}{S^2 C_d + C_a} \tag{15.2.4} $$

The combination of diaphragm compliance with the compliance of the air in the closed cabinet adds some stiffness to the loudspeaker system, raising the resonance frequency and giving rise to the popular term "acoustic suspension loudspeaker", owing to the fact that the air stiffness is the dominant part of the compliance.

To continue in our transducer analysis, we must convert the mechanical impedance to a mechanical mobility (inverse of impedance) as seen in part (c) of Figure 6 In a mobility analogy, the impedance of each element is inverted and the series circuit is replaced by a parallel circuit. Using a mechanical mobility analogy, a transformer with turns ratio Bl:1 represents a physical circuit device where the force in the mechanical system is a current proportional to the electrical input current i by the relation $f = Bli$ and the diaphragm velocity u is a voltage in the equivalent circuit directly to the physical applied voltage e by the relation $e = Blu$. Finally, part (d) of Figure 6 shows the equivalent electrical analogy circuit with the transformer proportionality reflected in the mechanical mobility elements. One literally could build this electrical circuit, using the physics-based element values for the mechanical parts, and measure exactly the same electrical input impedance as the actual loudspeaker. Figure 8 shows the modeled electrical input impedance for a typical 30 cm loudspeaker where $Re = 6$ Ω, $Le = 3$ mH, $Bl = 13.25$, $Md = 60$ gr, $Rd = 1.5$ mechanical Ω, the box volume V_b is 1 m³, and the driver resonance is 55 Hz. Figure 8 shows the electrical input impedance to the equivalent circuit for the loudspeaker in its closed cabinet.

The loudspeaker input impedance is a very interesting and useful quantity because it provides features indicative of many important physical parameters and it can be monitored while the loudspeaker is in use. One can simply monitor the voltage across a 1 Ω series resistor in series with the loudspeaker to get the current, and the voltage across the voice coil terminals and calculate a transfer function of voltage over current, or electrical impedance. In electrical impedance units, the input impedance is simply $Re + j\omega Le + (Bl)^2$ divided by the mechanical impedance of the loudspeaker. As easily seen in Figure 8 the impedance at 0 Hz is simply the voice coil resistance Re. At high frequencies, the slope of the impedance curve is Le. The total system mechanical resonance is also easily seen as the peak at

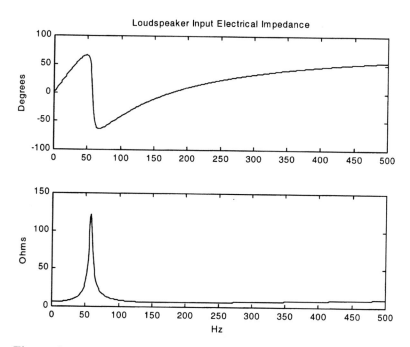

Figure 8 Electrical input impedance for a typical loudspeaker in a closed cabinet showing the total system resonance at 57.7 Hz, a dc resistance of 5 ohms, and a voice coil inductance of 3 mH.

57.7 Hz. The system resonance is slightly higher than the loudspeaker resonance of 55 Hz because of the added stiffness of the air inside the 1 m^3 cabinet. The total system resonance quality factor, Q_T, can be measured by experimentally determining the frequencies above and below resonance, where the impedance drops from Z_{res} (the value at resonance) to $Z(f_2^g, f_1^g) = \sqrt{Z_{res}R_e}$. This "geometric mean" is used because the total system Q_T is the parallel combination of the mechanical Q_{ms} and electrical Q_{es}. The system Q_T is correspondingly

$$Q_T = \frac{f_s}{f_2^g - f_1^g}\sqrt{\frac{Re}{Z_{res}}} \tag{15.2.5}$$

where f_2^g and f_1^g are the frequencies above and below resonance, respectively, where the impedance is $\sqrt{Z_{res}R_e}$.

One can measure the other loudspeaker parameters by simple experiment. To determine the diaphragm mass and compliance, one can simply add a known mass M' of clay to the diaphragm. If the resonance of the loudspeaker is f_0 and with the added mass is f_0', the mass and compliance of the diaphragm are simply

$$M_d = \frac{M'f_0^2}{f_0^2 - f_0'^2} \qquad C_d = \frac{1}{M_D 4\pi^2 f_0^2} \tag{15.2.6}$$

A similar approach would be to place the loudspeaker in a closed cabinet of precisely-known air volume and note the resonance change. To measure the magnetic motor force, one can apply a known dc current to the coil and carefully measure the deflection of the driver. The force is known given the compliance of the loudspeaker and measured displacement. Thus the "Bl factor" is obtained by direct experiment. However, it can also be indirectly measured by the following equation

$$Bl = \sqrt{\frac{\omega_s M_d R_e}{Q_T} - R_e R_d} \tag{15.2.7}$$

where R_d can be found from $R_d = (\omega_s M_D R_E / Q_T Z_{res})$.

Given all the physical parameters of a loudspeaker, one can design an efficient high fidelity loudspeaker system. We start by determining an analytical expression for the total system Q_T.

$$Q_T = \frac{2\pi f_s M_d Re}{B^2 l^2 + ReRd} \tag{15.2.8}$$

The Q_T factor is an important design parameter because is determines system response. If $Q_T = \frac{1}{2}$ the system is critically damped. If more low frequency response is desired, one may select a loudspeaker with $Q_T = 0.707$ which will align the pressure response to a maximally flat "2nd order Butterworth" low pass filter polynomial, as seen in Figure 9. The pressure response is analytically

$$\frac{p}{e} = \frac{Bl\, S\, R_a}{B^2 l^2 + (Re + j\omega Le)\left(j\omega M_d + \dfrac{1}{j\omega C_{da}'} + Rd + S^2 Z_a\right)} \tag{15.2.9}$$

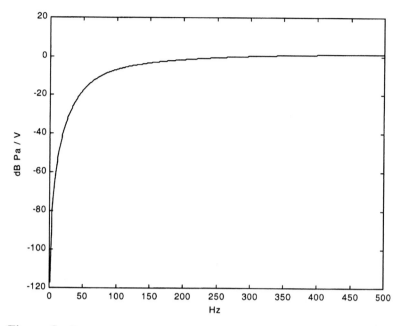

Figure 9 Pressure response for the loudspeaker example with $Q_T = 0.707$ for a maximally flat 2nd order low pass filter Butterworth alignment.

For a given loudspeaker, Q_T can be raised by raising the system resonance frequency f_s, which is easily done by using a smaller cabinet volume. However, as Q_T is raised above 0.707, the alignment is technically a Chebyshev, or "equal ripple" high-pass filter response. The response will peak at the system resonance as Q_T is increased above 1.0, but the effective low frequency bandwidth does not increase significantly. This suggests that for a given driver, there is an optimal cabinet size. Thiele and Small published a series of articles during the 1970s in the *Journal of the Audio Engineering Society* available today as part of an Anthology (3). There one can find similar 4th-order polynomial alignments for vented loudspeaker systems where an acoustical port is designed to extend the low frequency response.

While loudspeakers are definitely fun and certainly extremely common, one can also use this true transducer as a sensor. Since the diaphragm is relatively heavy compared to air, the motion of the diaphragm is caused mainly by pressure differences across the diaphragm. If the diaphragm is large, these forces can be quite significant. The force on the diaphragm due to an acoustic pressure wave is $F = Sp$, where S is the diaphragm area. The diaphragm velocity is this force divided by the mechanical impedance. Since $e = Blu$, we can simply write the voltage output of the loudspeaker due to an acoustic pressure on the diaphragm.

$$\frac{e}{p} = \frac{Bl\,S}{\dfrac{B^2 l^2}{Re + j\omega Le} + Rd + j\omega M_d + \dfrac{1}{j\omega C'_{da}}} \qquad (15.2.10)$$

Figure 10 shows the voltage response relative to acoustic pressure for our

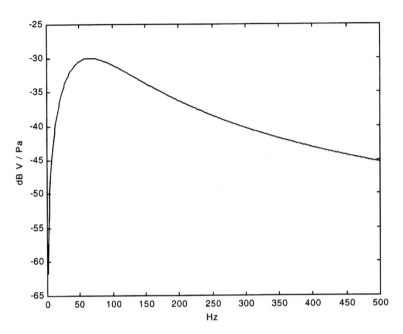

Figure 10   Voltage response of the loudspeaker when used as a "Dynamic Microphone".

loudspeaker example. Even though the loudspeaker Q_T is 0.707 the receiving response is anything but uniform. This is why loudspeakers are rarely used as microphones and vice-versa. One would either optimize the transducer design for a loudspeaker or optimize the design for a bandpass microphone rather than try to design a system that would work well for both.

However, moving coil microphones are quite popular in the music and communication industries mainly because of their inherent non-flat frequency response. For example, in aviation a microphone "tuned" to the speech range of 500 Hz to 3 kHz naturally suppresses background noise and enhances speech. The simple coil and magnetic design also provide an extremely robust microphone for the harsh environment (condensation, ice, high vibration, etc) of an aircraft. The "bullet shape" of aviator microphones is still seen today and the unique frequency response has become the signature of many blues harmonica players.

A variant of the dynamic microphone even more important to the broadcast and music industries is the ribbon microphone. These large microphones originally introduced in the 1930s are still in high demand today for very good reasons well beyond historical stylistic trends. The ribbon microphone is seen in a sketch in Figure 11 and works by direct sensing of acoustic velocity with a light-weight metal ribbon in a magnetic field. As the ribbon moves with the acoustic waves, an electrical current is created along the length of the ribbon from the magnetic field traversing across the face of the ribbon. A typical ribbon size is about 3 cm long and only about 1–2 mm wide. It is used above it's resonance and essentially has a perfectly flat frequency response. The deep rich sound of a voice from a ribbon microphone can make any radio announcer sound like a giant. Also contributing to the low frequencies, are nearfield acoustic velocity waves (refer to Section 6.3) which add to the volume

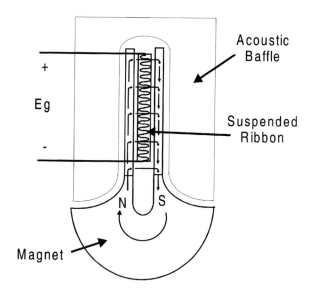

+

Eg

-

Acoustic
Baffle

Suspended
Ribbon

N   S

Magnet

Figure 11 A ribbon microphone is used to sense the acoustic velocity directly and also has a unique frequency response and dipole directivity response.

of wave energy in the proximity of the person talking into the microphone. Finally, since the ribbon has a dipole response pattern aided by the acoustic baffle around the edges of the ribbon, it can be oriented to suppress some unwanted sounds in the studio. However, because of its strong low frequency velocity response, the ribbon microphone is an extremely poor choice for outdoor use due to high wind noise susceptibility.

The ribbon microphone senses velocity directly, so the force on the ribbon is proportional to the ribbon mass times the acceleration due to the velocity $j\omega U/S$. The resulting mechanical velocity of the ribbon is this force divided by the mechanical impedance. Therefore, the velocity response is simply

$$\frac{e}{U} = \frac{Bl\,j\omega M_d}{\dfrac{B^2 l^2}{Re + j\omega Le} + Rd + j\omega M_d + \dfrac{1}{j\omega C'_{da}}} \qquad (15.2.10)$$

which can be seen in Figure 12 for a transducer widely known as a geophone. We also call this a "geophone" response because it represents the transducer's response to velocity. Geophones are the standard transducer used for seismic surveys and for oil exploration. Obviously, one would isolate a ribbon microphone well from vibration and design a geophone for minimal acoustic sensitivity by eliminating the diaphragm and completely enclosing the coil. A photograph of a commercially-available geophone is seen in Figure 13.

The electrostatic transducer is commonly used for wide frequency response condenser microphones, accelerometers for measuring vibrations over a wide frequency range, force gauges, electrostatic loudspeakers, and in sonar and ultrasonic imaging. The Greek word *piezen* means to press. The piezoelectric effect is found

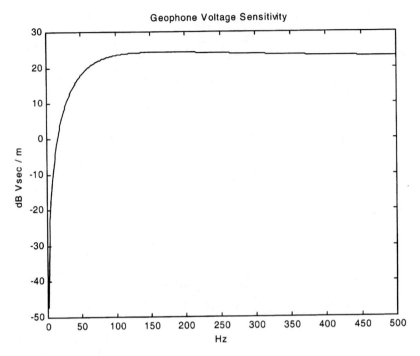

Figure 12 Voltage response of the loudspeaker as a pure velocity sensor such as a ribbon microphone or a geophone, commonly used for low frequency vibration measurements.

Figure 13 A cutaway in the case of a common geophone shows the enclosed coil which produces a high voltage sensitivity in response to low frequency velocity vibration (photo courtesy Oyo-Geospace).

in crystal and polymer transducers, in which when stress is applied a voltage is produced. Microphones and electrostatic loudspeaker transducers can be made based on the electrostatic charge on a lightweight movable plate capacitor (also known as a condenser). With one plate constructed as a lightweight diaphragm, movement of the diaphragm results in a corresponding change in capacitance, thus producing a subsequent change in voltage for a fixed amount of charge in the capacitor. Both piezoelectric and condenser-based transducers can be physically modeled using equivalent circuits. From Coulomb's law, we know that the force between two charges is directly proportional to their products and inversely proportional to the square of the distance between them $F = k\ q_1 q_2 / r^2$. This force can also be written as a charge times the electric field $F = qE$, where E has units Nt/C or V/m. For constant charge, there is a hyperbolic relation between the electric field and distance separating the charges. Therefore, one can linearize the relation between electric field and displacement for small changes, and to a greater extent, if a bias voltage or static force has been applied to the transducer. Since pressing into the piezoelectric crystal or on the capacitor plates causes a voltage increase, the linearized constitutive equations for small displacements and forces are

$$e = -\psi \xi \qquad f = \psi q \tag{15.2.12}$$

where ψ has physical units of Volts per m (V/m) or Newtons per coulomb (Nt/C). It can be seen that if one supplies charge to the device (input an electrical current $i = dq/dt$) it will produce an outward positive force. But, if the voltage potential is increased, there will be an inward displacement ξ. Assuming a sinusoidal excitation, the constitutive equations become

$$e = -\psi \frac{u}{j\omega} \qquad f = \psi \frac{i}{j\omega} \tag{15.2.13}$$

which can be seen as similar to the relations between mechanical and electrical impedances for the electromagnetic transducer ($f = Bli$ and $e = Blu$) with the exception of the sign.

Piezoceramic materials are materials which when a static electric field is applied, produce a dynamic voltage in response to stress waves. *Piezoelectric* materials are polarized either naturally or by manufacturing process so that only a small dc voltage bias is needed, mainly to insure no net charge leakage from the transducer. Piezoelectric materials require a little more detailed description of the physics. Natural piezoelectrics like quartz are crystals of molecules of silicon bonded to a pair of oxygen atoms denoted chemically as SiO_2. In the crystal lattice of atoms these molecules form helical rings. When the ring is compressed along the plane of the silicon atoms, an electric field is produced along the orthogonal direction from the extension of silicon's positive charge on one side (above the stress plane) and O_2's negative charge on the other side (below the stress plane). Clearly, there is a complicated relation between the various stress planes available in the crystal as well as shear planes. For our simplistic explanation, it is sufficient to say that the crystal can be cut appropriately into a parallelepiped such that the voltage measured across two opposing faces is proportional to the stress along an orthogonal pair of faces. Piezoelectric effects are also naturally found in Rochelle salt, but its susceptibility to moisture is undesirable. Man-made ceramics dominate the marketplace for piezoelectric transducers because they can be molded into desirable shapes and have their electrodes precisely bonded as part of the ceramic firing process. Polarization of the ceramic is done by heating the material above a temperature known as the Curie temperature where the molecular orientations are somewhat mobile, and then applying a strong electrostatic field while cooling the transducer. Polarization can thus be applied in the desired direction allowing the transducer to respond electrically to orthogonal stress (called the 1–3 mode) or even to longitudal stress (in-line or 1–1 mode). However, when the ceramic is stressed in one direction, the other two orthogonal directions also change, affecting the coupling efficiency. Some of the formulas for man-made piezoceramics are PZT (lead-zirconate-titanate), $BaTiO_3$ (barium-titanate), and a very unique plastic film called PVDF (polyvinylidene fluoride), which also has a pyroelectric (heat sensing) effect used commonly in infrared motion sensors.

With the exception of PVDF, piezoelectrics are generally very mechanically stiff and have very high electrical resistance due to the strong covalent bonds of the molecules. This means that a generated force has to overcome the high stiffness of the ceramic to produce a displacement. Thus piezoelectrics can produce a high force but only if very low displacements occur. Conversely, a force applied to a piezoelectric will produce a high voltage but very little current. If the electricity produced is harvested as a power source or dissipated as heat from a resistor, the work done is seen mechanically as damping. An example of electronic material damping can be seen in some high performance downhill skis, which have embedded piezoelectrics and light-emitting diodes to dissipate the vibrational energy. Another common application of piezoelectrics is a mechanical spark generator on for gas-fired barbeque grills. But perhaps the most technically interesting application of piezoelectric ceramics is as a force, strain, acceleration, or acoustic sensor, or as an acoustic pressure or mechanical force generator.

To analyze the transducer response, we introduce the *Mason equivalent circuits* for plane waves in a block of material with cross-sectional area S, thickness l, density ρ, and wave speed c in Figure 14 where the mechanical wavenumber k_m is ω/c. If material damping is significant, the mechanical wavenumber is complex. For relatively low frequencies where the wavelength λ is much larger than any dimension of the material block, the inductors and capacitor equivalents are shown in Figure 14. However, at high frequencies where the block becomes resonant, these parameters will change sign accordingly representing the material as a waveguide rather than as a lumped element. The equation for the "mass" components M_1 seen on the left in Figure 14 is

$$Z_1 = j\rho_m c_m S \tan\left(\frac{k_m l}{2}\right) \approx j\omega \frac{\rho_m S l}{2} \quad \lambda \gg l \tag{15.2.14}$$

such that M_1 in Figure 14 represents actually half the mass of the block of material. The complementary impedance for the left circuit is

$$Z_1^c + \frac{-j\rho_m c_m S}{\sin\left(\frac{k_m l}{2}\right)} \approx \frac{1}{j\omega \dfrac{l}{\rho_m c_m^2 S}} \quad \lambda \gg l \tag{15.2.15}$$

such that the mechanical compliance is seen to be $1/\{\rho_m c_m^2 S\}$, or the stiffness is

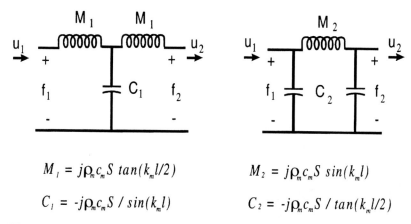

$$M_1 = j\rho_m c_m S \, tan(k_m l/2)$$

$$C_1 = -j\rho_m c_m S / sin(k_m l)$$

$$M_2 = j\rho_m c_m S \, sin(k_m l)$$

$$C_2 = -j\rho_m c_m S / tan(k_m l/2)$$

Figure 14 Mason equivalent circuits for a block of material of density ρ_m, plane wave speed c_m, area S, and length l showing the low frequency approximation symbols for capacitor and inductor.

$\rho_m c_m^2 S/l$ which has the proper units of Kg/sec. Either circuit in Figure 14 can be used, the choice being made based on which form is more algebraically convenient. The low frequency approximations given in Eqs (15.2.14) and (15.2.15) are quite convenient for frequencies where geometry of the block is much smaller than a wavelength. For high frequencies, the block behaves as a waveguide where for any given frequency it can be either inductive or capacitive depending on the frequency's relation to the nearest resonance.

Applying the Mason equivalent circuit to an accelerometer is straightforward. As seen in Figure 15, it is quite typical to mount a proof mass (sometimes called a seismic mass) on the transducer material (referred to here as PZT) using a compression bolt or "stud". The compression stud makes a robust sensor structure and is considerably less stiff than the PZT. Its presence does contribute to a high frequency resonance, but with clever design, this resonance can be quite small and well away from the frequency range of use.

The electrical part of the accelerometer is seen in Figure 15 as a capacitor C_0 connected through a transformer with turns ratio 1: $C_0\phi$. For our example here, we will assume the resistance of the piezoceramic crystal R_0 is similar to quartz at 75×10^{16} Ω or greater (750 peta-ohms, therefore we can ignore the resistance) and the electrical capacitance is 1600 pF. If the resistance R_0 is less than about 1 $G\Omega$, it will significantly cut the ultra low frequency response which can be seen as well for condenser microphones later in this section. One can appreciate the subtleties of piezoelectric transducers by considering that the crystal contributes both an electrical capacitance and a mechanical compliance which are coupled together (4). The unusual turns ratio in the transformer in Figure 15 provides a tool to physically model the piezoelectric circuit. For example, if one applies a current i to the electrical terminals, a voltage appears across the capacitor C_0 of $i/j\omega C_0$, which subsequently produces an open circuit force of $f = (\phi C_0) \, i/j\omega C_0 = \phi i/j\omega$, as expected assuming the mechanical velocity is blocked ($u = 0$). Going the other way, a mechanical velocity u_c on the right side of the transformer produces an open circuit current through C_0 of $-uC_0\phi$. This gen-

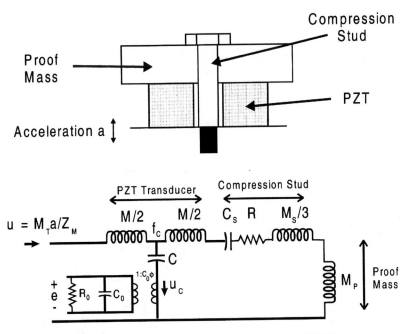

Figure 15 Physical sketch and block diagram of a typical PZT accelerometer constructed with a compression stud showing the electrical circuit interface through a transformer with a transformer with turns ratio $1:C_0\phi$.

erates an open circuit voltage of $e = -u_c C_0 \phi/j\omega C_0 = -\phi u_c/j\omega$, as expected. The effect of the electrical capacitance on the mechanical side of the transformer is a series compliance of value $1/(C_0 \phi^2)$. In other words, the mechanical impedance effect of the electrical capacitor is $C_0 \phi^2/j\omega$. For our example we assume a value of ϕ of 1×10^9 Nt/C or V/m. The coupling coefficient $d_{31} = 1/\phi$ is therefore 1000 pC/Nt. The coupling coefficient d_{31} is a common piezoelectric transducer parameter signifying the orthogonal relation between the voltage and stress wave directions. The crystal actually is sensitive to and produces forces in the same direction as the voltage giving a d_{11} coupling coefficient as well. Manufacturing process and mechanical design generally make the transducer most sensitive in one mode or the other.

The parameters M and C in Figure 15 represent the mass and compliance of the PZT material block. We assume a density of 1750 Kg/m^3, a plane longitudal wave speed of 1×10^7 m/sec, a cylindrical diameter of 1/4" and length of 1/4". This gives a mechanical compliance of 1.1458×10^{-15} and a mass of 3.5192×10^{-4} Kg. The compression stud is assumed to be stainless steel with a density of 7750 Kg/m^3 and wave speed of 7000 m/sec. The stud is cylindrical, 1/8" in diameter, and 1/4" long, giving a mass of 3.8963×10^{-4} Kg and a compliance of 2.112×10^{-9} s/Kg. It can be shown using kinetic energy that exactly 1/3 of the spring mass contributes to the mass terms in the mechanical impedance. A small mechanical damping of 400 Ω is used and the proof mass is 2.5 grams. As the base of the accelerometer accelerates dynamically, the force applied to the PZT is simply the total mass times the acceleration. The velocity into the base is this applied force divided by the total mechanical impedance Z_M.

$$u = \frac{F}{Z_M} = \frac{a\left(M + M_p + \dfrac{M_s}{3}\right)}{Z_M} \tag{15.2.16}$$

The mechanical impedance looking to the right at the node defined by f_c in Figure 15 is the parallel combination of the branch with C and the electrical mobilities and the right-hand branch.

$$Z_c = \frac{\left(\dfrac{1}{j\omega C} + \dfrac{C_0 \phi^2}{j\omega}\right)\left(j\omega\left[\dfrac{M}{2} + \dfrac{M_s}{3} + M_p\right] + R + \dfrac{1}{j\omega C_s}\right)}{\dfrac{1}{j\omega C} + \dfrac{\phi^2 C_0}{j\omega} + \dfrac{1}{j\omega C_s} + R' + j\omega\left[\dfrac{M}{2} + \dfrac{M_s}{3} + M_p\right]} \tag{15.2.17}$$

The total input mechanical impedance from the base of the accelerometer is $Z_M = j\omega M/2 + Z_c$. It can be seen that the velocity u_c in the PZT is

$$u_c = \frac{a\left[M + M_p + \dfrac{M_s}{3}\right]Z_c}{Z_M\left(\dfrac{1}{j\omega C} + \dfrac{C_0 \phi^2}{j\omega}\right)} \tag{15.2.18}$$

where a is the base acceleration. Using Eq. (15.2.13) we can solve for the voltage sensitivity in Volts per m/sec^2 (multiplying by 9810 gives the sensitivity in mV/g).

$$\frac{e}{a} = \frac{-\phi\left[M + M_p + \dfrac{M_s}{3}\right]Z_c}{j\omega Z_M\left(\dfrac{1}{j\omega C} + \dfrac{C_0 \phi^2}{j\omega}\right)} \tag{15.2.19}$$

Figure 16 shows the calculated voltage sensitivity for our example in mVper g.

Piezoelectric materials are also widely used as transmitters and receivers for underwater sound and as force sensors and actuators. To convert from force to acoustic pressure, one simply divides the force by the diaphragm area radiating the sound. For the underwater acoustic radiation impedance, apply Eq. (15.2.3) with $\rho = 1025$ Kg/m^3 and $c = 1500$ m/sec. It can be seen that a force gauge and hydrophone receiver are essentially the same transducer hydrophone transmitter and a force actuator will have a response depending on the load impedance. To convert from an accelerometer to a hydrophone or force transducer, replace the compression stud and proof mass with a general load impedance Z_L. This has the effect of shifting the transducer resonance much higher in frequency. If the transducer is blocked on the left, the PZT mass inductor on the left can be ignored. Otherwise, if the left side is free floating, the PZT mass inductor is attached to ground on the left. To simplify the algebra, we will assume we have a blocked transducer. Also to simplify

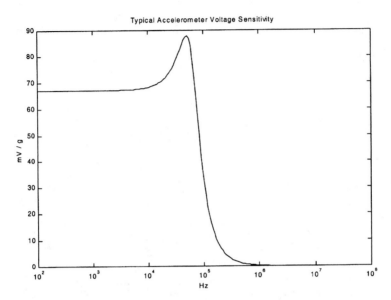

Figure 16 Typical accelerometer receiving voltage response in mV/g showing the resonance from the stud and proof mass and an extremely flat response at low frequencies.

our algebra, let

$$Z_M = \frac{1}{j\omega C} + j\omega \frac{M}{2} + R \tag{15.2.20}$$

where C is the PZT mechanical compliance, M is the PZT mass, and R is the transducer mechanical damping. Applying a current i, the force f_c in Figure 15 is seen to be

$$f_c = \frac{i\dfrac{\phi}{j\omega}(Z_M + Z_L)}{\dfrac{C_0\phi^2}{j\omega} + Z_M + Z_L} \tag{15.2.21}$$

The resulting mechanical velocity through the load is

$$U_M = \frac{i\dfrac{\phi}{j\omega}}{\dfrac{C_0\phi^2}{j\omega} + Z_M + Z_L} \tag{15.2.22}$$

The resulting force across the load impedance relative to the current is

$$\frac{f_L}{i} = \frac{Z_L}{C_0\phi + \dfrac{j\omega}{\phi}(Z_M + Z_L)} \tag{15.2.23}$$

Figure 17 shows the force actuation response when the load is a 1 Kg mass. As seen in

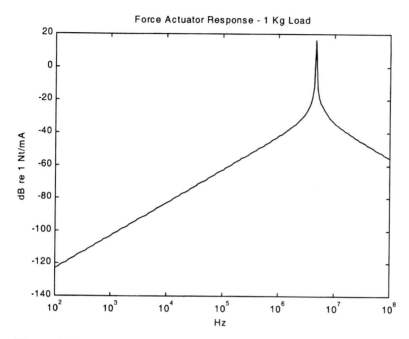

Figure 17 Force response relative to applied current for the PZT driving a 1 Kg mass.

the response plot, at resonance a substantial amount of force can be generated. For a hydrophone transmitter, we replace Z_L with Z_a/S^2 for water. However, it is more interesting to examine the pressure response relative to applied voltage. To convert current to voltage we need the total electrical input impedance.

$$Z_{ET} = \frac{1}{j\omega C_0} + \frac{\frac{\phi^2}{\omega^2}}{Z_M + Z_L + \frac{C_0\phi^2}{j\omega}} \qquad (15.2.24)$$

The third term in the denominator of Eq. (15..2.24) is relatively small and can be ignored. Actually, the input electrical impedance is dominated by the electrical capacitance C_0. The acoustic pressure radiated to the far field assuming a narrow beamwidth is

$$\frac{P}{e} \frac{R_a}{SZ_{ET}\left[C_0\phi + \frac{j\omega}{\phi}(Z_M + Z_a)\right]} \qquad (15.2.25)$$

where R_a is the real part of the acoustic radiation impedance defined in Eq. (15.2.3). Figure 18 presents a plot of the transmitted pressure response in water.

For a PZT force gauge or hydrophone receiver, one calculates to applied force on the PZT (across the load impedance) and solves for the velocity through the crystal by dividing by the mechanical impedance plus the mechanical compliance of the electrical capacitor. Given the velocity, it is straightforward to derive the

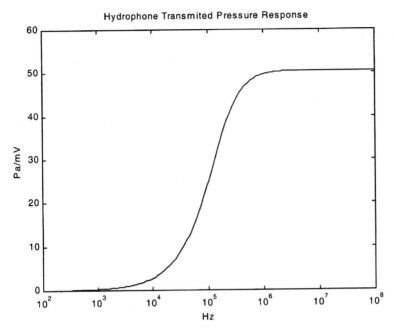

Figure 18 Transmitted acoustic pressure response relative to applied voltage for a PZT hydrophone.

voltage on the transducer's terminals due to the applied force. Further dividing by the diaphragm area S gives the voltage response to acoustic pressure, rather than mechanical force.

$$\frac{e}{p} = \frac{-\phi}{j\omega S\left[\dfrac{C_0\phi^2}{j\omega} + Z_M\right]}$$
(15.2.26)

Figure 19 shows the voltage response for our example PZT in a receiving hydrophone configuration. Clearly, one would want to add significant damping for the hydrophone receiver and force actuator if a broadband response is desired. The hydrophone transmitter does not need any additional mechanical damping because of the high resistive component of the radiation impedance into the water. Since the radiation resistance increases as ω^2, the transmitting pressure response of the hydrophone remains constant above the resonance, just as is the case with a common loudspeaker.

It is useful to discuss why the resonance of the accelerometer is so different from the receiving hydrophone and why adding a compression stud to a hydrophone or force transducer has little affect on the frequency response. Without the compression stud, the accelerometer's response would resemble its electrical input impedance, which is essentially a capacitor. The stud-less response would thus show the maximum response at 0 Hz and then fall as $1/\omega$ as frequency increases. This makes our accelerometer a very sensitive gravity sensor provided a dc-coupled amplifier is used which does not drain any net charge from the accelerometer.

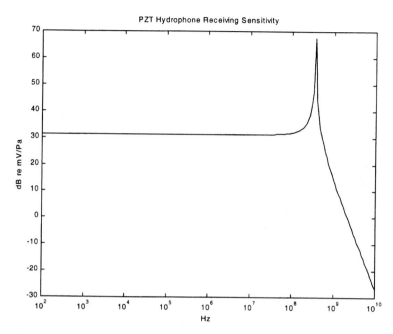

Figure 19 Receiving pressure sensitivity for our (undamped) PZT in a hydrophone configuration.

The compression stud, if infinitely stiff, would block any velocity through the crystal and if infinitely compliant would vanish from the equivalent circuit. Thus, a series capacitor between the PZT and the proof mass is the appropriate circuit model. The effect of the compression stud is to shift the accelerometer resonance up from 0 Hz while also bringing down the low frequency voltage sensitivity to a value that is constant for acceleration frequencies below the resonance. Note that for vibrations with a given velocity, at low frequencies the displacement will be large and at high frequencies the acceleration will be large. Generally, this makes force, displacement, and strain gauges the natural choice for dc and ultra low frequency measurements, geophones the good choice for measurements from a few Hz to a few hundred Hz, and accelerometers the best choice for almost high frequency measurements. However, accelerometer technology is a very rapidly developing field and new technologies are extending the bandwidth, environmental robustness, and dynamic range. For example, some "seismic" accelerometers are recently on the market with voltage sensitivities as high as 10 V/g with excellent response down to less than 1 Hz. Generally these low-frequency accelerometers configure the PZT into a layered cantilever sandwich with the proof mass on the free end of the cantilever. Stacking the PZT layers and constructively adding the voltages from compression and tension makes a very sensitive transducer.

   The condenser microphone is physically a close cousin to the piezoelectric transducer. It has a nearly identical equivalent electro-mechanical circuit as the piezoelectric transducer and in actuator form is known as an electrostatic loudspeaker. The "condenser" transducer is optimized for air acoustics by simply replacing the crystal with a charged electrical capacitor with one plate configured as a light-weight movable diaphragm. Configuring the movable diaphragm as one surface of a sealed air-filled chamber, the diaphragm will move from the applied force due to acoustic waves. As the diaphragm moves, the capacitance changes, so for a fixed amount of charge stored in the condenser one obtains a dynamic voltage which is linear and constant with the acoustic pressure for frequencies up to the resonance of the diaphragm. This is analogous to the response curve in Figure 19 for the receiving hydrophone or force gauge. The stiffness of the condenser microphone is found from the tension of the diaphragm and the acoustic compliance of the chamber. However, since 1 atmosphere is 101,300 Pa, and 1 Pa rms is equivalent to about 94 dB relative to 20 μPa, a small "capillary vent" is added to the microphone chamber to equalize barometric pressure changes and to allow for high acoustic sensitivity. We also include a shunt resistor in the circuit for a very good reason. Since the capillary vent is open to the atmosphere, the condenser microphone cannot be hermetically sealed from the environment. Water condensation or direct moisture can get into the microphone cavity and significantly change the electrical resistance. It will be seen that moisture, condensation, or damage to the vent or cavity can cause a very significant change to the condenser microphone's low frequency response, and in particular, its low frequency phase response. We note that temperature is also a significant environmental factor affecting diaphragm tension, air density, and resistance. Some scientific-grade condenser microphones are heated slightly above room temperature as a means to hold the sensitivity response constant during calibrations and measurements.

Figure 20 depicts a typical condenser microphone design and its equivalent circuit showing a thin tensioned membrane as the diaphragm and the back plate with an applied dc voltage bias. This back plate usually has small holes or annular slots to let the air easily move. Some felt or other damping material is used to suppress these and other high frequency acoustic resonances of the cavity structure,

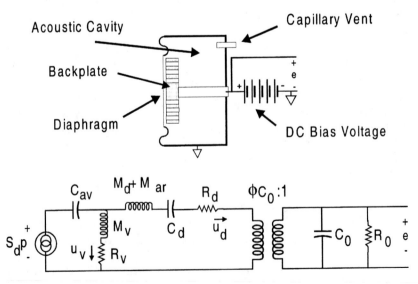

Figure 20 Condenser microphone and equivalent circuit showing acoustic cavity, capillary vent, bias voltage supply, and mechanical velocities.

and we limit the fidelity of our model to the basic physics. The small capillary vent in the cavity allows for barometric pressure equalization and also causes a low frequency sensitivity reduction. The force acting on the movable part of the cavity (the diaphragm and vent hole) is approximately $S_d p$, where p is the acoustic pressure and $S_d = 4 \times 10^{-6}$ m^2 is the diaphragm area for our example condenser microphone. This force results in a diaphragm velocity u_d, and vent velocity u_v, while a much smaller force $S_v p$, where $S_v = \pi a_v$ is the vent area, that causes a vent velocity u_v to be compensated slightly. The vent velocity takes away from the diaphragm velocity at low frequencies, which directly weakens voltage sensitivity e at the microphone terminals at low frequencies. The mechanical compliance of the cavity behind the diaphragm is $C_{av} = V/(\rho c^2 S_d^2)$ where V is 5×10^{-7} m^3 for our example. The mechanical response of the diaphragm contains the damping $R_d = 100$ Ω, diaphragm compliance $C_d = 6.3326 \times 10^{-8}$ (free space diaphragm resonance is 20 kHz), diaphragm mass $M_d = 1$ gram, and the mass of the air which moves with the diaphragm $M_{ar} = 0.6 \rho a_d S_d$, where $S_d = \pi a_d^2$ as seen from the radiation reactance in Eq. (15.2.3). The air-mass is included because its mass is significant relative to the lightweight diaphragm.

The capillary vent requires that we present some physics of viscous damping. For a cylindrical vent tube of length L_v and area $S_v = \pi a_v$, some of the fluid sticks to the walls of the tube affecting both the resistance and mass flow. When the tube area is small the frictional losses due to viscosity and the moving mass are seen in acoustic impedance units to be

$$Z_{va} = \left[\frac{8\eta L_v}{\pi a_v^4} \right] + j\omega \left[\frac{4\rho L_v}{3\pi a_v^2} \right] \tag{15.2.27}$$

where $\eta = 181$ μ Poise for air at room temperature. There are forces acting on both ends of the capillary vent, $S_v p$ on the outside and the much stronger force $S_d p$ through the cavity. We can simplify our analysis by ignoring the small counteracting force $S_v p$. The mechanical impedance of the vent is therefore

$$Z_v + R_v + j\omega M_v = \left[\frac{8\eta L_v S_d^2}{\pi a_v^4} \right] + j\omega \left[\frac{4\rho L_v S_d^2}{3\pi a_v^2} \right] \tag{15.2.28}$$

because the volume velocity through the vent is determined by the diaphragm, not the capillary vent area. The length of our example capillary tube is one mm and several vent areas will be analyzed. In our example, the electro-mechanical coupling factor is 360,000 V/m, which may sound like a lot, but its generated by a dc bias voltage of 36 V over a gap of 0.1 mm between the diaphragm and the back plate. This gives an electrical capacitor value of 0.354 pF since our diaphragm area is 4×10^{-6} m^2.

Figure 21 shows our example "typical" condenser microphone frequency response from 1 Hz to 100 kHz in units of mV/Pa and degrees phase shift for a range of shunt resistances. The 10 GΩ shunt is typical of a high quality professional audio condenser microphone where the resistance is provided by an audio line transformer. The 100 GΩ shunt is typical of a scientific-grade condenser microphone and is usually the gate impedance of an FET (field-effect transistor) preamplifier attached directly to the microphone terminals. The infrasonic condenser microphone requires an extraordinarily high shunt resistance to maintain its low frequency response.

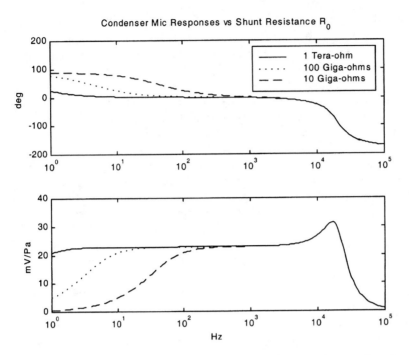

Condenser Mic Responses vs Shunt Resistance R_0

Figure 21   Magnitude and phase response of a typical condenser microphone showing the effect of input shunt resistance changes which can occur as the result of moisture of condensation.

Infrasonic microphones can be used to measure earthquakes, nuclear explosions, and even tornadoes from thousands of kilometers distance. This is because there is very little absorption in the atmosphere below 10 Hz. However, it is so difficult to maintain the required shunt resistance, alternative parametric techniques such as carrier wave modulation are used. To observe the diaphragm motions in the infrasonic range, one can measure the phase shift of a radio-frequency electromagnetic signal passed through the capacitor. This straightforward modulation technique is widely used in micro electro-mechanical systems (MEMS) as an indirect means to observe the motions of silicon structures.

Figure 22 shows our example condenser microphone frequency responses as a function of capillary vent size. The capillary vents are all 1 mm long and the cross-sectional area is varied to explicitly show the sensitivity of the magnitude and phase responses. The shunt resistance R_0 is taken as 1×10^{15} Ω to insure it is not a factor in the simulation. The most important observation seen in both Figure 21 and 22 is the very significant phase changes seen all the way up to about 1 kHz. If the condenser microphone is part of an acoustic intensity sensor, or part of a low frequency beamforming array, this phase response variation can be very detrimental to system performance. The effect of the phase error is compounded by the fact that the observed acoustic phase differences at low frequencies across an array might well already be quite small. A small phase difference signal with a large phase bias will likely lead to unusable results. Again, the most common threat to condenser microphones is moisture. But, the vent size sensitivity also indicates that damage

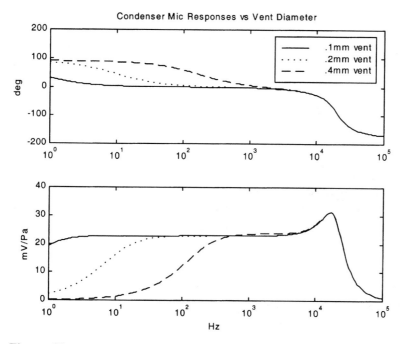

Figure 22 Typical condenser microphone response showing the effect of various capillary vent diameter changes to the sensitivity magnitude and phase.

to the microphone housing or vent can have a profound effect on the microphone response. Note that the shunt resistance and all sources of mechanical and acoustic damping contribute to the thermal "Johnson" noise of the transducer.

One of the major engineering breakthroughs in audio which led to widespread inexpensive condenser-type microphones was the development of the electret microphone (5). There are a number of possible configurations and respective patents, but the "electret" material is essentially a metallized foil over an insulating material which is permanently polarized. The metallic side of the electret sheet serves as an electrode. The electret sheet can be directly stretched over a backplate (see Figure 20) of a condenser microphone and the charge of the electret reduces the required bias voltage. The electret material can also be place on the backplate with the charged insulated side of the sheet facing the metal diaphragm of the condenser microphone in the so-called "back-electret" configuration. But the most inexpensive to manufacture configuration has the electret material placed directly on an insulating sheet attached to the backplate. As with the condenser microphone, holes in the backplate allow sound to enter a cavity behind the backplate, thus allowing the diaphragm to move in places. No tension is required in the diaphragm and this extremely simple arrangement can produce a reasonable sensitivity, signal to noise ratio, and frequency response at very low cost. Electret style condenser microphones can be found is almost every small microphone, particularly in telephones.

Micro Electro-Mechanical Systems (MEMS) are a fascinating product of advancements in silicon manufacturing at the nanometer scale during the last 20 years of the 20th century. Recall that integrated circuits are manufactured as layers of *p*-doped silicon, *n*-doped silicon, insulating glass (silicon dioxide), and wires (typically aluminum). The actual circuit weaves its way through the layers on the chip as part of the artful design which has brought us inexpensive microprocessors with literally hundreds of millions of transistors. Micro-machining is a variant of large-scale integrated circuits where 3-dimensional structures are created out of silicon. Micro-machining is not done using a microlathe any more than the first table-top radios had micro-sized musicians inside. Micro-machining is done by etching away layers of material to create 3-dimensional structures. Analogy would be to create a 3-dimensional casting of various metals by making a sand mold, digging out the desired shape, casting the metal, then stacking another sand mold on top, digging out the desired shape for that layer, casting the metal, and so-on. When the casting is complete, the sand is washed away to reveal the complex 3-dimensional structure. For a silicon structure, the layering is done by diffusion doping (kiln process), vacuum vapor deposition, electron beam epitaxy, to name just a few methods. In micro-machining, some of these layers are made from materials which can be dissolved using various acids. One can subsequently remove layers from the structure, analogous to removal of the sand from the metal casting, to produce remarkable structures on a nano-meter scale. Scientists have built small gears, pumps, electric motors, gyros, and even a turbine! One of the most widely used MEMS sensors in use today is the micro-machined accelerometer which is used as a crash sensor to deploy a driver air bag in an automobile.

The MEMS accelerometer essentially has a movable silicon structure where the motion can be sensed electronically. The electronic interface circuitry is integrated directly into the surrounding silicon around the MEMS sensor. The sensing of motion is typically done by either measuring changes in capacitance between the movable and non-movable parts of the silicon, or by sensing changes in resistance of the silicon members undergoing strain. Figure 23 shows a MEMS accelerometer which measures the changes in resistance in several thin members attached to a movable trapezoidal cross-section of silicon. By combining the resistance of the silicon members as part of a Wheatstone bridge, temperature compensation is also achieved. This MEMS sensor is a true dc device and therefore can measure gravity and orientation of an object relative to gravity. However, because it has relatively high resistance and requires a net dc current to operate, the intrinsic electronic noise (see Section 15.1) of the device can be a design concern. The more common micro-machined accelerometer is the capacitance sensing type, where the capacitance between the movable and non-movable silicon is measured in much the same way as the operation of a condenser microphone. An FET or high input impedance operational amplifier is used in current amplifier mode to produce a strong sensor voltage. Again, because of the electronic elements and high resistance, electronic background noise is a design concern.

MEMS sensors are going to be important for the foreseeable future even if they are not as good as a traditional transducer for reasons of cost, shock robustness, and obviously, their small size. Because the layering and etching processes can be highly automated, large numbers of MEMS sensors can be made for very low cost and with very high mechanical precision. Shock robustness is a slightly more elusive feature of MEMS, because it requires thoughtful mechanical design. However, if a material can withstand a certain stress before breaking, the amount of acceleration

Figure 23 A strain-type accelerometer cut-away showing the movable trapezoidal section and the thin resistance bridges used to detect the deflection due to gravity or motion due to acceleration (photomicrograph courtesy of EG&G IC Sensors).

tolerable can be significantly increased if the mass of the structure can be reduced. Therefore, very small stiff structures can be inherently very difficult to break. This is one reason why the micro-machined accelerometer is used as a crash impact detector in automobiles. The initial shock of the crash can conceivably crack the piezoelectric crystal before the deceleration signal of the collapsing car body is sensed to cause the proper deployment of the air bag. MEMS sensors will likely have a role in many unusual and extreme applications ranging from sports equipment to heavy manufacturing and construction process controls. Finally, the small size of MEMS sensors has several intriguing aspects worth noting. Small size allows for use in less invasive medical operations such as catheterization sensors, implant-able sensors, and cellular biology technology. Another artifact of small size is the ability to generate very high voltage fields as well as high frequency mechanical resonances. Very recently, material scientists have been able to grow piezoelectric materials on silicon MEMS structures which opens a whole new avenue for development.

Electronic sensor interface circuits are worthy of discussion especially for the capacitive-type sensors of piezoelectric transducers, condenser and electret microphones, and MEMS sensors. Because these sensors all have very high output impedance, they can deliver only a very small current. Under these conditions, a cable connecting the transducer to the data acquisition system is subject to electromagnetic interference. As noted in Section 15.1, magnetic fields will induce small currents into these high-impedance cables, which are comparatively large to the small sensor currents. There are two practical choices to correct the weak signal current: (a) use a transformer to increase the current (the sensor voltage will be proportionately stepped down), or; (b) convert the line impedance at the sensor using a junction field effect (JFET) transistor. The JFET has extremely high input resistance and thus will draw little current from the sensor. An n-channel JFET works by controlling the conductivity of a bar of n-type silicon by inducing a charge depletion region with an applied "gate" (p-type gate) voltage across the path of current flow along the bar. The current through the source and drain (at their respective ends of the bar) can be varied by the gate voltage between the gate and source. As the gate voltage is driven negative, the depleted charge carriers in the bar cause the n-type silicon to change from a conductor to an insulator, or more appropriately, behave as a semiconductor. By supplying a dc current and appropriately biasing the gate voltage of the JFET, small fluctuations of the gate voltage directly vary the drain current, which is of course much stronger than the current from the high impedance sensor. By locating the JFET as close as possible to the high impedance sensor, the size of the "antenna" which can pick up electromagnetic fields is greatly reduced making for inefficient reception of unwanted electromagnetic interference. Proper voltage biasing of a JFET is an art, made more difficult by the variations in transconductance and the sensitivity of the JFET to temperature. To avoid saturating or even damaging the JFET, the supply current must be limited to a fairly narrow range. This is usually done using a device known as a current-limiting diode.

When the sensor has a high output impedance which is predominantly capacitive, special care should be taken to provide an interface circuit which will not oscillate at high frequencies. Figure 24 depicts what is known as a charge amplifier. The charge amplifier is really just a modified current amplifier with the addition of a feedback capacitor. The negative feedback resistor and capacitor

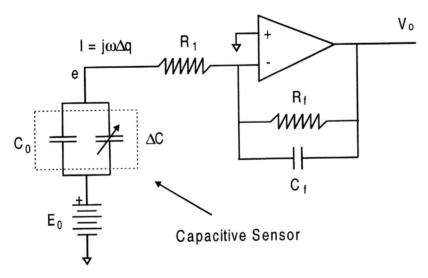

Figure 24 Charge amplifier circuit showing the capacitive sensor, sensor bias voltage E_0, and a current amplifier modified with a feedback capacitor C_f where $V_0 = -E_0 \Delta C / C_f$.

have the effect of defining a low frequency limit of

$$f_L = \frac{1}{2\pi R_f C_f} \tag{15.2.29}$$

where one make R_f and C_f as large as practical for extended low frequency response. Above this lower corner frequency the "charge gain" is $V_o/\Delta q = -1/C_f$. But since the dynamic voltage e of the transducer output is $e = \Delta q/C_0$, one can also write that the net voltage gain of the charge amplifier is $V_o/e = -C_0/C_f$. The resistor R_1 also serves the important purpose of creating a high frequency cutoff at

$$f_H = \frac{1}{2\pi R_1 C_0} \tag{15.2.30}$$

which helps keep the charge amplifier stable at very high frequencies. With R_1 nonexistent, the operational amplifier loses its ability to suppress oscillations at high frequencies. Note that for a given amount of amplifier response delay (slew rate), what is stable negative feedback at low frequencies can become unstable positive feedback at high frequencies. This is because the phase shift due to delay τ increases with frequency as $\theta = \omega\tau$. For the given sensor capacitance C_0, one simply increases R_1 until an acceptable amount of stability margin is implemented.

The Reciprocity Calibration Technique is used to obtain an absolute calibration of a sensor or transducer by using a reversible transducer and the associated reciprocity physics to obtain a receiving voltage sensitivity. It should be noted that there are really two types of calibration, relative calibration and absolute calibration. With relative calibration one transducer or sensor is compared to another calibrated transducer or sensor which has a traceable calibration to national metrology standards. The second type of calibration is called absolute calibration because the calibration can be derived from traceable non-transducer quantities, such as mass, length, density, volume, etc. For the absolute calibration to be traceable, these other quantities must also be traceable to the national standard.

Figure 25 depicts reciprocity calibration in a free field (top) and in a cavity (bottom). Both configurations require two setups to allow a measurement with transducer B used as a sensor (setup A) and as an actuator (setup B). For setup A, we require that both sensor A and transducer B receive the same pressure field from C. Co-locating transducer B with sensor A some distance from C or locating all 3 transducers in a small cavity (small compared to wavelength) is adequate insurance of a uniform received pressure for transducers A and B. In setup B, transducer B is used as a transmitter and one notes the distance d_0 for the free-field reciprocity case. For our example, we consider acoustic pressure and that transducer A is only used as a sensor for which we seek the receiving voltage sensitivity. But, the reciprocity technique is powerful and completely general for any type of transducers.

We define the receiving voltage sensitivity as M (V/Pa), the transmitting voltage sensitivity as S (Pa/A), and the reciprocity response function as $J = M/S$, which has the units of mobility (inverse impedance) in whatever domain the transducers are used (mechanical, acoustical, or electrical). For the case of electromagnetic transducers, recall that the transduction equations are $f = Bli$ and $e = Blu$ while for electrostatic or piezoelectric transducers the transduction equations are $f = \phi i/j\omega$ and $e = -\phi u/j\omega$, thus the "Bl" can be replaced by "$\phi/j\omega$" provided the minus sign is

Free Field Reciprocity

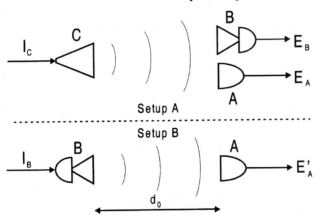

Cavity Reciprocity

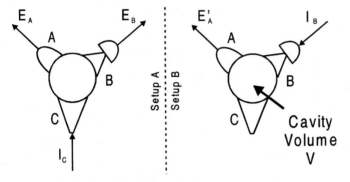

Figure 25 Reciprocity calibration in a free field or in a cavity requires two setups to exploit the reciprocal physics of transduction in transducer B to obtain an absolute voltage calibration for sensor A.

handled correctly for the piezoelectric/electrostatic transducer case. When a microphone diaphragm is subjected to acoustic pressure p, the applied force is pS_d, where S_d is the diaphragm area. The velocity of the diaphragm u is this applied force divided by the mechanical impedance of the diaphragm Z_M. The voltage generated by a velocity u is $e = Blu$. For an electromagnetic transducer, the receiving voltage is the velocity times Bl.

$$e = \frac{BlpS_d}{Z_M} \tag{15.2.31}$$

Therefore, the receiving voltage sensitivity is

$$M = \frac{e}{p} = \frac{BlS_d}{Z_M} \tag{15.2.32}$$

The transmitting sensitivity is found for a free space by noting the pressure generated by a point source with volume velocity Q a distance d_0 meters away

$$p = \frac{jk\rho c Q}{4\pi d_0} e^{-jkd_0} = \frac{\rho f Q}{2d_0} e^{-j(kd_0 - \frac{\pi}{2})} \tag{15.2.33}$$

where k is the wavenumber (ω/c) and the time harmonic dependance has been suppressed. To get the mechanical velocity of the diaphragm, we note the applied force due to the current i is $f = Bli$, the velocity is the applied force divided by the mechanical impedance, or $u = Bli/Z_M$. Multiplying the mechanical velocity u times the diaphragm area S_d gives the acoustic volume velocity Q in Eq. (15.2.33). The free-space transmitting voltage sensitivity is therefore

$$S = \frac{p}{i} = \frac{\rho f Bl S_d}{2d_0 Z_M} e^{-j(kd_0 - \frac{\pi}{2})} \tag{15.2.34}$$

and the free-space reciprocity response function is

$$J = \frac{M}{S} = \frac{2d_0}{\rho f} e^{+j(kd_0 - \frac{\pi}{2})} \tag{15.2.35}$$

When all three transducers are in a small closed cavity of volume V as seen in the bottom of Figure 25, we note that the acoustical compliance of the cavity C_A is $V/\rho c^2$, and that the mechanical equivalent $C_M = C_A/S_d^2$. Since the cavity is small, this compliance is low (the stiffness is high) and the transducer diaphragm motion is affected by the stiff cavity impedance. The receiving voltage sensitivity for the cavity is therefore

$$M = \frac{e}{p} = \frac{Bl S_d}{\left(Z_M + \frac{1}{j\omega C_M} \right)} \tag{15.2.36}$$

For an applied force due to a current i, the pressure in the cavity is the diaphragm velocity (applied force over total mechanical impedance) times the acoustic impedance of the cavity. The transmitting voltage sensitivity is

$$S = \frac{p}{i} = \frac{Bl S_d}{\left(Z_M + \frac{1}{j\omega C_M} \right)} \frac{1}{j\omega C_A} \tag{15.2.37}$$

where C_A is the acoustic compliance of the cavity.
The reciprocity response function for the cavity is simply

$$J = \frac{M}{S} = j\omega C_A \tag{15.2.38}$$

where the result is the acoustic mobility of the cavity. The reciprocity response in Eqs (15.2.35) and (15.2.38) are called absolute because they do not have any terms associated with the transducers, only dimensions (V and d_0), parameters of the medium (ρ and c), and frequency, which can be measured with great precision.

How does one make use of reciprocity to calibrate a sensor? Consider setup A in Figure 25 where both sensor A and transducer B receive the same pressure from transducer C. We measure a spectral transfer function of E_A over E_B for the frequency range of interest and can write the following equation for the receiving voltage sensitivity of sensor A.

$$M_A = \frac{E_A}{E_B} M_B \tag{15.2.39}$$

For setup B the voltage output of sensor A due to current I_B driving transducer B is

$$E'_A = pM_A = I_B S_B M_A \tag{15.2.40}$$

where S_B is the transmitting voltage sensitivity of transducer B. Therefore, we can write another equation for M_A, the receiving voltage sensitivity for sensor A, in terms of the transfer function E'_A over I_B for the frequency range of interest.

$$M_A = \frac{E'_A}{I_B} \frac{1}{S_B} \tag{15.2.41}$$

Independently, Eqs (15.2.39) and (15.2.41) may not appear useful, but if we multiply them together a take the square root (the positive root) we get an absolute expression for sensor A's receiving voltage sensitivity.

$$M_A = \sqrt{J\left(\frac{E'_A}{I_B}\right)\left(\frac{E_A}{E_B}\right)} \tag{15.2.42}$$

One simply uses the appropriate reciprocity response function J (free space or cavity) depending on the situation in which the reversible transducer B is employed, measures the required transfer functions, and calculates the receiving voltage sensitivity as given in Eq. (15.2.42). Given good transfer function estimates (i.e. linearity enforced, low background noise, and large numbers of spectral averages used) and an accurate reciprocity response function model, the calibration technique is both broadband and accurate. The calibration seen in Eq. (15.2.39) is referred to as a "relative" calibration where the sensitivity of one transducer is gauged against another. Usually, a calibration transducer traceable to national standards (i.e. NIST in the United States) is used to provide a relative calibration metric for a number of other transducers in a system. However, the reciprocity technique provides an absolute calibration, provided one can measure the needed dimensions, medium parameters, and frequency in an accurate and traceable way.

Reciprocity is the preferred technique for intelligent sensor systems to self-calibrate their transducers. It requires one or more reversible transducers in addition to the typical sensor. It requires some environmental sensing (such as temperature and static pressure for acoustical measurements), and some control over the environment (fixed dimensions or masses in a mechanical system). There are two or more transducers available to check measurements as well as check relative calibration. But if the reciprocity technique is employed, all the transducers can be calibrated absolutely and automatically by the intelligent sensor system. Sensor damage, poor electrical connections, or breakdowns in sensor mounting can be identified using calibration techniques. This "self-awareness" capability of an intelligent sensor system is perhaps the most intriguing aspect of what we define as a truly "smart" sensor system. The obvious benefit is that sensor data from a questionable or out-of-calibration sensor can be discounted automatically from the rest of the intelligent sensor system's pattern recognition and control algorithms allowing automatic re-optimization of the system's mission given a new situation.

15.3 NOISE CANCELLATION TECHNIQUES

Noise reduction is both straightforward and useful to a wide range of signal processing applications. We define *noise* simply as any unwanted signal. The noise may not be random, but rather some sort of interference which adversely effects the desired information processing of the signal. One can even describe an FFT as a noise reduction technique because is separates the signal into frequency bins and allows the magnitude and phases of many signals to be processed with little interference from each other as well as random background noise. But, if our *signal* is itself a random waveform and we wish to separate it from other signals, one has an interesting and challenging task.

Signal and noise separation can be achieved in one of two general approaches: by using a statistical whitening filter (see Sections 3.3 and 10.2) or by optimal Wiener filtering (Section 10.2). The optimal "Wiener" separation filter is defined from a knowledge of the signal one wishes to reject. "Knowledge" of the undesirable noise is defined here as either access to the waveforms or the ability to accurately synthesize either the signal or "noise" waveform coherently. On the other hand, a whitening filter is based on the approach of our "signal" waveform being periodic, thus, we can synthesize it coherently for a short time into the future (say one sample ahead), and use this linear prediction to suppress the "noise" waveform and therefore enhance our signal. This is done by first suppressing all the periodic components of our waveform, or "whitening" the waveform. The whitened waveform "residual" can then be subtracted from the original waveform to enhance the signal, which is assumed periodic. This is called *Adaptive Signal Enhancement* and is generally used to remove random noise components (or unpredictable noise transients) from a waveform with periodic (predictable) signal components. A good example of this are signal processing algorithm which remove "hiss", "clicks", and "pops" from old audio recordings in music and film archives. Using considerably more numerical processing, the same technique can be applied to images in film or video archives to remove "snow" or "specs" from damaged film.

Optimal Wiener filtering for noise suppression (and even complete cancellation) is more straightforward than whitening filtering. Suppose our desired signal is 62 Hz and our "noise" waveform is a 60 Hz sinusoid (typical of a ground-loop problem). The "optimal" separation filter has a narrow unity-gain passband at 62 Hz with a zero response at 60 Hz. Obviously, filtering our signal in this way rejects the 60 Hz interference. For separation of steady-state sinusoids of different frequencies, the FFT can be seen as having each bin as an optimal filter to allow separation and measurement of the amplitude and phase of the signal closest to the bin center frequency. Indeed, the convolution of a signal with a sinusoid representing a particular bin frequency can be seen as a FIR filter where the integral of the FIR filter output gives the FFT amplitude and phase for the bin. In general, optimal Weiner filtering is best suited for situations where a "reference signal" of interfering noise is available. This noise waveform is also a component of our original waveform of interest, but the net time delay and frequency response of the channel which introduces the noise into our original waveform is unknown. In Section 10.2, the optimal Wiener filter problem was discussed from the point of view of identification of this transfer function. In the system identification configuration, the "reference noise" is the system input, and our original waveform is the system output corrupted by another interfering waveform (our desired signal). For noise cancellation, one uses a Wiener filter to remove the "reference noise" which is coherent with the original waveform. The residual "error signal", which optimally has all components coherent with the reference noise removed, provides us our "desired signal" with this reference noise interference suppressed. This is sometimes called a *correlation canceler* in the literature because the optimal filter removes any components of our waveform which are correlated with the reference noise.

In this section we will describe in detail the adaptive signal processing applications of adaptive signal whitening, adaptive signal enhancement, adaptive noise cancellation, and the physical cancellation of unwanted noise known as active noise cancellation. In active noise cancellation, one must take into account the possibility of feedback from the physical actuator back into the reference signal. In addition, active noise control requires delays in the error plant to be accounted for in the adaptive controller updates. Finally, we will present a brief note on feedback control approaches for active noise control.

Adaptive Signal Whitening describes the technique of removing all predictable signal components, thus driving the processed signal towards spectrally-white Gaussian noise. One begins by assuming a model for our waveform y_t as the output of an infinite impulse response (IIR) filter with white noise input, or *innovation*. Using z-transforms, the waveform is

$$Y[z] = \frac{E[z]}{[1 + a_1 z^{-1} + a_2 z^{-2} + \ldots + a_M z^{-M}]} \tag{15.3.1}$$

where $E[z]$ is the z-transform of the white noise innovation sequence for the IIR process y_t. Rearranging Eq. (15.3.1) yields

$$Y[z][1 + a_1 z^{-1} + a_2 z^{-2} + \ldots + a_M z^{-M}] = E[z] \tag{15.3.2}$$

and taking inverse z-transforms one has

$$y_t + \sum_{i=1}^{M} a_i y_{t-i} = \epsilon_t \tag{15.3.3}$$

where ϵ_t is the innovation. To remove the periodicity in y_t, or in other words remove the predictability or correlation of y_t with itself, one can apply a FIR filter $A[z] = 1 + a_1 z^{-1} + a_2 z^{-2} + \ldots + a_M z^{-M}$ to our waveform y_t and the output approximated the innovation, or white noise. The filter $A[z]$ is said to be a "whitening filter" for y_t such that the residual output contains only the "unpredictable components" of our waveform. We can call our approximation to the innovation the "prediction error" and define the linear one-step-ahead prediction for y_t

$$\hat{y}_{t|t-1} = -\sum_{i=1}^{M} a_i y_{t-i} \qquad (15.3.4)$$

where the notation "$t|t-1$" describes a prediction for the waveform for time t using past waveform samples up to and including time $t-1$. This "linear predicted" signal by design does not contain very much of the unpredictable innovation. Therefore, the predicted signal is seen as the enhanced version of the waveform where the noise is suppressed.

To estimate the linear prediction coefficients of $A[z]$, a variety of adaptive algorithms can be used as seen in Section 10.2. The most straightforward adaptive algorithm is the least-mean square, or LMS algorithm. If we consider the innovation in Eq. (15.3.3) as an error signal, the gradient of the error is a positive function of the coefficients a_i; $i = 1, 2, \ldots, M$. Therefore, a gradient decent for the LMS algorithm will step the coefficient updates in the opposite direction of the gradient as seen in Eq. (15.3.5). Figure 26 graphically depicts a whitening filter.

$$a_{i,t} = a_{i,t-1} - 2\mu\varepsilon_t y_{t-i} \qquad \mu \le \frac{\mu_{rel}}{ME\{y_t^2\}} \qquad \mu_{rel} \approx \frac{1}{N} \qquad (15.3.5)$$

The step size μ in Eq. (15.3.5) is defined in terms of its theoretical maximum (the inverse of the model order times the waveform variance) and a relative parameter μ_{rel} which effectively describes the memory window in term of N waveform samples. As described in Section 10.2, the smaller μ_{rel} is, the slower the LMS convergence will be, effectively increasing the number of waveform samples influencing the converged signal whitening coefficients. This is a very important consideration for whitening filters because the filter represents a *signal model* rather than a *system model*. By slowing the adaptive filter convergence we can make the whitening filter ignore nonstationary sinusoids. This can be very useful for canceling very stationary sinusoids such as a 60 Hz power line frequency from a ground fault.

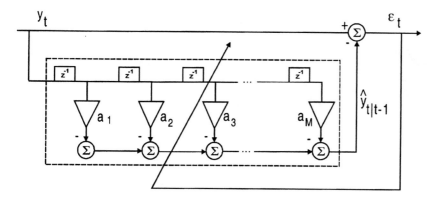

Figure 26   Flow diagram for an adaptive FIR whitening filter for the signal y_t where ε_t is the whitened "error" signal used to drive the adaptive filter, and $\hat{y}_{t|t-1}$ is the linear predicted or "enhanced" signal.

Adaptive Signal Enhancement is the technique of enhancing desired predict-able signal components by removal of the unpredictable noise components. Signal enhancement usually requires a whitening filter to produce the unpredictable com-ponents to be removed. To further illustrate whitening and enhancement filters, Figure 27 shows the results of the whitening filter output, and the enhanced (linear predicted) signal output for a case of a stationary 120 Hz sine wave in white noise along with a non stationary frequency in the 350 Hz range with a standard deviation of 10 Hz. The instantaneous frequency is found for the nonstationary sinusoid by integrating a Gaussian random variable over several thousand samples to obtain a relatively low frequency random walk for the sinusoid frequency. Since instan-taneous frequency is the time derivative of the instantaneous phase, one can synthesize the nonstationary spectra seen in Figure 27 on the left spectrogram. The center spectrogram shows the enhanced signal (from the linear prediction) and the right spectrogram shows the whitened error signal output. An LMS adaptive algorithm is used in the time domain with a memory window of about 200 samples. Figure 28 shows the time-averaged power spectra of the original and whitened signals for the time interval from 60 to 70 sec. The whitened signal clear shows over 50 dB of cancellation for the 120 Hz stationary sinusoid and about 20 dB cancel-lation for the nonstationary sinusoid. Figure 29 shows the time average spectra for the enhanced signal which clearly shows in detail a 10–20 dB broad band noise reduction while leaving the sinusoids essentially intact.

Adaptive Noise Cancellation significantly suppresses, if not completely removes, the coherent components of an available noise reference signal from the signal of interest through the use of optimal Wiener filtering. Figure 30 shows a block diagram for a straightforward Wiener filter used to remove the interference noise correlated with an available "reference noise", thus leaving the residual error containing mostly the desired signal. If coherence is high between the reference

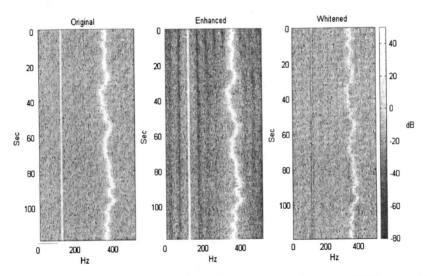

Figure 27 Adaptive whitening and enhancement results for a 120 Hz sinusoid in white noise plus a frequency-modulated sinusoid to show the effects of signal nonstationarity.

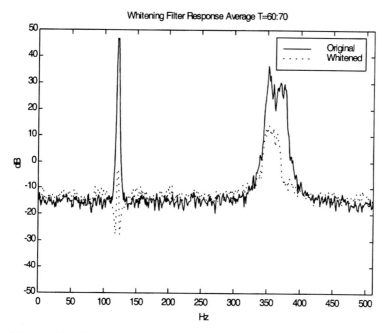

Figure 28 Ensemble-averaged original and whitened signal spectra showing the most significant whitening of the 120 Hz stationary sinusoid.

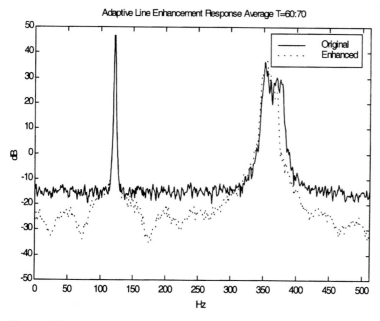

Figure 29 Ensemble-averaged original and enhanced spectra showing the suppression of unpredictable noise in the adaptively-enhanced signal produced by the linear prediction.

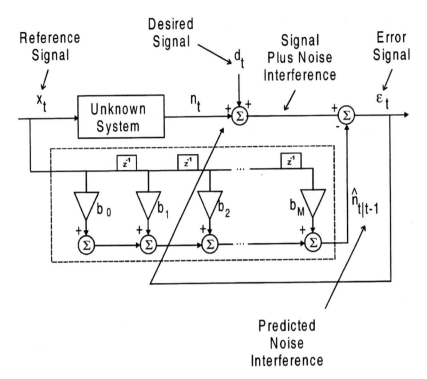

Figure 30   An adaptive Wiener filter for removing noise correlated with a known reference noise signal, thus driving the error signal output to contain mostly the desired signal components.

noise and the waveform with our desired signal plus the noise interference, the cancellation can be nearly total. This of course requires that we have a very accurate model of the transfer function between the reference noise and our waveform. The Wiener filtering in Section 10.2 is mainly concerned with the identification of this transfer function system, or "system identification". A valuable by-product of precise system identification is the capability of separating the reference noise components from the output waveform to reveal the "desired signal" with the noise interference removed.

For example, suppose the President, not wishing to answer reporters' questions while walking from his helicopter to the White House, instructs the pilot to race the turbine engines creating enough noise to mask any speech signals in the area. However, one gadget-wise reporter places one microphone near the helicopter and uses another to ask a question of the President. A real-time Wiener filter is used to remove the turbine noise from the reporter's microphone allowing the question to be clearly heard on the videotape report. This technique has also been used to reduce background noise for helicopter pilot radio communications, as well as many other situations where a coherent noise reference signal is available.

We begin our description of the adaptive noise cancellation seen in Figure 30 by assuming an FIR system model (the model could also be IIR if desired), with the following linear prediction of the noise interference n_t, given a reference signal

x_t, and system model last updated at time $t - 1$.

$$\hat{n}_{t|t-1} = \sum_{k=0}^{M} b_k x_{t-k} \tag{15.3.6}$$

The linear prediction error is then simply

$$\varepsilon_t = n_t + d_t - \hat{n}_{t|t-1} \tag{15.3.7}$$

where d_t is our desired signal.

The adaptive LMS update for the FIR coefficients is

$$b_{k,t} = b_{k,t-1} + 2\mu \varepsilon_t x_{t-k} \qquad \mu \le \frac{\mu_{rel}}{ME\{x_t^2\}} \qquad \mu_{rel} \approx \frac{1}{N} \tag{15.3.8}$$

where a plus sign is used to insure the step is in the opposite direction of the gradient of the error with respect to the coefficients.

Use of the LMS algorithm and a FIR system model makes the processing remarkably simple. If the transmission channel between the reference signal and the noise interference is essentially propagation delays and multipath, the FIR model is very appropriate. If the transmission channel contains strong resonances (the system transfer function has strong peaks in the frequency domain), an IIR filter structure is to be used because it requires fewer coefficients to accurately model the system.

If the reference noise contains signal components which are uncorrelated with our waveform, these components will unfortunately add to the noise interference. This is because the converged system filter only responds to the coherent signals between the reference noise and our waveform. The uncorrelated noise in the reference signal will be "filtered" by the system model, but not removed from the output waveform. This is a particular concern when the reference signal contains random noise from spatially uncorrelated sources, such as turbulence on a microphone. Such situations are problematic when acoustic wind noise is interfering with our desired signal. Because the turbules shed and mix with the flow, the pressure fluctuations are spatially uncorrelated after a given distance depending on the turbulence spectrum. The signal plus turbulence at one location (or generated by one microphone in air flow) is in general not correlated with the wind noise at another location. If one attempts to cancel the flow noise using a reference sensor a large distance away one would simply add more random noise to the output signal. However, wind noise on a microphone has been canceled using a co-located hot-wire turbulence sensor signal as a reference noise (6). When tonal acoustic noise is to be removed, one can design, or synthesize, a "noiseless" reference signal by using a non-acoustic sensor such as an optical or magnetic tachometer pulse train designed to produce all the coherent frequencies of the noise generating device. This frequency-rich reference signal is then adaptively-filtered so that the appropriate gain and phase are applied to cancel the coherent tones without adding more random noise. This is the best strategy for canceling fan noise in HVAC systems (heating, ventilation, and air conditioning), propellor noise in commercial passenger aircraft, and unwanted machinery vibrations.

As an example demonstrating the ease, power, and simplicity of adaptive noise cancellation, a pair of nonstationary sinusoids is seen in the left spectrogram of Figure 31. Nonstationary noise (around 100 Hz) and signal (around 300 Hz) sinusoids are used to emphasize that once the system transfer function is identified, cancellation occurs independent of the actual signals. For this case our reference noise is the lower sinusoid near 100 Hz plus white Gaussian noise. The SNR of the signals is 20 dB, the SNR gain of the 1024-point FFT is 30.1 dB, yet the observed SNR of about 40 dB is less than 50.1 dB because of the nonstationarity of the frequencies. The reference noise mean frequency is 105 Hz and its standard deviation is 10 Hz while our desired signal has a mean of 300 Hz and a standard deviation of 20 Hz. The Gaussian frequency estimates were integrated over 10,000 samples (sample rate for the simulated waveforms is 1024 Hz) to give a low-frequency randomness. The center spectrogram represents the predicted noise interference signal consisting of the system-filtered 105 Hz nonstationary sinusoid plus the random noise. The right spectrogram illustrate the nearly complete removal of the noise interference, even though the signals are nonstationary. Figure 32 depicts the time-average spectra of the original waveform and the cancellation filter error output averaged over records 40–45. One can clearly see the cancellation of the random noise as well as the interfering 105 Hz nonstationary sinusoid. Adaptive noise cancellation is a straightforward and impressively simple technique for separating signal and noise interference.

Active Noise Cancellation suppresses the unwanted noise in the physical medium (rather than on the signal wire or in a digital recording of the waveform). This physical cancellation requires that feedback from the actuator to the reference signal be accounted for, and that delays in the actuator to error sensor transfer func-

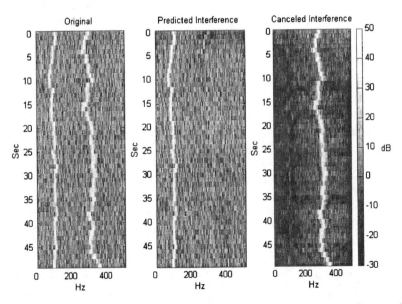

Figure 31 Signal separation using Wiener filtering on two nonstationary sinusoids where the approximately 100 Hz wave in white noise is available as a reference noise signal for cancellation.

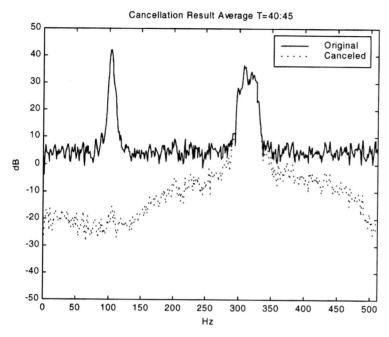

Figure 32 Average spectra from records 40–45 in Figure 31 showing the original signal plus interference (100 Hz sinusoid plus noise) and the output signal with the interference canceled.

tion be accommodated. There are significant complications due to the additional delays and control loop transfer functions and we emphasize here that *Active Noise Cancellation* and *Adaptive Noise Cancellation* refer to significantly different signal processing tasks.

Figure 33 provides a basic comparison between active noise cancellation and adaptive noise cancellation (ANC). While both algorithms are adaptive controllers, the word *active* signifies that a physical actuator produces a real wave which mixes the interfering noise to produce a physical cancellation. Where does the wave energy go? The answer depends on how the ANC system in configured. If one is canceling an acoustic wave outdoors, the cancellation simply occurs at a point in space from out-of-phase superposition. If the acoustic cancellation occurs in a duct where the wavelength is larger than twice the longest dimension in the duct cross section, the ANC actively causes an impedance zero at the active control source. This actually causes a reflection of the acoustic wave back towards the "primary" noise source, and zero power radiation downstream in the duct (7). The concept of active control of impedance control is extremely powerful because it allows one to control the flow of power and the efficiency of energy conversion. To reduce radiated energy, one drives radiation impedance to zero. To create a "zone-of-silence" at the location of a sensor, one is not concerned so much about impedance and uses ANC to allow sensor signal gains to be increased for a physical increase in SNR.

As seen in Figure 33, the ANC control system is considerable more complicated than a Wiener filter for adaptive noise cancellation "on the wire". The Block "$S[z]$" symbolizes an active source which could be electrical, mechanical, or

Adaptive Noise Cancellation

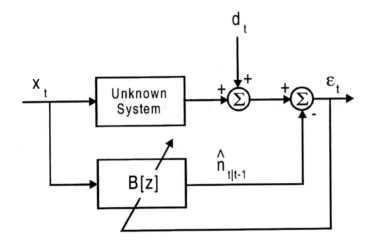

Active Noise Cancellation

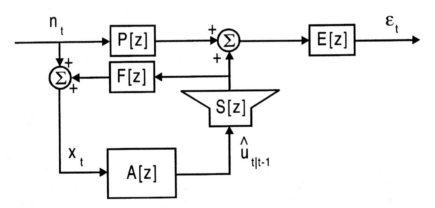

Figure 33   Comparison of Adaptive Noise Cancelation and Active Noise Cancelation Algorithms.

acoustical. However, once the ANC "secondary" source produces its canceling wave, it can have an effect on the reference sensor which is symbolized by the feedback block "$F[z]$." For adaptive noise cancellation seen in the top of Figure 33, the effect of the control signal output is seen immediately in the error signal after subtracting the predicted noise interference from the waveform. However, in the ANC system, this control output must first pass through the actuator transfer function "$S[z]$" and then through the error propagation channel and transducer symbolized by "$E[z]$." The error loop transfer function can be seen as "$S[z]E[z]$," or "SE" for short, and can have significant time delays due to the physical separation and the electro-mechanical transfer function of the transducers. Since in the least-squared error adaptive filter the error and reference signals are cross-correlated to provide the filter coefficient update (no matter which adaptive algorithm is used), any phase shift (or time delay caused by the response of *SE* will be significantly detrimental to the success of the cancellation. Thus, by filtering the reference signal with *SE* we make a new signal r_t, which we call the "filtered-x" signal which is properly aligned in time and phase with the error signal. Note that a small amplitude error in the error signal due to *SE* is not a significant problem since the adaptive algorithm will continue to update the control filter coefficients until the error is zero. But with a phase or time delay error, the coefficients of the controller will be updated with the wrong cross correlation. The "Filtered-X" adaptive ANC algorithm is seen in Figure 34.

To see the optimal converged control filter solution for ANC, we break down the system transfer functions into frequency-domain representative blocks which we can manipulate algebraically. These transfer functions could be of FIR or IIR type in terms of model paradigm. For now we will simply leave them as a letter block to simplify our analysis. The error signal can be seen to be

$$\varepsilon(f) = n(f)P(f)E(f) + x(f)A(f)S(f)E(f) \tag{15.3.9}$$

which $A(f)$, or $A[z]$ in the digital z-domain, can be solved for the control filter A which gives zero error signal explicitly as (dropping the f or z for brevity)

$$A = \frac{-P}{S(1 - FP)} \tag{15.3.10}$$

assuming the propagation and error sensor transfer function "E" is nonzero for all frequencies. If the error loop response is zero at some frequency(s), the ANC control system will not respond to that frequency(s), but otherwise is stable.

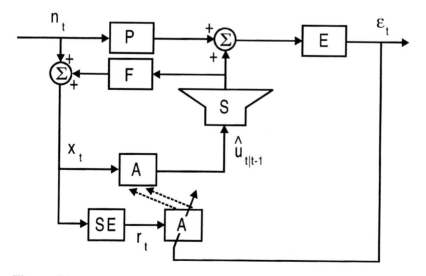

Figure 34 The signal "r_t" is the Filtered-X signal required for adaptive control filter convergence when a non-zero time delay exists in the actuator-to-error sensor control loop.

One can clearly see from the optimal ANC control filter in Eq. (15.3.10) that to be a stable filter, "S" must be non-zero at all frequencies and the forward path "P" times the feedback path "F" must be less than unity when the phase is near zero. While this makes A a stable filter, the feedback loop from the actuator S back into the reference sensor through F must also be stable. This is guaranteed if the loop gain through ASF is less than unity, or $|FP| < \frac{1}{2}$. *We call the requirement for $|FP|$ $< \frac{1}{2}$ the "passivity condition"* for ANC systems which must be met to insure system stability. In general, the feedback from the active control source back into the reference sensor is highly undesirable and robs the ANC system of performance and dynamic range. It is best eliminated by using a reference sensor which does not respond to the acoustic or mechanical waves from the actuator, such as an optical tachometer for canceling tonal noise interference. Directional sensor arrays can also be used to insure passivity. While eliminating the feedback is the best strategy, if present and meeting the passivity constraint, feedback will simply cause the ANC system to have to cancel much of its own signal, leading to poor overall performance. Clearly, applying a basic physical understanding of the system can significantly improve the performance of the applied adaptive signal processing.

We now turn our attention to measurement of the error loop transfer function SE. Clearly, one could measure this while the ANC system offline and apply the SE Filtered-X operation while hoping that this transfer function does not change during the ANC system operational time. But, the Filtered-X operation is important in that it is generally done concurrently with the ANC adaptive filtering in commercially-available systems (8). Figure 35 shows an "active" on-line system identification which injects a white noise signal into the error loop to simultaneously identify the SE transfer function while the ANC system is operating. Clearly, this

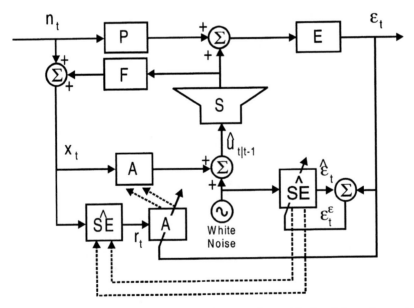

Figure 35   White noise is injected at low, but detectable, levels into the control signal for active identification of the error loop transfer function SE.

noise has to be strong enough that the *SE* transfer function can be measured in a reasonable amount of time. However, if the noise is too loud, the purpose of the ANC is defeated. Figure 36 shows a very clever (9) approach which allows both the *PE* and *SE* transfer functions to be estimated using only the available control signals. This is desirable because it adds no noise to the ANC system or ANC error output. It works by balancing the models for *PE* and *SE* to predict the error signal, which could be near zero during successful cancellation. The "error of the error", $\varepsilon_t^\varepsilon$, is then used to update the estimates for the *PE* and *SE* transfer functions. Only the *SE* transfer function is needed for the Filtered-X operation, and only its phase at the frequency components in the noise, reference, and error signals is needed to insure correct convergence of the ANC system. However, if the reference noise is narrowband, the estimated *SE* transfer function will only be valid at the narrowband frequency.

The Filtered-X algorithm can also be applied for multichannel ANC where one has p reference signals, q actuators, and k error sensors (10). One would choose to use multiple reference signals when several independently observable coherence reference noise signals are available, and as such, one can optimize the adaptive controllers independently for each type of reference signal summing the control outputs at the actuators. The ANC control filter A is represented in the frequency domain as a pxq matrix for each frequency. This way each reference signal has a path to each actuator. In general, when a number of actuators and error sensors are used, the ANC system is designed to produce spatial, or modal, control. One should note simultaneously minimizing the squared error at every error sensor location is but one spatial solution. One can apply a number of useful modal filters to the error sensor array for each frequency of interest in much the same way one could design a beampattern with nulls and directional lobes as desired. The issue of modal control

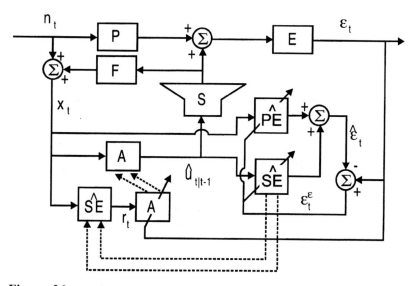

Figure 36 Passive system identification can be done by using both the reference and control signals to identify the time delay of the error loop which is all that is required for system convergence.

is central to ANC problems of sound radiation from large structures and high frequency sound cancellation in enclosures such as airplanes, automobiles, and HVAC ducts. Using modal ANC control, one can design the ANC converged response to minimize vibrational energy which radiates from a structure, or excites enclosure modes.

In general, each actuator "channel" should respond adaptively to every error signal, either independently or as part of a modal filter output. When the number of error sensors is greater than the number of actuators ($k > q$), this combining of error sensors allows each actuator signal to be optimized. When the number of actuators is larger than the number of error sensors ($q > k$), the error signals will be repeated for multiple actuators redundantly. This will in general still work, but there is a potential problem with linear dependence of actuator channels. Note that in Eq. (15.3.10) the actuator transfer function S is inverted. For multichannel ANC, S is a qxq matrix and the error propagation and transducer paths are modeled as a qxk matrix for each frequency. This allows each actuator a path to each error sensor. However, the actuator matrix S must be invertible, which means that the columns of S must be linearly independent and each element on the main diagonal of S must be nonzero for all frequencies. An example of linear dependence between ANC channels can be seen for a two channel system where both actuators and both error sensors are co-located. As such, the columns of the S matrix are linearly dependent. Channel 1 can interfere with channel 2 and the error sensors cannot observe the effect. Thus, the control effort can go unstable without correction instigated by the error signals. One can design around the actuator linear dependence problem by careful choice of actuator and error sensor location. Another method is to apply a singular value decomposition and "turn off" the offending actuators at selected frequencies. This is a significant source of complexity in robust design of a multichannel ANC system.

With p reference sensors, q actuators, and k error sensors the multichannel Filtered-X algorithm poses a problem: how does one filter p reference signals with the qxk SE transfer function estimate to produce p "filtered-x" (r_t in Figures 34–36) signals for use by the adaptive algorithm? We adopt a "copy/combine" strategy to support the matrix interfaces and the ANC design goals. If $p>q$ (more reference channels than actuators) the control filters for the extra reference channels can be calculated in parallel where the outputs are summed at the actuators. For $p<q$ the reference signals are simply copied into the extra channels. Now we have k filtered-x signals (one for each error channel) and we need to adapt the pxq ANC control filter frequency domain coefficients. For $k>q$ (more error channels than actuators) the extra filtered-x signals are combined for each actuator channel. In other words, more than q filtered-x signals and error signals contribute to the adaptive update for a particular control filter element. When $k<q$ (more actuators than error sensors) the extra actuator channels are updated with the same filtered-x signals and error signals as other channels at the risk of causing linear dependence. This is less risk than it sounds because the actual transfer functions between individual actuators and a particular error sensor may be different enough to support linear independence of the channels.

Active Noise Attenuation Using Feedback Control is best applied in situations where the closed-loop response has a time constant significantly shorter than the period of the highest frequency to be controlled. This is because the feedback loop time delay determines that available bandwidth for stable active attenuation. Generally speaking, active feedback control is well suited for "smart structures" where the vibration sensor and actuator are closely-located, the wave speed is very high, and therefore the control loop delay is very short compared to the period of the frequencies to be actively attenuated. This arrangement can be seen as a basic regulator circuit for very low frequencies. As one attempts to employ active feedback control at higher frequencies, one must limit the control bandwidth in accordance with the loop delay creating a "tuned regulator". Because of the devastating quickness of divergence in a high-gain tuned regulator, adaptive techniques are best executed off-line where one can apply stability gain margin analysis before updating the feedback controller.

Figure 37 depicts a basic feedback control system which can be optimized using what is known as ARMAX control. ARMAX stands for Auto Regressive Moving Average with auXiliary input and describes a modern digital adaptive control technique for feedback systems (the Filtered-X algorithm is a feedforward system). Before we get into the details of ARMAX control, let us examine the basic feedback control system where the plant ($H[z]$ in Figure 37) has some gain H_0 and delay τ. This delay is inherent to all transducers but is particularly large in electromagnetic loudspeakers and air-acoustic systems where the sound propagation speed is relatively slow. The delay response for most sensors in their passband is usually quite small if the frequency response is flat. For piezoelectric transducers embedded into structures, the delay between sensor and actuator is especially small. This delay determines the available bandwidth and amount of stable active attenuation.

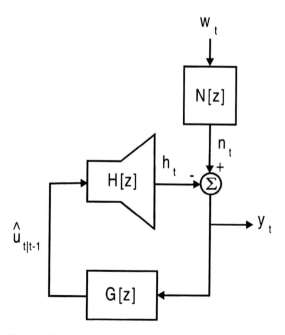

Figure 37 ARMAX Control can be used to optimized the design of a digital feedback control system for active attenuation of noise, but it is well-advised to operated the feedback controller with all filters non-adaptive to avoid instability with high-gains.

The delay in the plant will cause a linear phase shift where the phase becomes more negative as frequency increases. At some high frequency, what was designed as low frequency stable negative feedback become unstable positive feedback. Therefore, one typically designs a compensator filter ($G[z]$ in Figure 37) which attenuates the feedback out-of-band to maintain stability.

To examine the feedback control system stability problem in its most basic incarnation (a whole textbook has been devoted to this subject countless times), we'll consider a "perfect transducer" model where the net response from the electrical input to the actuator to the electrical output of the error sensor is simply $H(s) = H_0 e^{-j\omega\tau}$. Our compensator $G(s)$ is a simple 1st-order low pass filter defined by

$$G(s) = \frac{1}{1 + j\omega/\omega_c} \qquad \omega_c = \frac{1}{RC} \qquad (15.3.11)$$

which can be implemented a variety of ways. The compensator has a flat frequency response up to ω_c and then decreases at a rate of -6 dB per octave as frequency increases. Placing two such filters in series gives a 2nd-order (-12 dB/oct) filter, four filters in series yields a 4th-order (24 dB/oct), and eight filters in series yields a sharp cutoff of (48 dB/octave). Figure 38 compares the magnitude and phase of these filters designed with a 300 Hz cutoff frequency. While the higher order compensators provide a sharp cutoff and a lot of attenuation for high frequencies, there also is considerable phase change in the cutoff region. As will be seen, this

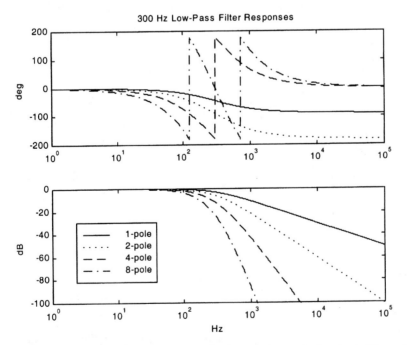

Figure 38 Comparison of the magnitude and phase of a 1st, 2nd, 4th, and 8th order low pass filter designed to attenuate signals above 300 Hz.

phase change is the source of considerable difficulty (and art) in designing stable feedback control systems.

The open-loop response is simple the $H(s)G(s)$ product, or

$$H(s)G(s) = \frac{H_0 e^{-j\omega\tau}}{\left(1 + j\dfrac{\omega}{\omega_c}\right)^n} \tag{15.3.12}$$

where "n" is the order of the filter (2nd order, etc) also denoted as the number of poles in the open-loop system. For the open-loop response to be stable, all of its poles must lie in the left half s-plane. However, if the open-loop system is unstable, the closed-loop response can still be stable, which is a useful and interesting aspect of feedback control. *Open-loop stability implies that there are no poles of* $H(s)G(s)$ *in the right half s-plane while closed-loop stability implies that there are no zeros of* $1 + H(s)G(s)$ *in the right half s-plane.* The closed-loop response of the feedback control system is seen to be

$$\frac{y(s)}{n(s)} = \frac{1}{1 + H(s)G(s)} \tag{15.3.13}$$

where the denominator of Eq. (15.3.13) is known as the *characteristic equation* for the closed-loop response. The closed-loop responses for our system with either 1st, 2nd, 4th, or 8th order low pass filters in seen in Figure 39. The 8th-order (8-pole) response is actually unstable, which will be explained shortly. The zeros of the

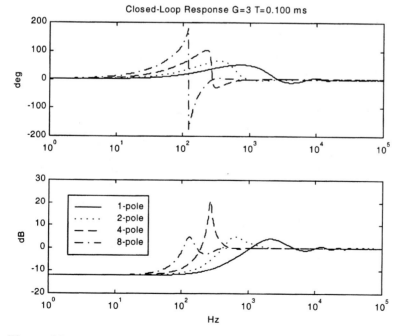

Figure 39 Closed-loop responses with a gain of 3 and 100 μs delay showing the 4-pole system marginally stable and the phase of the 8-pole system showing an unstable response.

characteristic equation are the closed-loop system poles. Therefore, the characteristic equation can have no zeros in the right half s-plane if the closed-loop response is stable.

Solving for the characteristic equation zeros and testing to determine their location is quite tedious, but can be done in a straightforward manner using numerical techniques. The Routh–Hurwitz criterion is a method for determining the location of the zeros (in either the left or right half planes) for a real-coefficient polynomial without actually having to solve for the zeros explicitly. However, the Nyquist criterion is a more appealing graphical solution to determining closed-loop system stability. For our simple system with no open-loop poles in the right half plane, one plots the real verses imaginary parts of the open loop response $H(s)G(s)$ for $s = + j\infty$ to $s = j0$ taking care to integrate around any poles of the characteristic equation on the $j\omega$ axis.

The "Nyquist plots" for our system can be seen in Figure 40. Since we have a low pass compensator $G(s)$, all the open-loop response plots start at the origin corresponding to $\omega = +j\infty$. As the contour integral (counter clockwise around the right-half s-plane) progresses down the $s = j\omega$ axis, the $H(s)G(s)$ plot spirals out in a general counterclockwise direction. Obviously, if it intersects the critical point $[-1, j0]$, the characteristic equation is exactly zero and the closed loop response diverges. What is less obvious is that if the Nyquist plot *encircles* (that is lies to the left of the critical point), the closed-loop system is unstable because there will be at least one zero of the characteristic equation in the right half s-plane. Figure 40 clearly shows the 4th order system stable, but nearly intersecting the

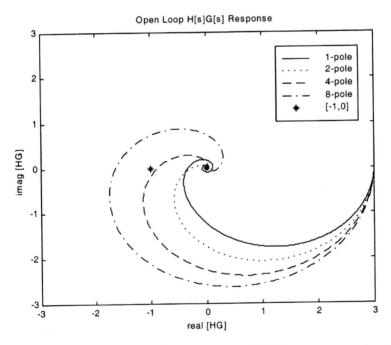

Figure 40 Nyquist plot of the open-loop $H[s]G[s]$ response showing the 8-pole system encirclement of the $[-1, 0]$ coordinate indicating instability.

critical point while the 8th order system encircles the critical point indicating instability.

But, the closed-loop response seen in Figure 39 shows no sign of instability in the magnitude of the 8th order compensator closed-loop response. Why? The answer can be seen in the fact that frequency response magnitudes provide no information about time causality. The phase response clearly shows a sizable phase shift of nearly 2π through the cutoff region. Recall from Section 3.3 that positive phase shifts imply a time advance (due to perhaps a time delay in the transfer function denominator). The 8th order response in Figure 39 and 40 is unstable because it is noncausal. The effect of increasing the gain is to expand the Nyquist plot of Figure 40 in all directions. The effect of increasing the time delay is to rotate the Nyquist plots clockwise, significantly increasing the chances of encirclement of the critical point. Lowering the compensator cutoff frequency will move the Nyquist plots towards the origin at high frequencies which buys some gain and phase margin. Phase margin is the angle between the point at high frequencies where the open-loop gain is unity and the real axis on the Nyquist plot. Gain margin is 1 over the magnitude of the open-loop gain where is crosses the real axis, hopefully to the right of the point $[-1, j0]$. If the compensator does not attenuate sharply the high frequencies, the Nyquist plot widely encircles the origin many times as frequency increases due to the plant delay τ. These wide encirclements cause ripples and overshoot to be seen in the closed-loop responses, all of which with a gain of 3.0 show a little over 12 dB active attenuation at low frequencies. Clearly, the design goal for robust active attenuation using feedback control is to have as short a time delay as possible in the control loop and to use a low-order compensator to avoid too much phase shift in the cutoff region. This is why embedded piezoelectrics with extremely short response delays work so well in smart structures designed to attenuate vibrations using active electronics. Even active hearing protectors, with nearly co-located loudspeakers and error microphones, get only 12–15 dB of active attenuation over a bandwidth of up to 1–2 kHz. An additional 20–25 dB of high frequency attenuation can happen passively if the hearing protectors are of the large sealed type.

Applying ARMAX control offers no relief from the stability physics of feedback control systems, only an approach which allows one to exploit (with some risk) the known closed loop response to gain greater attenuations over a wider bandwidth for periodic noise disturbances. If one knows the loop delay precisely, and can be sure that the loop delay will be unchanged, a periodic noise waveform can be predicted "d" samples into the future. This follows a whitening filter where one predicts the waveform one sample into the future to derive the whitened error signal. The compensator $G(s)$ is designed to invert the minimum phase response of $H(s)$ and provide a "d-step ahead" linear prediction of the control signal $u_{t+d|t-1}$. Clearly, this is risky business because if the control loop delay or noise disturbance suddenly change, the transient signals in the control loop can quickly lead to divergence and/or nonlinear response of the actuators. ARMAX control has been successfully applied to very low bandwidth problems in industrial process controls (11), among other applications. One of the nice features of ARMAX control is that one can, with least-squares error, drive the closed-loop response to produce any desired waveform or have any stable frequency response, assuming a high fidelity plant actuator is used.

ARMAX control starts by assuming an ARMA process for the noise disturbance n_t. The ARMA process has innovation w_t, which is zero mean Gaussian white noise.

$$N[z] = W[z]\frac{C[z]}{D[z]} \tag{15.3.14}$$

Taking inverse z-transforms one obtains

$$n_t = \sum_{k=0}^{K} c_k w_{t-k} - \sum_{m=1}^{M} d_m n_{t-m} \tag{15.3.15}$$

The plant transfer function is also assumed to be a pole-zero system, but we will separate the delay out so we can say that the pole-zero system is minimum phase. The delay τ corresponds to d samples in the digital system.

$$H[z] = U[z]\frac{B[z]}{A[z]}z^{-d} \tag{15.3.16}$$

Taking inverse z-transforms one gets

$$h_t \sum_{p=0}^{P} b_p u_{t-p-d} - \sum_{q=1}^{Q} a_q h_{t-q} \tag{15.3.17}$$

The ARMAX output y_t is the noise of Eq. (15.3.15) minus the plant output of Eq. (15.3.17) we can write a closed-loop system transfer function relating the ARMAX output to the noise innovation.

$$\frac{Y[z]}{W[z]} = \frac{C[z]A[z]}{A[z]D[z] + B[z]D[z]z^{-d}G[z]} \tag{15.3.18}$$

Solving for the compensator $G[z]$ which drives $Y[z]$ to $W[z]$ (in other words, completely whitens the ARMAX output), one obtains

$$G[z] = \frac{A[z]}{B[z]}z^{+d}\left(\frac{C[z]}{D[z]} - 1\right) \tag{15.3.19}$$

But, for active noise attenuation purposes, one can do slightly better by not completely whitening the noise, but rather just whitening the noise poles and leaving any zeros in the noise alone. The result is that ARMAX output will be driven to the MA process defined by $W[z]C[z]A[z]$.

$$\begin{aligned} G_{CA}[z] &= \frac{A[z]}{B[z]}z^{+d}\left(\frac{1}{A[z]D[z]} - 1\right) \\ &= z^{+d}\left(\frac{1 - B[z]D[z]}{B[z]D[z]}\right) \end{aligned} \tag{15.3.20}$$

The result in Eq. (15.3.20) becomes even more interesting one considers that the polynomial product $B[z]D[z]$ is just the AR part of the control signal u_t.

$$U[z] = \frac{W[z]C[z]A[z] - Y[z]A[z]D[z]}{B[z]D[z]} \qquad (15.3.21)$$

Equation (15.3.21) tells us that one can simply whiten the control signal u_t to obtain the necessary system identification to synthesize $G_{CA}[z]$. However, because $B[z]$ contains the net gain of the minimum phase plant $H[z]$, and a whitening filter operation to recover $B[z]D[z]$ will have a leading coefficient of zero, one must have an estimate of the plant gain and delay in order to properly scale $B[z]D[z]$ and employ the feedback control for active attenuation of periodic noise.

The ARMAX characteristic equation is

$$1 + H[z]G_{CA}[z] = \frac{1}{A[z]D[z]} \qquad (15.3.22)$$

which by design has no zeros to cause closed-loop instability. The polynomials in the denominator are also minimum phase, which helps stability. But, the closed-loop gain is very high if the plant has sharp resonances ($A[z]$ has zeros near the unit circle), or if the noise disturbance is nearly sinusoidal ($D[z]$ has zeros near the unit circle). With this high loop gain, any miscalculation of the plant delay, or change in the plant response, can lead to near instant divergence of the signals in the feedback loop. If we use the compensator given in Eq. (15.3.19), we have to be concerned that the noise is *persistently exciting* such that $C[z]$ has no zeros on the unit circle and that the poles of the plant $H[z]$ are inside the unit circle. This is another reason why the compensator given in Eq. (15.3.20) is desirable. Figure 41 shows an example ARMAX active attenuation result for a 50 Hz sinusoid noise disturbance and a slow-adapting LMS whitening filter. The control effort is deliberately slowed with initial small open-loop gain to prevent the start up transient from driving the feedback loop into instability. ARMAX active attenuation systems are specifically tuned to the plant transfer function and noise disturbance characteristics and not well-suited for nonstationary plants or disturbance signals.

15.4 SUMMARY, PROBLEMS, AND BIBLIOGRAPHY

This chapter presents some very important models for electrical noise, transducer performance, and noise cancellation techniques. These concepts are very much characteristic of intelligent sensor systems, which recognize signal patterns, have self awareness of system health, and have the ability to adapt to environmental changes such as changes in noise interference. Building a physical foundation for this requires an examination of how noise is generated, how transducers work, and how to minimize or even remove the effects of noise. In Section 15.1 we present well-accepted models for several types of electrical noise. Thermal noise, can also be generated mechanically in damping materials. This is quite interesting both practically and physically. An understanding of the signal features of the various types of noise (thermal, shot, contact, and popcorn) allows an intelligent sensor system to determine the cause of a given noise spectrum on one of its sensors. Section 15.2 provides a very brief overview of common sensor technology used in acoustic

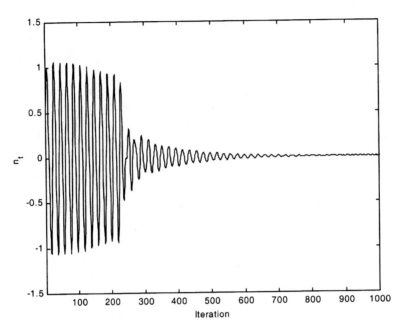

Figure 41 ARMAX Attenuation of a 50 Hz sinusoids sampled at 1024 Hz using a 10 coefficient LMS whitening filter and a slow-responding adaptive feedback control algorithm.

and seismic systems. These transducers are certainly more interesting than low-bandwidth sensors such as temperature, pressure, force, etc., because of their signal processing physical models for analysis of the frequency response. The transducer can be seen as a reversible device which can serve as both a sensor and actuator. One can model the transducer or sensor as a frequency response function which has a variable SNR over the bandwidth of interest. The variations in the SNR are the source of much interest to an intelligent sensor system. Section 15.3 presents some applications of adaptive signal processing in canceling or attenuating unwanted noise signals. Signals can be canceled adaptively "on-the-wire" in what we call adaptive noise cancellation, or physically in the medium which is being sensed in the more difficult, active noise cancellation. With feedforward adaptive controllers, the reference signal which is coherent with the error signal can be nearly completely canceled. Using feedback control systems, a reference signal is not required but the unwanted noise can only be attenuated by the feedback control, not completely canceled. For feedforward cancellers, one often seeks a reference signal which is well time-advanced relative to the error signal. For feedback control, the error sensor and actuator response must have as little delay as possible to allow significant cancellation over as wide a bandwidth as possible.

Active noise control is a fine example of an intelligent sensor system physically interacting with its environment by introducing cancellation waves to improve its performance. Such integrated adaptive control also provides the added benefit of transducer transfer functions, which can provide valuable information about system health. By identifying patterns in the electrical noise and system transfer functions, the intelligent sensor gains a level of "self-awareness", or even sentient (having the five senses) processing. Input impedances and reciprocity calibrations of transducers can also be used as part of an intelligent sensor systems ability to assess the health of the sensors and transducers attached to it. For example, the voltage sensitivity for a piezoelectric transducer will change depending on whether it is firmly attached to a massive base or if its loosely hanging from its connecting cable. By examining patterns in the background noise, the intelligent sensor system can determine if an electrical connection or semiconductor problem has occurred. Examination of the error and control spectra in an ANC system can determine the reachability of the cancellation, coherence of the reference signals to the noise, and the response of the ANC actuators. The intelligent controller can then apply some fuzzy syntax to decide whether the control emphasis should be on learning, avoiding instability, or ANC performance. For frequency-domain adaptive processing, the fuzzy control strategy can vary with frequency as needed.

Concluding Remarks

The purpose of this book has been to attempt to put together spatial, temporal, sensor information and adaptive signal processing algorithms towards the goal of developing meaningful sensor system intelligence. While it is not practical to cover every algorithm, or variant of algorithm which can be used as part of this lofty engineering quest, what we have done is to assemble the major topics in basic signal processing, sensor systems, adaptive processing, pattern recognition, and control to integrate the basis for intelligent sensor systems as a foundation. The reader is strongly encouraged to consult the many excellent texts currently available which provide far more rigor and detail than presented here. Our goal has been to provide the non-electrical engineer an entry to both basic and advanced topics in intelligent sensor signal processing as well as provide the electrical engineer some physical insight into practical applications of signal processing technology. There are even a few things here and there which are original in this book. The risk of doing this has been substantially reduced by validating every Figure plot using Matlab script. These scripts and data sets will eventually be made available on a CD ROM for interested readers who contact the author. This book is really a hierarchical conglomeration of many other texts with a few unconventional derivations. It represents a body of knowledge necessary to do work in the sensor signal processing area. The author and his students have frequently referred to its contents during the 5-year writing of the book. Hopefully, enough detail and references are included to allow rapid initial learning, resources for further more detailed research, and appropriate examples for continued reference. The writing process was both joyful and humbling.

PROBLEMS

1. Sketch the spectral shapes and typical time-domain plots of thermal, contact, and popcorn noise.
2. What is the amount of dB attenuation provided by a 1 mm thick cylindrical copper shield on a co-axial cable for a 120 Vrms 60 Hz cable located 0.1 m away?

3. Explain how two single-conductor co-axial cables can be used to provide a balanced differential sensor connection where the shield is floated out to a "two-wire" type sensor.

4. If one could replace the magnet on a loudspeaker with another which produces a magnetic field in the voice-coil gap 10 times greater, what happens to the driver resonance and damping?

5. A MEMS accelerometer has sensitivity of 0.1 V per g and an equivalent resistance of 10 MΩ. What is the SNR in dB for a 0.001 g 5 Hz sinusoid where the dc current is 10 ma?

6. A pair of microphone transducers in a 5 cm^3 air-filled cavity are exposed to an acoustic signal. Mic A responds with 10 times more voltage than Mic B. A current of 20 ma rms is applied to Mic B and Mic A responds with 1 vrms. What is the sensitivity of Mic A in V/Pa?

7. The coherence between a reference signal and error signal sensor is only 0.90 in a potential feedforward ANC system. Assuming no feedback to the reference signal and that the error plant is known or can be measured with great precision, what is the maximum dB cancellation possible?

8. For the "active SE plant identification" what would the "SE" estimate converge to if the error signal were completely driven to zero? Assuming the SE identification injected noise is uncorrelated with the reference signal, if the reference signal noise is completely canceled, what is the residual error signal level?

9. For a multichannel feedforward ANC system, explain why the transfer functions between actuators and error sensor of different channels should be as independent as possible.

10. A hydraulic actuator is used to stabilize a heavy gun barrel on a ship. The transfer function of the actuator to a vibration sensor on the gun barrel reveals a delay of 200 msec over a bandwidth from 0.1–3 Hz. Using feedback control, how much stable attenuation of the ships motion can be achieved over the 3 Hz bandwidth? How about 1Hz bandwidth?

BIBLIOGRAPHY

L. L. Beranek Acoustics, Chapter 3, New York: The American Institute of Physics, 1986.

L. J. Erikkson Active Noise Control, Chapter 15 in Noise and Vibration Control Engineering, L. L. Beranek, I. L. Vér (Eds), New York: Wiley, 1992.

J. Fraden AIP Handbook of Modern Sensors: Physics, Designs, and Applications, New York: AIP, 1993.

C. R. Fuller, S. J. Elliott, P. A. Nelson Active Control of Vibration, San Diego: Academic Press, 1996.

G. C. Goodwin, K. S. Sin Adaptive Filtering, Prediction, and Control, Englewood Cliffs: Prentice-Hall, 1984.

C. H. Hansen, S. D. Snyder Active Control of Noise and Vibration, New York: E & FN Spon, 1997.

W. H. Hayt Jr. Engineering Electromagnetics, 3rd ed., New York: McGraw-Hill, 1974.

F. V. Hunt Electroacoustical Transducers, New York: The American Institute of Physics.

Loudspeakers — An Anthology ..., Vol. I, The Audio Engineering Society, 60 East 42nd Street, New York, NY 10165.

P. A. Nelson, S. J. Elliott Active Control of Sound, New York: Academic Press, 1992.

H. W. Ott Noise Reduction Techniques in Electronic Systems, New York: Wiley, 1976.

G. M. Sessler, J. E. West "Condenser Microphones with Electret Foil," Microphones: An Anthology, The Audio Engineering Society, 60 E 42nd Street, New York, NY, pp. 111–113.

D. F. Stout, M. Kaufmann Handbook of Operational Amplifier Design, New York: McGraw-Hill, 1976.

J. G. Wessler (Ed.), Medical Instrumentation: Application and Design, New York: Wiley, 1998.

REFERENCES

1. J. B. Johnson "Thermal Agitation of Electricity in Conductors", Physical Review, Vol. 32, July, 1928, pp. 97–99.

2. H. Nyquist , "Thermal Agitation of Electrical Charge in Conductors", Physical Review, Vol 32, July, 1928, pp. 110–113.

3. Loudspeakers — An Anthology ..., Vol. I, The Audio Engineering Society, 60 East 42nd Street, New York, NY 10165.

4. L. L. Beranek Acoustics, Chapter 3, New York: The American Institute of Physics, 1986.

5. G. M. Sessler, J. E. West "Condenser Microphones with Electret Foil", Microphones: An Anthology, The Audio Engineering Society, 60 E 42nd Street, New York, NY, pp. 111–113.

6. R. S. McGuinn, G. C. Lauchle, D. C. Swanson "Low Flow-Noise Microphone for Active Control Applications", AIAA Journal, 35(1), pp. 29–34.

7. D. C. Swanson, S. M. Hirsch, K. M. Reichard, J. Tichy "Development of a Frequency-Domain Filtered-X Intensity ANC Algorithm", Applied Acoustics, 57(1), May 1999, pp. 39–50.

8. L. J. Erikkson Active Noise Control, Chapter 15. In: Noise and Vibration Control Engineering, L. L. Beranek, I. L. Vér (Eds.), New York: Wiley, 1992.

9. S. D. Sommerfeldt, L. Bischoff "Multi-channel Adaptive Control of Structural Vibration," Noise Control Eng. Journal, 37(2), Sept 1991, pp. 77–89.

10. D. C. Swanson, K. M. Reichard "The Generalized Multichannel Filtered-X Algorithm" In: 2nd Conf. Recent Advances in Active Control of Sound and Vibration, R. A. Burdisso (Ed.), Lancaster: Technomic, 1993, pp. 550–561.

11. K. J. Åstom, B. Wittenmark Adaptive Control, New York: Addison-Wesley, 1989.

Appendix

Answers to Problems

1. (a) Accelerometer output ± 0.25 volts. 11.5 dB $= 20$ log (gain), $\therefore$ gain $=$ 3.758. Voltage at A/D input is ± 0.9395 volts. 14-bit A/D has range $+2^{13}$ -1 to -2^{13}, or $+8191$ to -8192. Maximum A/D input voltage is $0.000305 \times 8192 = 2.5$ V. Numerical range of input is ± 3080.
    (b) A gain of 10 is needed, or 20 dB. The closest available gains are 20.5 dB and 19 dB. Since 20.5 dB would cause clipping, 19 dB is selected, or a linear gain of 8.9125. The A/D input is ± 2.228 giving a numerical range of ± 7305. The SNR is therefore 7305:1 or on a dB scale, 77.3 dB. Before gain adjustment it was 69.8 dB.

2. (a) Available signed 8-bit SNR is 127:1 or 42 dB. Therefore fc can be 25 kHz, half the Nyquist rate giving over 48 dB attenuation above 50 kHz.
    (b) A 16-bit signed A/D has ± 32767 range or 90.3 dB SNR. If fc is about 12.5 kHz, over 96 dB attenuation is possible above 50 kHz.
    (c) For a 100 kHz sample rate, the attenuation at 50 kHz is $48(50k - fc)/fc$ dB (i.e. 48 dB if $fc = 25$ kHz). Since the attenuation back at fc is twice that at the Nyquist rate, $48(100 - 2fc)/fc = 90.3$ is the minimum attenuation needed giving a maximum fc of 25,765 Hz.

3. (a) fs must be at least 1.5 Mhz
    (b) the Doppler-shifted frequencies can lie between 4.99667 Mhz and 5.00334 Mhz giving a minimum bandwidth of 6666.67 Hz. With the analog signal bandpass-filtered properly, a complex-sample A/D could gather 666.67 complex samples per sec. A real A/D system could sample at 20000 samples per sec to represent the data, but would not allow the distinction of positive and negative Doppler.

4. Assuming the sound is sinusoidal, the peak level is 1.41 times the rms, or 1.5 Pascals peak (conservatively). We therefore would expect a maximum voltage of, say, 20 mV from the microphone. A reasonable gain for the microphone would be 500, or about $+54$ dB.

5. The vertical distance of 1.0 m is sampled 525 times giving a spatial sample interval of 1.9 mm. At least two spatial samples per object dimension are needed to insure unaliased video, or 3.8 mm. A high contrast tweed jacket with dark and light weaves less than 1/8″ apart would look pretty confusing to viewers!

6. With 30° between the holes, the car will appear to be moving the fastest when the wheel turns only 15° between the 1/30 sec even or odd scans. This speed is 0.7π m/turn times (1/24) turns/sec, or 0.0916 m/sec which corresponds to about 0.2 mph. As the car moves faster than 0.2 mph, the holes appear to be moving backward slower and slower until the car is moving at about 0.4 mph, when the holes appear nearly stationary.

7. (a) Only 39.0625 samples per sec.
   (b) 1.445 Ghz.

8. (a) The offset binary data greater than 32767, say 32768 is misinterpreted as -32767 (complement the bits and add 1). The waveform plot looks pretty odd. Note that the largest offset binary value of 65535 maps to -1 in 2's complement.
   (b) The opposite situation produces a similar waveshape.

CHAPTER 2

1. Just follow through the proofs as given in section 2.1.

2. Note that $z = e^{sT}$ $s = \sigma + j\omega$. A stable signal is one where it stays bounded or decays as time increases. The real part of s is thus limited to be less than or equal to zero (left-half s-plane), which maps to the region on or inside the unit circle on the z-plane.

3. When $|\alpha| = 1$, the scaling on the z-plane amounts to a frequency shift in equation (2.1.13), which is the same as if a were purely imaginary in equation (3.1.13).

4. Note that functionally, $s \simeq (1 - z^{-1}/T)$. This functional similarity can also be seen by noting that $s = (1/T)\ln z = (1/T)\{(1 - z^{-1}) + \sum_{n=1}^{8}(1/n)(1 - z^{-1})^n\}$. Therefore, to a first approximation at least, the two operations can be seen as equivalent.

5. Computer model experiment – self explanatory.

CHAPTER 3

1. A 2nd-order FIR can be used where the two zeros are on the unit circle at the angles $\pm\pi(8000/22{,}050)$, or ± 65.3 degrees. The FIR coefficients are $b_0 = 1$, $b_1 = -0.8357$, and $b_2 = 1$. The output $y(n) = x(n) - 0.8357x(n-1) + x(n-2)$.

2. If 10 Hz resolution is sufficient, the length of the FIR must be at least 100 msec. With a sample rate of 50 kHz, about 5000 FIR coefficients are needed.

3. (c)

4. (a)

5. By writing the non-minimum phase ARMA filter as the product of a minimum phase filter (poles and zeros inside the unit circle) and an all pass filter, one is left with a pair of zeros at the origin (giving two unit delays) as the ARMA filter's non-minimum phase zeros move out to infinity. Therefore, the system delay increases the further zeros are outside the unit circle.

CHAPTER 4

1. Let $x_1[k+1] = x_2[k]$, $x_2[k+1] = u[k] - a_2 x_1[k] - a_1 x_2[k]$ as in Eq. (4.1.4.) The output $y[k]$ for input $u[k]$, is: $y[k] = b_1 x_1[k] + b_0 x_2[k]$ as in Eq. (4.1.6).

2. Using the guidelines in Eq. (4.1.21), $T < 2\zeta/\omega_0^2$, $\zeta = R/2L = 2500$, $\omega_0^2 = 5 \times 10^8$. Therefore, the minimum T for stablility is 10 μsec and the recommended T is 1 μsec.

3. α is 0.10 from the 10% requirement. Using 4.2.8 $\beta = 0.005267$ and γ is 0.0002774.

4. One could use a gradient along a $+45°$ direction such as:

$$-\nabla_{+45°} = \begin{bmatrix} 0 & -1 & -2 \\ 1 & 0 & -1 \\ 2 & 1 & 0 \end{bmatrix}$$

5. A convolution in the time domain of a filter impulse response with itself is equivalent to squaring the filter frequency response. This is evident viewing Figures 4.4.3 and 4.4.6.

CHAPTER 5

1. The difference between the two sinusoids is 72 Hz. For one bin to lie in-between the two peaks, the spectral resolution must be 36 Hz. This corresponds to an integration time of 1/32 or 31.25 msec. At 100,000 samples per sec, 3125 samples are the minimum needed in the DFT calculation to resolve the two sinusoids.

2. An FFT requires a theoretical minimum of $N\log_2 N$ operations. The total operations per FFT is $N(2 + \log_2 N)$, or 1152, 12,288, and 122,880, for $N = 128$, 1024, 8192, respectively. Given 25,000,000 operations available from the DSP, the maximum number of FFTs per sec is 21,701, 2,034, and 203, for $N = 128$, 1024, and 8192, respectively. The computational times are simply 46 μsec, 493 μsec, and 4.93 msec. (Note: actual DSP FFT algorithms will have much additional overhead depending on coding efficiency — looping vs. inline code. A typical 1024-point FFT on a 25 MFLOP DSP is several msec, not half a msec).

3. Given a real $x(t)$ placed in the imaginary part of the FFT input sequence with zeros in the real part of the input sequence, the imaginary part of the resulting complex spectrum would be symmetric in frequency while

the real part of the complex spectrum would be skew symmetric. Therefore, the true real spectrum is found by summing the positive and negative frequency bins of the imaginary bin of the FFT output. The true imaginary spectrum is found by subtracting the negative frequency bins from the positive frequency bins of the real part of the resulting FFT output.

4. The narrowband correction factor is found by matching the window integral to a rectangular window. The integral of $w(n) = 1 - n/N$ from $n = 0$ to N is simply $\frac{1}{2}N$, as opposed to N for the rectangular window. Therefore the narrowband correction factor is 2.00. The broadband correction factor is found by matching the square-root of window integral squared to the the same for the rectangular window. This result for our window is the square root of $\frac{1}{3}$ giving a correction factor of $3^{\frac{1}{2}}$ or 1.732. The broadband correction matches rms levels in the time domain to the sum of the rms bins in the frequency domain.

CHAPTER 6

1. The mean in the magnitude-squared bin is σ_t^2/N or 2.314 mv^2. The variance is $\sigma_t^4/(N^2 M)$ where $M = 15$, or $2 \mu v^4$ making the standard deviation ± 1.416 mV^2. The FFT resolution is 44,100/1024 or 43 Hz per bin. The noise level spectral density is therefore 53.81 $\mu v^2/$Hz.

2. The broadband correction factor for the Hanning window is about 1.65 from Table 5.1. Therefore the levels in question 1 would be too low by a factor of $1/(1.65)^2$. The observed noise spectral density would be approximately 19.765 $\mu V^2/$Hz.

3. Yes, the leakage errors requiring Hanning or other windows are most noticable for frequency responses with strong peaks and/or dips.

4. We seek a normalized variance of 0.0025. Using Eq. (6.2.21) we find 67 averages are needed.

5. Following Gauss's theorem, one computes the *normal* intensity to the surface at the available points, assigning a "patch" of the surface to each normal intensity. Multiplying each normal intensity by its respective patch area and summing gives the total radiated power. If the source is outside the surface, the measured power is zero, since what intensity enters the surface also leaves.

CHAPTER 7

1. With respect to the origin, let sensor 1 be at a distance r_1 and angle θ_1, sensor 2 be at a distance r_2 and angle θ_2, and sensor 3 be at a distance r_3 and angle θ_3. If the plane wave arrives from an angle θ moving at speed c, the propagation time from sensor k to the origin is simply $(r_k/c)\cos(\theta - \theta_k)$. The phase difference between sensor j and sensor k at frequency ω radians/sec is simply:

$$\Delta\phi_{jk} = \phi_j - \phi_k = \frac{\omega}{c}\left\{\cos\theta(r_j\cos\theta_j - r_k\cos\theta_k) + \sin\theta(r_j\sin\theta_j - r_k\sin\theta_k)\right\}$$

Using the phase difference between sensors 1 and 2 and the phase difference between sensors 1 and 3, we can solve two equations for the unknowns $\sin\theta$ and $\cos\theta$.

$$\begin{bmatrix} \cos\theta \\ \sin\theta \end{bmatrix} = \begin{bmatrix} (r_1\cos\theta_1 - r_2\cos\theta_2) & (r_1\sin\theta_1 - r_2\sin\theta_2) \\ (r_1\cos\theta_1 - r_3\cos\theta_3) & (r_1\sin\theta_1 - r_3\sin\theta_3) \end{bmatrix}^{-1} \begin{bmatrix} \dfrac{\Delta\phi_{12}c}{\omega} \\ \dfrac{\Delta\phi_{13}c}{\omega} \end{bmatrix}$$

The inverted matrix can be seen as an array shape weighting matrix, which when applied to the measured phase differences gives the sine and cosine of the plane wave. A simple arctangent calculation using the result conveniently gives the angle of arrival.

2. Let the array elements be numbered 1–16 from left to right where sensor 1 is our phase reference. The value of $kd\cos\theta = 0.1778$. The steering vector which makes the summed array output a maximum at $80°$ is

$$A(k,\theta) = \left\{1 + e^{-jkd\cos\theta} + e^{-j2kd\cos\theta} \ldots + e^{-j15kd\cos\theta}\right\}$$

For an approaching wave, all the sensors to the right of sensor 1 are slightly "ahead" in phase for the $80°$ arrival angle. Therefore, subtracting the appropriate phase from each sensor output puts all the array elements in-phase with sensor 1.

3. The radius is $r = 1$ m and the angular sensor spacing is $22.5°$. For the array with sensor 1 at $\theta_1 = 180$, sensor 2 at $\theta_2 = 157.5$, ... , sensor 9 at $\theta_9 = 0$, ... , sensor 13 at $\theta_{13} = 270$, ... , and sensor 16 at $\theta_{16} = 337.5$ degrees the phase relative to the origin for a particular sensor and a plane wave arriving from direction θ is

$$\phi_i = e^{jkr\cos(\theta_i - \theta)}$$

The steering vector for direction θ' is

$$A(k,\theta) = \left\{e^{-jkr\cos(\theta_1 - \theta')} + e^{-jkr\cos(\theta_2 - \theta')} \ldots + e^{-jkr\cos(\theta_{16} - \theta')}\right\}$$

The circular array response as a function of wavenumber k and direction θ with the main lobe of the beam steered to θ' is

$$X(k,\theta) = \sum_{i=1}^{N} A(k,\theta)Y(k,\theta) = \sum_{i=1}^{N} e^{-jkr\left(\cos\theta_i[\cos\theta - \cos\theta'] + \sin\theta_i[\sin\theta - \sin\theta']\right)}$$

4. W^s is symmetric 3×3 filter with a value of -7 in the center (2,2 position) surrounded by ones. A Fourier transform of the sequence $[+1 -7 +1]$ gives a wavenumber response of

$$I(m) = 2\cos\left(\frac{2\pi}{N}m\right) - 7$$

which is -5 at the zero wavenumber and -9 at $m = \pm N/2$. The larger amplitude at the higher wavenumbers increases the "sharpness" of the image.

CHAPTER 8

1. The best way to model the exponent is to use logarithms. Let $y = \log T$ and

$$H = (\bar{X}^H \bar{X})^{-1} \bar{X}^H \bar{y}$$

$$\begin{bmatrix} -0.02 \\ 5.3566 \end{bmatrix} = \begin{bmatrix} 0.0 & -0.005 \\ -0.005 & 1.1 \end{bmatrix} \begin{bmatrix} 60 & 120 & 180 & 240 \\ 1 & 1 & 1 & 1 \end{bmatrix} \begin{bmatrix} 4.15 \\ 2.95 \\ 1.75 \\ 0.55 \end{bmatrix}$$

$H(1)$ is -0.02 and $\beta = -1/H(1) = 50$. $T_0 = e^{5.356} = 212°C$.

2. No. But, orthogonal basis functions are the best choice because the covariance matrix is by definition diagonal. The basis functions do have to be linearly independent (full rank) to insure that the covariance matrix is invertible.

3. Yes. If the underlying error is not a linear function of the basis functions, the squared error will not be quadratic, thus a unique "least squared error" solution cannot be defined. Even with multiple error minima to contend with, one can obtain a solution, but one is not guaranteed that it is the "best" solution.

4. Substituting in the optimal H coefficients in the squared error sum we have

$$\bar{\varepsilon}^H \bar{\varepsilon} = \bar{y}^H \bar{y} - H^H (\bar{X}^H \bar{H}) H_0 - H_0^H (\bar{X}^H \bar{H}) + H^H (\bar{X}^H \bar{H}) H$$
$$= \bar{y}^H \bar{y} - (2H_0^H - H^H)(\bar{X}^H \bar{H}) H$$

where H_0 is the optimal (least-squared error) set of coefficients and we note that the squared error sum is a scalar. If $H = H_0$ then

$$\bar{\varepsilon}^H \bar{\varepsilon} = \bar{y}^H \bar{y} - H^H (\bar{X}^H \bar{H}) H_0$$
$$= \bar{y}^H \bar{y} - \bar{y}^H \bar{X}(\bar{X}^H \bar{H})^{-1} \bar{X}^H \bar{y}$$
$$= \bar{y}^H (I - P_X) \bar{y}$$

Dividing the sum of the squared error by N gives the desired result.

5. (a) For the linear moving average model stock markets use, one simply averages the previous 4 days of data to get an estimated value for Friday of 1212.75. However, one could also interpret a "linear model" as a

straight-line fit using 2 coefficients. Consider the following:

$$\begin{bmatrix} 1257 \\ 1189 \\ 1205 \\ 1200 \end{bmatrix} = \begin{bmatrix} 1 & 1 \\ 1 & 2 \\ 1 & 3 \\ 1 & 4 \end{bmatrix} \begin{bmatrix} h_0 \\ h_1 \end{bmatrix} \quad \text{The covariance is} \quad \begin{bmatrix} 4 & 10 \\ 10 & 30 \end{bmatrix}$$

$$\text{and its inverse is} \quad \begin{bmatrix} 1.5 & -0.5 \\ -0.5 & 0.2 \end{bmatrix}$$

giving coefficients of $h_0 = 1251.5$ and $h_1 = -15.5$. The Friday prediction is 1174. Sell!

(b) Using a basis function including a 5-day period sinusoid, plus an offset constant we have

$$\begin{bmatrix} 1257 \\ 1189 \\ 1205 \\ 1200 \end{bmatrix} = \begin{bmatrix} 1 & 0.9511 \\ 1 & 0.5878 \\ 1 & -0.5878 \\ 1 & -0.9511 \end{bmatrix} \begin{bmatrix} h_0 \\ h_1 \end{bmatrix} \quad \text{The covariance is} \quad \begin{bmatrix} 4 & 0 \\ 0 & 2.5 \end{bmatrix}$$

$$\text{and its inverse is} \quad \begin{bmatrix} 0.25 & 0 \\ 0 & 0.4 \end{bmatrix}$$

giving coefficients of $h_0 = 1212.8$ and $h_1 = 17.9$. The Friday prediction is 1212. Buy! Note how the sinusoid is an orthogonal basis function. Which model is best? Basis functions ought to be based on known physics or significant trending experience. Call a broker!

6. Self-explanatory.

7. Self-explanatory.

8. Self-explanatory.

9. Yes. The price will drop to $104 by closing time. The model is simply that tomorrow's price is a daily rate (weekly slope) plus some base price (intercept). Let the ordinate for Monday be 1, Tuesday 2, Wednesday 3, and Thursday 4, etc.

$$\boldsymbol{H} = (\bar{\boldsymbol{X}}^H \bar{\boldsymbol{X}})^{-1} \bar{\boldsymbol{X}}^H \bar{y}$$

$$\begin{bmatrix} -0.7 \\ 107.5 \end{bmatrix} = \begin{bmatrix} 0.2 & -0.5 \\ -0.5 & 1.5 \end{bmatrix} \begin{bmatrix} 1 & 2 & 3 & 4 \\ 1 & 1 & 1 & 1 \end{bmatrix} \begin{bmatrix} 110 \\ 100 \\ 108 \\ 105 \end{bmatrix}$$

Friday's price should be $107.5 - 0.7*5 = 104$. Sell at 105 and cut your losses. Note: computerized stock trading is extremely volatile at times due to false correlations, human emotion, and above all else, occasional misinformation. Least-squares modeling works best for very large observation sets and relatively simple models.

CHAPTER 9

1. Self-explanatory.

2. M^2 multiples and M^2 additions/subtractions, or simply M^2 MACs (multiply/accumulates).

3. The number of MACs and Divides for the RLS algorithm are $M^3 + 2M^2 + 2M$ plus 2 divisions, or 270,464 MACs and 2 DIVs, plus another 65 MACs to compute the filter output and error signals. The RLS Lattice requires $M^2 + 13M + 4$ MACs, plus $3M + 1$ divisions, or 4,832 MACs and 193 DIVs, plus another 64 MACs to compute the filter output. The LMS only requires $3M + 3$ MACs, plus 1 division, or 195 MACs plus 1 DIVs, plus another 65 MACs to compute the filter output and error signals.

4. The RLS and LMS algorithms are virtually the same in terms of operations for the Weiner filter problem (input and output available). However, the Weiner lattice requires about (depending how one codes the algorithm) 3M more MACs plus an additional M divides, plus an additional M^2 MACs for the Levinson recursion. Thus the total is seen to be $2M^2 + 16M + 4$ MACs, plus $4M + 1$ DIVs, or 9220 MACs, plus 257 DIVs, plus another 64 MACs to compute the filter output signal. The RLS Wiener lattice requires nearly twice the computation of the regular lattice when $M = 64$.

5. Use random noise as the unknown system input signal. It has the narrowest eigenvalue spread and will therefore allow the LMS to perform as efficiently as the RLS algorithm.

6. Okay, this one is not very obvious but damn useful. It is called a Schur Recursion. Start by post-multiplying the lattice error signals by y_{n-k} and taking expected values.

$$E\{\varepsilon_{p+1,n}y_{n-k}\} = E\{\varepsilon_{p,n}y_{n-k}\} - K^r_{p+1}E\{r_{p,n-1}y_{n-k}\}$$
$$E\{\varepsilon_{p+1,n}y_{n-k}\} = E\{r_{p,n}y_{n-k}\} - K^\varepsilon_{p+1}E\{\varepsilon_{p,n-1}y_{n-k}\}$$

The above can be written in terms of so-called "gapped" cross correlation functions.

$$g^+_{p+1}(k) = g^+_p(k) - K^r_{p+1}g^-_p(k-1)$$
$$g^+_{p+1}(k) = g^-_p(k-1) - K^\varepsilon_{p+1}g^-_p(k)$$

Note that $g^+_0(k) = R_k$, where $R_k = E\{y_n\ y_{n-k}\}$, the input signal autocorrelation to be whitened and, that $g^+_p(k) = 0$ for $k = 1, 2, ..., p$, and $g^+_p(k) = 0$, for $k = 0, 1, 2, ..., p-1$. It also follows from the forward and backward predictors that $g^-_p(k) = g^+_p(p-k)^H$. The PARCOR coefficients are solved using $g_{p+1}^+(p+1) = 0$ and $g_{p+1}^+(0) = 0$.

$$K_{p+1}^r = \frac{g_p^+(p+1)}{g_p^-(p)}$$

$$K_{p+1}^\varepsilon = \frac{g_p^-(-1)}{g_p^+(0)} = \left[\frac{g_p^+(p+1)}{g_p^-(p)}\right]^H = K_{p+1}^r{}^H$$

Schur Recursion summary:
(1) let $g_0^\pm(k) = R_k$, $k = 0,1,2,...,M$
(2) for $p = 0,1,...,M-1$
   (a) compute PARCOR coefficients

$$K_{p+1}^r = \frac{g_p^+(p+1)}{g_p^-(p)} = K_{p+1}^\varepsilon{}^H$$

   (b) update $g_{p+1}^+(k)$ for $p+2 \le k \le M$
   (c) update $g_{p+1}^+(k)$ for $p+1 \le k \le M$

7. Assume the mean square value for the input is 2.0.

$\varepsilon_{0,t}$	$r_{0,t-1}$	$R_{0,t}^\varepsilon$	$R_{0,t-1}^r$	$\Delta_{1,t}$	$K_{1,t}^\varepsilon$	$K_{1,t}^r$
2	0	6	2	0	0	0
1	2	7	6	2	2/7	2/6
0	1	7	7	2	2/7	2/7
−1	0	8	7	2	2/8	2/7
2	−1	12	8	0	0	0

8. (a) Assuming the mean square value for the signal is 2.0, let $\mu = 0.5$

ε_t	y_t	a_t	y_{t-1}	$K_{1,t}^\varepsilon$
2	2	0	0	0
1	1	0	2	2/7
−0.5	0	−0.5	1	2/7
−1	−1	−0.25	0	2/8
1.75	2	0.25	−1	0

(b) For the 5 given numbers the mean square is 2.4 and $\mu_{max} = 0.4$
(c) $\mu = 5.0$

ε_t	y_t	a_t	y_{t-1}	$K_{1,t}^\varepsilon$
2	2	0	0	0
1	1	0	2	2/7
−10	0	−10	1	2/7
−1	−1	+40	0	2/8
−38	2	+40	−1	0

Clearly, the large μ is leading to instability.

9. We know that for a converged lattice, the projection operator tends towards the identify matrix and the likelihood variable tends towards unity as given in equation (9.4.12). Equation (9.4.29) clearly shows that the forward and back ward error variances should tend to be equal under this circumstance.

10. No, the Cholesky factorization is only guaranteed to work for the backward error signal variances since the forward variances are not guaranteed to be orthogonal.

CHAPTER 10

1. The maneuverability index is given using (10.1.38) and is 0.10. Equations (4.2.8) and (4.2.11) give tracking gains of 0.3475 and 0.0739 for α and β, respectively.

2. The airplane heading 110 (20 degrees South of East) and the radar measurement error variance is $100\,\text{m}^2$ in all directions. The East–West, or x-direction component of measurement noise is $93.97\,\text{m}^2$ and the North–South, or y-direction noise is $34.20\,\text{m}^2$.

   Since the airplane is not maneuvering we can consider the x-y directions independently. Solving equation (10.1.25) directly using equation (10.1.30) for the updated (steady state) state error covariance one obtains

   $$p_{11} = (1 - \alpha)p'_{11} = \alpha\sigma_w^2$$

   $$p_{12} = (1 - \alpha)p'_{12} = \frac{\beta}{T}\sigma_w^2$$

   $$p_{22} = p'_{22} - \frac{\beta}{T}p'_{12} = \frac{\beta}{T^2}\frac{\alpha - \beta/2}{1 - \alpha}\sigma_w^2$$

   The above equations are very useful for a "covariance analysis" to test the underlying assumptions used to construct a tracking filter. If the observed state errors do not match the theoretical steady-state errors above, either the maneuverability index, measurement, or process noises need adjustment. The velocity state error in the x-direction is $(93.97)(0.03517) = 3.3\,\text{m}^2/\text{s}^2$ and in the y-direction is $1.2\,\text{m}^2/\text{s}^2$.

3. Self-explanatory.

4. A weighting factor of $0.01 = 1/100$ or 1% requires 100 times the observations of a recursive update with a weighting factor of unity. This has the effect of creating a sliding "memory window" where the data within is combined in the estimate. In the lattice recursions, an exponentially-decaying memory window of approximate length N sample is denoted as $\alpha = (N - 1)/N$, or 0.99 for N = 100.

5. Consider the symmetric nature of the embedding process. For the AR coefficients, the leading coefficient must be unity. The MA coefficients are scaled by the lattice except for the leading ones, b_0 and d_0.

6. Consider the prediction error for the Mth-order embedded model. The scaling for the lattice Levinson recursion is straightforward.

7. For a converged lattice the likelihood variable $(1 - \gamma_{p,n})$ approaches unity, thus making the a priori and a posteriori error signals equivalent.

8. Self-explanatory.

9. Since the reference signal spectrum is conjugate multiplied by the error signal spectrum, the oldest M error samples in the 2M sized buffer should be zeroed. One could also zero the oldest M samples of the reference and use a full error signal buffer. Noting that a spectral conjugate is the same as a time reversal, one could calculate $X_n^c(\omega)\{E_n^c(\omega)\}^*$ instead of $\{X_n^c(\omega)\}^* E_n^c(\omega)$, zero the newest M samples of either buffer prior to the FFTs, and extract the first M samples of the FIR response in reverse order if desired.

   The effect is that the cross correlation (from the inverse FFT of the conjugate frequency domain multiply) will have any circular correlation errors confined to the largest lags. These can be simply omitted when one wants an FIR filter.

10. The error spectrum is calculated from a 2M-point FFT. If one updated the FDLMS algorithm every sample the effect on the error signal spectrum would not be observed significantly until over M samples (over half an FFT buffer) is filled with the new error results. Therefore, the FDLMS algorithm would keep adapting leading to over-corrections. It is recommended that no more than 50% FFT buffer overlap be used and that data windows be avoided in the FDLMS algorithm.

CHAPTER 11

1. For both (a) and (b) the spectral SNR is 30.1 dBV (10 log N or the magnitude expressed on a dB scale). Why? The spectral density of the signal and noise is constant in a 1-Hz FFT bin (the spectral resolution of both FFTs). There is no improvement in SNR with the 8192 FFT because the bandwidth has been increased, allowing more of the white noise energy into the FFT data buffer. To increase the spectral SNR, one has to integrate longer in physical time, which decreases the noise level in the narrower FFT bins relative to the signal level.

2. Twenty trials per sec, 3600 sec/hr, 24 hr/day, 30.417 days/month means 52,560,000 trials per month with no more than 1 false alarm, or a P_{fa} of 0.000001902%. Using the approximation for low P_{fa} with a Gaussian magnitude detector in Eq. (11.2.6) we estimate the relative threshold using

$$T \approx \left[-\ln(P_{fa})\right]^{\frac{1}{1.6}} \quad P_{fa} < 0.2$$

or

$$T \approx \left[-\ln(0.00000001902)\right]^{\frac{1}{1.6}} \approx 6.042$$

Therefore, setting the absolute detection threshold at a level 6.042 times the background noise standard deviation (or an absolute level of 6.042×10^{-7} m/sec) should give the required false alarm rate.

3. The derivation is seen in equations (11.1.17–23). The result for high SNR is seen in equation (11.1.25).

4. Note the following: $\int xe^{-x^2} dx = -\frac{1}{2}e^{-x^2}$. A useful detection system would be a narrowband signal detector in the frequency domain. The Rayleigh pdf describes the magnitude spectrum in a given bin, while the integral of the pdf gives the probability of detection, which is conveniently Gaussian.

5. Using the characteristics in equation (11.1.17) it can be seen that the variance of M-averages is reduced by a factor of $1/M$. The standard deviation therefore is reduced by the square-root of $1/M$.

6. (a) No, the complex phasors for each FFT bin would add incoherently. Thus, the sum of the phasors in each FFT bin would tend to add to zero. (b) In this case, we are forcing coherence at the FFT bin corresponding to the sinusoid, whether a sinusoid is present or not, this bin would coherently add while all the other bins would tend to sum to zero. (c) Yes, in this case the SNR should increase so long as the phase and frequency of the sinusoid are stationary. This is the same effect as time-synchronous averaging except it is being accomplished by phase synchronization in the frequency domain.

7. The relative detection threshold T for 1% false alarm rate is about 2.597 times the noise magnitude. For 0.1% false alarm, T is increased to about 3.346. If the absolute detection threshold stays the same, the noise magnitude must decrease by about 0.776, or pretty close to 1 over the square-root of two. Therefore, simply averaging two signal magnitude observations together before applying detection would appear to reduce the false alarm rate by nearly a factor of 10.

8. Form an analytical model relating the mean path lengths, frequency, and wave speed to received signal amplitude. Determine the statistics by calculating the characteristic functions for the two path densities, multiplying in the frequency domain, and inverse transforming. Repeat the procedure with the Rician to obtain the signal plus noise probability density function. The detector's false alarm rate is determined from the noise statistics and detection threshold alone. The propagation path fluctuations affect the probability of detection by lowering it and broadening the density over a wider range.

CHAPTER 12

1. 100 samples. See equation (12.1.28). Note that the variance of the variance estimate is $\sigma^4/50$ for the case of 100 averages.

2. Assuming a very high SNR and an arrival angle near broadside, equation (12.2.11) provides a straightforward assessment of the best possible bearing error, which is 0.03105/SNR. An SNR of only 10 (10 dB) gives a bearing error variance of only 0.1779 degrees. This is because of the large number of sensors and the spacing of about 0.3λ.

3. (a) Sensor spacing is $2\pi/16$ m or 0.39267 m. To avoid spatial aliasing $\lambda/d > 2$. Therefore, $f < 1909.859$ Hz
   (b) Consider a symmetric spacing where 8 sensors are located at angles $(2n+1)\pi/16$, $n = 0, 1, 2, \ldots, 7$ relative to the x-axis, with a mirror image below the x-axis. The plane wave arrives along the x-axis passing sensors

(numbered counter-clockwise from the positive x-axis) 1 and 16 first, then 2 and 15, 3 and 14, and so on, passing sensors 8 and 9 last. The phases, relative to sensors 1 and 16 are determined by the distance on the x-axis between sensor pairs, the speed of the wave, and the frequency. The distance on the x-axis between the 1–16 and 2–15 sensor pairs is $\cos(\pi/16)$–$\cos(3\pi/16)$, or 0.1493 m. The wavelength of a 1 kHz wave in water in 1.5 m. The phase difference between sensors 1–16 and 2–15 is therefore 0.6254 radians, or 35.83 degrees.

4. With 16 sensors, one can theoretically solve the eigenfunctions to resolve up to 15 sources at the same frequency.

5. The array beamwidth is a function of aperture and wavelength, not the array shape or number of elements. Using the line array response in equation (12.2.16) one can approximate the $-3\,dB$ response as the angle where $\frac{1}{2}kL\sin\theta$ equals about 1.4, where L is the aperture and k is the wavenumber.

f_{max}	200 Hz	1 kHz	2 kHz	L	c
1909.859 Hz	NA	$\pm19.5°$	$\pm9.62°$	2 m	1500
828 Hz	$\pm8.844°$	$\pm1.76°$	$\pm0.88°$	5 m	345

Note that if $f > f_{max}$ grating lobes will be present in the beam response.

6. Using equation (12.3.11) we find the maximum distance for the scan plane to be 8.66 cm. Let's choose a scan distance of 5 cm giving a resolution at 120 Hz of 28.79 cm, or about $1/10$ of a wavelength.

7. Spherical wave spreading loss is $-20\log_{10}R$ or $-6\,dB$/range doubling. Circular wave spreading is $-10\log_{10}R$
No surface losses or scattering on the water surface
$100 - 10\log_{10}(1000) - 10\log_{10}X > 40$ dB, $10\log_{10}X = 30$ dB, $X = 10^3 = 1000$ m.

CHAPTER 13

1. Consider the following diagram

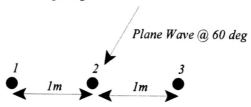

Assuming the wave passes sensor 3 first, then 2, and finally one, the phase of the wave at sensor 2 is $kd\cos\theta$ radians delayed relative to sensor 3, and the phase at sensor 1 is $2kd\cos\theta$ delayed relative to sensor 3 (k is the wavenumber, d is 1 m and θ is 60 degress). To "steer" the array response to 60 degrees, we need to add the appropriate phase to each array element in the frequency domain such that the wave from 60 degrees has the same phase at all 3 elements. Adding the "steered" array element outputs produces an array gain of 3 in the steered direction of 60 degrees. If the raw element outputs are denoted as $X = [x_1(\omega)x_2(\omega)x_3(\omega)]$ the steering vector is $S = [1 e^{-jkd\cos\theta} e^{-j2kd\cos\theta}]^T$ or $S = [e^{+j2kd\cos\theta} \quad e^{+jkd\cos\theta} 1]^T$, depending on the desired net phase shift. The steered array output is simply XS. Note that $k = \omega/10,000$ and $d = 1$ so that the steering works for frequencies up to 5 kHz, where the spacing is half a wavelength.

2. Assume the array elements are arranged counter-clockwise with element 1 on the x-axis at $< 1, 0 >$, element 5 on the y-axis at $< 0, 1 >$, element 9 at $< -1, 0 >$, etc. If the plane wave arrives from 0 degrees (along the x-axis in the -x direction), element 1 is passed first, followed by 2 and 16, then 3 and 15, and so on. The steering vector which allows the summed array output to be coherent in the $\theta = 0$ degrees direction is

$$S(\omega, \theta = 0) = [e^{-jk} e^{-jk\cos(\pi/16)} e^{-jk\cos(2\pi/16)} \ldots e^{jk\cos([n-1]\pi/16)} e^{-jk\cos(15\pi/16)}]^T$$

Note that the phases of the steering vector elements are relative to the phase of the plane wave at the origin. Because the wave is approaching from 0 degrees, the y-axis coordinates of the array elements are not a factor. Multiplying the array element spectral outputs by the steering vector above make each element output coherent with the phase of the wave at the origin, thus giving an array gain of 16 in the 0 degrees direction. For an arbitrary direction θ, the steering vector is

$$S(\omega,) = [e^{-jk[\cos\theta+\sin\theta]} e^{-jk[\cos(\pi/16)\cos\theta+\sin(\pi/16)\sin\theta]} \ldots$$
$$e^{-jk[\cos([n-1]\pi/16)\cos\theta+\sin([n-1]\pi/16)\sin\theta]} \ldots$$
$$e^{-jk[\cos(15\pi/16)\cos\theta+\sin(15\pi/16)\sin\theta]}]^T$$

To plot the beam response over $-\pi < \theta' < +\pi$, simply replace θ with $\theta - \theta'$. The wavenumber range for propagating waves is $k \leq \omega/c$ where ω is the radian frequency and c is the wave speed. For waves $k > \omega/c$ (it can happen in the nearfield of a complicated source), the wavelength is smaller than the plane wavelength and the wave collapses upon itself and is thus termed "evanescant". Waves traveling faster than c are (at least in acoustics) called supersonic, have wavelengths larger than the plane wave, and wavenumber smaller than $k = \omega/c$. An array beamformer coherently amplifies a given wavenumber, which usually, but not always, is associated with a specific temporal frequency.

3. This is done following equations (13.1.5–9) using an LMS step smaller than the sum of the squared array element outputs. The LMS calculated steering vector elements cause a null in the source direction, so inverting the filter gives a beampattern to amplify the source relative to the background.

4. Equation (13.3.1) shows the Vandenburg matrix for 3 coherent sources. Comparing this to the example in equations (13.1.1–4) it can be seen that the off-diagonal covariance elements will have source phase information. Under realistic conditions, the source phases are not exactly stationary, thus some time averaging will cause the source phases and background noise to appear uncorrelated. Under conditions with sources phases coher-

ent (Vandenburg matrix), the off-diagonal phase and amplitude is not solely due to the angles of arrival. Therefore, the spatial information is corrupted by the source phases, and the eigenvectors are not orthogonal. Thus, the "nulls" of the signal eigenvectors beam responses will not necessarily correspond to the other source directions.

5. When the MVDR steering vector is steered to one of the sources, the steering vector is a signal eigenvector of the covariance matrix. By definition, its nulls are in the direction of the other sources.

6. Assume the sensors are in a straight line but spaced $d = 2/15$ m apart with $\pm\sigma_d$ standard deviation, say along the x-axis. We can make a good approximation assuming $\sigma \ll d$ in that $(1/d \approx (1d)(1 - \sigma/d + \sigma^2/d^2 - \ldots)$. Thus equation (12.2.11) can be used

$$\sigma_\theta \approx \frac{\lambda}{16\sqrt{2\pi}SNRd}\left(1 - \frac{\sigma_d}{d}\right) = 1.81° \pm \frac{\sigma_d}{d}$$

7. The trick here is to represent the MLS sequence not as 0 and 1's, but as a -1 and $+1$ signal for autocorrelation purposes. Depending on how you initialize the shift registers, you get an MLS sequence like 0011101, or 11110100, etc. We started with bits 1–3 equal to 100, giving an output of 0. The output for the [3,1] MLS sequence is XOR'd with bit 1 to generate bit in the next set. The next set of bits is 110 with and output 0, then 111 with output 1, 011 with output 1, 101 with output 1, 010 with output 0, and finally 001 with output 1. Thus the [3,1] sequence is like 0011101, then repeats exactly.

   Representing the sequence as a bipolar signal gives $-1 - 1 + 1 + 1 + 1$ $-1 + 1$. It is straightforward to show that the 0th autocorrelation lag is 7 and all other lags are -1. All MLS sequence have this unique property where the 0th is $2^N - 1$ and the remain lags are -1. MLS sequences also have the nice property that any two sequences of the same length have a bounded crosscorrelation, thus allowing individual sequences to be recovered from a communications band with many MLS channels.

8. The spread spectrum signal can be seen, on average, to be the time domain product of the carrier sinusoid and a square-wave with period 50 ms. The spectrum is therefore the convolution of a delta function at the carrier frequency, and a sinc function about 40 Hz wide. The actual signals are randomized somewhat. But as the chip rate increases, so does the effective bandwidth or the spread spectrum signal.

9. As the chip rate approaches the sample rate, the corresponding sinc function in the frequency domain covers the entire bandwidth of the digital signal.

10. As the chip rate increases, so does the signal bandwidth and the ability to resolve multipath propagation in time and space. The longer the MLS sequence length, the larger the net time delay can be observed between source and receiver. Thus, to measure a travel times for a range R m, the sequence length should be longer than $2R/c$ seconds to insure that at most half a buffer overlap exists between the received and transmitted waves.

CHAPTER 14

1. Recall that the 2×2 covariance matrix can be diagonalized by a simple coordinate rotation. The square root of the diagonal matrix elements give the major and minor axis, and the rotation angle comes directly from the eigenvector matrix.

$$\Sigma_A = \begin{bmatrix} \cos\theta_A & -\sin\theta_A \\ \sin\theta_A & \cos\theta_A \end{bmatrix} \begin{bmatrix} \sigma_{11}^2 & 0.00 \\ 0.00 & \sigma_{22}^2 \end{bmatrix} \begin{bmatrix} \cos\theta_A & \sin\theta_A \\ -\sin\theta_A & \cos\theta_A \end{bmatrix}$$

$$= \begin{bmatrix} 2.5 & -3.2 \\ -3.2 & 5.5 \end{bmatrix}$$

Since the eigenvector rotation matrix is orthonormal

$$\begin{bmatrix} \sigma_{11}^2 & 0.00 \\ 0.00 & \sigma_{22}^2 \end{bmatrix} = \begin{bmatrix} \cos\theta_A & \sin\theta_A \\ -\sin\theta_A & \cos\theta_A \end{bmatrix} \begin{bmatrix} 2.5 & -3.2 \\ -3.2 & 5.5 \end{bmatrix} \begin{bmatrix} \cos\theta_A & -\sin\theta_A \\ \sin\theta_A & \cos\theta_A \end{bmatrix}$$

The following equation is then solved for θ_A.

$$0 = \cos\theta_A \sin\theta_A (5.5 - 2.5) + (-3.2)(\cos^2\theta_A - \sin^2\theta_A)$$

$$\theta_A = \frac{1}{2}\tan^{-1}\left\{ \frac{2(-3.2)}{2.5 - 5.5} \right\} = -57.56°$$

The diagonal elements are

$$\sigma_{11}^2 = (2.5)\cos^2\theta_A + (-3.2)2\cos\theta_A \sin\theta_A + (5.5)\sin^2\theta_A = 7.5341$$

$$\sigma_{22}^2 = (2.5)\sin^2\theta_A - (-3.2)2\cos\theta_A \sin\theta_A + (5.5)\cos^2\theta_A = 0.4659$$

2. The rotation angle for class B is $+85.62°$ while the standard deviations for the major and minor axis of the ellipse are 2.24 and 1.31, respectively.

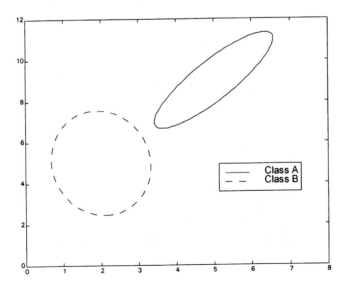

3. Recall equations (14.1–4). However, the features are not uncorrelated thus requiring the multichannel density in equation (14.1.9). Note that a capital "P" denotes a probability, while a lower case "p" denotes a probability density. The probability for class w_k given feature, $\bar{x}$ is

$$P(W_k|\bar{x}) = \frac{p(\bar{x}|w_k)P(w_k)}{\sum_k p(\bar{x}|w_k)P(w_k)}$$

where for a 2-element feature vector $\bar{x}$, the conditional density for $\bar{x}$ given class w_k is

$$p(\bar{x}|w_k) = \frac{1}{2\pi|\Sigma_k|^{\frac{1}{2}}} e^{-\frac{1}{2}[\bar{x}-\overline{m_k}]\Sigma_k^{-1}[x-\overline{m_k}]^H}$$

The covariance determinant for class A is $|\Sigma_A| = (1.8735)^2$ and class B is $(3.3634)^2$. The conditional density for our feature vector $[4.35 \ 5.56]^H$ is $p(x|A) = 0.00011745$ and $p(x|B) = 0.0097296$. The sum of the class conditional probability densities times their respective class probabilities is 0.0044664. Evaluating Bayes' theorem we have the probability of class A is 1.97% and class B is 98.03%.

It is important to carry out Baye's theorem in calculating the maximum liklihood class. While the above example is straightforward, with many classes and each class having unique probabilities, the correct class is often not determined by the minimum Mahalanobis distance (the maximum conditional density).

4. Task must be completed as a computer assignment.

5. Task must be completed as a computer assignment.

6. The crest factor is the peak relative to the rms voltage. The sinusoid is obviously 70.7 mV rms, thus the peak amplitude for a crest factor of 25 is 1.767 V.

7. The harmonic distortion at 200 and 234 Hz is 5%. The internodulation distortion is 10% at 17 Hz and 217 Hz (sum and difference frequencies). The total signal distortion assuming no other harmonics are detected is seen as 3%.

8. See Figure 14.2.16. The choice of segmentation order and direction is up to the reader. However, once the numbers are defined as mathematical lines, coordinates, and directions, simple vector algebra allows rotation, scaling, and translation, etc.

9. Using equation (14.3.1) assuming zero acceleration of the temperature, it can be seen that in 20 minutes the vessel temperature should reach 1000°C. With T equal 1 minute and zero acceleration, the 2-state transition matrix can be defined as

$$F(N+20) = \begin{bmatrix} 1 & 20 \\ 0 & 1 \end{bmatrix}$$

where 20 on the transition originates from 20 samples into the future. The process noise is

$$Q(N) = \begin{bmatrix} 400 & 20 \\ 20 & 1 \end{bmatrix}$$

Assuming the state error covariance is diagonal, equation (14.3.3) becomes

$$P'(N + 20|N) = \begin{bmatrix} 1 & 20 \\ 0 & 1 \end{bmatrix}\begin{bmatrix} 25 & 0 \\ 0 & 1 \end{bmatrix}\begin{bmatrix} 1 & 0 \\ 20 & 1 \end{bmatrix} + \begin{bmatrix} 400 & 20 \\ 20 & 1 \end{bmatrix} = \begin{bmatrix} 825 & 40 \\ 40 & 1 \end{bmatrix}$$

The covariance rotation angle is only a couple of degrees, so we can approximate the standard deviation of the temperature as 28.7 and temperature rate as 1. Therefore the ratio $\sigma_F/\sigma_R = 28.7$ in equation (14.3.6). The standard deviation of the time prediction is 1.836 times 28.7 or 52.7 minutes. In other words, we're predicting 20 minutes until the vessel reaches 1000°C where 68.4% of the time our (crude) prediction should be within ±52.7 minutes. While not a super precise prediction because of the variance of the raw temperature readings, it is still statistically useful.

10. The survivor rate is defined in equation (14.3.7) as the time integral from the current time to infinity of the time prediction density function. A plot of the survivor rate is

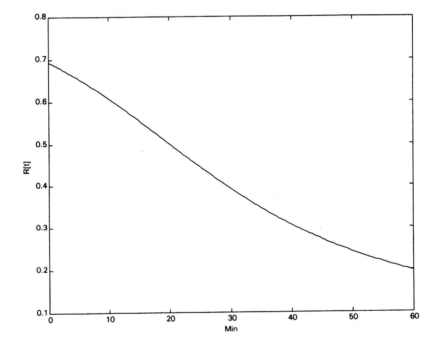

Note how the probability of surviving (vessel temperature below 1000°C) is at 50% at the predicted time of 20 minutes.

The hazard rate is found to be

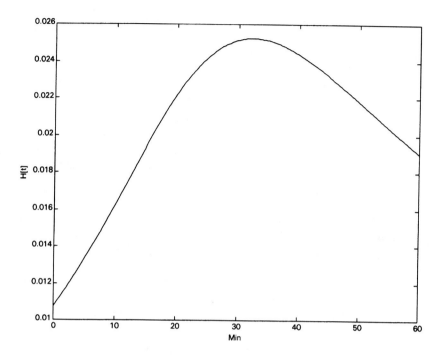

The maximum hazard rate actually occurs at 30 minutes, rather than the predicted failure time of 20 minutes. This is rather interesting and useful because the hazard rate is the ratio of the density function over the survivor rate. Depending on the shape of the two functions in the quotient, the peak hazard may not be at the predicted failure time.

CHAPTER 15

1. See Figures 15.1.1–3.

2. From equation 15.1.10, one finds 287 dB of electrical shielding. However, for magnetic shielding the skin depth is 8.517 mm (from equation 15.1.13–14 where $\mu\sigma = 73.136$). The absorption loss from equation (15.1.15) is only 1 dB while the magnetic reflection loss is about -1.2 dB, thus the net magnetic shielding can be seen as 0 dB! The current induced into our coaxial cable depends on the current flowing in the 120 vrms power cable and the differences in inductance between the coaxial and power cables.
   The external inductance of parallel wires is well known to be 10.95 $\cosh^{-1}$ (r/d), where d is the conductor diameter and r is the distance between them. If the power cable is 10 AWG (2.6 mm diameter with 3.15 mΩ/m resistance) and the coaxial cable center conductor is 24 AWG (0.51 mm diameter with 83 mΩ/m resistance), the inductance from the power cable to the coaxial center conductor is seen to be about 65 H, giving an impedance

at 60 Hz of 24.6 kΩ. If the two cables run along side each other for 100 m, an induced ac current in the coaxial center conductor which is 0.315/24600, or 12.8×10^{-6} times the current in the power cable. That may appear small until one considers, say 20 amperes in the power cable and 1 ma current from a sensor. The SNR in the coaxial center conductor is only 1/0.256 or +11.8 dB. Next to ground loops, magnetically induced noise is the most common noise problem to sensor cables. This is why a sensor should drive a low impedance coaxial line with as high a current as possible.

3. One simply ties the shields of each co-axial cable together at each end, grounding the shield preferably at the instrumentation end. The negative (grounded in a single-ended connection) terminal of the sensor is passed through the center connector of one co-axial cable while the positive (signal output) of the sensor is passed through the center conductor of the other co-axial cable.

4. Increasing the magnetic flux "B" in the motor factor "Bl" has no effect on the driver resonance frequency since the effect on diaphragm mass and stiffness is canceled in the resonance calculation. However, equation (15.2.8) clearly shows how an increase in Bl increases the system damping. Equation (15.2.9) also shows that the stronger magnet increases the acoustic pressure generated by the loudspeaker. Equation (15.2.11) shows that the voltage generated by a geophone in response to velocity also increases with a stronger magnet.

5. The signal voltage from a 0.001 g sinusoid (assume rms) is 100 μV. Thermal noise is small (the problem is not defined at high temperatures) at 387 nV in a 1-Hz band. Shot noise is seen as 56.56 pA, or 565 μV across a 10 MΩ resistance in a 1-Hz band. Contact and popcorn noise are ignored because we don't know the material constants. Therefore, in a 1-Hz band the SNR is 100/565 or −15 dB. The minimum detectable acceleration in a 1-Hz band is seen as around 0.006 g. For broadband signals it is much worse because the SNR can be seen as 20 log $(0.2/B^{\frac{1}{2}})$, where B is the bandwidth in Hz.

6. Equation (15.2.38) gives the reciprocity constant for a cavity $J = j\omega C_A$, where C_A is $V_A/(\rho c^2)$, and V_A is the cavity volume in m^3.

$$M_A = \sqrt{(5.45 \times 10^{-8})\left(\frac{1}{0.020}\right)(10)} = 5.22 \ mV/Pa$$

7. The "SE" plant is known and the "PE" plant can have error due to the low coherence. Recall from Section 6.2 for a coherence of 0.9, the "real-time" (n = 1) transfer function error magnitude is about 23% and phase error is 13 degrees. Both of these correspond to an error of about 23% for the cancellation wave. Therefore, the maximum cancellation is only about 12.7 dB – quite poor.

8. The SE plane estimate would also converge to zero – not good! With the reference noise completely canceled, the residual error noise would be wSE, there w is the injected noise.

9. For complete independence, the S matrix is diagonal and easily invertible. If two or more channels are highly correlated (i.e. cross transfer functions nearly identical to main-diagonal transfer functions), the channels are linearly dependent. Thus the "S" matrix is not invertible. Physically, the two channels can "fight each other" without any effect seen in the error signals. This is a significant concern for multichannel ANC and has recently been solved using Gram–Schmidt Orthogonalization (F. Asano, Y. Suzuki, and D. C. Swanson, "Optimization of Control Source Configuration in Active Control Systems Using Gram–Schmidt Orthogonalization," *IEEE Trans Speech and Audio Proc.*, 7(2), pp. 213–220, March, 1999.)

10. A time delay of 200 ms results in a phase shift of 3.77 radians at 3 Hz. This means that negative (canceling) feedback at very low frequencies can turn into positive (unstable or oscillating) feedback near 3 Hz. The maximum stable feedback gain is slightly less than unity (0 dB) in magnitude. This results in less than 6 dB cancellation at 0.1Hz tapering off to no cancellation near 3 Hz.

    If the bandwidth were limited to 0.1 Hz to 1.0 Hz, the 200 ms delay results in a maximum of 1.257 radians phase shift at 1 Hz. One could use a low-pass filter in the feedback path as a "compensator" and achieve some attenuation such that the net phase shift at, say 2 Hz, is less than π radians, and the net feedback gain is slightly less than unity at 2 Hz. Assuming a negative feedback gain of X below 1 Hz, the vibration cancellation in the 0.1 Hz to 1 Hz decade would be $-20\log(1/[1 + X])$ dB.

Index